AF182803

Artificial Intelligence and Machine Learning with R

Bernd Heesen

Artificial Intelligence and Machine Learning with R

Applications in the Field of Business Analytics

Springer

Bernd Heesen
Hochschule Ansbach
Ansbach, Bayern, Germany

ISBN 978-3-658-45391-6 ISBN 978-3-658-45392-3 (eBook)
https://doi.org/10.1007/978-3-658-45392-3

Translation from the German language edition: "Künstliche Intelligenz und Machine Learning mit R" by Bernd Heesen, © Der/die Herausgeber bzw. der/die Autor(en), exklusiv lizenziert an Springer Fachmedien Wiesbaden GmbH, ein Teil von Springer Nature 2023. Published by Springer Fachmedien Wiesbaden. All Rights Reserved.

This book is a translation of the original German edition "Künstliche Intelligenz und Machine Learning mit R" by Bernd Heesen, published by Springer Fachmedien Wiesbaden GmbH in 2023. The translation was done with the help of an artificial intelligence machine translation tool. A subsequent human revision was done primarily in terms of content, so that the book will read stylistically differently from a conventional translation. Springer Nature works continuously to further the development of tools for the production of books and on the related technologies to support the authors.

This Springer imprint is published by the registered company Springer Fachmedien Wiesbaden GmbH, part of Springer Nature.
The registered company address is: Abraham-Lincoln-Str. 46, 65189 Wiesbaden, Germany

Preface

Machine Learning and Artificial Intelligence are much-discussed and current topics because they can be associated with exciting opportunities and difficult-to-assess risks. CEOs in Germany and worldwide (Price Waterhouse Coopers 2019, p. 35–36) believe that Machine Learning and Artificial Intelligence will change the world more sustainably than the Internet has done. Despite this fact, Machine Learning and Artificial Intelligence are not yet widely used by many companies. Why is that?

Firstly, decision-makers in organizations are often not familiar with the principles, methods, algorithms, and application areas of these innovations. Many prefer to wait and see if the investments prove worthwhile in other organizations before spending money. Who knows if it's just another hype? Investing in Machine Learning and Artificial Intelligence primarily requires know-how. Powerful software solutions, like R presented in this book, are often available for free as open-source and the costs for hardware are anyway decreasing.

With this book, I want to contribute to presenting what "Machine Learning" and "Artificial" Intelligence actually are and where their limits lie. Of course, I also want to present where and in what way they can be beneficial. In particular, I introduce the three types of Machine Learning, Supervised, Unsupervised, and Reinforcement Learning, and selected, widely used algorithms, which form the basis for Artificial Intelligence.

Ultimately, machine learning can only function effectively if the underlying data for the learning process is of sufficient quality. Therefore, I also present best practices of data science, including data preprocessing and visualization.

Since a book can never be as current as the contents of a website, and I also want to present you with up-to-date applications and solutions of machine learning and artificial intelligence beyond the book, I invite you to visit the website for the book (URL: http://www.prescient.pro). In addition, you can find videos on application solutions on my YouTube channel profheesen (URL: https://www.youtube.com/profheesen).

The datasets used in this book are provided to you through my R-package "machine-learning" (see Sect. 4.6), so that you can execute the code examples presented in the book yourself. The package also includes interactive self-learning tutorials for learning the R programming language.

I would be delighted if you let yourself be inspired by the reading and apply the innovations of machine learning and artificial intelligence for the benefit of your organization.

Wolframs-Eschenbach, in June 2023

Bernd Heesen

P.S.$_1$: I would like to thank my editor Mareike Teichmann from Springer-Verlag for her support with this book project!

P.S.$_2$: Contents in sections 4 and 5 on the basics of R are excerpts taken from my book publication "Data Science and Statistics with R: Practical Application Solutions" (Heesen 2022).

References

Price Waterhouse Coopers. (2019). *PwC Global CEO Survey 2019.* Retrieved on September 30, 2019 from https://www.pwc.de/de/ceosurvey.html

Heesen, B. (2022). *Data Science and Statistics with R: Practical Application Solutions*. Heidelberg: Springer-Gabler.

Contents

List of Figures

Benefits of Machine Learning and Artificial Intelligence

1

Digitization is changing the world, the markets, the competitors, and the expectations of customers. The speed at which markets are changing has increased to such an extent that it is becoming increasingly important to have relevant information about one's own organization, the competition, and the customers available in a timely manner.

For a long time, this data has been analyzed and visualized in organizations using software, so-called data science tools. The more experienced the users of these systems are, the more they can utilize the capabilities of the software products. The software is developed by programmers, and over time, the programmers learn to develop more efficient and powerful software. At the same time, the users of the software learn to use the software more effectively. Thus, the learning takes place among the people. This is accompanied by a certain progress, which is nothing new.

What is new, however, in recent years, is that computers, machines, can learn independently. This is referred to as machine learning or also as artificial intelligence, as the machines develop the ability to solve problems largely on their own. In general, it is assumed that intelligence is needed to solve problems. Since it is now machines that can solve certain problems independently, this is referred to as artificial intelligence in these cases.

Machine learning and artificial intelligence are not entirely new, as both were used by large organizations already in the last century. What has changed in recent years, however, is the fact that computers have become more powerful, that more and more data is available as a basis for machine learning, and that the algorithms for machine learning have also evolved. Some powerful software tools for using machine learning are even available for free as open-source solutions, e.g., R, Python, or others. The time

B. Heesen, *Artificial Intelligence and Machine Learning with R*,
https://doi.org/10.1007/978-3-658-45392-3_1

for a more comprehensive use of machine learning and artificial intelligence has thus come. Everyone can benefit from this innovation, provided they have the basic knowledge for it.

Artificial intelligence, based on machine learning, has already arrived in many areas of our everyday life, without us perhaps being so aware of it. When searching on the internet browser, suitable, personalized search results are presented to us, the navigation device suggests the best route, cameras automatically select the best settings for a subject (Federal Government 2020), voice assistants answer questions, lawnmowers find their way alone, intelligent or so-called active prostheses pick up nerve impulses and improve people's mobility (Shaer 2014), intelligent traffic control helps avoid traffic jams, people are identified based on camera images, diseases like breast cancer are detected early, or the setting of prices is done by machines, so that profit can be optimized.

The medication of patients can be done with the help of artificial intelligence, which improves the medical care of patients and helps to avoid life-threatening situations (Müller and Massaron 2018). In the UK, an AI system was trained with data from nearly 300,000 patients with heart disease and heart attacks (Rittershaus 2020). It was found that diseases could be predicted significantly better by artificial intelligence than was possible by humans. In the context of machine learning, indicators were also used that were not found on the doctors' checklists, and thus valuable new correlations could be identified.

Potholes in roads can now be detected based on vehicle data with the help of artificial intelligence and fixed in a timely manner (bitkom 2019). Caries can be proactively detected, so that treatment with medication may be sufficient. The buying behavior of customers can be predicted and marketing campaigns can be made more effective through customer segmentation.

How powerful AI is, is shown time and again when innovations are reported, as in the following example: "When AlphaGo defeated the world's best Go players (Go is a highly complex strategic board game) in 2016, it sent a shockwave through the AI community. Because Go is considerably more complex than chess and cannot be mastered by brute computing power. In the analyses of the competition games between AlphaGo and its human opponents, it made various maneuvers that were described by observing experts as completely surprising. These were maneuvers that a human would never play. But AlphaGo obviously discovered new maneuvers that work during its millions of training games, which it played against itself" (Rittershaus 2020).

Of course, voice assistants like Alexa, Siri, and Cortana also use artificial intelligence in speech processing.

All this and much more is possible through machine learning and artificial intelligence. Artificial intelligence is therefore considered one of the future technologies that will massively influence our lives, both professionally and privately. As with all innovations, this innovation also has both advantages and disadvantages, opportunities and risks. The weighing of the pros and cons largely depends on the situation and the interests of the individual. This will not be discussed in depth in this book. The focus of this book is to present the concepts and basics of machine learning and artificial intelligence and to impart competencies to the readers through examples of applications, which are useful to assess the possibilities of using machine learning and artificial intelligence in their own organization, to apply this and hopefully to benefit from it.

This is also the case in organizations. It is necessary to check whether and possibly at which points in the organization data science, machine learning, and artificial intelligence can provide a benefit. In order to be able to decide and weigh this up, a basic knowledge of these future technologies is required. The following explanations are intended to contribute to this.

1.1 Artificial Intelligence

The term Artificial Intelligence (AI) can lead to misunderstanding. Artificial Intelligence is ultimately based on algorithms that were programmed by humans and differs in one essential aspect from conventional computer programs, namely that the programs can learn by themselves with the help of the algorithms, which is referred to as Machine Learning. So it is not conventionally programmed which instruction a program should execute, but it is programmed how a program should learn independently. The fact that a human still has to create a program for this means that the machine cannot learn entirely on its own.

The program is given instructions on how it should learn. Subsequently, it is provided with data from which it can learn, e.g., by trying to recognize a pattern or determine the correlation between attributes and behavioral alternatives. Thus, the computer learns more with each additional data set. Autonomous driving, for example, is learned by a computer by initially having a car driven by a human and the program recording and analyzing the relevant data. Once "enough" has been learned, this experiential knowledge can be applied to mimic the driver's driving behavior. Due to the quality and number of sensors, it is even possible that autonomous driving is superior to a human driver in certain situations due to better sensory abilities, e.g., when reversing into a parking space with multiple cameras and distance sensors. With the help of Artificial Intelligence, some tasks can therefore be solved better and other tasks worse than by a human.

1.2 Human Intelligence

It is easier to estimate where the so-called Artificial Intelligence promises advantages and where one should not expect any contribution from it after the term intelligence is better understood. Howard Gardner (2008) from Harvard University distinguishes intelligence into:

- Linguistic-linguistic intelligence
- Logical-mathematical intelligence
- Musical-rhythmic intelligence
- Visual-spatial intelligence
- Bodily-kinesthetic intelligence
- Naturalistic intelligence
- Interpersonal intelligence or also Social intelligence
- Intrapersonal intelligence

Artificial Intelligence has little to do with human intelligence. In any case, human intelligence is a combination of innate intelligence via inheritance and learned intelligence. In the case of computers, their intelligence results solely from the technical capabilities, the performance of the programmed software, and recently the Artificial Intelligence learned through Machine Learning.

But how does learning take place and into which categories can learning be divided? The taxonomy model widely used for learning by Bloom (Anderson and Krathwohl 2000) distinguishes the three learning areas of cognitive, affective, and psychomotor learning. **Cognitive learning objectives** refer to knowledge and intellectual skills (Ruhr University Bochum 2022). **Affective learning objectives** refer to changes in interests, feelings, attitudes, and values. Values are derived from the integration of values in personality, the establishment of a value order, attributing an emotional value to things and actions, acting according to recognized or known value ideas, and recognizing the evaluation of things or behaviors by other people. **Psychomotor learning objectives** refer to learning movement sequences by executing and coordinating movement sequences and imitating movement sequences.

Since the consideration of Artificial Intelligence particularly concerns the question of where the cognitive abilities of humans can possibly be supplemented or replaced by Artificial Intelligence, the following focuses on this area. The **Cognitive abilities** can be subdivided into remembering, understanding, applying, analyzing, evaluating, and creating (see Fig. 1.1).

Bloom assumes that the cognitive ability of one level is always the prerequisite for reaching the next higher level in the taxonomy pyramid. The cognitive potential increases steadily upwards. The cognitive competence of decision-making is approximately at the level between evaluating and creating (bitkom 2017, p. 13) and especially in supporting decisions or even making autonomous decisions, Artificial Intelligence can be particularly helpful.

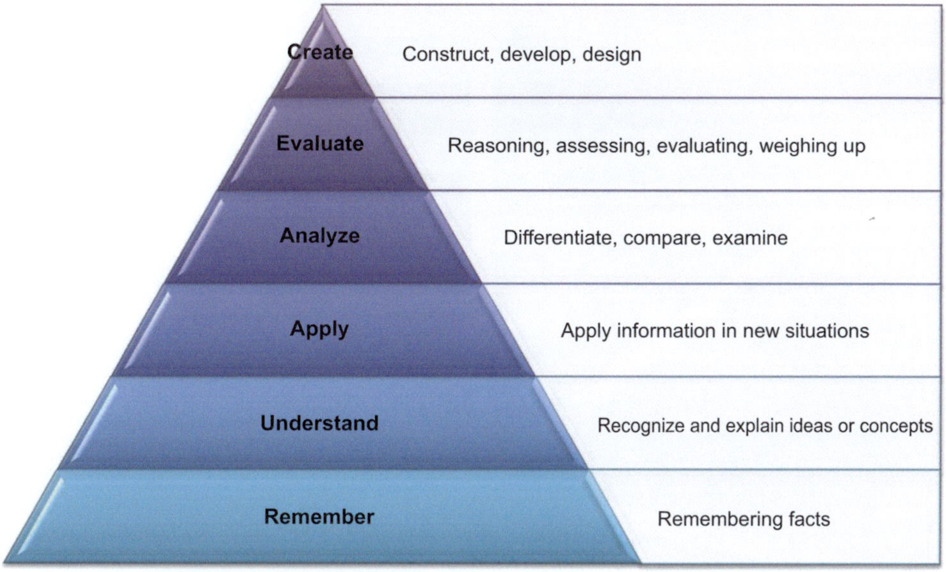

Fig. 1.1 Bloom's taxonomy of learning

1.3 Rational Decisions

It is clear that many decisions are made in organizations every day and that these decisions should be made as well as possible. Decisions made rationally and comprehensibly seem advisable for this. In Fig. 1.1, the decision-making process is presented according to the rational model of decision-making (Fig. 1.2).

The decision-making process consists of the following steps (Heesen 2016, p. 113):

1. Identifying the problem and assessing the situation:
   Whether a situation is considered a problem depends on a person's assessment and can vary greatly. For example, is a productivity increase of 2% a problem if the competition was able to increase it by 5%?
2. Identification of decision criteria:
   Is the aim of a 2% productivity increase to reduce the workforce and thus personnel costs by 2%, because the decision criterion and goal is cost savings? Or should the personnel available through the productivity increase be used in research and product development, because future growth opportunities are to be realized through new, innovative products? In the latter case, the proportion of profits invested in research and development would offer itself as a decision criterion.

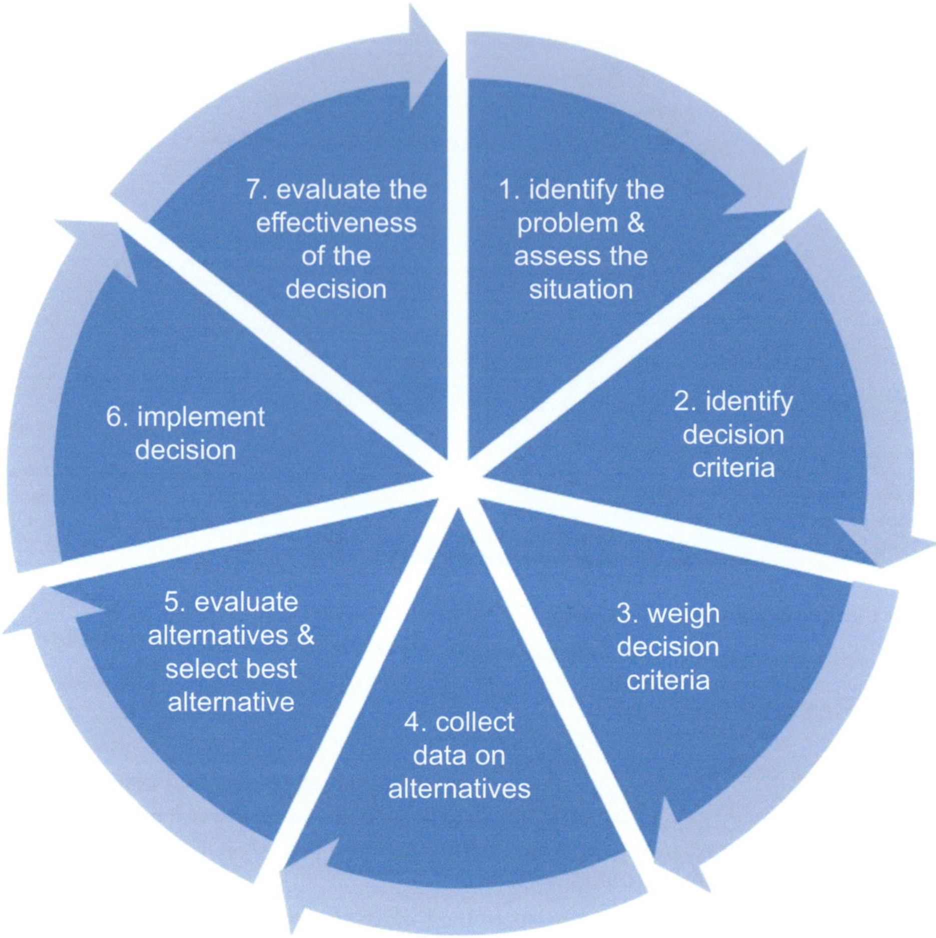

Fig. 1.2 Rational decision-making process

3. Weighting of decision criteria:
   Often, several goals are to be achieved at the same time, e.g., cost reduction and increase in research and development quota or cost reduction and sales growth. Which of these goals has which priority and should therefore be weighted to what extent?
4. Development of alternatives:
   For recurring problems, a set of already known alternatives can be used here. For new problems or when new alternatives exist, creativity is required.
5. Evaluation of alternatives and selection of the best alternative:
   If all decision criteria for each of the alternatives can be determined and these are objectively quantifiable, then the evaluation and selection of the best alternative is justifiable. However, it is a challenge for decision-makers to predict as perfectly as

possible which alternative will achieve the optimal goal. This requires comprehensive knowledge about the interrelationships of many variables, e.g., the variables costs, sales, profit, reputation, quality, customer loyalty. In ignorance of the expected short-term and long-term effects of a decision on these variables, the selection of the best alternative is impossible.
6. Implementation of the alternative
7. Evaluation of the effectiveness of the decision:
   An ex-post evaluation of the effects of past decisions, whether made by people or machines, is the basis for continuous improvement through learning.

If artificial intelligence can now be used in the decision-making process, the question arises as to which decisions should continue to be made by people and which decisions should in future be made by machines, i.e., computers, based on machine learning and artificial intelligence? This results in the degree of decision automation.

1.4 Decision Automation

The appropriate degree of automation depends on the circumstances. The extent of the use of artificial intelligence can be chosen in stages, as shown in Fig. 1.3 (bitkom 2017, p. 21–23).

At level 0, the human decides alone without a machine to assist him. At level 1, the human uses a machine, for example, to analyze data in advance of the decision and thus

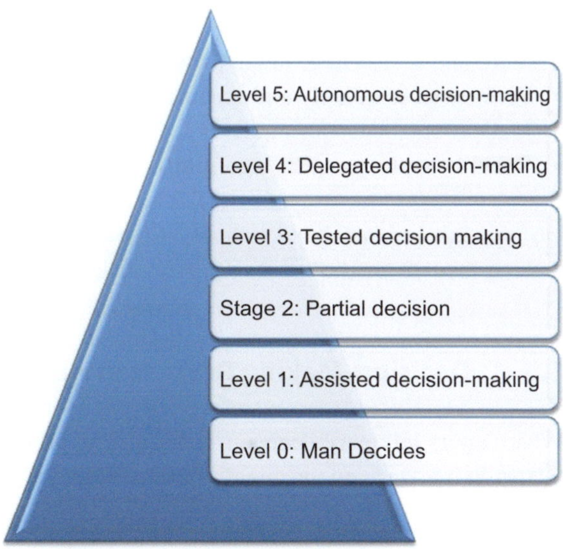

Fig. 1.3 Staged model of decision automation

better understand the context. This is also described as evidence-based management (EBM). At level 2, the machine carries out certain actions, but control and responsibility remain with the human, for example, when Alexa completes a purchase transaction at the human's call. At level 3, the machine generates decision proposals, for example, order suggestions. The decision whether to accept or reject the proposals remains in the hands of the human. At level 4, the human delegates decisions within certain limits to a machine, for example, the automated control of the switching on and off of the heating and the required heating power, after the desired room temperature has been set by the human. At level 5, the machine also takes over complex decisions and no longer needs a human to complete a task, for example, in autonomous driving. However, the human should always be given the option to withdraw the decision-making authority from the machine and take over control and steering himself.

To decide which decisions are suitable for which degree of automation, the strengths of humans and machines in decision-making should be considered in advance. The following cognitive strengths of humans and machines can be mentioned, among others (bitkom 2017, p. 61):

Human

- Perceive unexpected stimuli
- Creatively develop new solutions to problems
- Make difficult decisions with incomplete data

Machine

- Store and access large amounts of data
- Perform tasks quickly and execute multiple tasks in parallel
- Make routine decisions quickly
- Reliably apply complex algorithms

Do humans and machines actually follow the previously presented rational decision-making process and how does decision-making differ between individuals and machines?

1.5 Decision-Making Process in Humans and Machines

What factors influence decision-making? The following list distinguishes the influencing factors in humans and machines:

Human

- Situation (perception)
- Conflicts of interest between person and organization (what is best for the decision-maker, what is best for the organization):
  Based on the Principal-Agent Theory (Heesen 2016, p. 31), there can be conflicts of interest between the interests of the commissioned agent (e.g., employee) and the interests of the commissioning organization, e.g., when an employee advocates for a decision alternative from which he expects more individual benefits, although another decision variant would bring greater benefits to the organization.
- Conflicts of interest between goals (when improving one goal variable leads to a deterioration of other goal variables):
  If a reduction in customer service staff leads to a short-term cost reduction and profit increase, but in the medium term to lower customer satisfaction and thus to lower customer loyalty and in the long term to lower sales and profit, then such a conflict of interest between goals exists. The question arises as to which goal variable should be optimized in the short term or in the long term and which goals have which priority (Heesen 2016, pp. 118–119).
- Interpersonal conflicts of interest (when a conflict between people influences decision-making).
- Values
- Feelings
- Time of day (Leone et al. 2017)
- Time until a decision must be made
- Estimating the consequences (based on experiential knowledge and intuition)

Machine

- Situation (available data on the situation)
- Estimating the consequences (optimization of the defined goal variables)

Obviously, decisions made by humans are usually not made rationally and often even ignore available data (Selection Bias), as conflicts of interest, values, feelings, and other aspects also influence a decision. In contrast, machines make their decisions based on the available data rationally and optimize the goal variables according to the algorithms used (Müller and Massaron 2018).

Suppose a machine is to derive from data which recommendations for action should be given to the population and it is about minimizing the risk of shark attacks. Assuming that the machines with the ability of artificial intelligence initially only have the two variables of consumed ice cream in liters and the number of shark attacks over the course of the last 10 years with information per hour. The machine's conclusion would be that people should not go into the water when many people are seen eating ice cream. This would suggest that one should go swimming at night or in winter. If the temperature variable is also provided, the new recommendation would be that people should not go into the water when it is warm. If the location of the shark attacks is also provided as a variable, it would be derived that one should not go into the water at beaches that are very popular and therefore well attended (if more shark attacks are usually observed there) and that it is safer to go into the water at lonely, less visited beaches. All these recommendations would result as optimizations according to the algorithms and data and yet they would be more than questionable and definitely not optimal. However, if the variable of the number of people seeking water per location were also provided, a more sensible result would be obtained, because the number of shark attacks depends, among other things, significantly on how many people go into the water. If no person goes into the water in winter, at night, on cold days, and at lonely beaches, then the probability of a shark attack is naturally lower there. This example is intended to show that machines, despite fast computing power and the best algorithms, depend very much on which independent variables (context variables) as well as dependent (target variables) variables are provided for machine learning. Artificial intelligence is therefore not a panacea, but must be used sensibly and contextually appropriate.

1.6 Speed of Decision Making

Another factor in decision making is time. Due to the variety of influencing factors of a decision in humans presented in the preceding section, their decision is usually slower. A delay can occur, for example, due to the coordination of several people involved in the decision. The value of a decision is determined not only by its quality but also by the duration, as visible in Fig. 1.4 (Heesen 2016, p. 88).

If machines require less time for decision making, then the automation of decisions represents an additional advantage. This applies not only to the speed of decision making (step 4) but also to prompt data availability (step 2), data analysis (step 3), and implementation of the decision (step 5).

What data basis should be provided to decision-makers for the best possible decision making? Reliable data bases, both quantitative and qualitative, are essential for both people and machines. The example with the shark attacks made this clear. Perfect information availability does not exist in any decision, as the search for information will always be limited for reasons of time, effort, and selection bias. Ultimately, it must be accepted that decisions, both in humans and machines, are always made based on incomplete information, which is also referred to as bounded rationality (Heesen 2016, p. 154).

Due to the special importance of the data basis for the quality of decisions, many companies are developing into so-called data-driven companies, where data science, machine learning, and artificial intelligence take on a whole new, central importance. To ensure the best possible data availability and data quality, it is necessary to apply best practices of data science, which are presented in section 3.

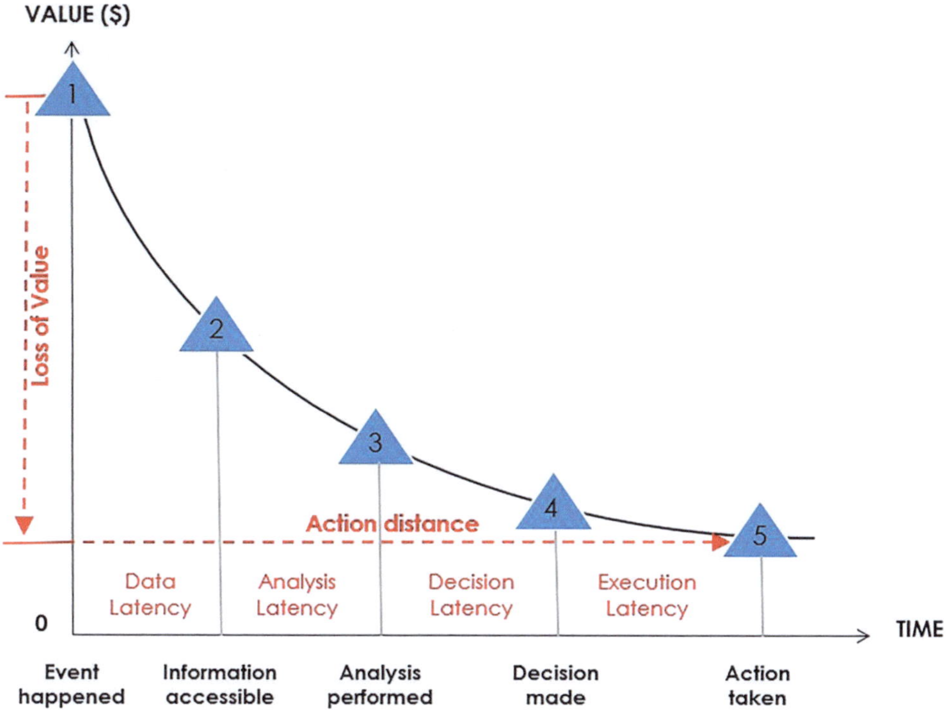

Fig. 1.4 Loss of value of a decision over time

1.7 Data Basis for Decisions

The data situation describes the availability and condition of data sources needed for a decision. The following aspects play a role here, among others: completeness, structure, ambivalence, real-time availability or delayed availability. Machine learning is based on the available data situation. If the quantity or quality of the data is insufficient, the quality of the decisions also suffers (bitkom 2017, p. 17).

Hardware and software are becoming increasingly powerful, thus enabling increasingly easier use of the advantages of analytics, even with larger amounts of data, machine learning and artificial intelligence. The global data volumes are constantly increasing (**Volume**) and thus also the possibilities to evaluate these data (see Fig. 1.5). The proportion of data from social media and in particular the proportion of text, image and video data, which are usually unstructured compared to business data, e.g. accounting, personnel or logistics data from ERP systems, is increasing significantly. Therefore, the analysis of unstructured data is becoming increasingly important (statista 2020).

From the perspective of an organization, however, the amount, the so-called volume, of the analyzed data does not play the primary role, but the added value that is created by

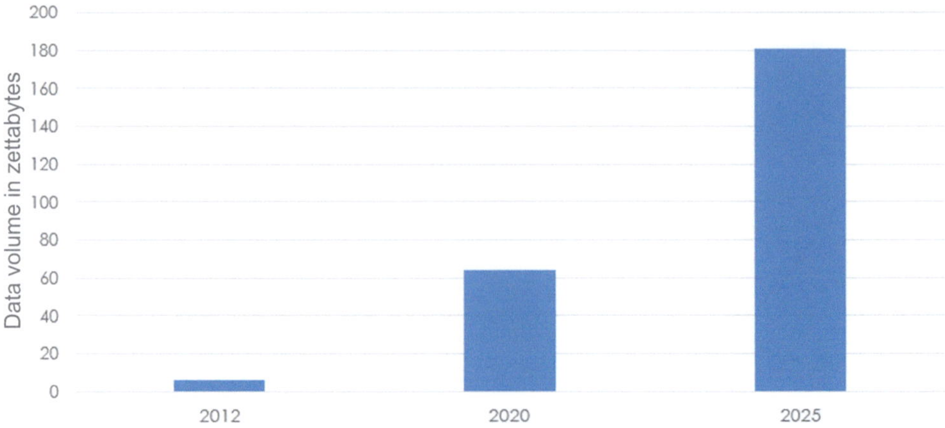

Fig. 1.5 Annual digital data volume generated worldwide from 2012–2020 and forecast until 2025. (Statista 2022)

the analytics, the so-called **Value** (see Fig. 1.6). By analyzing data, better decisions can be made if decision-relevant facts are provided.

In addition to volume and value, the new age of many data is characterized by other properties, which are collectively referred to as the 5 V's of Big Data: Volume, Velocity, Variety, Veracity and Value (see Fig. 1.6).

The increasing speed at which new data is generated is referred to as **Velocity**. The constant availability of new data in large quantities means that the insights gained from data analysis have a shortened half-life, as more current data is constantly available and the "old" data and analyses lose value. This results in the necessity and pressure to repeat the analyses regularly to stay up to date. Even the more powerful computers are once again reaching their limits. Efficient analysis is therefore still relevant with growing data volumes and more frequent execution of analyses despite the increased performance of computers. This also applies to machine learning.

Variety refers to the heterogeneity of the available data formats. Unstructured data is usually extracted from the source systems via a so-called ETL process before the analysis (E), then transformed into the target format or data model (T) and finally loaded into a system (L) where the data can be analyzed and displayed. In machine learning, this process is also often referred to as **preprocessing**.

Afterwards, the data is available in a format suitable for analysis (see Transformed Data in Fig. 3.2). For example, audio data, e.g. the transcript of a YouTube video, is converted into text so that it can then be analyzed. However, if a preprocessing process is too time-consuming and the timeliness of the evaluation takes precedence, the primary data can alternatively be analyzed directly, unchanged and without conversion, provided the data quality of the primary data is considered sufficient for this purpose (see Original Data in Fig. 3.2).

Fig. 1.6 The Five V's of Big Data

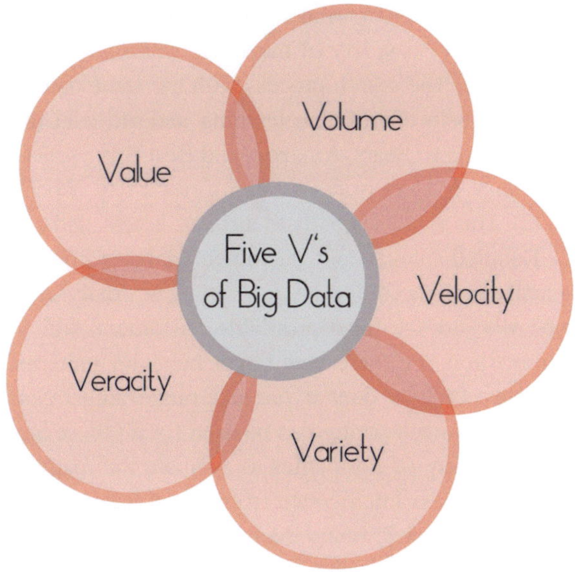

Veracity refers to data quality. For example, texts in social media are often grammatically incorrect, formulated in colloquial language and peppered with spelling errors or abbreviations. To ensure that text analysis, e.g. sentiment analysis of positive and negative feelings, does not lead to inaccurate results, a cleaning of the data is carried out if necessary before an analysis, in which process spelling mistakes and other errors are eliminated. This step is part of the preprocessing or transformation of the data. Also significant for data quality is, among other things, the handling of missing data.

1.8 Business Intelligence

The last section already explained the importance of data for decision-making and that its value can vary greatly for an organization. Data in isolation, considered on its own, has no value. Only when it serves to make better decisions does it provide benefit. For example, a better understanding of customer preferences can contribute to targeted and effectively tailored marketing measures. In such a case, the effort of data storage and data analysis is worthwhile to increase business intelligence. Machine learning can be usefully employed for this purpose.

To compare the expected benefit of an analysis with the corresponding costs, a corresponding business case should be created. If one compares a decision without the presence of the information with the decision when the information is present and the resulting consequences, it can be weighed which of the two scenarios is better for the organization. If, for example, one knows the probability that a customer is willing to pay a certain price for a product (e.g., a ticket for a concert), one can adjust this variably, person-specifically (e.g., depending on age, marital status, income, past purchase transactions …) and time-dependently (time of day, day of the week, season, temporal distance to the event, temporal distance to the last search process with the same date and product) and thus maximize profit. With the help of machine learning and artificial intelligence, this is possible in a timely manner, and so Sears, Amazon, and Best Buy use a **Dynamic Pricing** via machine learning to change product prices several times a day depending on market parameters (Campbell 2022). This applies equally to airlines, hotels, and e-commerce sites.

Detailed market and customer analyses contribute to new insights that can subsequently lead to competitive advantages. Analyses can help answer important questions, e.g., who are the most profitable customers, which customer groups have turned to the competition, are there new buyer behavior patterns, what influence does the price have on sales, or how does a product range adjustment or change in sales channels affect sales. Machine learning can be used for all these analyses.

The ability of companies to evaluate the many available data in a timely manner and derive business intelligence from it can hardly keep pace with the rapid data growth. As shown in Fig. 1.7, many decisions are not made based on rational decision criteria and analytics, and far from all available data is used. This creates the so-called Critical Gap and a Business Intelligence Gap (Strategy and Transformation Consulting 2020).

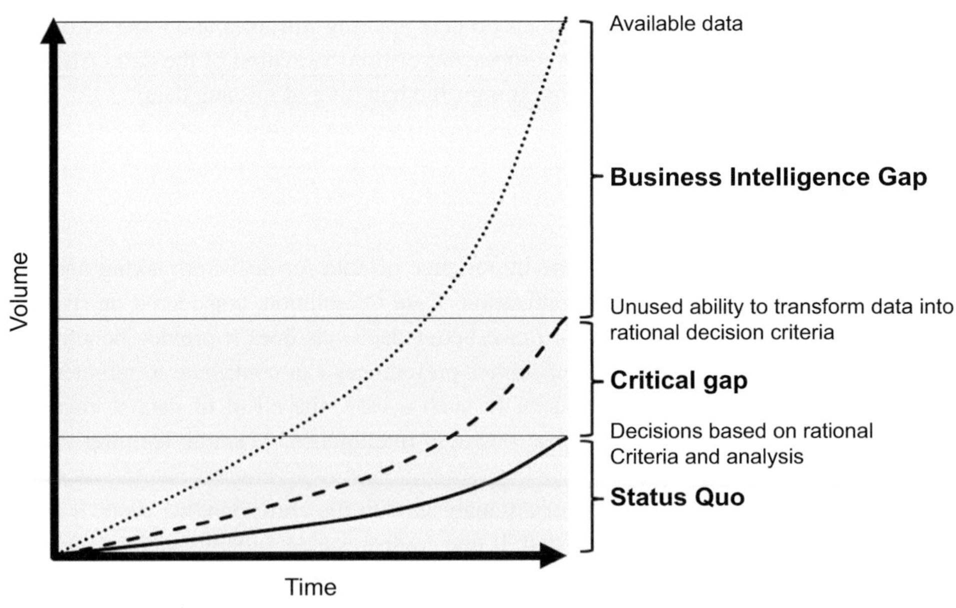

Fig. 1.7 Critical Gap and Business Intelligence Gap. (Own illustration based on Strategy and Transformation Consulting 2020)

1.9 Innovation

Obviously, it makes sense to examine the innovation of machine learning and artificial intelligence to see if and where their use in an organization is appropriate. Basically, changes are only advisable if they can solve a problem that exists. The perception of a problem is therefore the basis, also for the use of an innovation.

Empirical sciences, which also include social and economic sciences, are based on observing, measuring, and collecting data and conducting experiments to better understand the subject of study. Knowledge is gained through observation, description, explanation, and prediction of phenomena.

Questions or problems in reality motivate questioning things to derive explanations, predictions, or theories from them. Data science, including machine learning and artificial intelligence, can be valuable based on data for solving problems and answering questions (see Fig. 1.8).

Especially in a time of constant change and transformation, it is increasingly important for organizations to quickly recognize and solve challenges, find innovative answers to questions and problems, and data science, machine learning, and artificial intelligence can contribute to this.

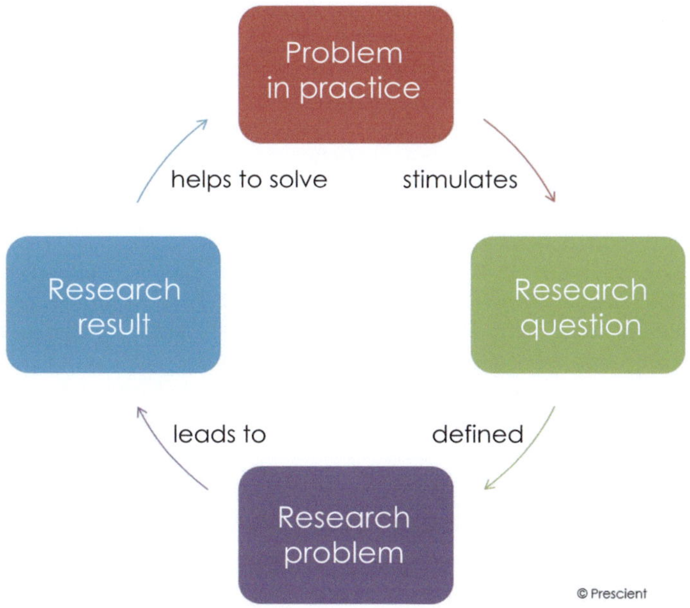

Fig. 1.8 Problem-solving process. (Heesen 2021, p. 28)

References

Anderson, L. W., & Krathwohl, D. R. (2000). *Taxonomy for Learning Teaching: A Revision of Bloom's Taxonomy of Educational Objectives.* München: Langenscheidt ELT.

bitkom. (2017). *Künstliche Intelligenz verstehen als Automation des Entscheidens: Leitfaden.* Abgerufen am 04. Dezember 2022 von https://www.bitkom.org/sites/default/files/file/import/Bitkom-Leitfaden-KI-verstehen-als-Automation-des-Entscheidens-2-Mai-2017.pdf.

bitkom. (2019). *Konkrete Anwendungsfälle von KI & Big-Data in der Industrie.* Abgerufen am 27. Dezember 2020 von https://www.bitkom.org/sites/default/files/2020-02/200203_lf_ki-in-der-industrie_0.pdf.

Bundesregierung (Hrsg.). (2020). *Schwarzrotgold 03/2019: Wie uns künstliche Intelligenz hilft.* Abgerufen am 25. Dezember 2020 von https://www.bundesregierung.de/breg-de/suche/schwarzrotgold-03-2019-wie-uns-kuenstliche-intelligenz-hilft-1643006.

Campbell, P. (28. April 2022). *How does dynamic pricing work? Examples, strategies, and models.* Abgerufen am 05. Dezember 2022 von https://www.priceintelligently.com/blog/bid/198355/how-to-implement-a-dynamic-pricing-strategy-without-the-pr-backlash.

Gardner, H. (2008). *Intelligenzen: Die Vielfalt des menschlichen Geistes.* Stuttgart: Klett-Cotta.

Heesen, B. (2016). *Effective Strategy Execution: Improving Performance with Business Intelligence* (2. Ausg.). Berlin: Springer.

Heesen, B. (2021). *Wissenschaftliches Arbeiten: Methodenwissen für Wirtschafts-, Ingenieur- und Sozialwissenschaftler* (4. Ausg.). Heidelberg: Springer-Gabler.

Leone, M. J., Slezak, D. F., Golombek, D., & Sigman, M. (2017). Time to Decide: Diurnal Variations on the Speed and Quality of. *Cognition*(158), 44–55.

Müller, J. P., & Massaron, L. (2018). *Artificial Intelligence For Dummies.* Hoboken: John Wiley & Sons.

Rittershaus, A. (5. Mai 2020). *Was Sie zum Thema KI wissen müssen.* Abgerufen am 26. Dezember 2020 von https://www.computerwoche.de/a/was-sie-zum-thema-ki-wissen-muessen,3544140.

Ruhr-Universität Bochum. (2022). *Typen und Stufen von Lernzielen.* Abgerufen am 04. Dezember 2022 von https://dbs-lin.ruhr-uni-bochum.de/lehreladen/planung-durchfuehrung-kompetenzorientierter-lehre/lehr-und-lernziele/typen-und-stufen/.

Shaer, M. (2014). *Is This the Future of Robotic Legs?* Abgerufen am 26. Dezember 2020 von https://www.smithsonianmag.com/innovation/future-robotic-legs-180953040/.

statista. (2020). *Internet Traffic.* Abgerufen am 25. Dezember 2020 von https://de.statista.com/statistik/daten/studie/266885/umfrage/prognose-zum-datenvolumen-des-privaten-und-geschaeftlichen-ip-traffics-weltweit/.

statista. (2022). *Volumen der jährlich generierten/replizierten digitalen Datenmenge weltweit in den Jahren 2012 und 2020 und Prognose für 2025 (in Zettabyte).* Abgerufen am 05. Dezember 2022 von https://de.statista.com/statistik/daten/studie/267974/umfrage/prognose-zum-weltweit-generierten-datenvolumen/.

Strategy & Transformation Consulting. (2020). *Fachkompetenzen: Business Analytics.* Abgerufen am 27. Dezember 2020 von https://www.strategy-transformation.com/kompetenzen-business-analytics/.

2 Machine Learning

There exist three methods of machine learning (see Fig. 2.1):

- **Supervised Learning**
  Learning based on training data, whose input (independent variables) and output (dependent variable) are known. The algorithm compares its predicted output of the dependent variable based on the independent variables with the real value and adjusts its approach accordingly.
  - Advantage: Traceability, comparability.
  - Disadvantage: Effort in data preparation.
  - Application: Predictions, chatbots, translators, detection of spam emails, recommendation for proactive machine maintenance, dynamic price management.
- **Unsupervised Learning**
  Identification of unknown structures in data, without knowing the target output in advance. It recognizes unknown patterns from the data and derives rules.
  - Advantage: Little effort, high flexibility.
  - Disadvantage: Complexity, manual evaluation of the results required.
  - Application: Clustering, pattern mining, personalized advertising, market-basket analysis (who buys this, also buys that …).
- **Reinforcement Learning**
  Identification of solutions and strategies for complex questions without initial material simply based on the trial-and-error method and a reward system. Knowledge and intelligence arise from as many repetitions of the trial-and-error method as possible.

B. Heesen, *Artificial Intelligence and Machine Learning with R*,
https://doi.org/10.1007/978-3-658-45392-3_2

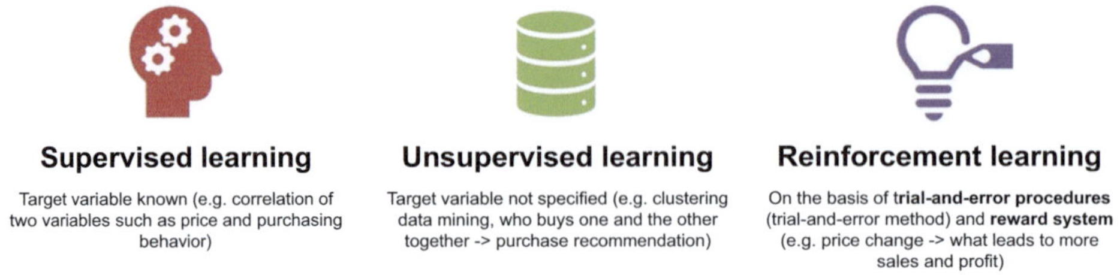

Fig. 2.1 Machine Learning Methods

- Advantage: Little effort, optimization of the target variable without intervention.
- Disadvantage: Mistake/error, explanation of the decision is limited transparent.
- Application: Dynamic price adjustment on online stores (which price optimizes sales/profit), autonomous robots (vacuuming, lawn mowing), autonomous driving.

Supervised machine learning is based on the concept that a machine learns from data and transfers this to other data and situations after the learning phase has ended. Algorithms build a statistical model in machine learning that is based on training data. This means that the model does not simply remember the data with which it was trained, but it recognizes patterns and regularities in the training data. Thus, the model can possibly also learn something new from data or apply learned (**training data**) to other data (current data), e.g. make predictions. Before a model is considered "trained" (**trained model**) and applied to other data, the patterns and regularities are evaluated with the help of **test data**. If the quality of the model (**model quality**) is considered sufficient, then the trained model is applied. The results of the trained model should also be continuously checked to recognize when, for example, old patterns can no longer be sensibly applied to the new data and thus a new training of the model (**re-training**) should be initiated.

A question that often arises is whether one can really trust the machine. Does the machine make good decisions and produce good results, e.g. reliable predictions? People trust a machine more when it works reliably and people understand how and why the machine behaves in one way or another (bitkom 2017, p. 10). It is therefore important to understand how a machine uses machine learning and how machine learning works.

There is often the misconception that machine learning happens completely autonomously and without human influence, but in fact machine learning is also controlled by software developers. Classic programming of software processes data based on an explicit sequence of instructions in a programming language, which lets the machine execute specific processing steps and algorithms (Fig. 2.2).

In machine learning, this is not fundamentally different, as here too an explicit sequence of instructions is given in a programming language. In the case of machine

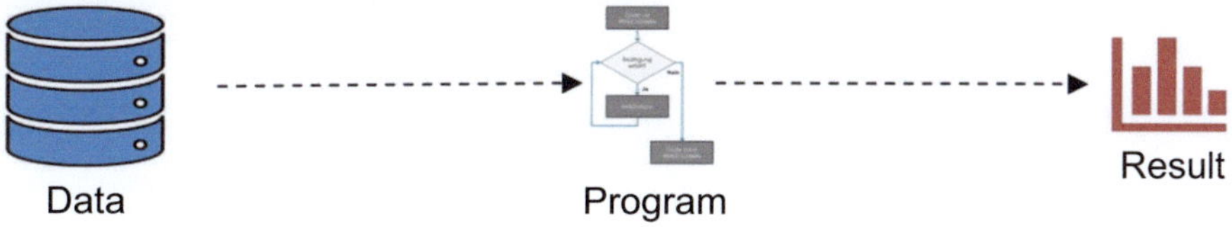

Fig. 2.2 Programming

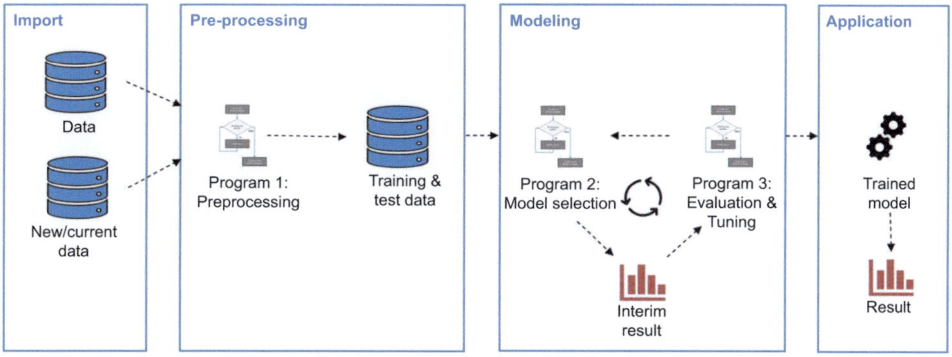

Fig. 2.3 Machine Learning

learning, this is the preprocessing of the data, the generation of training and test data, the model selection, the generation of intermediate results, the validation of the intermediate results, and the parameter tuning. The examination and decision as to whether a model meets the quality requirements before it is actually applied is also the responsibility of humans. Even after a machine learning model is applied, the quality is regularly checked and, if necessary, the model is re-trained (Fig. 2.3).

Validation should also be carried out repeatedly in the application phase, as machine learning is based on statistics and probability theory. Accordingly, the probability of an error by machines can be minimized, but not excluded.

To effectively use machine learning, it is important to follow an established process, as shown in Fig. 2.4.

After import, the data is preprocessed. This includes the selection, validation, fusion, transformation, supplementation, and summarization of the data (for details, see section 6). Subsequently, it is important to gain a better understanding of the data with the help of exploratory data analysis. This can contribute to further data being removed or added.

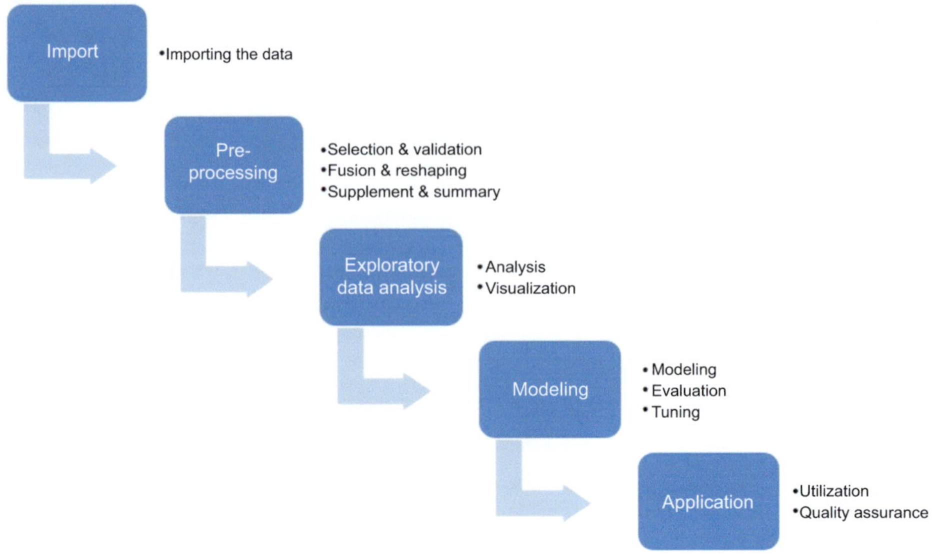

Fig. 2.4 Machine Learning Process

2.1 Supervised Learning

In supervised learning, the two most important applications are regression and classification. Classification, for example, involves dividing mail into the two classes of spam and non-spam. Regression is about predicting a numerical target variable as accurately as possible. This could be the arrival time, the price with the optimal transaction probability in an online store, or the value of a property or a used car.

To train machine learning models, a set of training data is provided. This training data includes both the target variable (dependent variable), which is to be predicted (spam = yes/no, value of a property), and descriptive variables (independent variables). The model then analyzes the relationship of these variables and remembers significant relationships in the so-called trained model. In a next step, based on test data, it is checked what prediction the trained model generates for the test data and to what extent this corresponds to the real target variables from the test data. The percentage of prediction error is considered a key quality criterion to decide whether the model is suitable for practical application. The application then allows the target variable to be predicted based on the availability of the independent variables.

2.2 Unsupervised Learning

In unsupervised learning, key applications are clustering, dimension reduction, outlier detection, and market basket analysis, which is based on association rules. Unlike supervised learning, there is no target variable that can be used for training. Machine learning tries to recognize completely new relationships in the data.

Clustering involves assigning all data to a certain number of clusters that should be as similar as possible, and should be as close as possible in the n-dimensional space (n = number of variables used) to the cluster centers assigned to them. The number of clusters does not have to be defined in advance, but can be determined by the machine learning model itself. In organizations, clustering is used, among other things, for customer and market segmentation, where customers with similar purchasing behavior are grouped together and then addressed with marketing campaigns tailored to them.

Dimension reduction involves machine learning models that investigate the relevance of factors to a dependent variable and eliminate those factors that are of low relevance. This can be used well, among other things, to preselect and reduce the variables that are important when training machine learning models of supervised learning.

Market basket analysis involves creating customer profiles based on the analysis of previous product purchases. The purchase probability for a product is predicted depending on the purchase of another product or a customer type assignment. In retail, customer cards are used, among other things, to collect data in order to be able to combine additional personal and demographic data as independent variables with purchasing behavior. In online retail, purchasing behavior is recorded using cookies and user accounts. The method of association analysis is used for the evaluation of the data and the creation of association rules. The results of this machine learning method are used, among other things, for recommendations on websites, such as on Amazon under "Related products to this article" or "Similar products". These hints have been proven to increase sales in online stores. The information is also used to determine the placement of products on websites and in stores, thereby also increasing sales.

2.3 Reinforcement Learning

In reinforcement learning, an agent is supposed to independently learn a strategy (policy) based on rewards through trial and error in order to achieve a maximum reward. A so-called interpreter analyzes the environmental variables and determines both the next state and the amount of the reward based on an action taken. This is repeated until the agent has completed its task. In the case of an autonomous vehicle, for example, the task would be completed when the destination has been reached. Rewards could be, for example, for a vacuum cleaner robot, if it could locate and eliminate a lot of dirt. Punishments for a vacuum cleaner robot could be, for example, if it repeatedly goes to a place where there is little dirt or repeatedly hits a wall or table leg. Without prior knowledge of the room, the vacuum cleaner robot learns over time all by itself, where walls and furniture are and that there is more dirt to be found in the entrance area. An application in an online store would be the dynamic adjustment of the sales price depending on the customer profile, country of call, time, date and scarcity of the goods and learning via reward when a transaction has been successful.

Many other applications for machine learning are conceivable.

Reference

bitkom. (2017). *Künstliche Intelligenz verstehen als Automation des Entscheidens: Leitfaden.* Abgerufen am 04. Dezember 2022 von https://www.bitkom.org/sites/default/files/file/import/Bitkom-Leitfaden-KI-verstehen-als-Automation-des-Entscheidens-2-Mai-2017.pdf

3 Best Practices

Data analysis, including machine learning, ideally takes place systematically and transparently, thus following a standardized process. Standards that have been established for the management of data science projects include the Cross Industry Standard Process for Data Mining (Crisp-DM) and the Team Data Science Process (TDSP), which was established by Microsoft (Data Science Project Management 2021).

The **CRISP-DM** process consists of six process phases:

1. Business Understanding—Defining objectives.
2. Data Understanding—Collecting, describing, exploring, validating data.
3. Data Preparation—Selecting, cleaning (cleansing), transforming, integrating, formatting data.
4. Modeling—Selecting modeling techniques, creating models and testing their suitability.
5. Evaluation—Determining results.
6. Deployment—Monitoring operation, maintenance and updating, communicating insights.

The **TDSP** process consists of five process phases:

1. Business Understanding
2. Data Acquisition and Understanding
3. Modeling
4. Deployment
5. Customer Acceptance

B. Heesen, *Artificial Intelligence and Machine Learning with R*,
https://doi.org/10.1007/978-3-658-45392-3_3

Both process models support a structured approach to data science projects, which may include machine learning based on artificial intelligence algorithms, statistical methods, or established methods of business intelligence, corporate performance management, data warehousing, data mining, and text mining. A process model specifically adapted for machine learning looks as follows (see Fig. 3.1).

Before data can be used for machine learning (see Frontend Applications layer in Fig. 3.2), appropriate preprocessing may be necessary (see Data Management layer in Fig. 3.2). As shown in the Business Analytics Framework, both structured and unstructured data and data streams (see Data Supply layer in Fig. 3.2) are used. In the Data Management layer, both the master data required for system integration and the harmonization of heterogeneous data structures, as well as preprocessed data (base tables or OLAP cubes or Online Analytical Processing Cubes), which are already optimized and aggregated for evaluation through transformation processes, are stored. However, if access to real-time data that has not yet undergone a transformation process, so-called original data, is required, access can be direct, either physically, via the cloud, embedded Application Programming Interface (API), or databases. A variety of frontend applications can then use the data for Past (Historical Data), Realtime (Real-time Data), Predictive (Forecasts/Planning), or Prescriptive (Prescribing Actions) analyses.

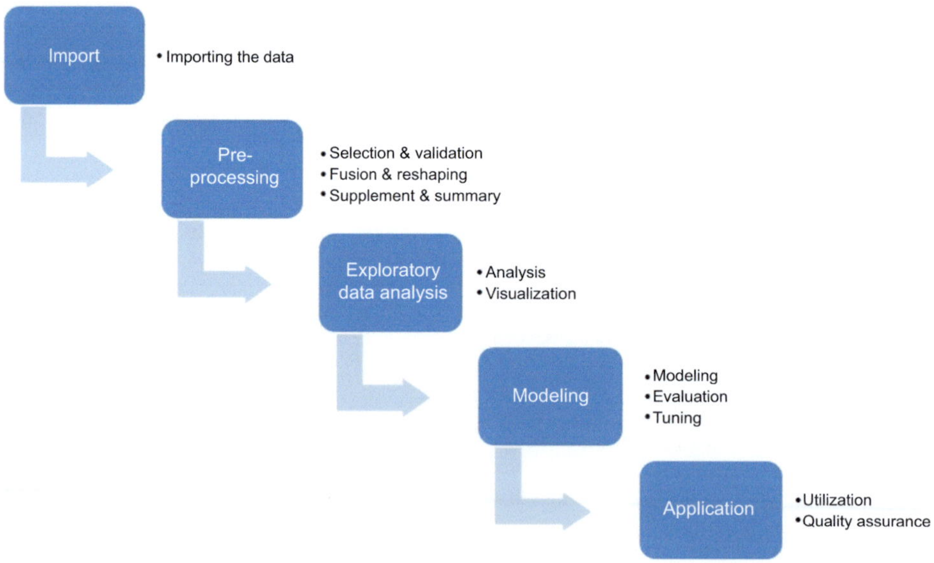

Fig. 3.1 Machine Learning Process

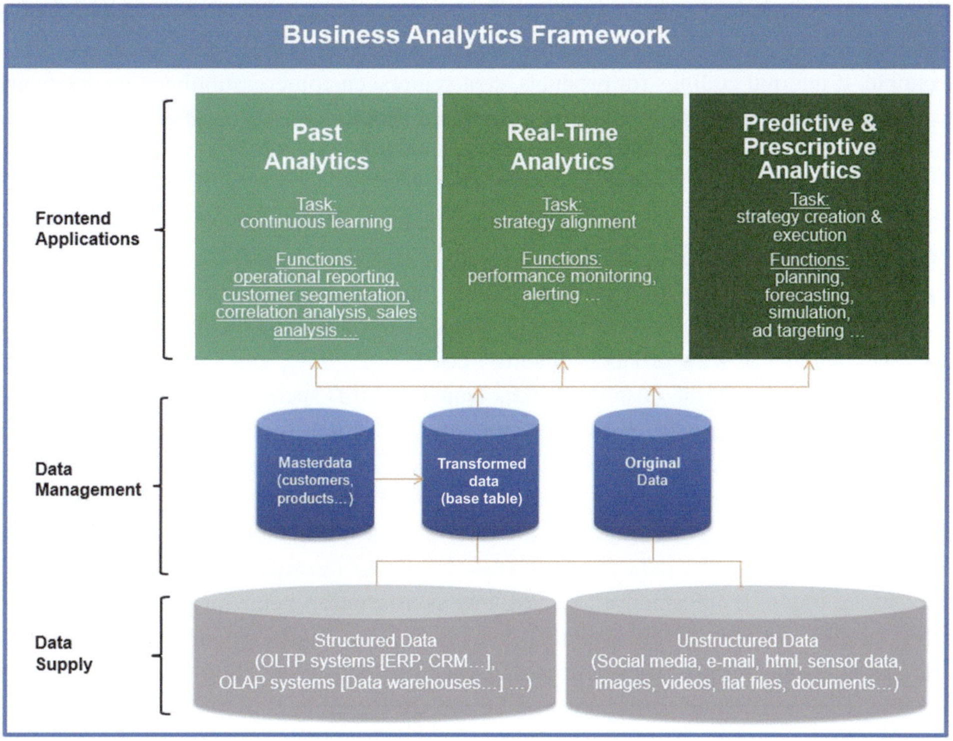

Fig. 3.2 Business Analytics Framework

3.1 Data Quality

When data is shared with other people and organizations, it pays off to use established standards for data exchange. Those who practice data science often use data from authorities, organizations, suppliers, multiple organizational units, etc. Not all of this data necessarily uses globally uniform encodings of countries, industries, products, currencies, and much more. If these basics do not match, data integration is required in preprocessing before a comprehensive evaluation is possible.

The ISO standard **SDMX** (Statistical Data and Metadata Exchange, ISO 17369:2013) (Stahl and Staab 2017) has been established to define data structures for any phenomena of economic life, i.e., finance and real economy, and to develop data exchange processes, data collections, and corresponding data analysis products based on them. The initiative was launched in 2001 by organizations including the IMF, OECD, UN, World Bank, Eurostat, and ECB.

Data often promises objectivity and provides security, but it is also important to trust the data and as a data scientist to ensure that the data is correct and meaningful. For example, if internal codings for products, cost centers, customers, suppliers, locations are used in the data, then these codings (master data) should be used in all datasets used for data validation (transaction data). Through this process of validation, for example, data records with sales data can be filtered out that refer to invalid product numbers or invalid countries. For this reason, many organizations make great efforts towards **data integration**, by integrating the data from various data sources into a unified, high-quality database. This integration enables systematic and reliable analysis of the database.

For example, scanner cash registers in stores store data (**transaction data**) regarding customer purchases with product, quantity, price, time, date, and location. If this data can be integrated with the customer's data (**master data**) such as name, address, email, phone number, age, gender, profession, income, and much more, then the information value can be increased through the linkage (integration). This additional customer information can be used, among other things, for customer segmentation, for predicting sales per age group, gender, profession, and income, or for sending targeted advertising to customers. The value of the data can be further increased through additional data integration, for example, if information on customer consumption behavior at other locations, at gas stations, airlines, etc. is available. Such data can be collected, among other things, via customer bonus point program cards. However, data integration is only efficiently feasible if certain requirements are met by the data sources, for example, the uniform coding of the transaction data based on the master data such as customer number, currency, product number (see Fig. 3.3).

For the interpretation of a customer's measured sales, it requires, among other things, uniform information on the customer number, product coding, and price with a currency coding. If the data is available in such a format as transaction data for each purchase, the data can be evaluated independently of the location of the recording and the data source (supermarket, gas station, pharmacy …).

SDMX provides a data model that corresponds to the widespread star or snowflake schema. The actual information (facts in the star schema) such as the number and price of products are at the center and are surrounded in a star shape by the identifiers or key fields (dimensions in the star schema, key field to a master data table) (Fig. 3.4). This data model is also referred to as a multidimensional data cube or **OLAP cube**. It

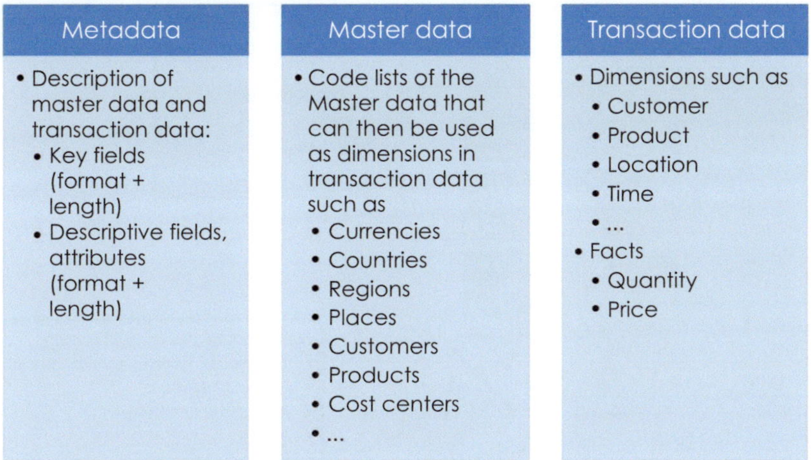

Fig. 3.3 Requirements for data sources

Fig. 3.4 Star schema with facts and dimensions

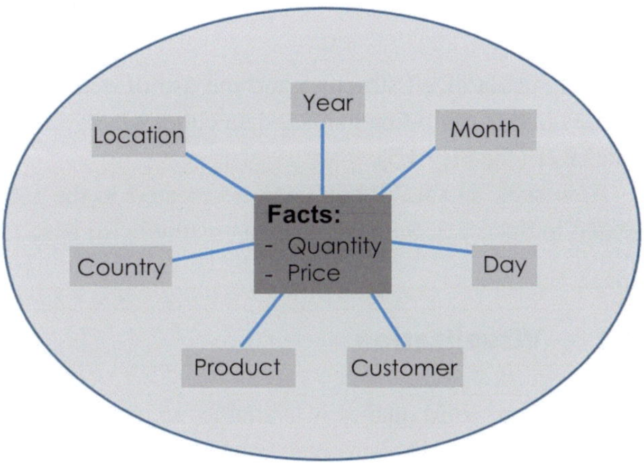

is important that standardized code lists are available for the dimensions in the master data, such as ISO country keys, product keys, postal codes, GPS coordinates (see Fig. 3.5).

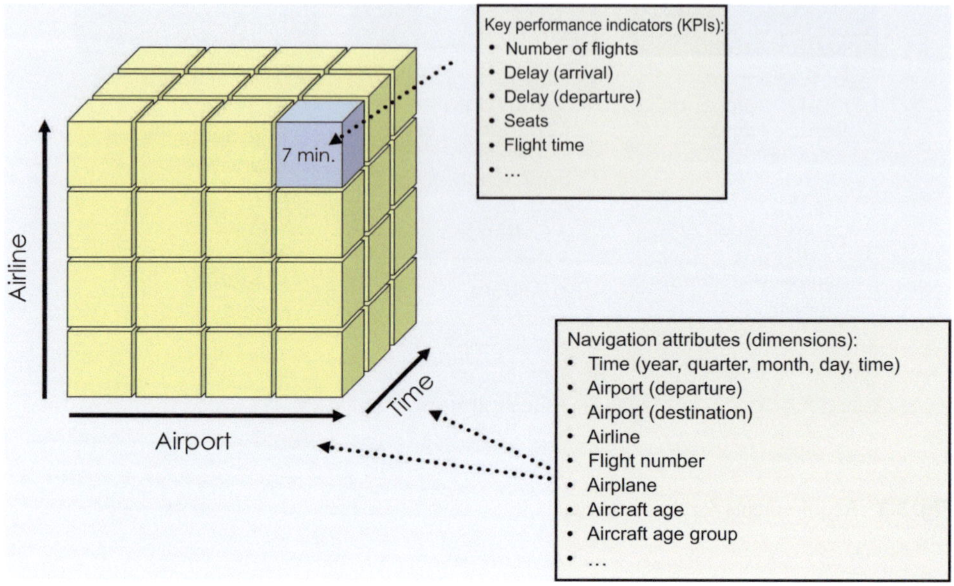

Fig. 3.5 Data cube flight data

This standardized structure and the use of code lists (dimensions) then allows search queries in this multidimensional data cube, which can be imagined as a multidimensional data table (see Fig. 3.5).

How such an OLAP cube can be created in the form of a base table with R is presented in Sect. 6.2. Suitable analysis methods for base tables are described in Sect. 6.3.1.

3.2 Visualization

When data of solid quality is available, an exploratory data analysis can initially be carried out. Visualizations are very valuable for recognizing relationships in data, especially with large amounts of data. Visualizations also serve to communicate new insights from the data to the relevant target groups. Visualization facilitates the intuitive understanding of relationships better than any other form of representation and is therefore an essential part of effective communication. Information visualization combines the strengths of interactive data analysis with the human ability to quickly grasp patterns and trends, such as in a dashboard or management cockpit with navigation functions (see Fig. 3.6).

In many disciplines such as music (notes …), traffic (traffic signs …), construction (construction drawings), navigation (maps) etc., standards have proven themselves in visualization. What does this look like in the world of business? The organization **Inter-**

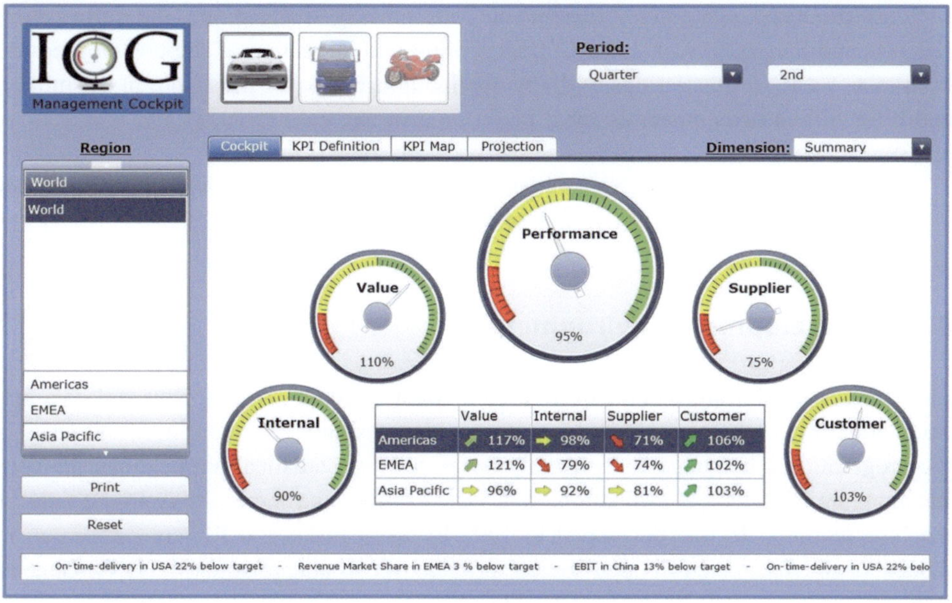

Fig. 3.6 Management Cockpit

national Business Communication Standards (IBCS Association 2017) describes valuable standards that data scientists can follow. It describes how labels, times, scenarios, deviations, highlights, and font sizes can be adequately represented. The design and formatting of tables are also described there. Recommendations are given on how their position and format should be designed and that abbreviations and units should be used uniformly (Hichert and Faisst 2019, p. 67). Among other things, it is recommended to always depict time axes horizontally (Hichert and Faisst 2019, p. 74). Scenarios such as the comparison of actual data, plan data, and forecast data should be visually differentiated. The recommendation is to depict actual data filled with color, plan data framed without fill color, and forecast data framed and filled with a hatching (Hichert and Faisst 2019, p. 79). When comparing plan, previous year, or other deviations, positive deviations should always be green and negative deviations always red (Hichert and Faisst 2019, p. 89). Columns or bars are suitable for absolute deviations and for better differentiation, relative or percentage deviations can be depicted as needles with a green or red head. It is recommended to make highlights recognizable through colored difference markings, trend indicators, or similar (Hichert and Faisst 2019, p. 93). For table design, the recommendation is that both row and column headers should be right-aligned and calculation results such as sums should be displayed in bold (Hichert and Faisst 2019, p. 103). For attributes such as the operational functions production, sales, R&D etc., it is recommended to use symbols or icon sets, just like for the differentiation of categories such as products, customers, regions etc., to quickly differentiate the information (Hichert and Faisst 2019, p. 131). Correct scaling and indication of the unit of measure such as €, $ etc. in illustrations with the applied unit like k for thousand, m for million and b for billion or comparable must be ensured in any case (Hichert and Faisst 2019, p. 138).

Recommendations for the implementation of organization-wide visualization standards are presented in Sect. 6.3.3.1.

3.3 Development Environment

Extracting valuable information from large amounts of data requires powerful data science tools. A data science tool that has become increasingly important in recent years is the programming language R, originally developed for statistical applications. Its core functions lie in statistical evaluation, data visualization, and increasingly in the field of machine learning. R was developed in 1992 by Ross Ihaka and Robert Gentleman in Auckland and distributed as open-source software free of charge under the GNU General Public License by the R Foundation for Statistical Computing (R Core Team 2022).

R is among the top 15 most popular programming languages in the TIOBE Index (Tiobe 2022), Redmonk Ranking (O'Grady 2022), O'Reilly Popularity Index (Brust 2020) and is on the list of most popular data science programming languages (Muenchen 2020). Among the analysis tools for Big Data Analytics, R is considered one of the most popular (Guru99 2022). As a statistical application, for data mining, machine learning, and analytics, R has established itself as one of the popular tools alongside Python (Muenchen 2020).

The free open-access journal "The R Journal" regularly reports on current publications on R and newly available packages/functions (The R Foundation 2022). The free journal "Journal of Statistical Software" (Foundation for Open Access Statistics 2022) is also recommended. Valuable information about R can also be found on the Bloggers platform (R Bloggers 2022) and on Stack Overflow (Stack Overflow 2022).

In R, everything revolves around objects, as entire expressions of an R script, variables, functions, and even operators are considered objects. Nevertheless, R is not a classic, fully object-oriented language that works with concepts such as classes, inheritance, etc.

3.4 Programming

Best practices should also be applied as developers and programmers, including, for example, consistent naming conventions. When data is read in, these objects should be named and transformed according to one's own conventions.

The following best practices are recommended for programming with R:

- **Naming conventions**:
  - Object names should be as short as possible.
  - Object names should consist of lowercase letters if the programming language distinguishes between upper and lower case, as is the case with R. This helps to avoid unnecessary errors. Delimiters such as "_" or "." are suitable for indicating details such as the year, e.g. "sales_2021" or "sales.2021". Variables (columns) in a table (dataframe, tibble, matrix, array) can either be uniformly defined in lowercase or starting with a capital letter. It should be considered that the column names will later also be displayed with their name in the tabular or graphical output.
- **Structure conventions**:
  - **Base tables** (dataframe, tibble, matrix, array), which form the basis for evaluations, should consist of columns, each representing a variable (key field or descriptive feature), and rows, each containing an observation (data record). The manifestations or values of the variables are in the cells of the table. This format is also referred to as "long format" or "narrow format" in R. What should be avoided for base tables, on the other hand, is the so-called "wide format", in which different manifestations of values of a variable are in different columns, which makes their evaluation more difficult.
  - Key fields in tables, which serve to identify data records, e.g. customer number or order number, should always be in the first columns (left in the table) as they are of particular importance.
- **Selection conventions**:
  - Columns (variables) not required for the analysis should be deleted or deselected from base tables.
  - Rows (data records, observations) not required for the analysis should be deleted or deselected from base tables.
- **Coding conventions**:
  - Date information
    - If only a year is given, it should be coded as an integer.
    - If a year and month are given, the day should be indicated with 01 and this should be coded as a date variable.
    - If a year, month, and day are given, this should be coded as a date variable.
  - Nominal variables should be coded as factor variables. In the case of binary variables that have Yes/No as an option, coding as a logical variable (TRUE/FALSE) is also possible.
  - Ordinal variables should be coded as factor variables with ordered=TRUE.
  - Metric variables should be coded either as integer or numeric.

How these best practices are applied in R during data transformation is described in more detail in Sect. 6.2.

References

Brust, A. (18. Februar 2020). *Python "preeminent" in O'Reilly learning platform usage analysis*. Abgerufen am 29. Januar 2021 von https://www.zdnet.com/article/python-preeminent-in-oreilly-platform-usage-analysis/#:~:text=In%20a%20report%20released%20by,usage%2C%20for%20the%20same%20period

Data Science Project Management. (2021). *What is CRISP DM?* Abgerufen am 01. June 2021 von https://www.datascience-pm.com/effective-data-science-process/

Foundation for Open Access Statistics. (2022). *Journal of Statistical Software*. Abgerufen am 30. Apri 2022 von https://www.jstatsoft.org/

Guru99. (2022). *10 Best Data Analytics Tools for Big Data Analysis (2022)*. Abgerufen am 30. April 2022 von https://www.guru99.com/big-data-analytics-tools.html

Hichert, R., & Faisst, J. (2019). *Gefüllt, gerahmt, schraffiert: Wie visuelle Einheitlichkeit die Kommunikation mit Berichten, Präsentationen und Dashboards verbessert.* München: Vahlen.

IBCS Association. (2017). *International Business Communication Standards.* Charleston, VA: CreateSpace

Muenchen, R. A. (2020). *The Popularity of Data Science Software*. Abgerufen am 28. Dezember 2020 von http://r4stats.com/articles/popularity/

O'Grady, S. (28. March 2022). *The RedMonk Programming Language Rankings: January 2022*. Abgerufen am 30. April 2022 von https://redmonk.com/sogrady/2022/03/28/language-rankings-1-22/

R Bloggers. (2022). *R-bloggers*. Abgerufen am 30. April 2022 von https://www.r-bloggers.com/

R Core Team. (2022). *The R Project for Statistical Computing*. Abgerufen am 15. November 2022 von https://www.r-project.org/

Stack Overflow. (2022). *Questions tagged [r]*. Abgerufen am 30. April 2022 von https://stackoverflow.com/questions/tagged/r

Stahl, R., & Staab, P. (2017). *Die Vermessung des Datenuniversums: Datenintegration mithilfe des Statistikstandards SDMX.* Berlin: Springer-Vieweg

The R Foundation. (2022). *The R Journal*. Abgerufen am 30. April 2022 von https://journal.r-project.org/

Tiobe. (2022). *TIOBE Index for April 2022*. Abgerufen am 30. April 2022 von https://www.tiobe.com/tiobe-index//

Setting up the R Development Environment

4

How an R environment is designed depends largely on which tools are to be used. The following presents how the installation is done when R, R-Tools, and RStudio are to be used as an Integrated Development Environment (**IDE**) or development environment. How you can install additional functions, which are available in the form of more than 15,000 R packages, is presented in section 4.5.

4.1 Installing R

Download the current version of R or another desired version from the R-Project website (R Core Team 2022) (see Fig. 4.1).

First, click on the link "download R". You will then be shown a list of the CRAN network servers from which R can be downloaded. Select a server near you by clicking on the link "https://ftp.fau.de/cran" of the FAU University (see Fig. 4.2).

The R software will then be downloaded from the selected server when you click on the link "Download R for …" (Choose your operating system) (see Fig. 4.3).

Next, click on "install R for the first time" (see Fig. 4.4).

You will automatically be offered the current version of R for download. Click on the corresponding link, e.g., "Download R 4.2.2 for Windows" (see Fig. 4.5).

Next, click on "Open" to continue the installation. Then select the language in which the installation should be done. The license terms will then be displayed. Confirm these by clicking on "Next" to continue the installation (see Fig. 4.6).

Select the target folder and confirm this with "Next" (see Fig. 4.7).

Select the components you want to install. It is recommended to install the settings as suggested (see Fig. 4.8).

You can leave the startup options unchanged (see Fig. 4.9).

B. Heesen, *Artificial Intelligence and Machine Learning with R*,
https://doi.org/10.1007/978-3-658-45392-3_4

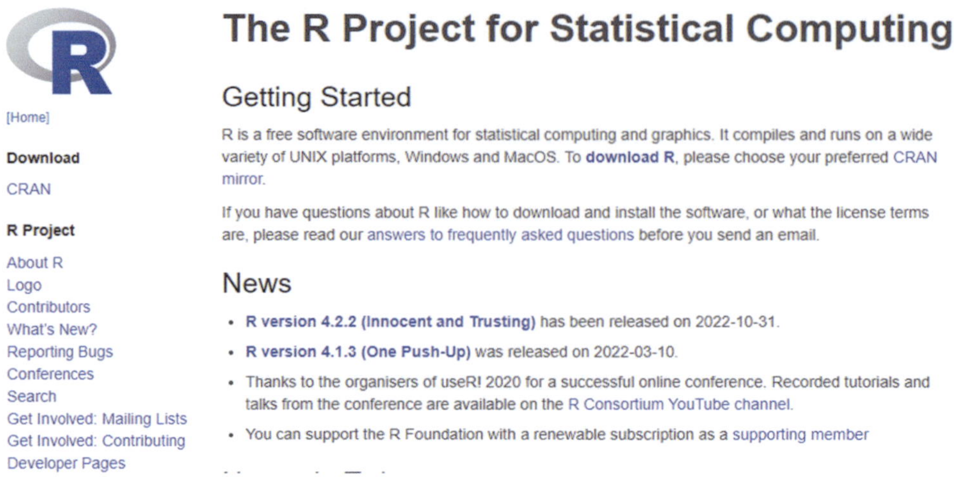

Fig. 4.1 The R Project

Fig. 4.2 Choosing the CRAN Server

To create a folder in the start menu, click on the "Next" button (see Fig. 4.10).

In addition, you can also have a desktop icon created. The version number should be stored in the registry and the files with the .RData extension should be linked with R, so that they can be opened with R later (see Fig. 4.11).

The successful installation will be confirmed to you and you can confirm this with "Finish" (see Fig. 4.12).

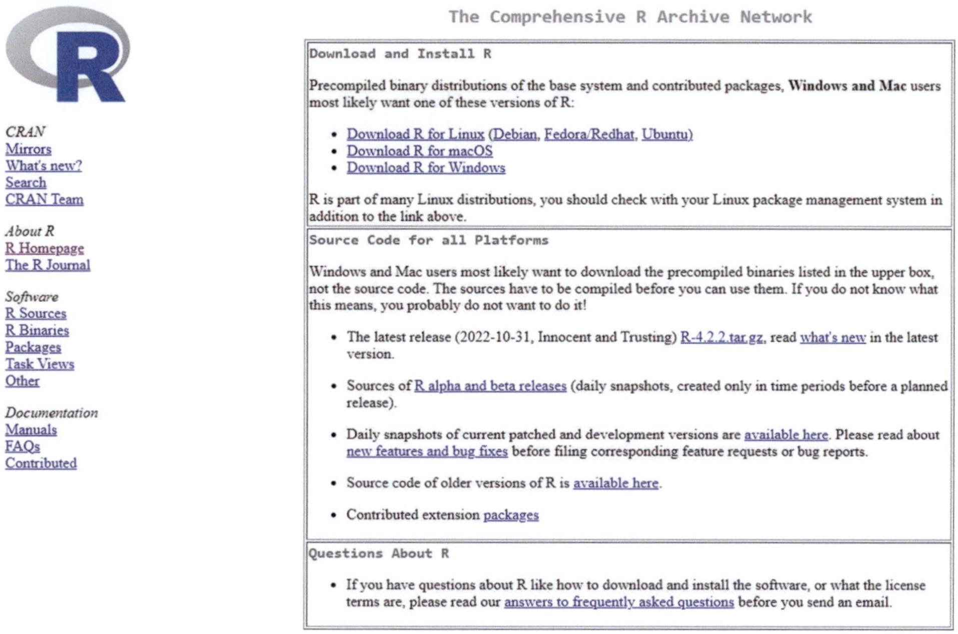

Fig. 4.3 Download R for …

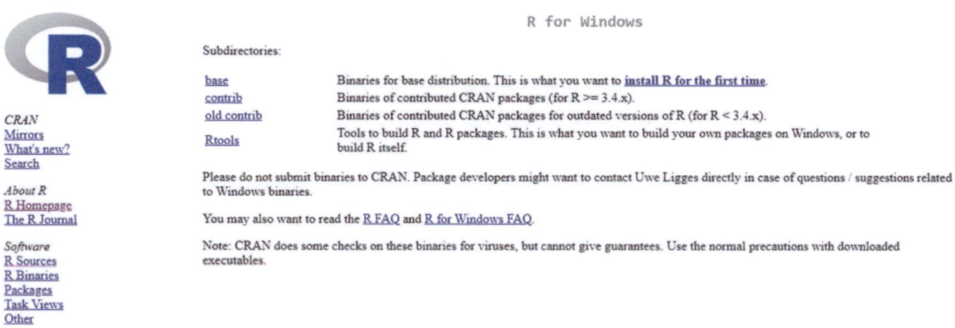

Fig. 4.4 Installing R

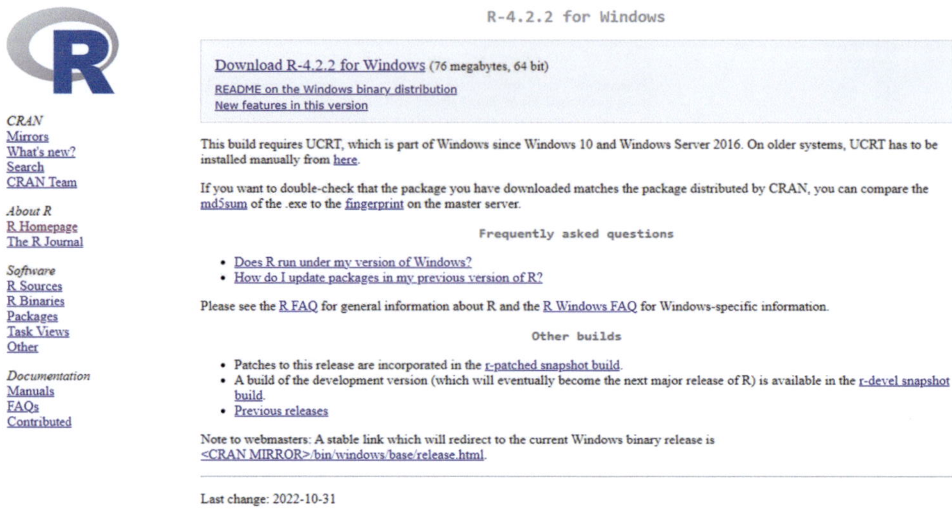

Fig. 4.5 Start download

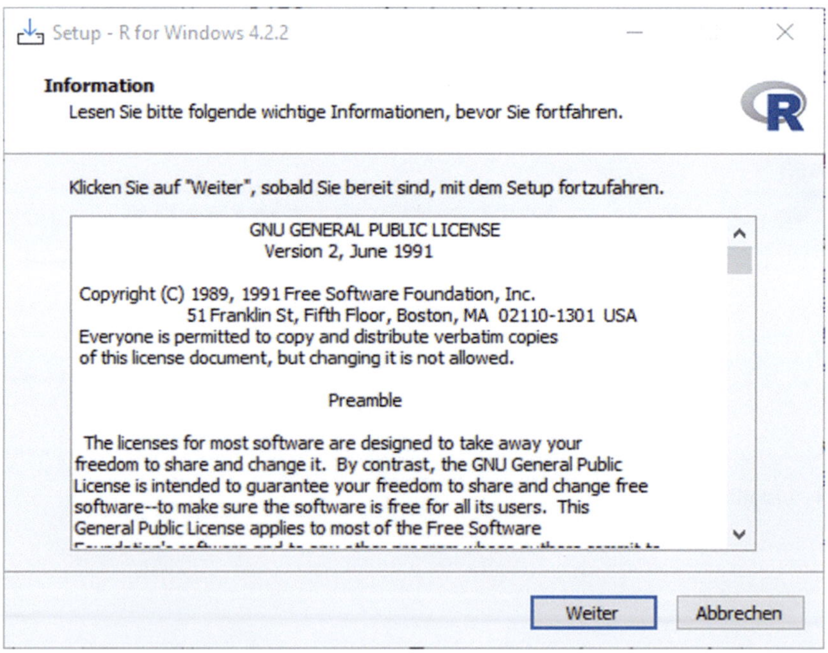

Fig. 4.6 Accepting the license terms

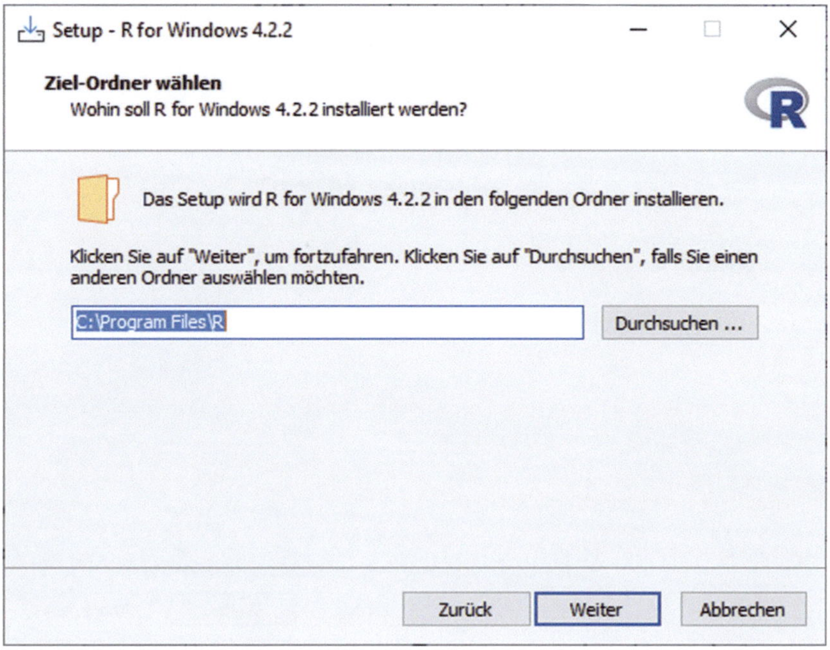

Fig. 4.7 Target folder for installation

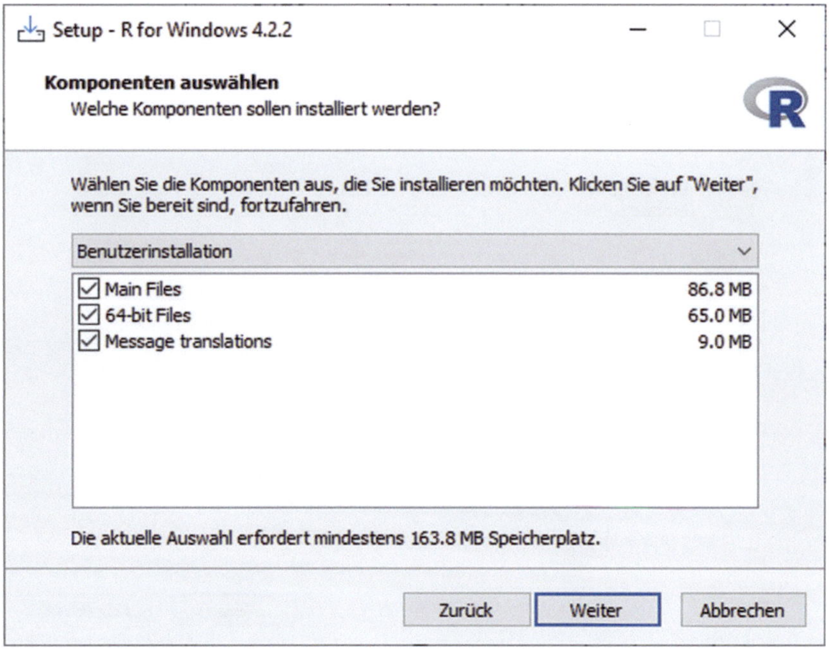

Fig. 4.8 Selecting components

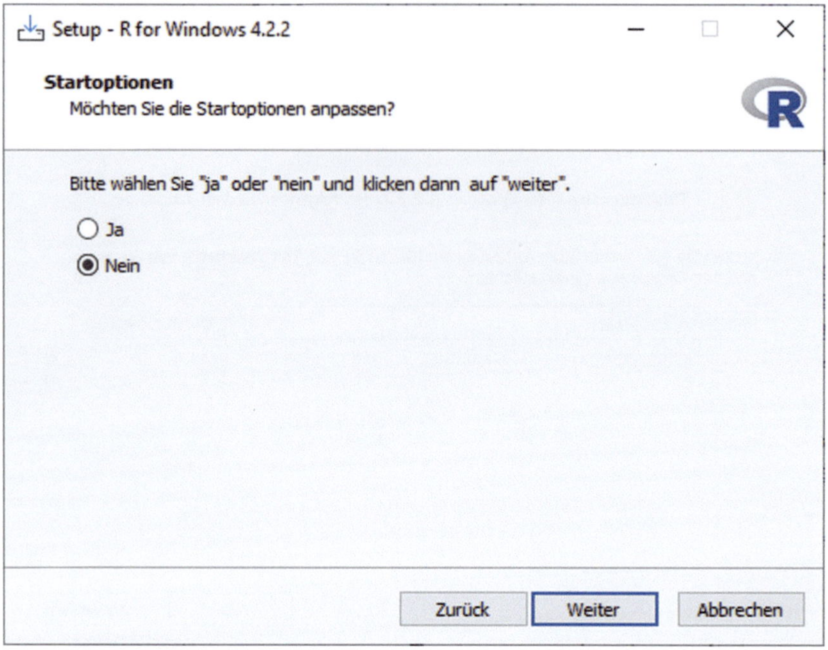

Fig. 4.9 Startup options

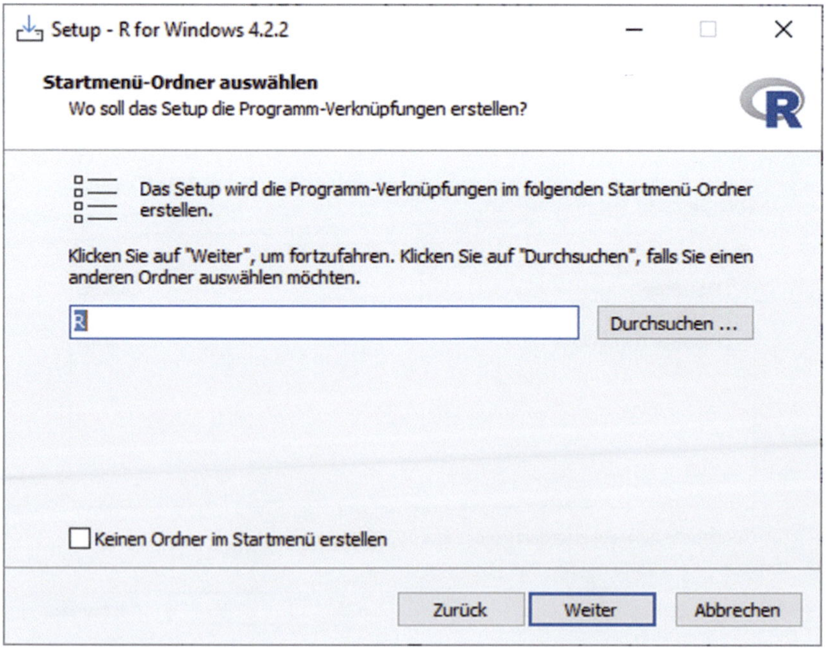

Fig. 4.10 Entry in the start menu

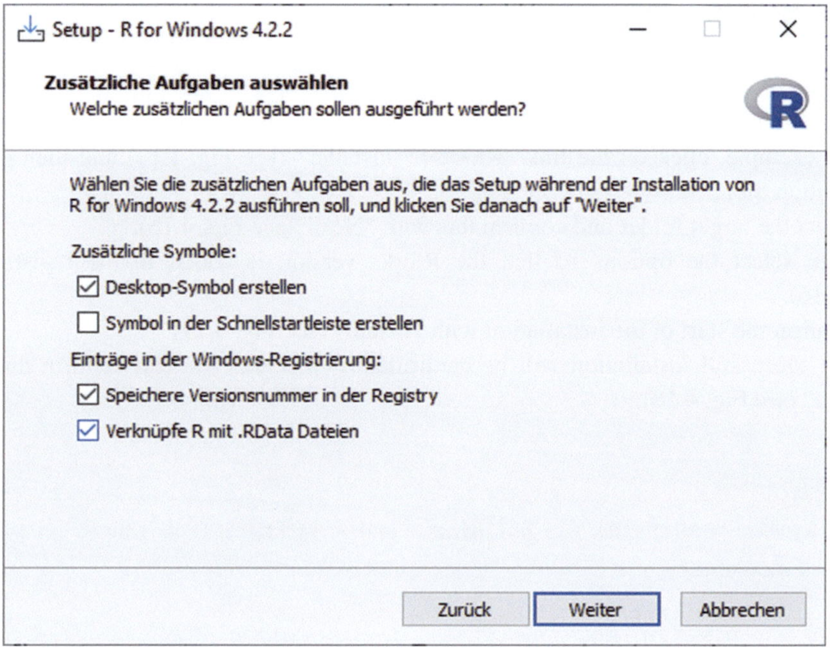

Fig. 4.11 Option for desktop icon

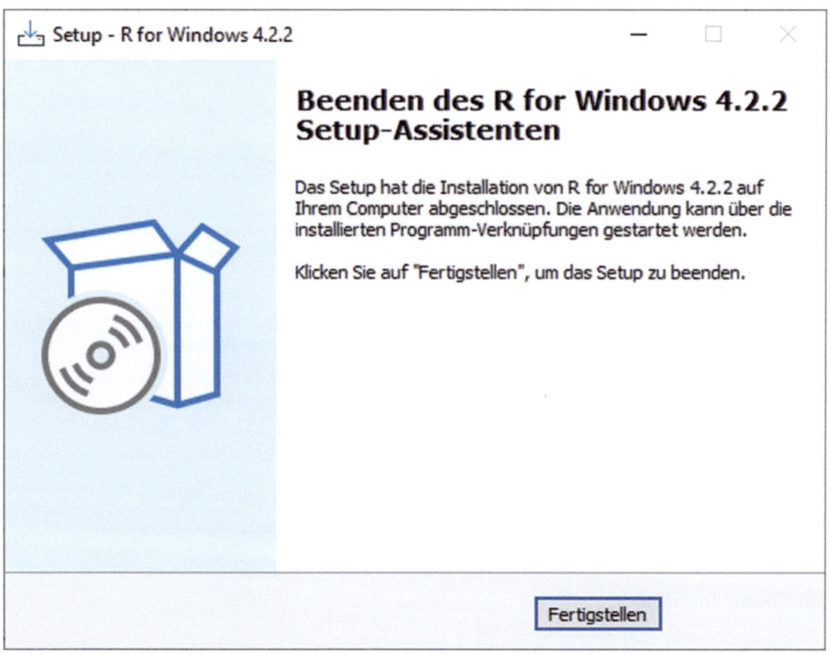

Fig. 4.12 R successfully installed

4.2 Installing Rtools

For some of the functionalities of R and its packages, special tools are required. Therefore, download the current version of Rtools (R Core Team 2022) (Fig. 4.13).

For example, click on the link "Rtools42_installer" (see Fig. 4.14) and then run the installation file.

Select the target folder and confirm this with "Next" (see Fig. 4.15).

Then select the options so that the Rtools version is added to the registry (see Fig. 4.16).

Confirm the start of the installation with "Install" (see Fig. 4.17).

The successful installation will be confirmed to you and you can confirm this with "Finish" (see Fig. 4.18).

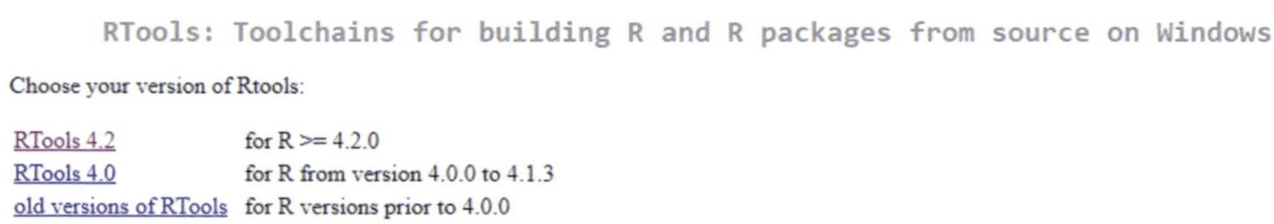

RTools: Toolchains for building R and R packages from source on Windows

Choose your version of Rtools:

RTools 4.2	for R >= 4.2.0
RTools 4.0	for R from version 4.0.0 to 4.1.3
old versions of RTools	for R versions prior to 4.0.0

Fig. 4.13 R-Tools Versions

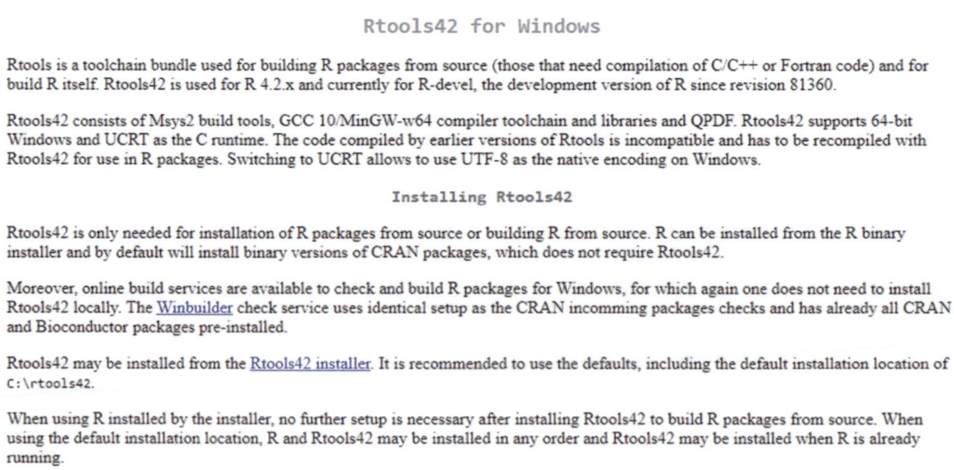

Rtools42 for Windows

Rtools is a toolchain bundle used for building R packages from source (those that need compilation of C/C++ or Fortran code) and for build R itself. Rtools42 is used for R 4.2.x and currently for R-devel, the development version of R since revision 81360.

Rtools42 consists of Msys2 build tools, GCC 10/MinGW-w64 compiler toolchain and libraries and QPDF. Rtools42 supports 64-bit Windows and UCRT as the C runtime. The code compiled by earlier versions of Rtools is incompatible and has to be recompiled with Rtools42 for use in R packages. Switching to UCRT allows to use UTF-8 as the native encoding on Windows.

Installing Rtools42

Rtools42 is only needed for installation of R packages from source or building R from source. R can be installed from the R binary installer and by default will install binary versions of CRAN packages, which does not require Rtools42.

Moreover, online build services are available to check and build R packages for Windows, for which again one does not need to install Rtools42 locally. The Winbuilder check service uses identical setup as the CRAN incomming packages checks and has already all CRAN and Bioconductor packages pre-installed.

Rtools42 may be installed from the Rtools42 installer. It is recommended to use the defaults, including the default installation location of `C:\rtools42`.

When using R installed by the installer, no further setup is necessary after installing Rtools42 to build R packages from source. When using the default installation location, R and Rtools42 may be installed in any order and Rtools42 may be installed when R is already running.

Fig. 4.14 Rtools Installation

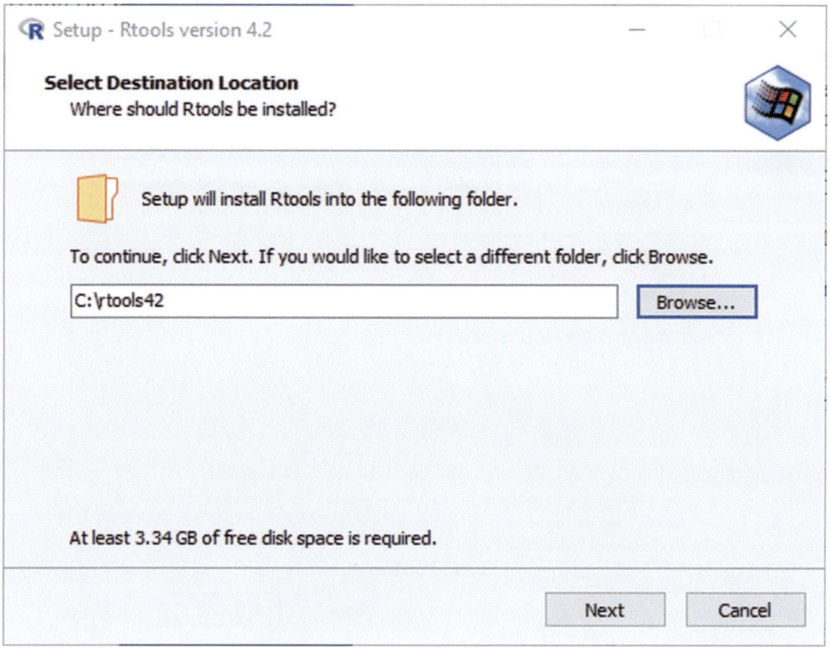

Fig. 4.15 Target folder for installation

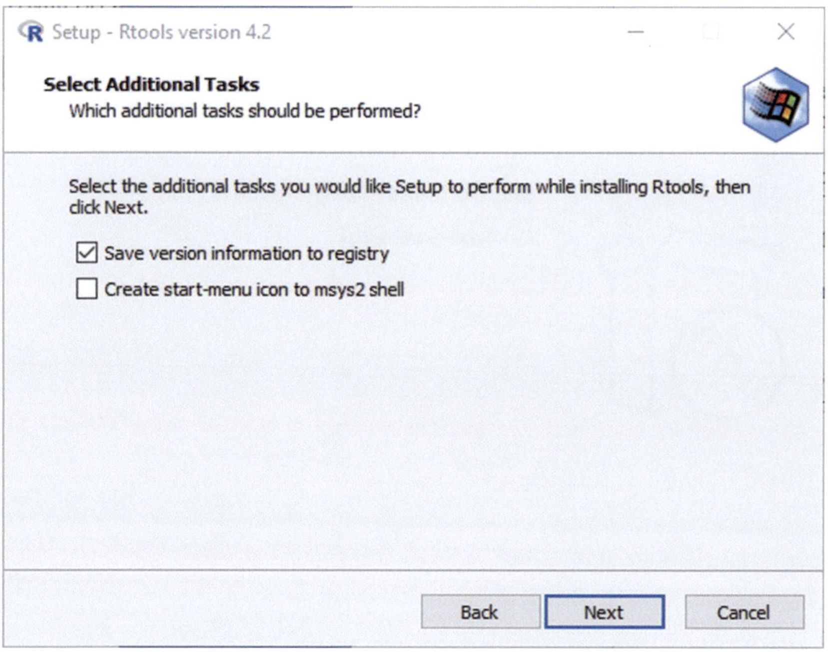

Fig. 4.16 Add Rtools version to registry

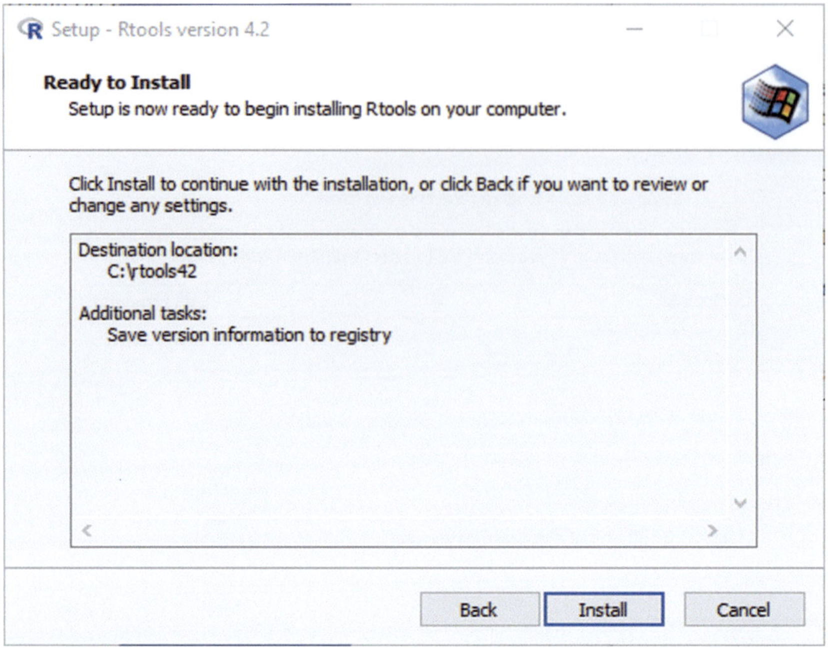

Fig. 4.17 Execute installation

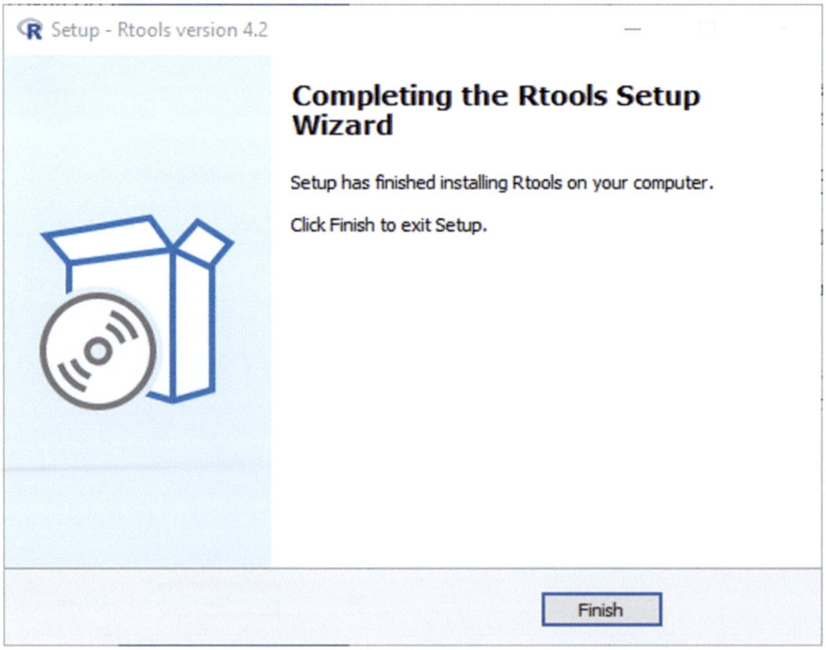

Fig. 4.18 Rtools successfully installed

4.3 Installing RStudio

An established user interface for users of R and Python is RStudio, which you can download from the RStudio website (posit 2022). Click on Products and then select "RStudio IDE" (see Fig. 4.19).

On the next screen, scroll down and then select "Download RStudio Desktop". This application is suitable for single-user computers (see Fig. 4.20).

The open-source edition of RStudio Desktop is free. Click on "Download" to start the download (see Fig. 4.21).

On the following screen, select the version for your operating system and click on the corresponding link, e.g. "RStudio-2022.07.2-576.exe" (see Fig. 4.22).

Click on the "Next" button to start the installation (see Fig. 4.23).

Select the target directory for the installation (see Fig. 4.24).

Next, select the start menu folder where RStudio should be added and then click on "Install" (see Fig. 4.25).

The successful installation will be confirmed to you and you can confirm this with "Finish" (see Fig. 4.26).

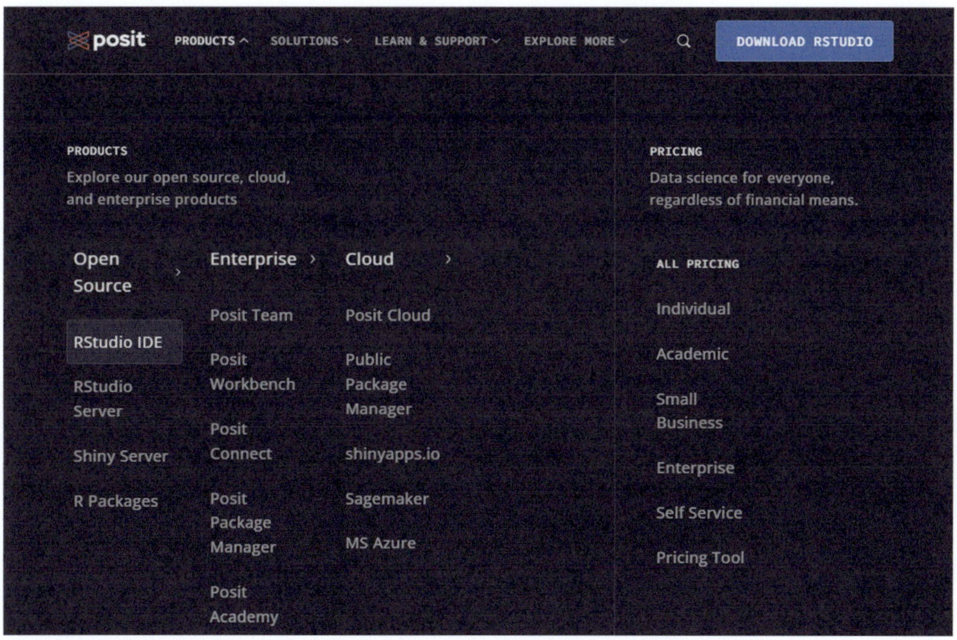

Fig. 4.19 RStudio

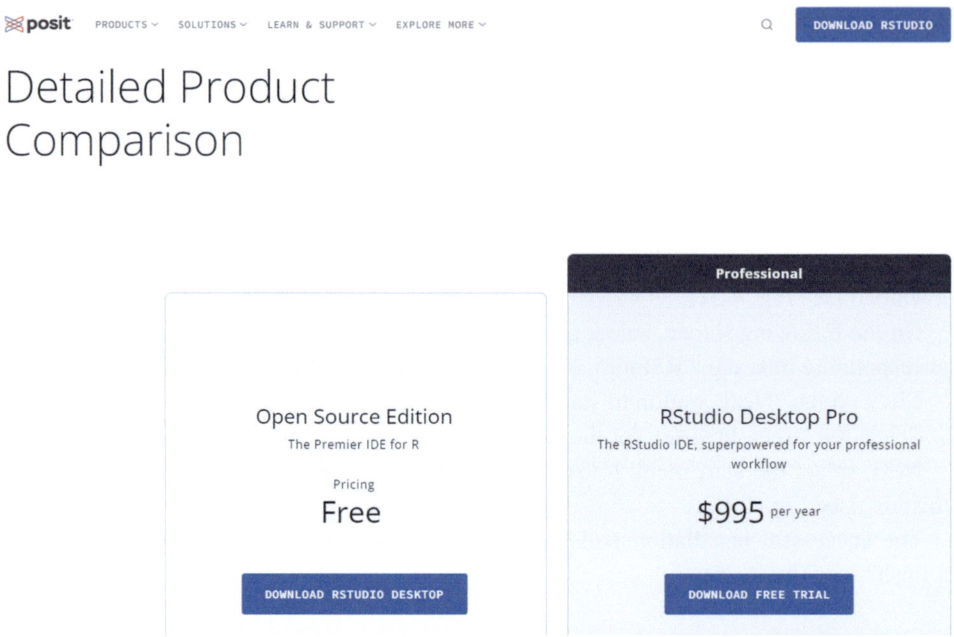

Fig. 4.20 RStudio Desktop

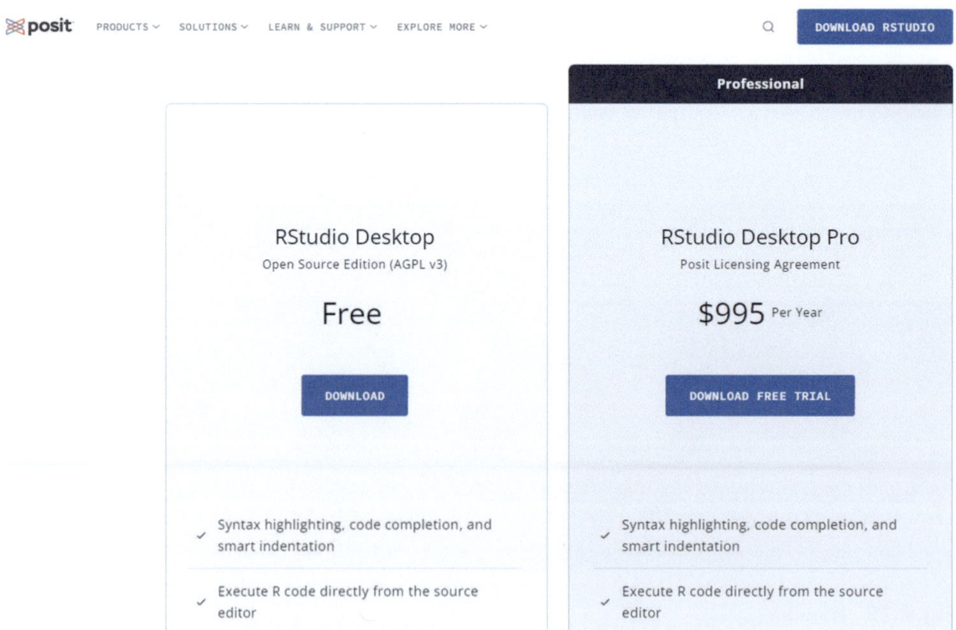

Fig. 4.21 Download RStudio Desktop

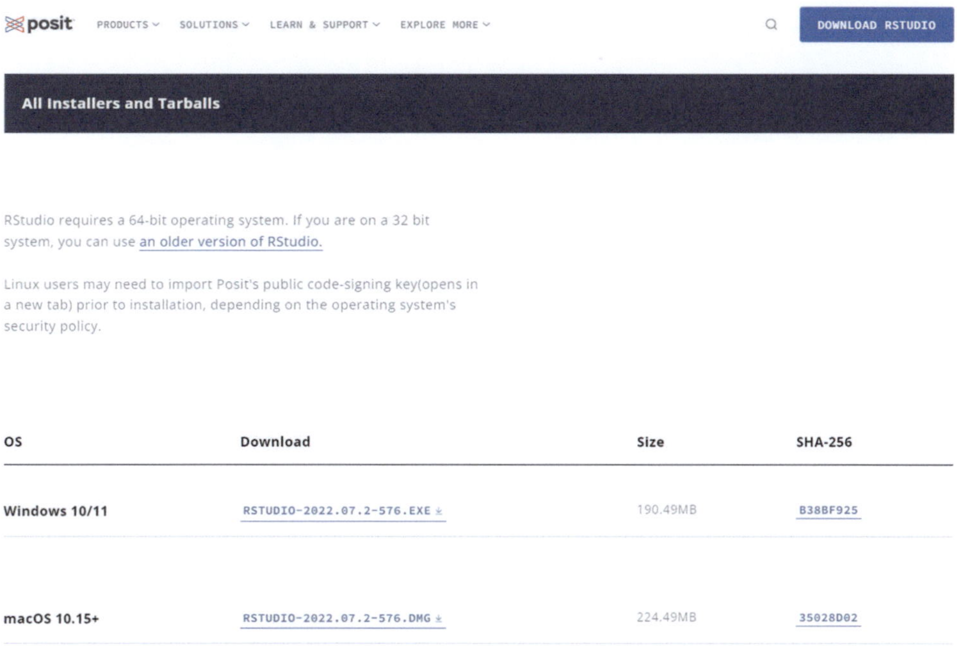

Fig. 4.22 Selection of RStudio by operating system

Fig. 4.23 Execute RStudio installation

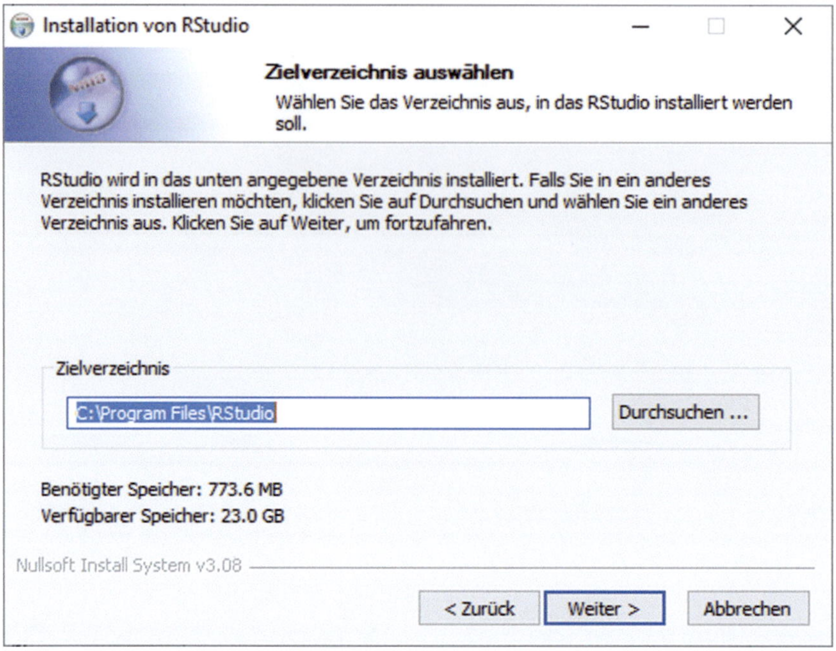

Fig. 4.24 Target directory for RStudio

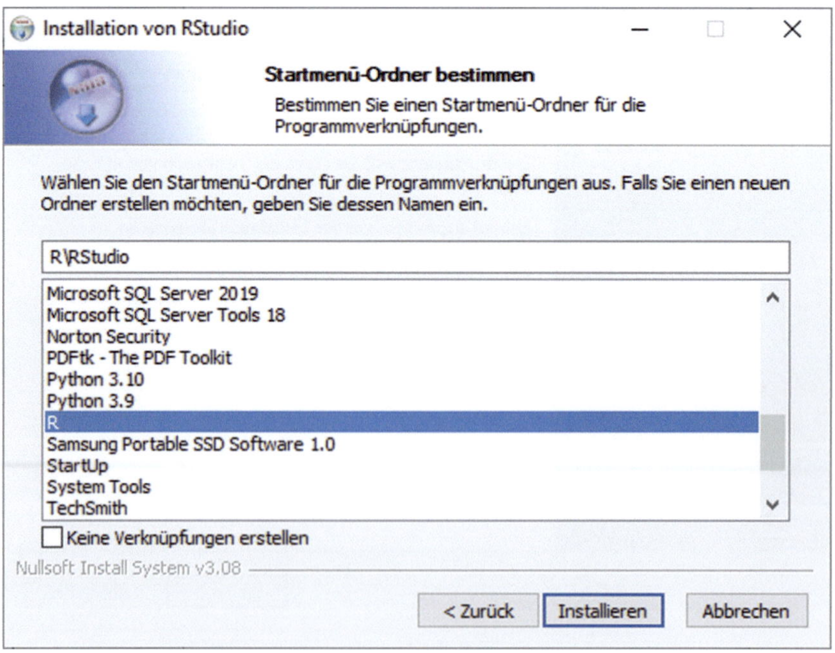

Fig. 4.25 Start menu for RStudio

Fig. 4.26 RStudio successfully installed

4.4 User Interface

The following will introduce you to how you can start and stop R and make use of advanced functionalities with the help of packages.

When you start the basic version of R, the RGui (Gui = Graphical User Interface) appears with the console and a simple menu (see Fig. 4.27).

The simplest way to work with R is to enter commands into the console. After executing a command, its output, if the command outputs something, is displayed in the console.

A more comfortable user interface than the Gui of the basic version of R is RStudio. Therefore, RStudio will be used in the following (see Fig. 4.28).

In RStudio, commands are entered either in the console, a window for command input, or in a script file. A script file has the advantage that the commands contained in it can be saved for later re-execution and thus reused. If a command is entered in the console (quadrant on the left in Fig. 4.28), its execution takes place after pressing the Enter key. If the command cannot be completed after pressing Enter, e.g. because a bracket or a parameter is missing, R shows a + instead of the output because the command cannot be executed in this way. To execute a command from a script file, it is first necessary to create a script file via the menu **File/New File/R Script** (Fig. 4.29).

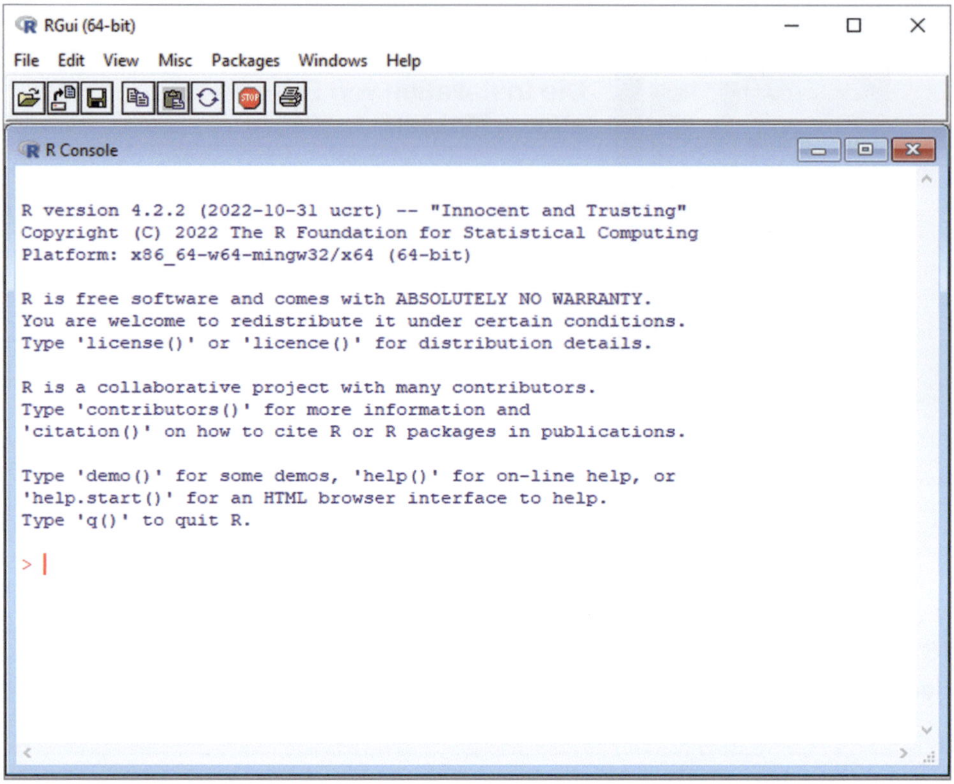

Fig. 4.27 Gui of the basic version of R

In the newly opened window with the script, a command, e.g. 1 + 2, can be executed by positioning the cursor on the line (quadrant top left in Fig. 4.30) and then pressing Ctrl+Enter or alternatively by pressing the **Run** button. After executing a command, its output, if the command outputs something, is displayed in the console. To cancel an R instruction if it takes too long to execute, the Esc key can be pressed.

If one of the last used commands should be executed again, the ↑ key can be used. To repeatedly execute a sequence of process steps in data analysis, it is recommended to work with script files. The commands from the script window can be saved as a script file via the menu **File/Save As …** and reloaded at a later time via the menu **File/Open File …** (Fig. 4.31).

In the context of the script window, the name of the script file is displayed after saving (Fig. 4.32). It is possible to open several script files at the same time. These are then displayed side by side (Fig. 4.32).

An R session can be ended in RStudio via the menu with **File/Quit Session** or with the function `q()`. When R is terminated, the default is to ask whether the workspace (all objects/variables/functions in the Global Environment) should be saved. Each time R is

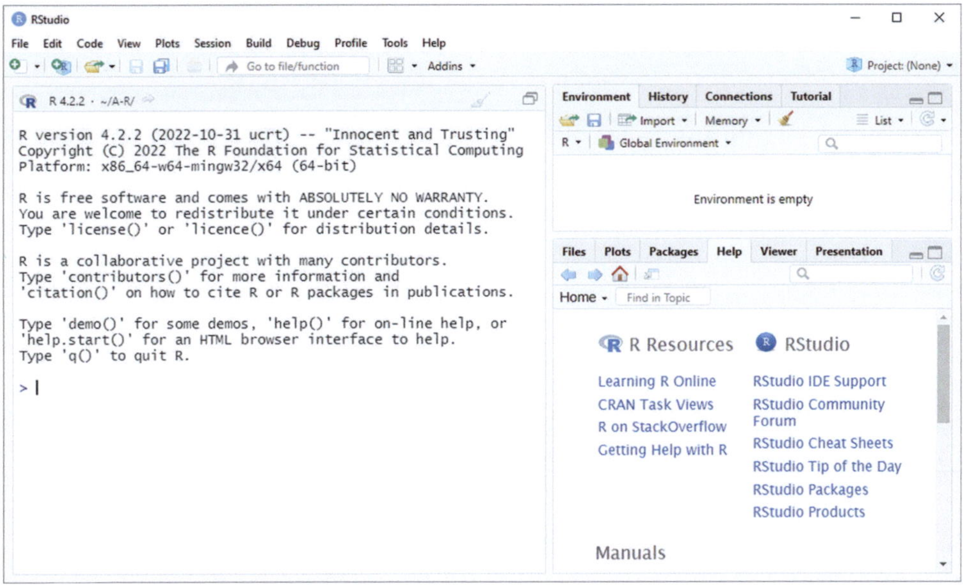

Fig. 4.28 Gui of RStudio

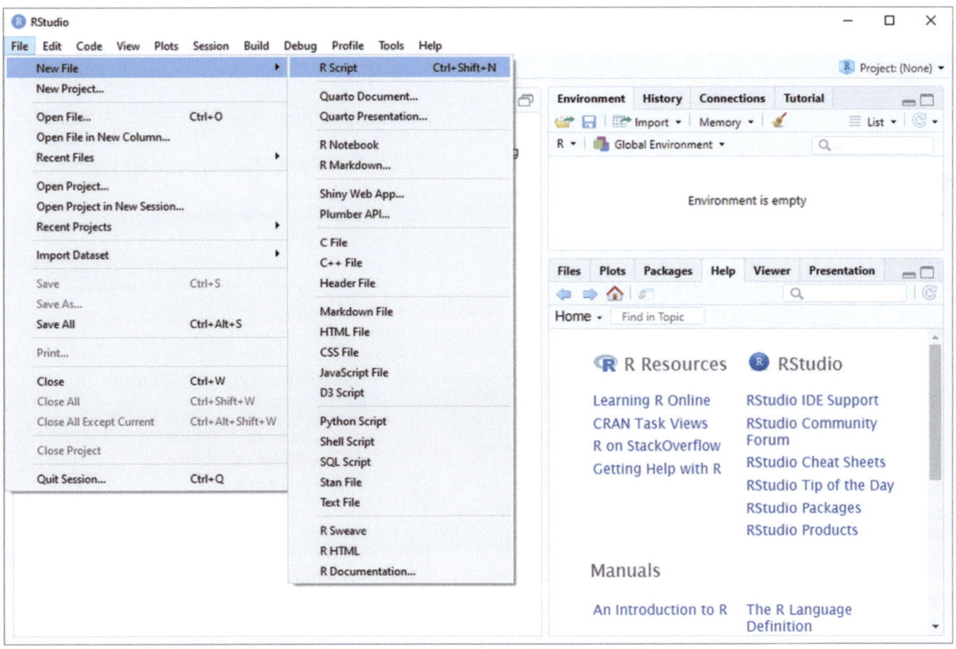

Fig. 4.29 Create new script file

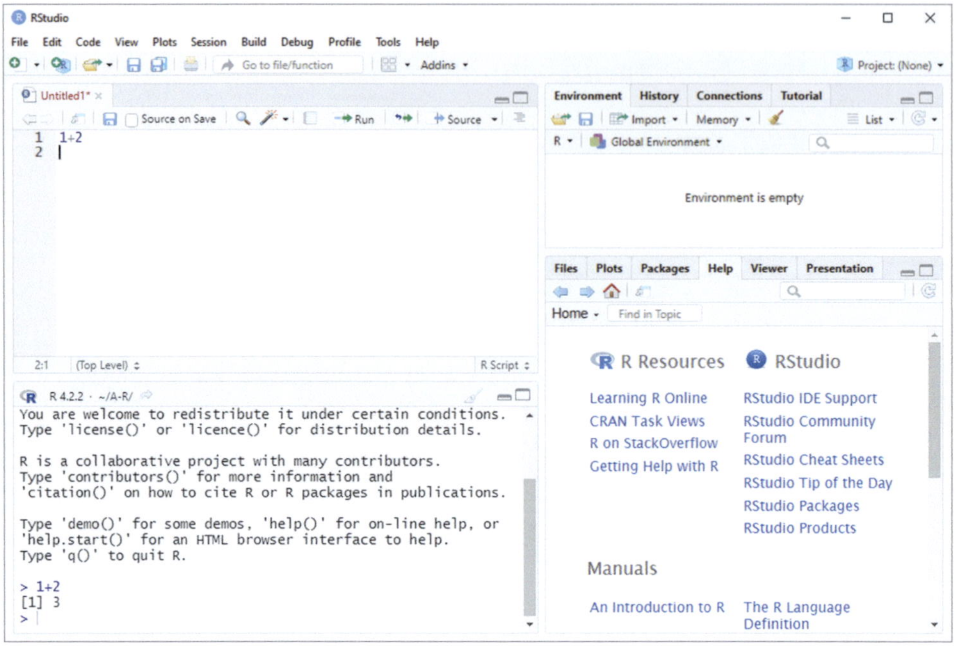

Fig. 4.30 RStudio with script window

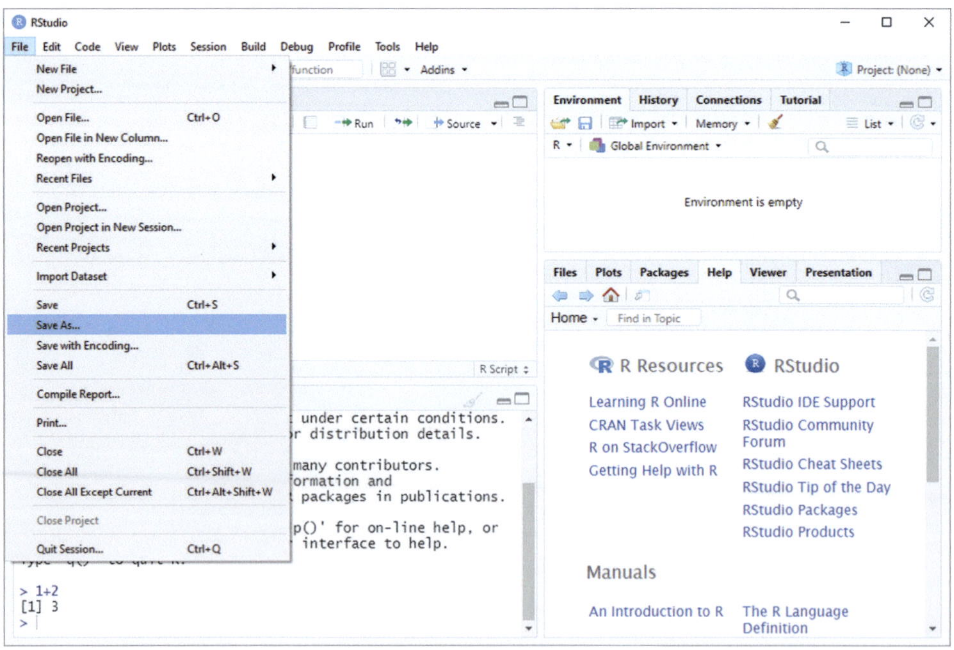

Fig. 4.31 Save script file as Demo-1and2.R

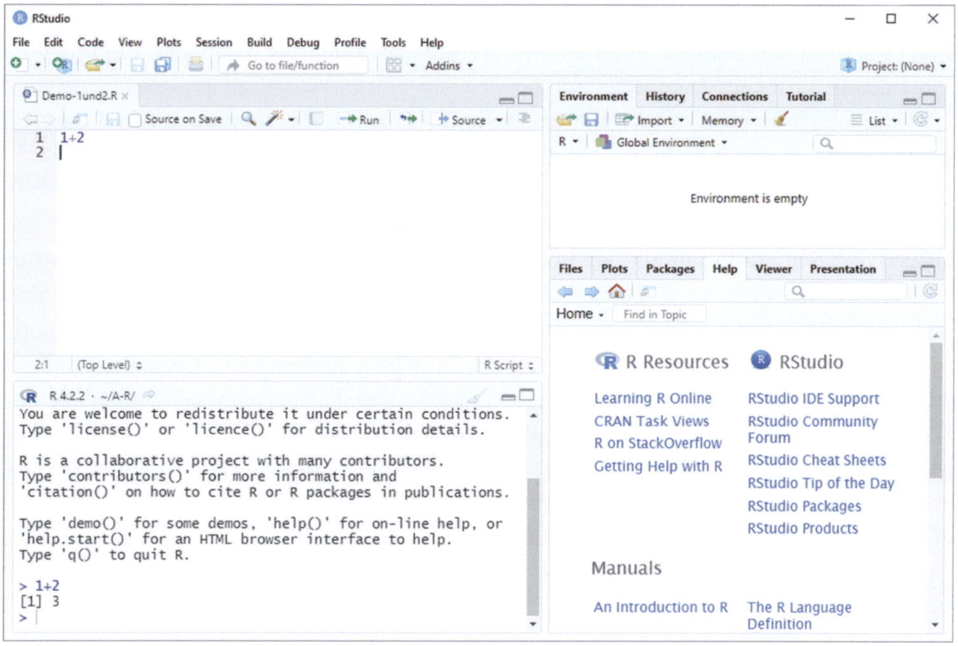

Fig. 4.32 Save script file as Demo-1and2.R

started, the workspace is loaded. So if it is desired that the work should be continued later at the same point where it was stopped, it makes sense to save the workspace. If you do not want to be asked every time you quit R whether the workspace should be saved, you can also permanently set this in the menu **Tools/Global Options/General** (options: Always, Never, Ask).

4.5 Advanced Functions

There are more than 15,000 so-called packages with advanced functions for R, which are usually available via the CRAN and can be retrieved from there and copied into the R-Library directory on the local computer with the function `install.packages("package")`. The package name must be in quotation marks for this function. Installation is only required once on a computer. It is recommended to use the parameter `dependencies=TRUE` so that all other packages on which it relies for functions and content are automatically installed as well.

A list of already installed packages can be displayed with the function `installed.packages()`.

After each start of R, the function `library(package)` must be called again to load or activate the package into memory. Before executing this command, the functions contained in a package cannot be used.

If a package is required for the execution of a script, the function `require()` is suitable for initiating both the installation and the loading of a package as needed.

The package `tidyverse` (RStudio 2022) is a so-called composite package, which includes the packages ggplot2, dplyr, tidyr, readr, purr, tibble, stringr, and forcats. Instead of having to install and load these individually, this is achieved with one instruction via the package `tidyverse`. It is recommended to load this package for data science projects. For essential functions of machine learning, the installation of the composite package `tidymodels` (RStudio 2022), which includes the packages rsample, parsnip, recipes, workflows, tune, yardstick, broom, and dials, is also recommended.

If conflicts are displayed when loading a package, as shown below when loading `tidyverse`, that the function `filter` exists not only in the package `dplyr`, but also in the package `stats`, then it is necessary to precede the function call of filter from the respective package with the package name and two colons, e.g., `dplyr::filter()` or `stats::filter()`. This ensures that the desired filter function is executed by the system.

Installing and activating packages, help with packages

```
#- Pakete---------------------------------------------------------------------------
#install.packages("tidyverse") # Installation von Package (einmalig)
library(tidyverse)            # Aktivieren von Package (nach jedem Start von R)

── Attaching packages ─────────────────────────────────────── tidyverse 1.3.2 ──
✓ ggplot2 3.3.6     ✓ purrr   0.3.4
✓ tibble  3.1.7     ✓ dplyr   1.0.9
✓ tidyr   1.2.1     ✓ stringr 1.4.1
✓ readr   2.1.3     ✓ forcats 0.5.2
── Conflicts ────────────────────────────────────────── tidyverse_conflicts() ──
✗ dplyr::filter() masks stats::filter()
✗ dplyr::lag()    masks stats::lag()

#?tidyverse                   # Hilfe zu Package
#help(tidyverse)              # Hilfe zu Package

#- Pakete---------------------------------------------------------------------------
#install.packages("tidymodels")# Installation von Package (einmalig)
library(tidymodels)           # Aktivieren von Package (nach jedem Start von R)

── Attaching packages ─────────────────────────────────────── tidymodels 0.2.0 ──
✓ broom        1.0.1     ✓ rsample      0.1.1
✓ dials        1.0.0     ✓ tune         0.2.0
✓ infer        1.0.0     ✓ workflows    0.2.6
✓ modeldata    0.1.1     ✓ workflowsets 0.2.1
✓ parsnip      1.0.0     ✓ yardstick    1.0.0
✓ recipes      1.0.3
── Conflicts ────────────────────────────────────────── tidymodels_conflicts() ──
✗ scales::discard() masks purrr::discard()
✗ dplyr::filter()   masks stats::filter()
✗ recipes::fixed()  masks stringr::fixed()
✗ dplyr::lag()      masks stats::lag()
✗ yardstick::spec() masks readr::spec()
✗ recipes::step()   masks stats::step()
• Learn how to get started at https://www.tidymodels.org/start/
```

When a package is loaded or activated, the functions contained in it can be called. With `?Package` or `help(Package)` you can call up the help for a package or with `?Function` or `help(Function)` the help for a function from the package, e.g., `?tidyverse` for help with the package `tidyverse` or `?geom_boxplot` for help with the function `geom_boxplot()` from the package `ggplot2`.

Since the functionality of the R packages is constantly evolving, it is occasionally useful to bring the installed packages up to date. This is possible with the function `update.packages()`.

If certain packages are no longer needed, they can be removed at any time with the function `remove.packages()`.

The installation, updating, activation, or removal of packages is of course also possible via RStudio (quadrant bottom right in Fig. 4.33).

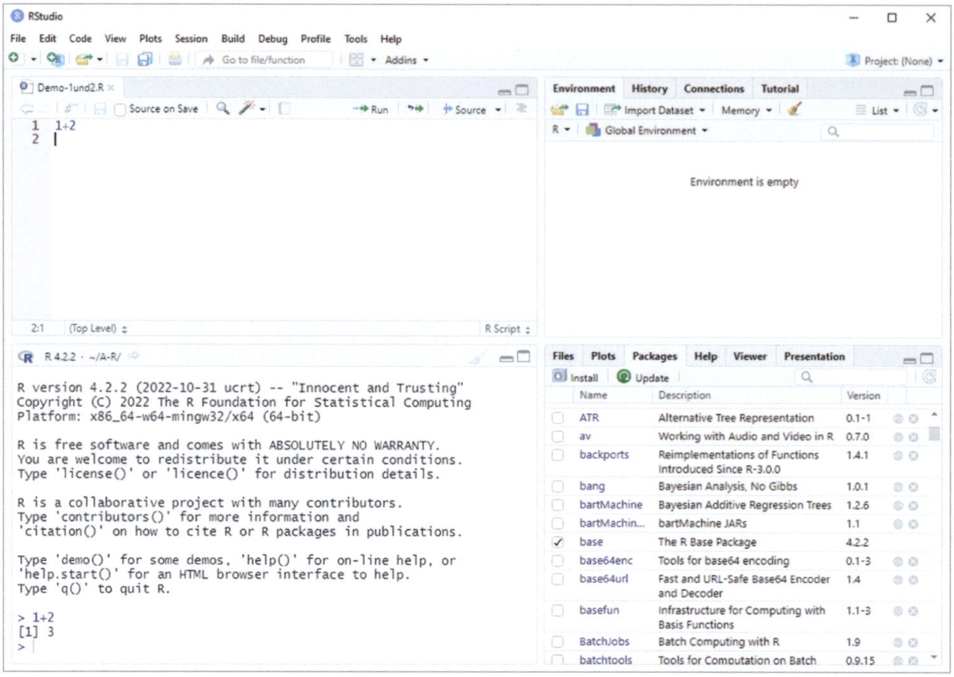

Fig. 4.33 Installation, activation … of packages

4.6 The machinelearning Package

The datasets used in this book, unless explicitly referred to other packages, are located in the R package `machinelearning` that I created. You can install this package in R with the following instruction lines:

- install.packages("devtools")
- library(devtools)
- devtools::install_github("bheesen/machinelearning")
- library(machinelearning)

After successful installation, you can display the included data and functions in the "Environment" section after selecting the machinelearning package from the dropdown (see Fig. 4.34). Even if you do not display the data in the Environment, you can still work with them.

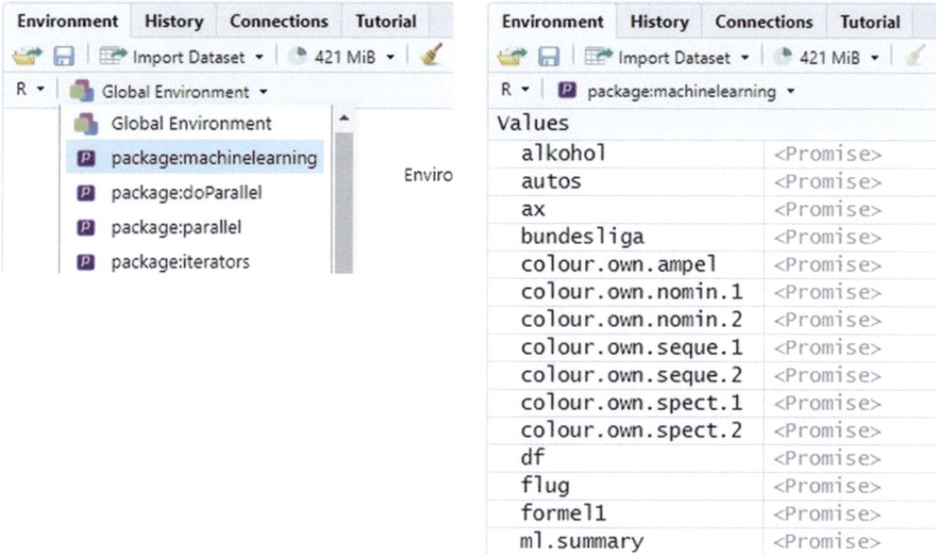

Fig. 4.34 Machinelearning Package

References

posit. (2022). *posit*. Abgerufen am 15. November 2022 von https://posit.co/

R Core Team. (2022). *The R Project for Statistical Computing*. Abgerufen am 15. November 2022 von https://www.r-project.org/

RStudio. (2022). *Tidymodels*. Abgerufen am 04. November 2022 von https://www.tidymodels.org/

RStudio. (2022). *Tidyverse packages*. Abgerufen am 13. Juli 2022 von https://www.tidyverse.org/packages/

5 Basics of the R Programming Language

R instructions consist of expressions, i.e., **assignments** (arithmetic or logical operators) or **function calls**. Function calls can be recognized by the fact that brackets follow the name of the function, in which parameters can optionally be passed to a function, e.g., `mean(employee)`, to calculate the average for the variable "employee".

R is case-sensitive, so the names of functions and objects distinguish between upper and lower case, so `mean(employee)` only works if there is a function `mean()` and an object `employee` that both do not contain capital letters. The command `Mean(Employee)` would cause an error message if either the function `Mean()` or the object `Employee` does not exist in this combination of upper/lower case. In addition to alphanumeric characters, the period and underscore are also allowed for the names of functions and variables.

Assignments are made in R using the assignment operator `<-`. Therefore, the command `employee.germany <- 3237` would be correct to assign the value 3237 to a variable employee.germany. Alternatively, the equal sign could also be used as an assignment operator. However, the arrow operator has the advantage of indicating the direction of the assignment. The name of the variable and the assigned value could therefore also be swapped as follows `3237 -> employee.germany`. To increase the readability of the instructions, however, it is advisable to always place the receiver variable at the beginning of a line, so that it is uniformly and clearly recognizable there.

The content of a variable can be displayed in the console by entering the name of the variable and then executing this command. Elements of a vector variable can be accessed by indexing in square brackets, with indexing starting at 1, e.g., `employee[2]` outputs the second value of the vector variable named employee.

B. Heesen, *Artificial Intelligence and Machine Learning with R*,
https://doi.org/10.1007/978-3-658-45392-3_5

Code blocks are enclosed with curly braces and can be used for simple groups of R instructions or loops. The advantage of code blocks is that they can be executed together with the Run Selected Line(s) command in the menu or the Ctrl+Enter keyboard shortcut.

To comment on the commands in script files, comments can be initiated with a #. All characters following in the respective line are then not considered as an R instruction, but as a comment.

Assignment, Comment, Code Block, Function Call

```
#- Kommentare zu Wertzuweisungen--------------------------------------------------
mitarbeiter.deutschland <- 3237  # gutes Beispiel  Wertzuweisung
mitarbeiter.deutschland          # Ausgabe der Variable

## [1] 3237

(mitarbeiter.holland <- 259)     # gutes Beispiel  Wertzuweisung inkl. Ausgabe

## [1] 259

(193 -> mitarbeiter.italien)     # schlechtes Beispiel  Wertzuweisung inkl. Ausgabe

## [1] 193

#- Funktion c() fügt Variablen zu einem Vektor zusammen (c="combine").
(mitarbeiter<-c(mitarbeiter.deutschland, mitarbeiter.holland,mitarbeiter.italien))

## [1] 3237  259  193

#- Codeblock einfache R-Anweisungen-----------------------------------------------
{
  (mitarbeiter.spanien <- 1244)  # Wertzuweisung inkl. Ausgabe
  (mitarbeiter<-c(mitarbeiter, mitarbeiter.spanien))# Vektorergänzung inkl. Ausgabe
}

## [1] 3237  259  193 1244

#- Codeblock Schleife-------------------------------------------------------------
for (i in 1:2) {           # For-Schleife von 1 bis 2
  print(mitarbeiter[i])  # Gibt Inhalt von mitarbeiter mit Index i aus
}

## [1] 3237
## [1] 259

#- Funktionsaufrufe---------------------------------------------------------------
mean(mitarbeiter)          # Funktion mean() berechnet den Mittelwert

## [1] 1233.25

Mean(mitarbeiter)          # Funktion mit Großbuchstabe M existiert nicht

## Error in Mean(mitarbeiter): could not find function "Mean"

mean(Mitarbeiter)          # Variable mit Großbuchstabe M existiert nicht

## Error in mean(Mitarbeiter): object 'Mitarbeiter' not found
```

5.1 Data Types

In many programming languages, a variable must be declared by assigning a name and a specific data type before a value can be assigned to it. In R, variables do not need to be declared in advance, but they are dynamically created at the moment of assignment and R automatically determines the data type, unless it is specifically indicated. This means fewer lines of code and facilitates programming, but is also a possible source of error. The fact that R can automatically determine the data type means that R is weakly typed.

In the automatic determination of the data type, R applies the mechanism of coercion. This mechanism tries to ensure that data corresponds to the type necessary for the operation being performed. For example, if text and numbers are merged in a vector, the new vector automatically gets the type character, so that both types of data can be stored. R thus chooses the smallest common type denominator.

The most important elementary data types in R are:

- **numeric** are numbers, which are usually stored and displayed as floating-point numbers and consist of the mantissa m and the exponent e with $x=m*10^e$. Historically, R also worked with the data type double, which is identical to numeric. If necessary, a number of the data type double can be converted to numeric using the function `as.numeric()` if the result of the function `is.numeric()` should yield a TRUE. The decimal point is input and output in R by default with a dot. R values of the type numeric follow the standard IEC 60559:2011 (also known as IEEE 754-2008) or Binary64 format, which works with a precision of 53 bits and can represent absolutely exact values from $2*10^{-308}$ to $2*10^{308}$.
- **integer** are integer numerical values. These are indicated with a subsequent L, e.g., `33L`, so that they are stored in the data type integer and consume less storage space than the data type numeric.
- **logical** are logical values, TRUE and FALSE or T and F.
- **Date** are date specifications.
- **character** are character strings or strings and are either set in single ' ' or double " " quotation marks.

After a value has been assigned to a variable, the function `class()` can be used to display the data type automatically assigned by R. The function `str()` also displays the value of a variable in addition to the data type. The data types are then possibly abbreviated as num instead of numeric, logi instead of logical, and chr instead of character. The function `unclass()` can be used to show how a variable is viewed and stored internally by the computer (uncoded by the data type of type raw/raw). However, this is rarely needed.

The logical functions `is.numeric()`, `is.logical()`, `is.character()` and `is.Date()` can be used to check whether a variable is of a certain data type. The result of these functions is TRUE if the variable is of this data type, otherwise FALSE.

If the data type is to be converted, this is possible with the functions `as.numeric()`, `as.logical()`, `as.character()` and `as.Date()`. Conversion into a date variable is possible for any date formats via the parameter `format`, e.g., `as.Date("05.05.2022", format= "%d.%m.%Y")`.

R has a special way of handling date and time variables. The data type of date variables is Date and that of time variables is POSIXct and POSIXlt. All time information is stored as the time in days, hours, or seconds that has passed since 01.01.1970. For times before this date, a negative number indicates the days or seconds until 01.01.1970. The function `Sys.Date()` can be used to display the current date. With the command `unclass(Sys.Date())` you will be shown the internal representation of the current date and this is the number of days that have passed since 01.01.1970. The function `Sys.time()` can also be used to display the current time. The command `unclass(Sys.time())` then shows the number of seconds that have passed since 01.01.1970. To avoid paying attention to this rather technical circumstance, the lubridate package can be used. It offers many valuable functions for date and time variables such as `date()`, `year()`, `month()`, `day()`, `hour()`, `minute()`, `second()`, `week()`, `quarter()`, `am()` and `pm()`.

When working with variables of the Character type, the package `stringr` offers some functions, e.g., the function `str_detect()`, `str_locate()`, to search for character strings or words, or `str_count()`, to determine the frequency of a character string. The package `stringr` is part of the package `tidyverse` and therefore does not need to be installed and activated separately, provided the package `tidyverse` is already installed and activated.

Variables can have attributes in addition to the value. These can be displayed with the function `attributes()` and attributes can be changed or added with the function `attr()`, e.g., the variable a is given the attribute Meaning with the command `attr(a, "Meaning") <- "Number sold"`.

Data Types: Numeric, Integer, Logical, Date, String

```
#- Numerische Datentypen-------------------------------------------------------------
a <-  12                # Numerische Variable
a                       # Inhalt der Variable

## [1] 12

class(a)                # Datentyp der Variable

## [1] "numeric"

str(a)                  # str() zeigt beides: Datentyp und Inhalt

##  num 12

a <- as.integer(a)      # Konvertierung zu Integer
str(a)

##  int 12

b <- 13.5               # Numerische Variable mit Nachkommastellen
str(b)

##  num 13.5
c <- 48000000           # Numerische Variable: Gleitkommazahl
str(c)

##  num 4.8e+07

d <- 46000000L          # Integer Variable: Zahl
str(d)

##  int 46000000

#- Logische Datentypen---------------------------------------------------------------
e <- TRUE               # Boolsche Variable
str(e)

##  logi TRUE

#- Datum Datentypen------------------------------------------------------------------
f <- Sys.Date()         # Datumsvariable
str(f)

##  Date[1:1], format: "2023-01-08"

g <- "1992-12-23"       # Character Variable
str(g)

##  chr "1992-12-23"

h <- as.Date(g)         # Konvertierung in Datumsvariable
str(h)

##  Date[1:1], format: "1992-12-23"

i <- as.Date("01.09.1994", format= "%d.%m.%Y") # Konvertierung jeden Formates
str(i)

##  Date[1:1], format: "1994-09-01"

j<-Sys.time()           # Zeitvariable
str(j)

##  POSIXct[1:1], format: "2023-01-08 17:27:02"

require(lubridate)      # Funktion require(): Installiert und aktiviert Paket
date(j)                 # Datum
```

```
## [1] "2023-01-08"
year(j)                   # Jahr
## [1] 2023
month(j)                  # Monat
## [1] 1
day(j)                    # Tag
## [1] 8
hour(j)                   # Stunde
## [1] 17
minute(j)                 # Minute
## [1] 27
second(j)                 # Sekunde
## [1] 2.082957
week(j)                   # Woche
## [1] 2
quarter(j)                # Quartal
## [1] 1
am(j)                     # AM-Uhrzeit TRUE/FALSE
## [1] FALSE
pm(j)                     # PM-Uhrzeit TRUE/FALSE
## [1] TRUE
#- Character Datentypen----------------------------------------------------------------
k <- 'Erste Zeichenkette'     # Character Variable in einfachen Anführungszeichen
str(k)
##  chr "Erste Zeichenkette"
l <- "Machine Learning hilft"# Character Variable in doppelten Anführungszeichen
str(l)
##  chr "Machine Learning hilft"
str_detect(k,"hilft")         # Prüfung, ob String enthalten ist
## [1] FALSE
str_detect(l,"hilft")         # Prüfung, ob String enthalten ist
## [1] TRUE
str_locate(l,"hilft")         # Start und Endeposition von String
##      start end
## [1,]    18  22
m <- "27.62"                  # Character Variable
str(m)
##  chr "27.62"
```

```
n <- is.character(m)          # Logische Frage ob Datentyp == character
str(n)

##  logi TRUE

o <- as.numeric(m)            # Konvertierung zu numerisch
str(o)

##  num 27.6

#- Attribute von Datenobjekten beliebigen Typs-----------------------------------
str(a)                                      # Anzeige Typ, Wert, Attribute

##  int 12

attributes(a)                               # Attribute anzeigen

## NULL

attr(a, "Bedeutung") <- "Stückzahl verkauft"   # Attribut zuweisen
attributes(a)                               # Attribute anzeigen
## $Bedeutung
## [1] "Stückzahl verkauft"

str(a)                                      # Anzeige Typ, Wert, Attribute

##  int 12
##  - attr(*, "Bedeutung")= chr "Stückzahl verkauft"
```

5.2 Data Structures

If only a single value is stored in a variable, this is referred to as a **simple variable**. If multiple values are stored in a variable, this is referred to as a **vector**. There are two types of vectors in R, so-called **atomic vectors** (one data type) and **lists** (multiple data types).

To display variables or data structures, the function `View()` can be used. The function `fix()` supports, if necessary, the editing and modification of variables and data structures.

5.2.1 Vector

A new vector without content can be created with the function `new()`, e.g. `text<-new("character")` or `a<-new("numeric")`, to create an empty vector of type character or numeric. Multiple values of a data type can be stored in a vector by combining them into a vector using the function `c()` (c stands for the English word "combine"). The following function can be used to create a vector: `a<-c(1.2, -13, 46, 1880.12, -2.5)`. Note: In R, decimal places are separated by a point as usual in English, not by a comma as in German. However, commas are used in R to distinguish the arguments of a function from each other.

The entire vector can be output by entering the name. Individual elements of a vector can be accessed by indexing in square brackets, with indexing starting at 1, e.g. `a[2]` outputs the second value of the vector variable named a, in this case –13. If no index is specified in the brackets, all elements are output. Multiple values of a vector can also be output, e.g. elements 2 to 4 with the command `a[2:4]`. To display all elements except certain ones, a minus sign preceding the index can be used, e.g. `a[-2]` displays all elements except the 2nd element. A list of indices can be specified with the function `c()`, e.g. `a[c(1,4:6)]` outputs element 1 and 4 to 6. The output of a sequence of elements can also be combined arbitrarily, including repetitions of individual elements, e.g. to output elements 1, 3, 6, 1, 1, 5 `a[c(1,3,6,1,1,5)]`. Specifying an invalid index results in the output NA, e.g. `a[3:8]` results in NA for index 7 and 8.

Calculation of indices is also possible. The function `length()` returns the number of elements of a vector. If you want to access the penultimate element, you can use the command `a[length(a)-1]`.

Vectors can also contain a name for each element, e.g. `b <- c(anna = 2700, rolf = 1800, heike = 4000)`. The content of the vector can be displayed again by entering the name or with the function `str()`. Access to element 2 can be achieved through `b[2]` or `b["rolf"]`.

Logical indexing is also possible, where conditions determine which elements should be selected, e.g. `b[b>2000]`, `b[b==1800 | b>3000]` or `b[b>3000 & b>2000]`. Another option is to specify TRUE (should be output) or FALSE (should not be output) for each element, e.g. `b[c(FALSE, TRUE, FALSE)]`, to output only element 2. The design of logical operations is detailed in section 0.

A vector does not necessarily have to be numeric, but can be of any data type, e.g. `c <- c("Siemens","Bosch")` as a vector of data type character.

If you apply a command to a vector, R executes this command for each individual element of the vector, e.g. when converting a temperature from Celsius to Fahrenheit. This does not require loop programming as in many other programming languages, which keeps the number of required R commands low. A vector `d<-c(20,40,60,80,100)` can realize a conversion from degrees Fahrenheit to degrees Celsius for each element of the vector with the command `d<-(d-32)*5/9`.

Vector

```
#- Vektor numerisch---------------------------------------------------------------
a <- new("numeric")                                 # Vektor ohne Inhalt
str(a)

##  num(0)

a <- c(1.2, -13, 46, 1880.12, -2.5, 77)             # Vektor Typ numerisch
str(a)

##  num [1:6] 1.2 -13 46 1880.1 -2.5 ...

a[]                                                 # Ausgabe

## [1]    1.20  -13.00   46.00 1880.12   -2.50   77.00

a[2]                                                # Ausgabe 2. Element

## [1] -13

a[2:4]                                              # Ausgabe 2.-4. Element

## [1]  -13.00   46.00 1880.12

a[-2]                                               # Ausgabe ohne 2. Element

## [1]    1.20   46.00 1880.12   -2.50   77.00

a[-2:-4]                                            # Ausgabe ohne 2.-4. Element

## [1]  1.2 -2.5 77.0

a[c(1,4:6)]                                         # Ausgabe 1. & 4.-6. Element

## [1]    1.20 1880.12   -2.50   77.00

a[c(1,3,6,1,1,5)]                                   # Ausgabe beliebiger Elemente

## [1]  1.2 46.0 77.0  1.2  1.2 -2.5

a[3:8]                                              # Ausgabe ungültiger Indizes

## [1]   46.00 1880.12   -2.50   77.00      NA      NA

length(a)                                           # Anzahl Elemente von Vektor

## [1] 6

a[length(a)-1]                                      # Ausgabe vorletztes Element

## [1] -2.5

#- Vektor numerisch mit Namen-----------------------------------------------------
b <- c(anna = 2700, rolf = 1800, heike = 4100)# Vektor mit Namen und Gehalt
str(b)

##  Named num [1:3] 2700 1800 4100
##  - attr(*, "names")= chr [1:3] "anna" "rolf" "heike"

b[2]                                                # Ausgabe 2. Element

## rolf
## 1800

b["rolf"]                                           # Ausgabe 2. Element

## rolf
## 1800

#- Vektor logische Indizierung----------------------------------------------------
b[b>2000]                                           # Ausgabe wenn >2000
```

```
## anna heike
## 2700 4100

b[b==1800 | b>3000]                     # Ausgabe wenn 1800 oder >3000

## rolf heike
## 1800 4100

b[b>2000 & b<3000]                      # Ausgabe wenn >2000 und <3000

## anna
## 2700

b[c(FALSE, TRUE, FALSE)]                # Ausgabe 2. Element

## rolf
## 1800

#- Vektor character-------------------------------------------------------------
c <- new("character")                   # Vektor ohne Inhalt
str(c)                                  # Ausgabe mit Typ und Länge
## chr(0)

c <- c("Siemens","Bosch")               # Vektor Typ character
str(c)

## chr [1:2] "Siemens" "Bosch"

#- Vektor Operationen-----------------------------------------------------------
d<-c(20,40,60,80,100)                   # Vektor Temperaturen Fahrenheit
d<-(d-32)*5/9                           # Temperaturen Celsius
d

## [1] -6.666667  4.444444 15.555556 26.666667 37.777778
```

5.2.2 Factor Variable

A factor variable is an atomic vector of the data type factor. Multiple values of a data type can be stored in a factor variable. These are **categorical variables** with a predefined set of characteristics. Categorical variables are distinguished between **nominal variables** such as color or country (without order) and **ordinal variables** such as customer satisfaction (with order from very satisfied to very dissatisfied).

For each occurring variant, a so-called **Label** is assigned as a combination of the value of the variant and a counter, so that each variant only needs to be stored once. The labels are standardly numbered either from the smallest number to the largest number (numerically) or alphabetically (character) with an integer from 1 to n. This is fine for a nominal factor variable. However, for an ordinal factor variable, it is important that the integer assigned to a variant is assigned either in ascending or descending order according to the order, so that the integer values can later be used for correlation analyses and machine learning, for example. Such an assignment of the integer numbers is achieved by using the parameters order and levels, e.g., `kundenzufriedenheit.ordinal-factor(feedback, order=TRUE, levels=c("very satisfied","satisfied","neutral","dissatisfied","very dissatisfied"))`.

If a value occurs in the factor variable, not the value itself, but the assigned integer number is stored for the element. Since integer numbers require minimal storage space, this reduces the storage requirement for a vector of the type factor when values occur repeatedly. A factor variable can be created with the function `factor()`, e.g., for the nominal variable `employeeofthemonth<-factor(c("Robert","Kim"," Heike","Kim","Robert","Robert"))`. Alternatively, the function `as.factor()` can create a factor variable from an already existing vector a, e.g., `a<-as.factor(a)`. In both cases, R did not have to automatically recognize the data type, but the function `factor()` or `as.factor()` determined that a factor variable should be created. The parameter `order=TRUE` determines whether the function `class()` returns factor or ordered factor. For ordinal variables, the parameter `order=TRUE` should be used.

The entire factor variable can be output by entering the name or with the function `str()`, and the function `levels()` displays the variants of a factor variable.

If you create a factor variable in which you record the days of the week on which the production quantity targets were exceeded, you can do this with the command `day<-factor(c("Wednesday", "Monday", "Wednesday", "Tuesday", "Wednesday"))`.

With the function `table()`, you can display the frequencies of the occurring values of a vector as a contingency table, e.g., with `table(day)`. The order in which the frequencies are listed is in the order in which the levels occur in the factor variable (ascending for numerical content, alphabetically sorted for strings).

But if you prefer a specific order, you can also change or specifically specify the order. For this purpose, there are the functions `fct_inorder()` (sorted by order of occurrence), `fct_infreq()` (sorted by frequency of occurrence) and `fct_inseq()` (sorted numerically) from the package `forcats`. The package `forcats` is part of the package `tidyverse` and therefore does not need to be installed and activated separately, provided the package `tidyverse` is already installed and activated.

The order of the levels can also be specified individually, e.g., the sequence of the weekdays within a week, by using the parameter `levels` in the function `factor()`, e.g., `weekdays <-factor(day, levels = c("Monday","Tuesday","Wednesday","Thursday","Friday"))`. Now the order of the levels corresponds to the sequence of the days within a week. The output of the frequencies then also corresponds to the weekdays. If levels not used in the data should not remain stored in the variable, they can be removed with the function `droplevels()`, e.g., `weekdays.withoutmonday<-droplevels(weekdays,"Monday")`.

If a factor variable is to be used as the basis of an arithmetic operation, it is necessary to transform the categorical values into a numerical value beforehand with the function `as.numeric(Factorvariable)`. This transformation often needs to be carried out in advance of the use of factor variables for machine learning.

Factor

```
#- Faktor character-------------------------------------------------------------------
mitarbeiterdesmonats <- factor(c("Robert","Kim","Heike","Kim","Robert","Robert"))
str(mitarbeiterdesmonats)

##  Factor w/ 3 levels "Heike","Kim",..: 3 2 1 2 3 3

mitarbeiterdesmonats.num <-                    # Konvertierung numerisch
  as.numeric(mitarbeiterdesmonats)             # -> Vorbereitung für MachineLearning
str(mitarbeiterdesmonats.num)                  # Zahlwerte unverändert

##  num [1:6] 3 2 1 2 3 3

levels(mitarbeiterdesmonats)                   # Levels, Standard=nach Alphabet

## [1] "Heike"  "Kim"    "Robert"

#- Faktor numerisch--------------------------------------------------------------------
a <- c(1.2, -13, 46, 1880.12, -2.5, 77)        # Vektorvariable numerisch
str(a)

##  num [1:6] 1.2 -13 46 1880.1 -2.5 ...

a <- as.factor(a)                              # Faktor numerisch aus Vektor a
str(a)

##  Factor w/ 6 levels "-13","-2.5","1.2",..: 3 1 4 6 2 5

levels(a)                                      # Levels, Standard=klein->groß

## [1] "-13"     "-2.5"    "1.2"     "46"      "77"      "1880.12"

#- Faktor character-------------------------------------------------------------------
tag <- c("Mittwoch","Montag","Mittwoch",       # Vektorvariable character
         "Dienstag","Mittwoch","Dienstag")
tag.nominal <- factor(tag)                     # Faktor character aus Vektor tag
levels(tag.nominal)                            # Levels, Standard=nach Alphabet

## [1] "Dienstag" "Mittwoch" "Montag"

table(tag.nominal)                             # Häufigkeit

## tag.nominal
## Dienstag Mittwoch   Montag
##        2        3        1

#- Vektorvariable in nominale oder ordinale Faktorvariable konvertieren-----------
feedback<-c("sehr unzufrieden",NA,             # Vektorvariable character
            "unzufrieden","neutral",
            "zufrieden","zufrieden",NA)
feedback.nominal <- factor(feedback)           # Nominale Variable
class(feedback.nominal)

## [1] "factor"

str(feedback.nominal)

##  Factor w/ 4 levels "neutral","sehr unzufrieden",..: 2 NA 3 1 4 4 NA
table(feedback.nominal)                        # Kontingenztabelle ohne NAs

## feedback.nominal
##          neutral sehr unzufrieden      unzufrieden        zufrieden
##                1                1                1                2

table(feedback.nominal, useNA = "always")      # Kontingenztabelle mit NAs

## feedback.nominal
##          neutral sehr unzufrieden      unzufrieden        zufrieden
```

```
##                 1                 1                 1                 2
##              <NA>
##                 2

feedback.ordinal <- factor(feedback,          # Ordinale Variable
  order=TRUE,
  levels=c("sehr zufrieden","zufrieden","neutral",
          "unzufrieden","sehr unzufrieden"))
class(feedback.ordinal)

## [1] "ordered" "factor"

str(feedback.ordinal)

##  Ord.factor w/ 5 levels "sehr zufrieden"<..: 5 NA 4 3 2 2 NA

table(feedback.ordinal)                       # Kontingenztabelle ohne NAs

## feedback.ordinal
##   sehr zufrieden        zufrieden           neutral       unzufrieden
##                0                2                 1                 1
## sehr unzufrieden
##                1

table(feedback.ordinal, useNA = "always")     # Kontingenztabelle mit NAs

## feedback.ordinal
##   sehr zufrieden        zufrieden           neutral       unzufrieden
##                0                2                 1                 1
## sehr unzufrieden             <NA>
##                1                2

#- Levels mit Funktionen fct_inorder und fct_infreq aus Paket forcats festlegen---
str(tag)

##  chr [1:6] "Mittwoch" "Montag" "Mittwoch" "Dienstag" "Mittwoch" "Dienstag"

tag.inorder <- fct_inorder(tag)               # Levels nach Reihenfolge
str(tag.inorder)

##  Factor w/ 3 levels "Mittwoch","Montag",..: 1 2 1 3 1 3

levels(tag.inorder)

## [1] "Mittwoch" "Montag"   "Dienstag"

table(tag.inorder)

## tag.inorder
## Mittwoch   Montag Dienstag
##        3        1        2

tag.infreq <- fct_infreq(tag)                 # Levels nach Häufigkeit
str(tag.infreq)
##  Factor w/ 3 levels "Mittwoch","Dienstag",..: 1 3 1 2 1 2

levels(tag.infreq)

## [1] "Mittwoch" "Dienstag" "Montag"

table(tag.infreq)

## tag.infreq
## Mittwoch Dienstag   Montag
##        3        2        1
```

```
#- Reihenfolge der Levels individuell festlegen----------------------------------
wochentage <- factor(tag, levels = c("Montag","Dienstag","Mittwoch",
                                     "Donnerstag","Freitag"))
str(wochentage)

##  Factor w/ 5 levels "Montag","Dienstag",..: 3 1 3 2 3 2

levels(wochentage)

## [1] "Montag"     "Dienstag"    "Mittwoch"    "Donnerstag" "Freitag"

table(wochentage)

## wochentage
##     Montag   Dienstag   Mittwoch Donnerstag    Freitag
##          1          2          3          0          0

wochentage.ohnemontag<-droplevels(wochentage,"Montag")
str(wochentage.ohnemontag)

##  Factor w/ 2 levels "Dienstag","Mittwoch": 2 NA 2 1 2 1

table(wochentage.ohnemontag)

## wochentage.ohnemontag
## Dienstag Mittwoch
##        2        3
```

5.2.3 List

A link of several vectors is possible in a list, which can be created with the function `list()`. This does not necessarily have to be a representation of a table with rows and columns, because the linked vectors can have different lengths, unlike matrices, arrays, dataframes, and tibbles.

In the following example, three vectors of different data types are combined into a list and the elements of the list are accessed via index (syntax: `Object[Index]`) and name (syntax: `Object$Elementname` or `Object[["Elementname"]]`). The output when accessed via index creates a list while the output when accessed via name creates a vector.

List

```
#- Liste-------------------------------------------------------------------
x1 <- c("BMW", "Audi")                          # Vektor x1 (Kunde)
x2 <- c(42, 54992, 7, 356)                      # Vektor X2 (Auftragsnummer)
x3 <- c(TRUE, FALSE, FALSE, FALSE, TRUE)        # Vektor X3 (Zielerreichung)
liste <- list(Kunde=x1, Auftrag=x2, Ziel=x3)   # Liste
str(liste)

## List of 3
##  $ Kunde  : chr [1:2] "BMW" "Audi"
##  $ Auftrag: num [1:4] 42 54992 7 356
##  $ Ziel   : logi [1:5] TRUE FALSE FALSE FALSE TRUE

a <- liste[2]                                   # Ausgabe über Index erzeugt Liste
str(a)

## List of 1
##  $ Auftrag: num [1:4] 42 54992 7 356

b <- liste$Auftrag                              # Ausgabe über Name erzeugt Vektor
str(b)

##  num [1:4] 42 54992 7 356

c <- liste[["Auftrag"]]                         # Ausgabe über Name erzeugt Vektor
str(c)

##  num [1:4] 42 54992 7 356
```

5.2.4 Matrix and Array

Matrices, arrays, dataframes, and tibbles are multidimensional vectors. Matrices, dataframes, and tibbles are like tables and have two dimensions, namely rows and columns, while arrays can have any number of dimensions. Matrices and arrays are atomic vectors that can only contain elements of one data type. Dataframes and tibbles, on the other hand, can contain any number of data types.

A matrix can be created by assigning the dimension attributes (columns, rows) to a one-dimensional vector with the function `dim()`. Another option is to merge several one-dimensional vectors into a matrix using the functions `cbind()` (column bind connects the vectors as columns side by side) or `rbind()` (row bind connects the vectors as rows one below the other). In addition, it is possible to create a matrix using the function `matrix()`.

With the function `dim()`, the dimensions (rows, columns) can also be displayed for all multidimensional vectors. If the function `dim()` is applied to one-dimensional vectors, NULL is returned as a result.

Matrix

```
#- Matrix---------------------------------------------------------------------
zahlen <- 1:15                                # Vektor
zahlen                                        # Ausgabe

##  [1]  1  2  3  4  5  6  7  8  9 10 11 12 13 14 15

dim(zahlen)                                   # Dimensionen

## NULL

class(zahlen)                                 # Datentyp

## [1] "integer"

dim(zahlen) <- c(3,5)                         # Matrix
zahlen                                        # Ausgabe

##      [,1] [,2] [,3] [,4] [,5]
## [1,]    1    4    7   10   13
## [2,]    2    5    8   11   14
## [3,]    3    6    9   12   15

dim(zahlen)                                   # Dimensionen

## [1] 3 5

class(zahlen)                                 # Datentyp

## [1] "matrix" "array"

str(zahlen)                                   # Ausgabe mit Typ, Dimension und Länge

##  int [1:3, 1:5] 1 2 3 4 5 6 7 8 9 10 ...

matrix1 <- matrix(1:15, 3, 5)                 # Matrix
matrix1                                       # Ausgabe
##      [,1] [,2] [,3] [,4] [,5]
## [1,]    1    4    7   10   13
## [2,]    2    5    8   11   14
## [3,]    3    6    9   12   15

dim(matrix1)                                  # Dimensionen

## [1] 3 5

str(matrix1)                                  # Ausgabe mit Typ, Dimension und Länge

##  int [1:3, 1:5] 1 2 3 4 5 6 7 8 9 10 ...

zahlen1 <- 1:3                                # Vektor
zahlen2 <- 4:6                                # Vektor
zahlen3 <- 7:9                                # Vektor
zahlen4 <- 10:12                              # Vektor
zahlen5 <- 13:15                              # Vektor
matrix2 <- cbind(zahlen1, zahlen2, zahlen3, zahlen4, zahlen5) # Matrix
matrix2                                       # Ausgabe

##      zahlen1 zahlen2 zahlen3 zahlen4 zahlen5
## [1,]       1       4       7      10      13
## [2,]       2       5       8      11      14
## [3,]       3       6       9      12      15

str(matrix2)                                  # Ausgabe mit Typ, Dimension und Länge

##  int [1:3, 1:5] 1 2 3 4 5 6 7 8 9 10 ...
##  - attr(*, "dimnames")=List of 2
##   ..$ : NULL
```

```
##    ..$ : chr [1:5] "zahlen1" "zahlen2" "zahlen3" "zahlen4" ...

zahlen6 <- 1:5                                  # Vektor
zahlen7 <- 6:10                                 # Vektor
zahlen8 <- 11:15                                # Vektor
matrix3 <- rbind(zahlen6, zahlen7, zahlen8) # Matrix
matrix3                                         # Ausgabe

##         [,1] [,2] [,3] [,4] [,5]
## zahlen6    1    2    3    4    5
## zahlen7    6    7    8    9   10
## zahlen8   11   12   13   14   15

str(matrix3)                                    # Ausgabe mit Typ, Dimension und Länge

##  int [1:3, 1:5] 1 6 11 2 7 12 3 8 13 4 ...
##  - attr(*, "dimnames")=List of 2
##   ..$ : chr [1:3] "zahlen6" "zahlen7" "zahlen8"
##   ..$ : NULL
```

An array can be created with the function `array()` and differs from a matrix only in the arbitrary number of dimensions. Arrays are suitable for creating so-called Data-Cubes, which can be used to perform evaluations very efficiently. In the following example, an array with the three dimensions Quarter 1–4, Product Group 1–9, and Sales Region for Country 1–5 is used to store the sales figures per product group and country for each of the quarters. For simplicity, the sales figures were set at 1 to 180.

Array

```
#- Array---------------------------------------------------------------------
array1 <- array(1:180, dim=c(4,9,5))       # Array: Quartal, Produktgruppe, Land
array1                                     # Ausgabe

## , , 1
##
##      [,1] [,2] [,3] [,4] [,5] [,6] [,7] [,8] [,9]
## [1,]    1    5    9   13   17   21   25   29   33
## [2,]    2    6   10   14   18   22   26   30   34
## [3,]    3    7   11   15   19   23   27   31   35
## [4,]    4    8   12   16   20   24   28   32   36
##
## , , 2
##
##      [,1] [,2] [,3] [,4] [,5] [,6] [,7] [,8] [,9]
## [1,]   37   41   45   49   53   57   61   65   69
## [2,]   38   42   46   50   54   58   62   66   70
## [3,]   39   43   47   51   55   59   63   67   71
## [4,]   40   44   48   52   56   60   64   68   72
##
## , , 3
##
##      [,1] [,2] [,3] [,4] [,5] [,6] [,7] [,8] [,9]
## [1,]   73   77   81   85   89   93   97  101  105
## [2,]   74   78   82   86   90   94   98  102  106
## [3,]   75   79   83   87   91   95   99  103  107
## [4,]   76   80   84   88   92   96  100  104  108
##
## , , 4
##
##      [,1] [,2] [,3] [,4] [,5] [,6] [,7] [,8] [,9]
## [1,]  109  113  117  121  125  129  133  137  141
## [2,]  110  114  118  122  126  130  134  138  142
## [3,]  111  115  119  123  127  131  135  139  143
## [4,]  112  116  120  124  128  132  136  140  144
##
## , , 5
##
##      [,1] [,2] [,3] [,4] [,5] [,6] [,7] [,8] [,9]
## [1,]  145  149  153  157  161  165  169  173  177
## [2,]  146  150  154  158  162  166  170  174  178
## [3,]  147  151  155  159  163  167  171  175  179
## [4,]  148  152  156  160  164  168  172  176  180

str(array1)                                # Ausgabe mit Typ, Dimension und Länge

##  int [1:4, 1:9, 1:5] 1 2 3 4 5 6 7 8 9 10 ...

dim(array1)                                # Dimensionen

## [1] 4 9 5
array1[1,,]                                # Dim 1=1

##       [,1] [,2] [,3] [,4] [,5]
##  [1,]    1   37   73  109  145
##  [2,]    5   41   77  113  149
##  [3,]    9   45   81  117  153
##  [4,]   13   49   85  121  157
##  [5,]   17   53   89  125  161
##  [6,]   21   57   93  129  165
##  [7,]   25   61   97  133  169
##  [8,]   29   65  101  137  173
```

```
## [9,]   33   69  105  141  177
array1[1,2,]                                # Dim 1=1, Dim 2=2
## [1]   5  41  77 113 149
array1[1,2,4]                               # Dim 1=1, Dim 2=2, Dim 3=4
## [1] 113
array1[,1,]                                 # Dim 2=1
##      [,1] [,2] [,3] [,4] [,5]
## [1,]    1   37   73  109  145
## [2,]    2   38   74  110  146
## [3,]    3   39   75  111  147
## [4,]    4   40   76  112  148
array1[,,1]                                 # Dim 3=1
##      [,1] [,2] [,3] [,4] [,5] [,6] [,7] [,8] [,9]
## [1,]    1    5    9   13   17   21   25   29   33
## [2,]    2    6   10   14   18   22   26   30   34
## [3,]    3    7   11   15   19   23   27   31   35
## [4,]    4    8   12   16   20   24   28   32   36
```

5.2.5 Dataframe and Tibble

Perhaps the most widely used data structures in R are the dataframe and the tibble. Often, tables are used to store multiple dimensions and different data types of elements. For this purpose, a dataframe or a tibble is suitable.

A dataframe can be created with the function `data.frame()`. Individual columns of a dataframe can be addressed via index or with the help of the $ operator with the syntax `Dataframename$Columnname` or `Dataframename[["Columnname"]]`. In connection with the pipe operator (`%>%`), which passes the result of an R statement to the following R statement, a point must be prefixed to the $ or [[. If an element of a dataframe is addressed via indices, a notation of the form `dataframe[row,column]` is possible. Indices can also be omitted and if this is done, all data will be selected for this index. For example, the command `df[,3]` causes the third column of the dataframe, i.e., the variable Age, to be output for all records. Depending on whether a dataframe is accessed via index (syntax: `Object[Index]`) or name (syntax: `Object$Elementname` or `Object[["Elementname"]]`), the result will have a different object type. The output when accessed via the index is a dataframe and the output when accessed via the name is a vector.

In addition to dataframes, tibbles also allow, as a newer form of a dataframe, the storage of different data types in a two-dimensional table form. A tibble is of the data type tbl_df and the installation of the package `tibble` is required to use this data type. The package `tibble` is a component of the package `tidyverse` and therefore does not need to be installed and activated separately, provided the package `tidyverse` is already installed and activated.

Unlike dataframes, tibbles do not convert the data types of variables into vectors when only one column remains (this happens with dataframes) and they are somewhat more comfortable to output. With the function `as_tibble()`, dataframes, lists, matrices, or tables can be converted into a tibble. To create a new tibble, the function `tibble()` is also available. If a function only works with the data type dataframe, the function `as.data.frame()` allows a tibble to be converted into a dataframe.

The following functions are applicable to both dataframes and tibbles. The dimensions (rows, columns) can be output with the function `dim()`. To determine only the number of rows, the function `nrow()` can be used. Similarly, the function `ncol()` works to determine the number of columns. The names of the columns (variables) of a dataframe can be displayed with the function `names()`. With the functions `colnames()` and `rownames()`, names can be assigned to columns and rows (note: assigning row names does not work for tibbles). To get a quick overview of a dataframe or a tibble, the function `head()` is handy, which displays the first 6 data records, or with a specific number, any other number, e.g., `head(df, 10)`, to display the first ten data records. To display the last n data records, the function `tail(df, n)` can be used. The function `summary()` is also helpful to get an overview of the data. It provides for each variable, depending on its type (num, char, factor …) a different output. For numerical (num, int) data, it shows the minimum, lower quartile, median, arithmetic mean, upper quartile, and maximum. This information helps to understand how the data is distributed. For factor variables (categorical variables), the function shows the absolute frequency of each level, i.e., each manifestation.

Dataframe and Tibble

```
#- Dataframe-----------------------------------------------------------------------
x1 <- c("Berlin","Bremen","München","Frankfurt") # Vektor x1 (Ort Niederlassung)
x2 <- c(42, 356, 123, 200)                       # Vektor X2 (Stückzahl Verkauf)
x3 <- c(53, 22, 31, 26)                          # Vektor X3 (Warenbestand)
df <- data.frame(Stadt=x1,Verkauf=x2,Bestand=x3) # Dataframe
df.kopie <- df                    # Kopie des Dataframes
df

##       Stadt Verkauf Bestand
## 1    Berlin      42      53
## 2    Bremen     356      22
## 3   München     123      31
## 4 Frankfurt     200      26

class(df)

## [1] "data.frame"

str(df)

## 'data.frame':    4 obs. of  3 variables:
##  $ Stadt  : chr  "Berlin" "Bremen" "München" "Frankfurt"
##  $ Verkauf: num  42 356 123 200
##  $ Bestand: num  53 22 31 26

dim(df)                           # Dimension (Zeilen,Spalten)

## [1] 4 3

nrow(df)                          # Zeilenanzahl

## [1] 4

ncol(df)                          # Spaltenanzahl

## [1] 3

a <- df[2]                        # Ausgabe Spalte 2 über Index -> Dataframe
str(a)

## 'data.frame':    4 obs. of  1 variable:
##  $ Verkauf: num  42 356 123 200

b <- df$Verkauf                   # Ausgabe Spalte 2 über Name -> Vektor
str(b)

##  num [1:4] 42 356 123 200

c <- df[["Verkauf"]]              # Ausgabe Spalte 2 über Name -> Vektor
str(c)

##  num [1:4] 42 356 123 200

d <- df %>% .$Verkauf             # Ausgabe Spalte 2 mit Name und Pipe -> Vektor
str(d)

##  num [1:4] 42 356 123 200

e <- df %>% .[["Verkauf"]]        # Ausgabe Spalte 2 mit Name und Pipe -> Vektor
str(e)

##  num [1:4] 42 356 123 200

#                                 # Anders als Tibble: Single Spalte->Vektor
df[2]                             # Ausgabe Spalte 2 -> Vektor
```

```
##   Verkauf
## 1      42
## 2     356
## 3     123
## 4     200

df[3,2]                           # Ausgabe Zeile 3 Spalte 2 mit Index

## [1] 123

df[3,]                            # Ausgabe Zeile 3 mit Index
##     Stadt Verkauf Bestand
## 3 München     123      31

df[df$Stadt == "Bremen",]         # Ausgabe Zeile 2 mit logischem index

##    Stadt Verkauf Bestand
## 2 Bremen     356      22

object.size(df)                   # Größe des Dataframe in Bytes

## 1320 bytes

names(df)                         # Namen der Spalten

## [1] "Stadt"   "Verkauf" "Bestand"

colnames(df)[2] <- "Stück verkauft"# Spalte 2 umbenennen
colnames(df)[names(df)=="Stadt"] <-# Spalte "Stadt" umbenennen
  "Niederlassung"
rownames(df) <- c("Zeile-1",      # Zeilen benennen
                  "Zeile-2",
                  "Zeile-3",
                  "Zeile-4")
df                                # Ausgabe

##         Niederlassung Stück verkauft Bestand
## Zeile-1        Berlin             42      53
## Zeile-2        Bremen            356      22
## Zeile-3       München            123      31
## Zeile-4     Frankfurt            200      26

head(df)                          # Ausgabe der ersten Datensätze (Standard=5)

##         Niederlassung Stück verkauft Bestand
## Zeile-1        Berlin             42      53
## Zeile-2        Bremen            356      22
## Zeile-3       München            123      31
## Zeile-4     Frankfurt            200      26

head(df, 3)                       # Ausgabe der ersten 3 Datensätze

##         Niederlassung Stück verkauft Bestand
## Zeile-1        Berlin             42      53
## Zeile-2        Bremen            356      22
## Zeile-3       München            123      31

tail(df, 2)                       # Ausgabe der letzten 2 Datensätze

##         Niederlassung Stück verkauft Bestand
## Zeile-3       München            123      31
## Zeile-4     Frankfurt            200      26

summary(df)                       # Zusammenfassungsstatistik

##  Niederlassung       Stück verkauft     Bestand
```

```
## Length:4          Min.   : 42.0   Min.   :22.0
## Class :character  1st Qu.:102.8   1st Qu.:25.0
## Mode  :character  Median :161.5   Median :28.5
##                   Mean   :180.2   Mean   :33.0
##                   3rd Qu.:239.0   3rd Qu.:36.5
##                   Max.   :356.0   Max.   :53.0

table(df$Niederlassung)            # Häufigkeit der Niederlassungen
##
##    Berlin    Bremen Frankfurt   München
##         1         1         1         1

df$Niederlassung <-                # Konvertierung Niederlassung von chr->factor
  as.factor(df$Niederlassung)
summary(df)                        # Zusammenfassung (factor jetz mit Anzahl)

##    Niederlassung Stück verkauft     Bestand
## Berlin   :1     Min.   : 42.0   Min.   :22.0
## Bremen   :1     1st Qu.:102.8   1st Qu.:25.0
## Frankfurt:1     Median :161.5   Median :28.5
## München  :1     Mean   :180.2   Mean   :33.0
##                 3rd Qu.:239.0   3rd Qu.:36.5
##                 Max.   :356.0   Max.   :53.0

#- Tibble-----------------------------------------------------------------------
tbl <- as_tibble(df.kopie)         # Tibble als Kopie des Dataframes
tbl                                # Ausgabe

## # A tibble: 4 × 3
##   Stadt     Verkauf Bestand
##   <chr>       <dbl>   <dbl>
## 1 Berlin         42      53
## 2 Bremen        356      22
## 3 München       123      31
## 4 Frankfurt     200      26

class(tbl)                         # Datentyp

## [1] "tbl_df"     "tbl"        "data.frame"

tbl2 <- tibble(Stadt=x1, Verkauf=x2, Bestand=x3)  # Tibble neu
tbl2                               # Ausgabe

## # A tibble: 4 × 3
##   Stadt     Verkauf Bestand
##   <chr>       <dbl>   <dbl>
## 1 Berlin         42      53
## 2 Bremen        356      22
## 3 München       123      31
## 4 Frankfurt     200      26

class(tbl2)                        # Datentyp

## [1] "tbl_df"     "tbl"        "data.frame"

str(tbl)                           # Ausgabe mit Typ, Dimension und Länge

## tibble [4 × 3] (S3: tbl_df/tbl/data.frame)
##  $ Stadt  : chr [1:4] "Berlin" "Bremen" "München" "Frankfurt"
##  $ Verkauf: num [1:4] 42 356 123 200
##  $ Bestand: num [1:4] 53 22 31 26

a <- tbl[2]                        # Ausgabe Spalte 2 über Index -> Tibble
str(a)
```

```
## tibble [4 × 1] (S3: tbl_df/tbl/data.frame)
##  $ Verkauf: num [1:4] 42 356 123 200

b <- tbl$Verkauf                   # Ausgabe Spalte 2 über Name -> Vektor
str(b)

##  num [1:4] 42 356 123 200
c <- tbl[["Verkauf"]]              # Ausgabe Spalte 2 über Name -> Vektor
str(c)

##  num [1:4] 42 356 123 200

d <- tbl %>% .$Verkauf             # Ausgabe Spalte 2 mit Name und Pipe -> Vektor
str(d)

##  num [1:4] 42 356 123 200

e <- tbl %>% .[["Verkauf"]]        # Ausgabe Spalte 2 mit Name und Pipe -> Vektor
str(e)

##  num [1:4] 42 356 123 200

# Anders als Dataframe: Single Spalte->Tibble
tbl[2]                             # Ausgabe Spalte 2 mit Index -> Tibble

## # A tibble: 4 × 1
##   Verkauf
##     <dbl>
## 1      42
## 2     356
## 3     123
## 4     200

tbl[3,2]                           # Ausgabe Zeile 3 Spalte 2 mit Index

## # A tibble: 1 × 1
##   Verkauf
##     <dbl>
## 1     123

tbl[3,]                            # Ausgabe Zeile 3 mit Index

## # A tibble: 1 × 3
##   Stadt   Verkauf Bestand
##   <chr>     <dbl>   <dbl>
## 1 München     123      31

tbl[tbl$Stadt == "Bremen",]        # Ausgabe Zeile 2 mit logischen index

## # A tibble: 1 × 3
##   Stadt  Verkauf Bestand
##   <chr>    <dbl>   <dbl>
## 1 Bremen     356      22

object.size(tbl)                   # Größe des Tibble in Bytes

## 1456 bytes

object.size(df.kopie)              # Größe des Dataframes in Bytes

## 1320 bytes

names(tbl)                         # Namen der Spalten

## [1] "Stadt"   "Verkauf" "Bestand"

colnames(tbl)[2] <-"Stück verkauft"# Spalte 2 umbenennen
```

```
colnames(tbl)[names(tbl)=="Stadt"] <- # Spalte "Stadt" umbenennen
  "Niederlassung"
rownames(tbl) <- c("Zeile-1",       # Zeilen in Tibbles OHNE Namen
                   "Zeile-2",
                   "Zeile-3",
                   "Zeile-4")

## Warning: Setting row names on a tibble is deprecated.

tbl                                 # Ausgabe

## # A tibble: 4 × 3
##   Niederlassung `Stück verkauft` Bestand
## * <chr>                   <dbl>   <dbl>
## 1 Berlin                     42      53
## 2 Bremen                    356      22
## 3 München                   123      31
## 4 Frankfurt                 200      26

head(tbl)                           # Ausgabe der ersten Datensätze

## # A tibble: 4 × 3
##   Niederlassung `Stück verkauft` Bestand
##   <chr>                   <dbl>   <dbl>
## 1 Berlin                     42      53
## 2 Bremen                    356      22
## 3 München                   123      31
## 4 Frankfurt                 200      26

head(tbl, 3)                        # Ausgabe der ersten 3 Datensätze

## # A tibble: 3 × 3
##   Niederlassung `Stück verkauft` Bestand
##   <chr>                   <dbl>   <dbl>
## 1 Berlin                     42      53
## 2 Bremen                    356      22
## 3 München                   123      31

tail(tbl, 2)                        # Ausgabe der letzten 2 Datensätze

## # A tibble: 2 × 3
##   Niederlassung `Stück verkauft` Bestand
##   <chr>                   <dbl>   <dbl>
## 1 München                   123      31
## 2 Frankfurt                 200      26

summary(tbl)                        # Zusammenfassung

##  Niederlassung      Stück verkauft      Bestand
##  Length:4           Min.   : 42.0   Min.   :22.0
##  Class :character   1st Qu.:102.8   1st Qu.:25.0
##  Mode  :character   Median :161.5   Median :28.5
##                     Mean   :180.2   Mean   :33.0
##                     3rd Qu.:239.0   3rd Qu.:36.5
##                     Max.   :356.0   Max.   :53.0

table(tbl$Niederlassung)            # Häufigkeit der Niederlassung

##
##    Berlin    Bremen Frankfurt   München
##         1         1         1         1

tbl$Niederlassung <-                # Konvertierung Niederlassung von chr->factor
  as.factor(tbl$Niederlassung)
```

Data structures	Vector	Factor	List	Matrix	Array	Dataframe	Tibble
Data type	numeric double integer logical Date character	factor	list	matrix	array	data.frame	tbl_df
# Dimensions	1	1	n	2	n	2	2
# Data types	1	1	n	1	1	n	n

Fig. 5.1 Data types and data structures

```
summary(tbl)                        # Zusammenfassung (factor jetzt mit Anzahl)

##     Niederlassung Stück verkauft      Bestand
##  Berlin    :1     Min.   : 42.0   Min.   :22.0
##  Bremen    :1     1st Qu.:102.8   1st Qu.:25.0
##  Frankfurt:1      Median :161.5   Median :28.5
##  München   :1     Mean   :180.2   Mean   :33.0
##                   3rd Qu.:239.0   3rd Qu.:36.5
##                   Max.   :356.0   Max.   :53.0

#-  Tibble in Dataframe konvertieren-------------------------------------------
df3 <- as.data.frame(tbl)           # Konvertiert Tibble in Dataframe
class(df3)                          # Datentyp

## [1] "data.frame"
```

5.2.6 Overview Table

Figure 5.1 summarizes all data types again.

5.3 Operations

R supports the most essential operations, including arithmetic, logical, selective operations as well as text and date operations.

5.3.1 Arithmetic Operations

R can also be used as a calculator for arithmetic operations. If the result is to be saved, it can be assigned to a variable. Such an assignment is done in R using the operation `<-` (less-than sign and minus sign), i.e., an arrow pointing to a variable, e.g., `a<-6+8`.

Fig. 5.2 Arithmetic Operators and Functions

Operator	Description
+	Addition
-	Subtraction
*	Multiplication
/	Division
^	Potency
sqrt(x)	Root
abs(x)	Absolute amount
exp(x)	Exponential function
log(x)	Natural logarithm

In RStudio, the content of the variable is then displayed in the Global Environment. The following operators can be used in R (see Fig. 5.2).

Operations can also be applied to a vector. If a vector c was created with the command `c<-c(10,15,40,200,5000)` and subsequently the operation `c+8` is executed, R adds the number 8 to each element of the vector c. The result remains a vector of the same length as the original vector.

When R uses two vectors of the same length for arithmetic operations, the respective operations (`+`, `–`, `*`, etc.) are processed element by element. If the vectors have different lengths, the shorter vector is reused until all calculations have been performed. If the shorter vector cannot be fully multiplied to the length of the other vector when reusing, R issues a warning.

Arithmetic operations can also be applied well to matrices, arrays, dataframes, and tibbles. To add the columns with the names x and y, for example, the command `df$x + df$y` or the function `with(df,x+y)` can be used for a dataframe named df.

If the calculation results in a number with many decimal places, the function `round()` often proves helpful to round the number to a specified number of decimal places, e.g., `round(b,2)` rounds the number b to two decimal places.

Arithmetic Operations

```
#- Arithmetische Operationen---------------------------------------------------------
6+8                                       # Addition

## [1] 14

a <- 6+8                                  # Ergebnis in Variable speichern
a                                         # Ausgabe

## [1] 14

6-4                                       # Subtraktion

## [1] 2

5*3                                       # Multiplikation

## [1] 15

b <- 30/167                               # Division
b                                         # Ausgabe

## [1] 0.1796407

round(b,2)                                # Ausgabe gerundet auf 2 Nachkommastellen

## [1] 0.18

10^3                                      # Potenz
## [1] 1000

sqrt(16)                                  # Wurzel

## [1] 4

abs(-87)                                  # Betrag

## [1] 87

exp(3)                                    # Exponentialfunktion e hoch 3

## [1] 20.08554

2.718281828459^3                          # e hoch 3

## [1] 20.08554

c <- exp(3)                               # Exponentialfunktion e hoch 3
log(c)                                    # Logarithmus

## [1] 3

d <- c(10,15,40,200,5000,-1000)           # Vektor numerisch Länge 6
d                                         # Ausgabe

## [1]    10    15    40   200  5000 -1000

e <- d+8                                  # Operation auf Vektor
e                                         # Ausgabe

## [1]   18   23   48  208 5008 -992

f <- d*e                                  # Op Vektoren gleiche Länge
f                                         # Ausgabe

## [1]      180      345     1920    41600 25040000   992000

e                                         # Ausgabe

## [1]   18   23   48  208 5008 -992
```

```
g <- c(3, -10)                             # vektor numerisch Länge 2
g                                          # Ausgabe

## [1]   3 -10

h <- e-g                                   # Op Vektoren ungleiche Länge vielfaches
h                                          # Ausgabe

## [1]   15   33   45  218 5005 -982

e                                          # Ausgabe

## [1]   18   23   48  208 5008 -992

i <- c(3, -10, -20, 40)                    # vektor numerisch Länge 4
i                                          # Ausgabe

## [1]   3 -10 -20  40

j <- e-i                                   # Op Vektoren ungleiche Länge vielfaches

## Warning in e - i: Longer object Length is not a multiple of shorter object
## Length

j                                          # Ausgabe
## [1]   15   33   68  168 5005 -982

df <- data.frame(x= c(53,2,100), y=c(299,20,30)) # Dataframe
df                                         # Ausgabe

##     x   y
## 1  53 299
## 2   2  20
## 3 100  30

df$x + df$y                                # Addition auf Spalten eines Dataframes

## [1] 352  22 130

with(df,x-y)                               # Subtraktion auf Spalten eines Dataframes

## [1] -246  -18   70
```

5.3.2 Logical Operations

Logical operations (see Fig. 5.3) can also be applied to various objects. When R performs a logical operation on a vector, it is executed for each element of the vector individually and produces a result that generates as many logical values as there are elements in the vector. The individual values represent TRUE if an element meets the condition and FALSE if the condition is not met for this element.

In addition to a simple logical operation, a logical OR (symbol for disjunction in R: `|`), logical AND (symbol for conjunction in R: `&`) and logical NOT (symbol for negation in R: `!`) can also be used in conditions.

Fig. 5.3 Logical Operators

Operator	Description
==	Equal
!=	Unequal
<	Smaller
<=	Smaller equal
>	Larger
>=	Larger equal
!	Not
&	Logical AND
|	Logical OR

Logical Operations

```
#- Logische Operationen-----------------------------------------------------------
a <- 48                                 # numerische Variable
a                                       # Ausgabe

## [1] 48

a == 48                                 # Operator == mit Ergebnis TRUE

## [1] TRUE

a == 84                                 # Operator == mit Ergebnis FALSE

## [1] FALSE

a != 84                                 # Operator != mit Ergebnis TRUE

## [1] TRUE

a < 84                                  # Operator < mit Ergebnis TRUE

## [1] TRUE
a <= 84                                 # Operator <= mit Ergebnis TRUE

## [1] TRUE

a > 84                                  # Operator > mit Ergebnis FALSE

## [1] FALSE

a >= 84                                 # Operator >= mit Ergebnis FALSE

## [1] FALSE

class(a)                                # Datentyp

## [1] "numeric"

is.numeric(a)                           # Datentyp is.numeric TRUE
```

```
## [1] TRUE
! is.numeric(a)                       # Datentyp ! is.numeric FALSE
## [1] FALSE
is.numeric(a) & a > 84                # Datentyp is.numeric AND a>84 FALSE
## [1] FALSE
is.numeric(a) | a > 84                # Datentyp is.numeric OR a>84 TRUE
## [1] TRUE
#- Logische Operationen auf Datentypen mit mehreren Elementen------------------
i                                     # Ausgabe
## [1]   3 -10 -20  40
truefalse1 <- i > 0                   # Operator > auf Vektor
truefalse1                            # Ausgabe
## [1]  TRUE FALSE FALSE  TRUE
df                                    # Ausgabe
##     x   y
## 1  53 299
## 2   2  20
## 3 100  30
truefalse2 <- df > 50                 # Operator > auf Dataframe
truefalse2                            # Ausgabe
##          x     y
## [1,]  TRUE  TRUE
## [2,] FALSE FALSE
## [3,]  TRUE FALSE
```

5.3.3 Selective Operations

With the help of selective operations, it is possible to select specific elements from vectors, dataframes, and other data structures.

The vector with the age of the team members, which can be created with the command `age <- sample(c(floor(runif(16, min=0, max=100)), rep(NA, 4)))`, contains 20 numerical elements with positive numbers between 0 and 100 years or the indication NA for a missing value, if the person asked did not specify their age. The function `rep()` repeatedly generates an element (1st parameter) n times (2nd parameter). Similarly, a logical vector truefalse with 20 elements can be generated by the command `truefalse<-rep(c(TRUE, FALSE),10)`, which alternately contains a TRUE and FALSE. The function `age[truefalse]` then only displays the elements where the value TRUE is in the vector truefalse. Thus, a selection can be made using TRUE and FALSE.

The function `is.na()` returns TRUE for the parameter (vector) in brackets if the element is NA and FALSE if the element is NOT NA. The function returns a vector of the same length as the parameter vector and indicates for each element the position (index) with TRUE where NA is in the vector. The function is used to identify NA values and their position. For example, to display all values of the vector age that are not NA, you can use the function `age[!is.na(age)]`. Reminder: The `!` means the negation of a logical condition.

If you want to display only the elements from this vector that have an age between 5 and 15 years, you can specify the logical condition in square brackets after the vector name, e.g. `age[!is.na(age) & age>= 5 & age<= 15]`.

To limit future analyses only to the elements that are not NA, you can create a new vector age2 that no longer contains NAs with the function `age2<-age[!is.na(age)]`. Then the display of people aged between 5 and 15 years can be achieved with the function `age2[age2>=5 & age2<=15]`. The adults from this group can be displayed with the command `age2[age2 > 17]`.

Selective Operations

```
#- Selektive Operationen--------------------------------------------------------------
alter <- sample(c(floor(runif(16,min=0,max=100)),rep(NA,4))) # Vektor numerisch
alter                                                 # Ausgabe

##  [1] NA  1 81  4 99 NA 89 92 57 70  9 NA 37 28 56 67  3 62 87 NA

truefalse <- rep(c(TRUE, FALSE),10)                   # Vektor logisch Länge 20
truefalse                                             # Ausgabe

##  [1]  TRUE FALSE  TRUE FALSE  TRUE FALSE  TRUE FALSE  TRUE FALSE  TRUE FALSE
## [13]  TRUE FALSE  TRUE FALSE  TRUE FALSE  TRUE FALSE

alter[truefalse]                                      # Ausgabe wo TRUE

##  [1] NA 81 99 89 57  9 37 56  3 87

is.na(alter)                                          # Ausgabe logisch Länge 20

##  [1]  TRUE FALSE FALSE FALSE FALSE  TRUE FALSE FALSE FALSE FALSE FALSE  TRUE
## [13] FALSE FALSE FALSE FALSE FALSE FALSE FALSE  TRUE

alter[is.na(alter)]                                   # Ausgabe wo TRUE

## [1] NA NA NA NA

alter[!is.na(alter)]                                  # Ausgabe wo FALSE

##  [1]  1 81  4 99 89 92 57 70  9 37 28 56 67  3 62 87

alter2 <- alter[!is.na(alter)]                        # Vektor wo FALSE
alter2                                                # Anzeige

##  [1]  1 81  4 99 89 92 57 70  9 37 28 56 67  3 62 87

alter[!is.na(alter) & alter>=5 & alter<=15]           # Ausgabe wo FALSE & >=5 & <=15

## [1] 9

alter2[alter2>=5 & alter2<=15]                        # Ausgabe wo FALSE & >=5 & <=15

## [1] 9

alter2[alter2 > 17]                                   # Ausgabe wo FALSE & >17

##  [1] 81 99 89 92 57 70 37 28 56 67 62 87
```

5.3.4 Text Operations

Many functions can be performed with variables of the character type, including the function `paste()` which allows multiple individual words to be connected. With the command `satz<-paste("Ich", "teste", "die", "paste()-Funktion")`, a new variable named satz is created.

You can also generate combinations of strings with the function `paste()` when the result consists of multiple elements of different types, e.g. `namen<-paste("Name",3:7,collapse=", ",sep="-")`. The parameter `collapse` specifies how the newly created strings should be separated from each other and the parameter `sep` specifies if a character other than a space should be used to separate the first parameter (here: Name) and the second parameter (here: 3 to 7). The function `paste0()` can be used if strings should be joined without spaces.

Text Operations

```
#- Textoperationen-----------------------------------------------------------
satz <- paste("Ich","teste","die","paste()-Funktion.") # Vektor character(Länge 1)
satz                                                   # Ausgabe

## [1] "Ich teste die paste()-Funktion."

str(satz)                                              # Ausgabe mit Typ und Länge

##  chr "Ich teste die paste()-Funktion."

namen <- paste("Name",3:7,collapse=", ",sep="-")       # Vektor character(Länge 1)
namen                                                  # Ausgabe

## [1] "Name-3, Name-4, Name-5, Name-6, Name-7"

str(namen)                                             # Ausgabe mit Typ und Länge

##  chr "Name-3, Name-4, Name-5, Name-6, Name-7"
```

Other functions for text analysis include the functions `str_detect()` for detecting strings in a text variable, `str_locate()` for detecting the position at which a string is located in a text variable, the function `str_extract()` and `str_extract_all()` to extract strings of a specific format from a text variable. The function `str_replace_all()` replaces a string in a text variable with another string. Another valuable function is the function `str_count()`, which counts the number of words when the parameter `boundary("word")` is used. This will be demonstrated in the following example of analyzing a tweet.

Analysis of a Tweet

```
#-Exemplarische Analyse eines Tweets-------------------------------------------
tweet<-"Trotz #Corona-Beschränkungen sind für den morgigen #1Mai zahlreiche rechte \
n           und/oder verachwörungsideologische Demos in #Sachsen geplant. Ein Ü
bersichtsthread\n                mit den Top10 von @johannesgrunert: https://t.co/PO
tZkJRpJQ"
str_detect(tweet, "[:digit:]")                    # Frage, ob Ziffer enthalten ist

## [1] TRUE

str_locate(tweet, "[:digit:]")                    # Frage, an welcher Position Ziffer

##      start end
## [1,]    53  53

str_extract(tweet, "[:digit:]")                   # Ziffer extrahieren

## [1] "1"

str_extract_all(tweet, "[:digit:]")               # Ziffern extrahieren

## [[1]]
## [1] "1" "1" "0"

str_extract_all(tweet, "[:digit:]{2}")            # Ziffern extrahieren, auch Länge 2

## [[1]]
## [1] "10"

hashtag<-str_extract_all(tweet, "#[:alnum:]+")# Hashtags # mit nachfolgenden alpha
hashtags<-unlist(hashtag)                         # Hashtags in einfachen Vektor
hashtags<-as_tibble(table(hashtags))              # Hashtags in Tibble
str_extract_all(tweet, "@[:alnum:]+")             # User @ mit nachfolgenden alpha

## [[1]]
## [1] "@johannesgrunert"

str_extract_all(tweet, "https?://[:graph:]+") # URLs extrahieren

## [[1]]
## [1] "https://t.co/POtZkJRpJQ"

str_count(tweet, boundary("word"))                # Anzahl Tokens (Wörter) zählen
## [1] 27

str_extract_all(tweet, "[:alpha:]+")              # Alle Tokens mit wenigstens einem

## [[1]]
##  [1] "Trotz"                      "Corona"
##  [3] "Beschränkungen"             "sind"
##  [5] "für"                        "den"
##  [7] "morgigen"                   "Mai"
##  [9] "zahlreiche"                 "rechte"
## [11] "und"                        "oder"
## [13] "verachwörungsideologische" "Demos"
## [15] "in"                         "Sachsen"
## [17] "geplant"                    "Ein"
## [19] "Übersichtsthread"           "mit"
## [21] "den"                        "Top"
## [23] "von"                        "johannesgrunert"
## [25] "https"                      "t"
## [27] "co"                         "POtZkJRpJQ"

                                                  # Buchstaben extrahieren
str_replace_all(tweet, "[^[:alpha:]+]", "")       # Alle Leerzeichen entfernen
```

```
## [1] "TrotzCoronaBeschränkungensindfürdenmorgigenMaizahlreicherechteundoderverachw
örungsideologischeDemosinSachsengeplantEinÜbersichtsthreadmitdenTopvonjohannesgruner
thttpstcoPOtZkJRpJQ"
```

5.3.5 Date Operations

The date format varies by region, and the time zone also varies regionally. The function `Sys.getlocale("LC_ALL")` can be used to display the system settings, and `Sys.setlocale("LC_TIME")` can be used to set the region. The time zone can be queried with the function `Sys.timezone()`.

There are special functions that can be applied to date variables. In Sect. 5.1, it was already introduced that R has a special way of handling date and time variables. The data type of date variables is Date and that of time variables is POSIXct and POSIXlt. All time information is stored as the time in days, hours, or seconds that has passed since 01.01.1970. For times before this date, a negative number indicates the days or seconds until 01.01.1970. The function `Sys.Date()` can be used to display the current date and the function `Sys.time()` to display the current time and date. With the command `unclass(Sys.Date())` or `unclass(Sys.time())`, the internal representation of the current date or time is displayed, and this is the number of days or seconds that have passed since 01.01.1970.

If you want to display the day of the week of a date, you can do this with the function `weekdays()`. The display of the month can be achieved with the function `months()`. A more comprehensive option, however, is the conversion of date and time into the POSIXlt format with the function `as.POSIXlt()`. If you then convert the result with the function `unclass()`, you can access the variables wday, mday, mon, year, zone, hour, min, and sec.

Sometimes it is also necessary to convert a text into a variable of type POSIXlt. The function `strptime()` supports this, e.g., `strptime("March 09, 1964 14:23","%B %d, %Y %H:%M")`.

The most clever way to get the details of date and time, however, is after converting a variable into the POSIXct format with the function `as.POSIXct()` and then using the function `format()` to display day, month, year, hour, minute, and second, e.g., with the command `format(Variable, format="%d.%m.%Y, %H:%M:%S")`.

There are also useful calculation functions (+, -) and comparisons that work with date and time variables. The function `difftime(Time-1, Time-2, units="secs")` allows the difference between the time variables Time-1 and Time-2 to be determined. The parameter `units` can be used to specify in which unit the difference should be calculated (weeks=weeks, days=days, hours=hours, minutes=mins, seconds=secs).

The lubridate package allows many other essential operations with date and time variables.

Date Operations

```
#- Standardeinstellungen----------------------------------------------------------------
Sys.getlocale("LC_ALL")                              # Systemeinstellungen ALL anzeigen

## [1] "LC_COLLATE=English_World.utf8;LC_CTYPE=English_World.utf8;LC_MONETARY=Englis
h_World.utf8;LC_NUMERIC=C;LC_TIME=English_World.utf8"

Sys.setlocale("LC_TIME", "C")                        # Windows Betriebssystem, Zeit=US

## [1] "C"

Sys.setlocale("LC_TIME", "German")                   # Windows Betriebssystem, Zeit=D

## [1] "German_Germany.1252"

Sys.getlocale("LC_TIME")                             # Systemeinstellungen TIME anzeigen

## [1] "German_Germany.1252"

Sys.timezone()                                       # Aktuelle Zeitzone

## [1] "Europe/Berlin"

#- Datumsoperationen--------------------------------------------------------------------
heute <- Sys.Date()                                  # Aktuelles Datum
heute                                                # Anzeige

## [1] "2023-01-08"

str(heute)                                           # Anzeige Typ

##  Date[1:1], format: "2023-01-08"

zeit <- Sys.time()                                   # Aktuelle Zeit
zeit                                                 # Anzeige

## [1] "2023-01-08 18:54:50 CET"

str(zeit)                                            # Anzeige Typ

##  POSIXct[1:1], format: "2023-01-08 18:54:50"

unclass.heute <- unclass(heute)                      # Tage seit dem 1.1.1970
unclass.heute                                        # Anzeige

## [1] 19365

str(unclass.heute)                                   # Anzeige Typ

##  num 19365

unclass.zeit <- unclass(zeit)                        # Sekunden seit dem 1.1.1970
unclass.zeit                                         # Anzeige

## [1] 1673200490
str(unclass.zeit                                     # Anzeige Typ

##  num 1.67e+09

#- Datumsoperationen für Datentyp Date--------------------------------------------------
heute.konv <- as.POSIXlt(heute)                      # Konvertiert in Typ POSIXlt
str(heute.konv)                                      # Anzeige Typ

##  POSIXlt[1:1], format: "2023-01-08"

unclass.heute.konv <- unclass(heute.konv)            # Konvertiert mit unclass()

unclass.heute.konv                                   # Anzeige

## $sec
```

```
## [1] 0
##
## $min
## [1] 0
##
## $hour
## [1] 0
##
## $mday
## [1] 8
##
## $mon
## [1] 0
##
## $year
## [1] 123
##
## $wday
## [1] 0
##
## $yday
## [1] 7
##
## $isdst
## [1] 0
##
## attr(,"tzone")
## [1] "UTC"
unclass.heute.konv$wday                        # Wochentag

## [1] 0

unclass.heute.konv$mday                        # Tag im Monat

## [1] 8

1+unclass.heute.konv$mon                       # Monat (mon vorbei + 1)

## [1] 1

1900 + unclass.heute.konv$year                 # Jahr  (year + 1900)

## [1] 2023

weekdays(heute)                                # Wochentag

## [1] "Sonntag"

months(heute)                                  # Monat

## [1] "Januar"

#- Datumsoperationen für Datentyp POSIXct (Zeit + Datum)-------------------------
zeit.konv <- as.POSIXlt(zeit)                  # Konvertiert in Typ POSIXlt
str(zeit.konv)                                 # Anzeige Typ

##  POSIXlt[1:1], format: "2023-01-08 18:54:50"

unclass.zeit.konv <- unclass(zeit.konv)        # Konvertiert mit unclass()

unclass.zeit.konv                              # Anzeige

## $sec
## [1] 50.07266
##
```

```
## $min
## [1] 54
##
## $hour
## [1] 18
##
## $mday
## [1] 8
##
## $mon
## [1] 0
##
## $year
## [1] 123
##
## $wday
## [1] 0
##
## $yday
## [1] 7
##
## $isdst
## [1] 0
##
## $zone
## [1] "CET"
##
## $gmtoff
## [1] 3600
##
## attr(,"tzone")
## [1] ""      "CET"  "CEST"
unclass.zeit.konv$wday                        # Wochentag
## [1] 0
unclass.zeit.konv$mday                        # Tag im Monat
## [1] 8
1+unclass.zeit.konv$mon                       # Monat (mon vorbei + 1)
## [1] 1
1900 + unclass.zeit.konv$year                 # Jahr  (year + 1900)
## [1] 2023
unclass.zeit.konv$zone                        # Zeitzone CET=Central European Time
## [1] "CET"
unclass.zeit.konv$hour                        # Stunde
## [1] 18
unclass.zeit.konv$min                         # Minute
## [1] 54
unclass.zeit.konv$sec                         # Sekunde
## [1] 50.07266
weekdays(zeit)                                # Wochentag
```

```
## [1] "Sonntag"
months(zeit)                                    # Monat
## [1] "Januar"
#- Konvertierung von character in Datentyp POSIXlt (Zeit + Datum)-----------------
a <- strptime("März 09, 1964 14:23","%B %d, %Y %H:%M") # Konvertiert Text in Zeit
a                                               # Anzeige CET=Central European Time
## [1] "1964-03-09 14:23:00 CET"
str(a)                                          # Anzeige Typ
##  POSIXlt[1:1], format: "1964-03-09 14:23:00"
#- Konvertierung von character in Datentyp POSIXct (Zeit + Datum)-----------------
b <- as.POSIXct("2020-03-09 11:12:54")          # Konvertiert Text in Zeit
b                                               # Anzeige CET=Central European Time
## [1] "2020-03-09 11:12:54 CET"
str(b)                                          # Anzeige Typ
##  POSIXct[1:1], format: "2020-03-09 11:12:54"
format(b, format="%d")                          # Tag
## [1] "09"
format(b, format="%m")                          # Monat
## [1] "03"
format(b, format="%Y")                          # Jahr
## [1] "2020"
format(b, format="%H")                          # Stunde
## [1] "11"
format(b, format="%M")                          # Minute
## [1] "12"
format(b, format="%S")                          # Sekunde
## [1] "54"
format(b, format="%d.%m.%Y, %H:%M:%S")          # Tag, Monat, Jahr, Stunde, Min, Sek
## [1] "09.03.2020, 11:12:54"
c <- as.POSIXct("11.04.2019, 14:52:31", format="%d.%m.%Y, %H:%M:%S") #Konv T->Zeit
format(c, format="%d")                          # Tag
## [1] "11"
format(c, format="%m")                          # Monat
## [1] "04"
format(c, format="%Y")                          # Jahr
## [1] "2019"
format(c, format="%H")                          # Stunde
## [1] "14"
```

```
format(c, format="%M")                          # Minute
## [1] "52"
format(c, format="%S")                          # Sekunde
## [1] "31"
format(c, format="%d.%m.%Y, %H:%M:%S")          # Tag, Monat, Jahr, Stunde, Min, Sek
## [1] "11.04.2019, 14:52:31"
#-- Datumsoperationen für die Berechnung von Zeitdifferenzen---------------------
heute                                           # Ausgabe
## [1] "2023-01-08"
weekdays(heute)                                 # Wochentag
## [1] "Sonntag"
übermorgen <- heute+2                           # Datumsberechnung mit +
übermorgen                                      # Ausgabe
## [1] "2023-01-10"
weekdays(übermorgen)                            # Wochentag
## [1] "Dienstag"
vorgestern <- heute-2                           # Datumsberechnung mit -
vorgestern                                      # Ausgabe
## [1] "2023-01-06"
weekdays(vorgestern)                            # Wochentag
## [1] "Freitag"
übermorgen - vorgestern                         # Berechnung Zeitdifferenz in Tagen
## Time difference of 4 days
d <- as.POSIXct("11.04.2021, 14:52:31", format="%d.%m.%Y, %H:%M:%S") #Konv T->Zeit
d                                               # Ausgabe
## [1] "2021-04-11 14:52:31 CEST"
e <- as.POSIXct("11.04.2021, 15:54:31", format="%d.%m.%Y, %H:%M:%S") #Konv T->Zeit
e                                               # Ausgabe
## [1] "2021-04-11 15:54:31 CEST"
difftime(e,d,units="weeks")                     # Zeitdifferenz in Wochen
## Time difference of 0.006150794 weeks
difftime(e,d,units="days")                      # Zeitdifferenz in Tagen
## Time difference of 0.04305556 days
difftime(e,d,units="hours")                     # Zeitdifferenz in Stunden
## Time difference of 1.033333 hours
difftime(e,d,units="mins")                      # Zeitdifferenz in Minuten
## Time difference of 62 mins
difftime(e,d,units="secs")                      # Zeitdifferenz in Sekunden
## Time difference of 3720 secs
```

5.4 Functions

The most essential functions are provided by R in the standard installation. In addition, further functions can be made available by installing packages. The following will first introduce how functions can be called. Then it will explain how to create or program your own functions.

5.4.1 Calling Functions

Functions follow a uniform basic structure. They have a name and are called with brackets. Within the brackets, arguments/parameters can be passed to the function, i.e., `FunctionName(ArgumentName1 = Value1, ArgumentName2 = Value2, ...)`, e.g., `data.frame(Friend=x1, Facebook=x2, Age=x3)`. The arguments in the brackets are separated by a comma. The order in which the arguments are passed when calling the function is arbitrary, as long as the names of the arguments are specified. If the names of the arguments are not specified, one must adhere to the standard order specified in the function definition. The names and order of the arguments as well as default values can be taken from the help page of a function, e.g., `?data.frame`. In the command `mean()`, default values for the arguments `trim=0` and `na.rm=FALSE` are preset, which means that the default value is always used when the function is called without these arguments. The specification of `na.rm= FALSE` means that the calculation of the mean returns NA as a result if an element of the vector does not contain a value, e.g., with `mean(x3)` NA will appear as a result if the vector x3 contains an element without a numerical value. If, on the other hand, the function is called with `mean(x3, na.rm=TRUE)`, elements without a value are ignored and the mean is calculated from all other elements and reported as a result.

If a function from a package that is not activated is to be called, the package name and two colons must be prefixed to the function name, e.g., `dplyr::filter()`. The same applies if a function with the same name exists in several installed packages.

The arguments, also called parameters, can be concrete values (numbers, strings …), variables of any data type, data structures of any type, or functions. It is therefore possible for a function to be passed as a parameter to another function, e.g., `summary(tail(df,2))`.

The arguments or parameters of a function can be displayed with the function `args()`.

Parameters are usually passed in the brackets after the function name. An alternative to this is the use of the so-called pipe operator `%>%` from the package `dplyr`, which allows the output of one function to be used as the input of the following function. The package `dplyr` is part of the package `tidyverse` and therefore does not need to be installed and activated separately, provided the package `tidyverse` has already been installed and activated.

To reduce the number of function calls when executing repeatedly, the functions `map()`, `map_dbl()`, `map2()` and `map2_dbl()` from the purrr package are suitable, which execute a function for all values of a vector (map functions) or two vectors (map2 functions). The functions with the _dbl suffix produce a vector as a result, while the functions without the suffix produce a list.

Calling Functions

```
#- Funktion aufrufen----------------------------------------------------------
x1 <- c("Tobi","Silke","Klaus","Tobi")                    # Funktion c()
x2 <- c(42, 356, 123, 200)                                # Funktion c()
x3 <- c(53, 22, 31, NA)                                   # Funktion c()
df <- data.frame(Freund=x1, Facebook=x2, Alter=x3)# Funktion data.frame()
#?data.frame                                              # Hilfe zur Funktion
#?mean                                                    # Hilfe zur Funktion
mean(x3)                                                  # Standardparameter na.rm=FALSE

## [1] NA

mean(x3,na.rm=TRUE)                                       # Parameter na.rm=TRUE

## [1] 35.33333

dplyr::filter(df, Freund=="Silke")                        # Funktion aus Paket dplyr

##   Freund Facebook Alter
## 1  Silke      356    22

summary(tail(df,2))                                       # Funktion tail als Parameter

##     Freund             Facebook          Alter
##  Length:2           Min.   :123.0   Min.   :31
##  Class :character   1st Qu.:142.2   1st Qu.:31
##  Mode  :character   Median :161.5   Median :31
##                     Mean   :161.5   Mean   :31
##                     3rd Qu.:180.8   3rd Qu.:31
##                     Max.   :200.0   Max.   :31
##                                     NA's   :1

#- Parameterübergabe mit Pipe-Operator(%>%)-----------------------------------
df                                                        # Ausgabe

##   Freund Facebook Alter
## 1   Tobi       42    53
## 2  Silke      356    22
## 3  Klaus      123    31
## 4   Tobi      200    NA

mean(df$Facebook)                                         # Parameter in Klammern

## [1] 180.25

df$Facebook %>% mean()                                    # Parameter über Pipe

## [1] 180.25

df$Facebook %>% mean() %>% round()                        # Parameter über Pipe

## [1] 180

#- Aufruf einer Funktion für alle Werte von Vektoren--------------------------
sqrt(df$Facebook[1])                                      # 1 Wert

## [1] 6.480741

sqrt(df$Facebook[2])                                      # 1 Wert

## [1] 18.86796

sqrt(df$Facebook[1:4])                                    # 4 Werte

## [1]  6.480741 18.867962 11.090537 14.142136

map(df$Facebook,sqrt)                                     # 1 Vektor   -> 1 Liste

## [[1]]
```

```
## [1] 6.480741
##
## [[2]]
## [1] 18.86796
##
## [[3]]
## [1] 11.09054
##
## [[4]]
## [1] 14.14214
map_dbl(df$Facebook,sqrt)                       # 1 Vektor    -> 1 Vektor
## [1]  6.480741 18.867962 11.090537 14.142136
map2(df$Facebook,df$Alter,~ .x+.y)              # 2 Vektoren -> 1 Liste
## [[1]]
## [1] 95
##
## [[2]]
## [1] 378
##
## [[3]]
## [1] 154
##
## [[4]]
## [1] NA
map2_dbl(df$Facebook,df$Alter,~ .x+.y)          # 2 Vektoren -> 1 Vektor
## [1]  95 378 154  NA
```

5.4.2 Creating Functions

Often, one wants to create their own functions when there are no suitable functions in any of the packages or one does not know such a function. For orientation, it can be helpful to refer to already known, similar functions. To display the syntax of a function, simply enter its name without the brackets that would be required for a function call, for example, `difftime`.

Functions are also objects in R, which are stored in the working memory, and they consist of the function head and the function body. The content of the round brackets of the function `function()`, in which the arguments or parameters of the function are determined, is considered the function head. In accordance with the weak typing of R, the arguments are given without a data type. The function body is enclosed by curly brackets and contains the instructions to be executed by the function. The function `return()` determines what the function returns and appears in the console and also ends the execution of the function, for example, `function1<-function(x, y){z=x+y;return(z/y)}`. If a function does not use `return()`, then the value of the last value assignment within the function is returned, without being displayed in the console. Additional text and number outputs can be defined with the function `print("Text output")` or `print(Number output)`. If a combination of text and numbers is to be output, this can be achieved with the function `cat()`.

It is also possible to provide the arguments of a function with a default value, for example `function2<-function(x, y=6){z=x+y;return(z/y)}`. If the function is then called without the parameter y, i.e., `function2(15)`, the number 6 is used for y as a default. The call of a function can always be made with specified parameter names, for example, `function2(x=15, y=3)`.

A function can also be established for an indefinite number of arguments when there are three dots in the bracket, for example, `function9 <-function(...){x<-as.numeric(list(...));return(sum(x))}`.

All arguments of a function are basically passed as in a "call by value" and not a "call by reference", which means that the values are unchanged after the function is completed, i.e., their value can only be changed locally within the function. Also, all variables newly created within the function no longer exist after the function is finished. If there is a variable in the Global Environment with the same name as a local variable, its value also remains unchanged. All variables of the Global Environment can be used in a function, but remain unchanged after the function is finished.

Creating Functions

```
#- Code und Argumente einer Funktion anzeigen lassen------------------------------
difftime                                         # Code der Funktion anzeigen

## function (time1, time2, tz, units = c("auto", "secs", "mins",
##     "hours", "days", "weeks"))
## {
##     if (missing(tz)) {
##         time1 <- as.POSIXct(time1)
##         time2 <- as.POSIXct(time2)
##     }
##     else {
##         time1 <- as.POSIXct(time1, tz = tz)
##         time2 <- as.POSIXct(time2, tz = tz)
##     }
##     z <- unclass(time1) - unclass(time2)
##     attr(z, "tzone") <- NULL
##     units <- match.arg(units)
##     if (units == "auto")
##         units <- if (all(is.na(z)))
##             "secs"
##         else {
##             zz <- min(abs(z), na.rm = TRUE)
##             if (!is.finite(zz) || zz < 60)
##                 "secs"
##             else if (zz < 3600)
##                 "mins"
##             else if (zz < 86400)
##                 "hours"
##             else "days"
##         }
##     switch(units, secs = .difftime(z, units = "secs"), mins = .difftime(z/60,
##         units = "mins"), hours = .difftime(z/3600, units = "hours"),
##         days = .difftime(z/86400, units = "days"), weeks = .difftime(z/(7 *
##             86400), units = "weeks"))
## }
## <bytecode: 0x000001b000a400b8>
## <environment: namespace:base>

args(difftime)                                   # Argumente der Funktion

## function (time1, time2, tz, units = c("auto", "secs", "mins",
##     "hours", "days", "weeks"))
## NULL

#- Funktion erstellen-------------------------------------------------------------
funktion1 <- function(x, y){z=x+y;return(z/y)}  # Neue Funktion
funktion1                                        # Zeigt den Code der Funktion

## function(x, y){z=x+y;return(z/y)}

funktion1()                                      # Aufruf ohne Argument

## Error in funktion1(): argument "x" is missing, with no default

funktion1(15)                                    # Aufruf mit Argument x

## Error in funktion1(15): argument "y" is missing, with no default

funktion1(x=15)                                  # Aufruf mit Name und Argument x

## Error in funktion1(x = 15): argument "y" is missing, with no default

funktion1(y=3)                                   # Aufruf mit Name und Argument y
```

```
## Error in funktion1(y = 3): argument "x" is missing, with no default
funktion1(15,3)                              # Aufruf Argument xy
## [1] 6
funktion1(3,15)                              # Aufruf Argument xy
## [1] 1.2
funktion1(x=15,y=3)                          # Aufruf mit Name und Argument xy
## [1] 6
funktion1(y=3,x=15)                          # Aufruf mit Name und Argument yx
## [1] 6
funktion2 <- function(x, y=6){z=x+y;return(z/y)}# Neue Funktion mit Standard y=6
funktion2                                    # Zeigt den Code der Funktion
## function(x, y=6){z=x+y;return(z/y)}
funktion2()                                  # Aufruf ohne Argument
## Error in funktion2(): argument "x" is missing, with no default
funktion2(15)                                # Aufruf mit Argument x
## [1] 3.5
funktion2(x=15)                              # Aufruf mit Name und Argument x
## [1] 3.5
funktion2(y=3)                               # Aufruf mit Name und Argument y
## Error in funktion2(y = 3): argument "x" is missing, with no default
funktion2(15,3)                              # Aufruf Argument xy
## [1] 6
funktion2(3,15)                              # Aufruf Argument xy
## [1] 1.2
funktion2(x=15,y=3)                          # Aufruf mit Name und Argument xy
## [1] 6
funktion2(y=3,x=15)                          # Aufruf mit Name und Argument yx
## [1] 6
funktion3 <- function(x){return(mean(x))}    # Neue Funktion
x2                                           # Ausgabe
## [1]  42 356 123 200
funktion3(x2)                                # Aufruf mit x2
## [1] 180.25
funktion4 <- function(x){                    # Neue Funktion
  print("Mittelwert von:");                  # Print
  print(x2);
  return(mean(x))}
funktion4(x2)                                # Aufruf mit x2
## [1] "Mittelwert von:"
## [1]  42 356 123 200
```

```
## [1] 180.25
funktion5 <- function(x){                           # Neue Funktion
  print("Mittelwert von:",quote=FALSE);             # Print mit quote=FALSE
  print(x2); print(mean(x))
  return(mean(x))}
rückgabewert5 <- funktion5(x2)                      # Aufruf mit x2

## [1] Mittelwert von:
## [1]  42 356 123 200
## [1] 180.25

str(rückgabewert5)                                  # Ausgabe mit Typ Anzahl

##  num 180

funktion6 <- function(x){                           # Neue Funktion
  cat("Mittelwert von:",x2,"= ", mean(x));          # Cat
  Letzterwert = mean(x)                             # Ohne return()
  }
rückgabewert6 <- funktion6(x2)                      # Aufruf mit x2

## Mittelwert von: 42 356 123 200 =  180.25

str(rückgabewert6)                                  # Ausgabe mit Typ Anzahl

##  num 180

funktion7 <- function(x){                           # Neue Funktion
  cat("Mittel, Max, Min von:",x2,"=","\n");         # Cat
  return(c(mean(x),max(x),min(x)))}                 # Return von Vektor
rückgabewert7 <- funktion7(x2)                      # Aufruf mit x2

## Mittel, Max, Min von: 42 356 123 200 =

str(rückgabewert7)                                  # Ausgabe mit Typ Anzahl

##  num [1:3] 180 356 42

funktion8 <- function(x){                           # Neue Funktion
  cat("Mittel, Max, Min von:",x2,"=",mean(x),max(x),min(x),"\n");  # Cat
  return(data.frame(mean(x),max(x),min(x)))}        # Return von Dataframe
rückgabewert8 <- funktion8(x2)                      # Aufruf mit x2

## Mittel, Max, Min von: 42 356 123 200 = 180.25 356 42

str(rückgabewert8)                                  # Ausgabe mit Typ Anzahl

## 'data.frame':    1 obs. of  3 variables:
##  $ mean.x.: num 180
##  $ max.x. : num 356
##  $ min.x. : num 42

#- Funktion erstellen mit beliebiger Anzahl Parameter-----------------------------
funktion9 <- function(...){                         # Neue Funktion
  x<-as.numeric(list(...));return(sum(x))}
rückgabewert9 <- funktion9(1,3,"7",100)             # Aufruf beliebige Parameter
str(rückgabewert9)                                  # Ausgabe mit Typ Anzahl

##  num 111

rückgabewert9 <- funktion9()                        # Aufruf ohne Parameter
str(rückgabewert9)                                  # Ausgabe mit Typ Anzahl

##  num 0
#- Funktion erstellen: Argumentübergabe "by value"-------------------------------
```

```
zahl <- 2                                              # Globale Variable
str(zahl)

##  num 2

funktion10 <- function(x){                             # Neue Funktion
  cat("Zahl in Funktion vor Addition:",zahl,"\n");  # Lokale Variable
  zahl <- zahl + 50                                    # Addition
  cat("Zahl in Funktion nach Addition:",zahl,"\n"); # Lokale Variable verändert
  return(zahl)}
rückgabewert10 <- funktion10(zahl)                     # Aufruf by value

## Zahl in Funktion vor Addition: 2
## Zahl in Funktion nach Addition: 52

str(rückgabewert10)

##  num 52

str(zahl)                                              # Globale Variable unverändert

##  num 2
```

5.5 Control Structures

In R, the usual control structures if/else and loops of the type for, while, and repeat exist to represent program logic. However, it is also possible to structure instructions by forming blocks of instructions with curly brackets. The advantage is that blocks of instructions can be executed interactively in RStudio with a click on Ctrl-Enter.

An if/else statement follows this syntax: `if(condition) {instruction block-if} else {instruction block-else}` or for more than two case distinctions `if(condition) {instruction block-if} else if (condition-2) {instruction block-2} else {instruction block-else}` (see Fig. 5.4).

A for loop follows this syntax: `for (variable in list) {instruction block}` (see Fig. 5.5).

A while loop follows this syntax: `while (condition) {instruction block}` and runs as long as the condition is met. The repeat loop follows this syntax: `repeat {instruction block}` and repeats the instructions of the instruction block until the function `break()` is executed, which is tied to a condition. If the condition is not met or no break() function is integrated into the repeat loop, this leads to an infinite loop (see Fig. 5.6).

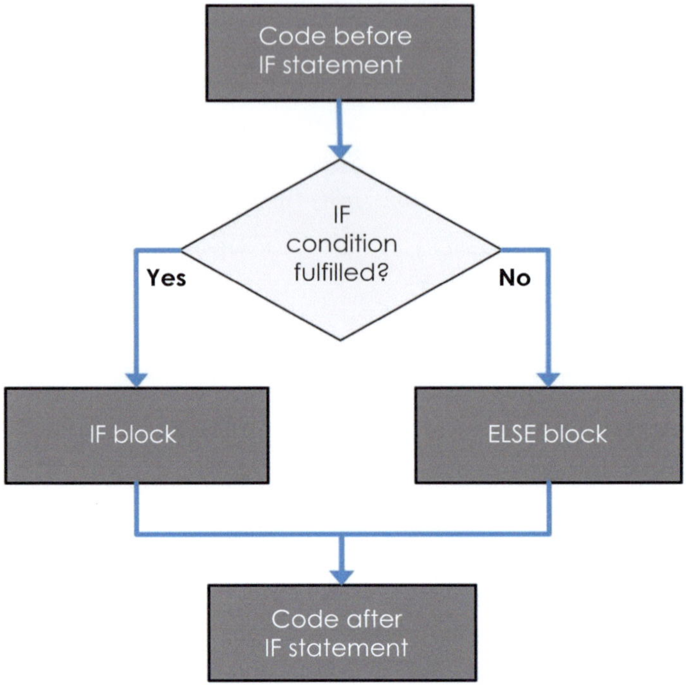

Fig. 5.4 IF Statement

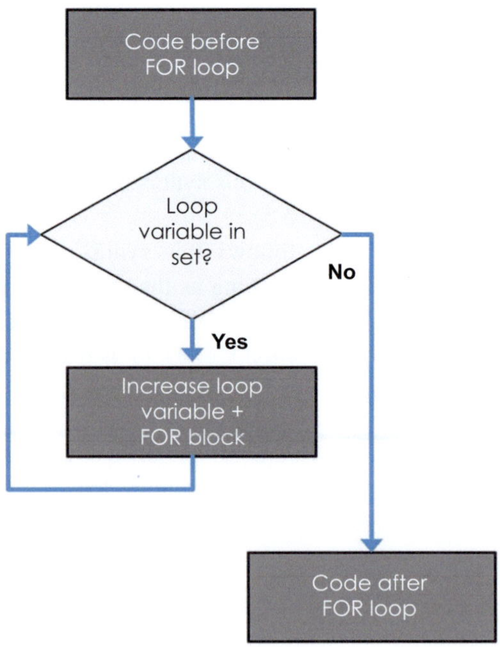

Fig. 5.5 FOR Loop

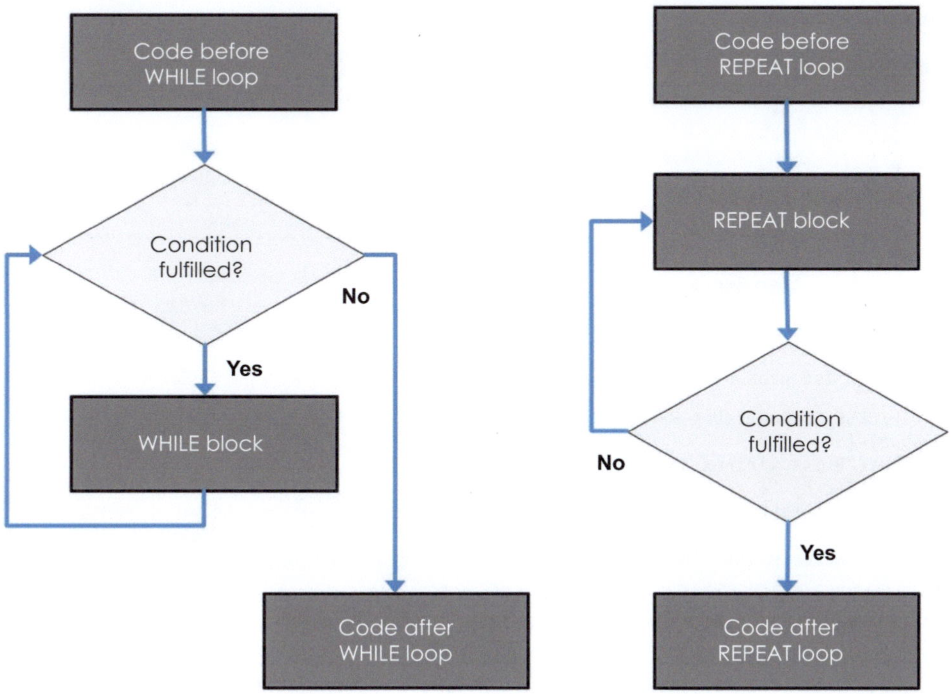

Fig. 5.6 While Loop and Repeat Loop

Control Structures

```
#- Block----------------------------------------------------------------------
{                                                    # Block von Anweisungen
  a = 1
  b = 2
  c = 3
  d = a + b + c
}
#- IF-Anweisung---------------------------------------------------------------
if(b>a) {
  print("B ist größer als a")
} else {
  print("B ist nicht größer als a")
}
## [1] "B ist größer als a"

#- IF/ELSE IF Anweisung-------------------------------------------------------
if(b<a) {
  print("B ist kleiner als a")
} else if(c<b)  {
  print("C ist kleiner als b")
} else {
  print("Weder B ist kleiner als a noch ist c kleiner als b")
}

## [1] "Weder B ist kleiner als a noch ist c kleiner als b"

#- FOR-Schleife---------------------------------------------------------------
x1 <- c("Tobi","Silke","Klaus","Tobi")   # Vektor x1 (Freunde)
for (i in x1){
  print(i)
}

## [1] "Tobi"
## [1] "Silke"
## [1] "Klaus"
## [1] "Tobi"

x2 <- c(42, 356, 123, 200)               # Vektor X2 (Facebook Freunde)
x3 <- c(53, 22, 31, 26)                  # Vektor X3 (Alter)
df <- data.frame(Freund=x1, Facebook=x2, Alter=x3)  # Dataframe
df                                       # Ausgabe

##   Freund Facebook Alter
## 1   Tobi       42    53
## 2  Silke      356    22
## 3  Klaus      123    31
## 4   Tobi      200    26

for (z in 1:nrow(df)){
  print(df[z,])
}

##   Freund Facebook Alter
## 1   Tobi       42    53
##   Freund Facebook Alter
## 2  Silke      356    22
##   Freund Facebook Alter
## 3  Klaus      123    31
##   Freund Facebook Alter
## 4   Tobi      200    26

for (s in 1:ncol(df)){
  print(df[,s])
```

```
}
## [1] "Tobi"  "Silke" "Klaus" "Tobi"
## [1]  42 356 123 200
## [1] 53 22 31 26

for (s in 1:ncol(df)){
  for (z in 1:nrow(df)){
    print(paste("Spalte/Zeile[",s,",",z,"]=",df[z,s]))
  }
}
## [1] "Spalte/Zeile[ 1 , 1 ]= Tobi"
## [1] "Spalte/Zeile[ 1 , 2 ]= Silke"
## [1] "Spalte/Zeile[ 1 , 3 ]= Klaus"
## [1] "Spalte/Zeile[ 1 , 4 ]= Tobi"
## [1] "Spalte/Zeile[ 2 , 1 ]= 42"
## [1] "Spalte/Zeile[ 2 , 2 ]= 356"
## [1] "Spalte/Zeile[ 2 , 3 ]= 123"
## [1] "Spalte/Zeile[ 2 , 4 ]= 200"
## [1] "Spalte/Zeile[ 3 , 1 ]= 53"
## [1] "Spalte/Zeile[ 3 , 2 ]= 22"
## [1] "Spalte/Zeile[ 3 , 3 ]= 31"
## [1] "Spalte/Zeile[ 3 , 4 ]= 26"

#- WHILE-Schleife-------------------------------------------------------------------
index <- 1
while (index <= 5){
  cat("Schleifendurchlauf",index,"\n")
  index <- index+1
}

## Schleifendurchlauf 1
## Schleifendurchlauf 2
## Schleifendurchlauf 3
## Schleifendurchlauf 4
## Schleifendurchlauf 5

index <- 1
while (index <= 5){
  cat("Schleifendurchlauf",index,"\n")
  if(index == 3){break}
  index <- index+1
}

## Schleifendurchlauf 1
## Schleifendurchlauf 2
## Schleifendurchlauf 3

#- REPEAT-Schleife------------------------------------------------------------------
index <- 1
repeat{
  cat("Schleifendurchlauf",index,"\n")
  index <- index+1
  if(index > 5){break}
}

## Schleifendurchlauf 1
## Schleifendurchlauf 2
## Schleifendurchlauf 3
## Schleifendurchlauf 4
## Schleifendurchlauf 5
```

6 Basics of Machine Learning with R

In order to use data for machine learning, it must first be imported as raw data from a source. Subsequently, the data must be prepared or processed for the subsequent analysis as part of a transformation. A data transformation includes, among other things, the selection of the data required for the analysis, validation, fusion with other data, reshaping, supplementation, and summarization. Once the data is cleaned, it needs to be better understood, analyzed, and with the help of visualizations, peculiarities can also be identified. It can also be determined how to deal with outliers and missing data in further analyses. After these preparatory activities, the methods and models of machine learning can be applied and their suitability evaluated before finally communicating the resulting insights (see Fig. 6.1).

6.1 Import and Export of Data

Before data can be processed with R, it must first be read into memory or imported. This can be done from various sources and in different formats, e.g., from files, databases, via web scraping, or through direct data collection.

If the data is to be collected directly in R, this can be done with the function `scan()`, e.g., `new<-scan()` allows the collection of elements that are combined into a vector named new. The new object is saved as soon as an empty element is present and the Enter key has been pressed.

Often, files are used as a source for data. Files can always be uniquely described by the combination of filename and pathname, in which the file is or should be saved. If no explicit pathname is given, R assumes that the file is located in the so-called working directory or should be saved there. The working directory can be displayed with the function `getwd()` or redefined with the function `setwd()`. The working directory can also be

B. Heesen, *Artificial Intelligence and Machine Learning with R*,
https://doi.org/10.1007/978-3-658-45392-3_6

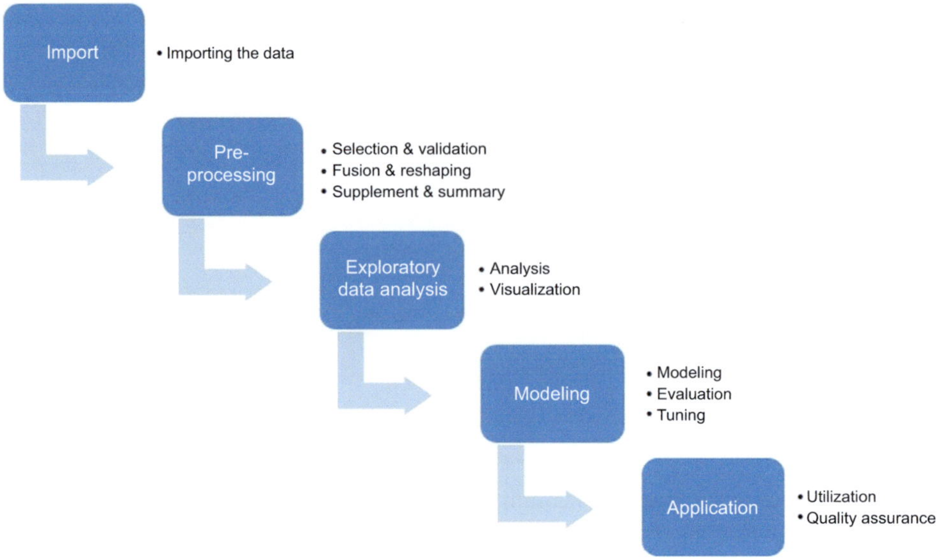

Fig. 6.1 Machine Learning Process

determined in RStudio via the menu Session/Set Working Directory/Choose Directory. The function `list.files()` or `dir()` can be used to display all files in a directory. If no directory is specified as a parameter, the working directory is used. With the argument `recursive=TRUE`, the files of the subdirectories are also displayed. New directories can be created with the function `dir.create()` and new files with the function `file.create()`. To check whether a file exists, the function `file.exists()` can be used. The function `file.info()` displays information (size = size, isdir = whether it is a directory, mtime = last change, atime = last access, exe = executable file yes/no) about a file. To rename a file, the function `file.rename()` can be used, to copy the function `file.copy()` and to delete the function `file.remove()`. To delete an entire directory, the function `unlink()` is used.

Data can also be sourced from objects that have already been loaded into memory, the so-called workspace. The workspace contains all objects in memory. The objects can be displayed with the function `ls()`. If you want to save the entire workspace with all objects for later use, this is possible with the function `save.image("filename.RData")`. If you want to explicitly save specific objects, this can be done with the function `save(objectname1, objectname2..., file="filename.RData")`. The objects saved in this way can then be read back into the workspace later with the function `load("filename.RData")`. Both file extensions .RData and .Rda can be used interchangeably for these file types.

6.1.1 Data Import

Data can be imported from various sources, including FlatFiles, such as CommaSeparatedFiles (CSV), text and HTML files, Excel files, software from other statistical packages, databases, or via APIs (Application Programming Interfaces). This section will only introduce the most basic import options for reading RData files and the following R-external file formats: TXT, CSV, XLSX, JSON, XML, and HTML.

Data from the SPSS statistical package can be read using the function `read.spss()` from the foreign package. The same package also includes the function `read.dta()` for reading files from the Stata statistical package. The sas7bdat package contains the function `read.sas7bdat()` for reading files from the SAS statistical package.

If data is to be imported from databases, the RMySQL package is available, which allows data to be read from a MySQL database using the functions `dbConnect()`, `dbSendQuery()`, and `dbWriteTable()`. For the non-relational MongoDB database, the RMongo package with the functions `mongoDbConnect()` and `dbGetQuery()` is available. To load data from an Oracle database, the ROracle package can be used (Oracle 2021).

If data is to be loaded from the web via WebScraping, the xml2 package in combination with the rvest package is a good solution. The read_html() function from the xml2 package reads the data from a URL. The html_table() function from the rvest package extracts a table from the website.

The function `readLines()` allows .txt files to be loaded into a dataframe. The functions `read_delim()` with a dot as a decimal separator and `read_delim2()` with a comma as a decimal separator support reading files whose variables are separated by a special character, if the columns are not separated by a tab as in a .txt file, but for example with a comma or semicolon. The parameter `delim=","` or `delim=";"` or similar is used to specify the separator. In addition to the function `read_delim()`, the functions `read_csv()` for comma-separated and `read_csv2()` for semicolon-separated .csv files can also be used. The parameter `header=TRUE` or `FALSE` indicates whether the first line contains the column names and the parameter `col_types` defines the target data format of the columns, e.g., `gebrauchtwagen<-read_delim("autos.csv",col_names=TRUE,delim=";",col_types=cols(Marke=col_factor(),Zulassung=col_integer(),PS=col_integer(),Kilometerstand=col_character(), Preis=col_character()))`.

Excel files can be read using the function `read_excel()` from the readxl package. The syntax of the function is `read_excel(PATH,sheet=NULL,range=NULL,col_names=TRUE)` with the following arguments:

- PATH: Directory path where the file can be found and filename.
- sheet: Name of the worksheet to be imported. If not specified, the first worksheet is imported.

- range: Range with row and column indication to be imported. If not specified, all cells are imported.
- col_names: TRUE if column names should be imported from the first row (default). FALSE if the column names are not in the first row.
- n_max: Number of rows to be read. If not specified, all rows are read.

To read files in JSON format, the function `fromJSON()` from the rjson package can be used. Similarly, files in XML format can be read using the function `xmlTreeParse()` from the XML package.

Data can also be contained in R packages. Importing data from packages is done using the command `data(dataset,package =„package")`.

6.1.2 Data Export

After data has been imported into the workspace, it can be helpful to save this data for later processing. To secure individual objects, the function `save(Objectname1, Objectname2…, file="Filename.RData")` can be used. The objects saved in this way can then be read back in later with the function `load("Filename.RData")`.

To export data to a statistical package, the functions `write_sav()` for SPSS, the function `write_sas()` for SAS, and `write_dta()` for Stata can be used.

Saving in the cloud, e.g., on Google Drive or Dropbox, is also supported by R. With the function `drive_upload()` in the googledrive package (to load the function `drive_download()`) and with the functions `drop_auth()`, `drop_create()` and `drop_upload()` in the rdrop2 package (to load the function `drop_read_csv()`), data exchange with these two platforms is possible.

Writing a dataframe to XLS files is possible with the function `write.xlsx()` from the openxlsx package, e.g., `write.xlsx(formula1,"formula1-copy.xlsx")`.

Results that you want to save in a file do not necessarily have to be objects. In R, it is also possible to redirect the output, which is usually displayed in the console, to a text file. For this purpose, the function `sink("Filename")` is used. This function does not affect outputs that are output in the Plots area. To have future outputs displayed again in the console, enter `sink()` again without parameters.

Import and Export of Data

```
#- Verzeichnisse und Dateien-----------------------------------------------------
setwd("~/Dateien")                                     # Arbeitsverzeichnis festlegen
getwd()                                                # Arbeitsverzeichnis anzeigen

## [1] "C:/Users/bernd/Documents/Dateien"

list.files()                                           # Dateien anzeigen

## [1] "autos.csv"              "Basis"                  "datasets.xlsx"
## [4] "flugdatenmodell.RData" "formel1.xlsx"           "ml.funktionen.RData"
## [7] "ml.tunepar.RData"       "Wizard_Of_Oz.txt"

list.files(recursive = TRUE)                           # Dateien inkl. Unterverzeichnisse

##  [1] "autos.csv"               "Basis/kfz1000.csv"
##  [3] "Basis/wflset.RData"      "Basis/wflsetmap2a.RData"
##  [5] "Basis/wflsetmap2b.RData" "Basis/wflsetmap3.RData"
##  [7] "datasets.xlsx"           "flugdatenmodell.RData"
##  [9] "formel1.xlsx"            "ml.funktionen.RData"
## [11] "ml.tunepar.RData"        "Wizard_Of_Oz.txt"

#- Erforderliche Pakete----------------------------------------------------------
library(readxl)                                        # Paket erforderlich
library(XML)
library(rvest)
library(openxlsx)
#- Import von Dateien------------------------------------------------------------
head(kfz, 4)

## # A tibble: 4 × 6
##   Marke         PS Preis Alter Kategorie Kilometer
##   <fct>      <int> <int> <int> <ord>         <int>
## 1 peugeot      158  5000    12 Alt             150
## 2 bmw          286  8500    19 Alt             150
## 3 volkswagen   102  8990     7 Mittelalt        70
## 4 smart         71  4350     9 Mittelalt        70

oz<-readLines("Wizard_Of_Oz.txt")                           # Importiere TXT-Datei
head(oz, 20)

##  [1] "The Wonderful Wizard of Oz"
##  [2] ""
##  [3] ""
##  [4] "by"
##  [5] ""
##  [6] "L. Frank Baum"
##  [7] ""
##  [8] ""
##  [9] ""
## [10] " Contents"
## [11] ""
## [12] "        Introduction"
## [13] "    1.  The Cyclone"
## [14] "    2.  The Council with the Munchkins"
## [15] "    3.  How Dorothy Saved the Scarecrow"
## [16] "    4.  The Road Through the Forest"
## [17] "    5.  The Rescue of the Tin Woodman"
## [18] "    6.  The Cowardly Lion"
## [19] "    7.  The Journey to the Great Oz"
## [20] "    8.  The Deadly Poppy Field"

autos <- read_csv2("autos.csv",col_names = TRUE) # Importiere CSV-Datei

## Rows: 49999 Columns: 5
```

```
## ── Column specification ────────────────────────────────────────────────
## Delimiter: ";"
## chr (3): Marke, Kilometerstand, Preis
## dbl (2): Zulassung, PS
##

head(autos, 4)

## # A tibble: 4 × 5
##   Marke      Zulassung    PS Kilometerstand Preis
##   <chr>          <dbl> <dbl> <chr>          <chr>
## 1 peugeot         2004   158 150,000km      $5,000
## 2 bmw             1997   286 150,000km      $8,500
## 3 volkswagen      2009   102 70,000km       $8,990
## 4 smart           2007    71 70,000km       $4,350

autos <- read_delim("autos.csv",col_names =TRUE,delim=";",
                  col_types=cols(Marke=col_factor(),Zulassung=col_integer(),
                                PS=col_integer(),Kilometerstand=col_character(),
                                Preis=col_character()))

head(autos, 4)

## # A tibble: 4 × 5
##   Marke      Zulassung    PS Kilometerstand Preis
##   <chr>          <dbl> <dbl> <chr>          <chr>
## 1 peugeot         2004   158 150,000km      $5,000
## 2 bmw             1997   286 150,000km      $8,500
## 3 volkswagen      2009   102 70,000km       $8,990
## 4 smart           2007    71 70,000km       $4,350

formel1 <- read_excel("formel1.xlsx")          # Importiere XLS-Datei
head(formel1, 4)

## # A tibble: 4 × 9
##   Rennen Ort          Startposition Rennwagen Fahrer   Team  Platz…¹ Punkte WRPun…²
##    <dbl> <chr>                <dbl>     <dbl> <chr>    <chr>   <dbl>  <dbl>   <dbl>
## 1      1 Australia                1        44 Lewis H… Merc…       2     18    50
## 2      1 Australia                2         7 Kimi Rä… Ferr…       3     15    33.3
## 3      1 Australia                3         5 Sebasti… Ferr…       1     25   100
## 4      1 Australia                4        33 Max Ver… Red …       6      8    16.7
## # … with abbreviated variable names ¹Platzierung, ²WRPunkte

#- Import von Excel im Detail---------------------------------------------------
file <- readxl_example("datasets.xlsx")        # PFAD mit Dateiname
excel_sheets(file)                             # Arbeitsblätter

## [1] "iris"     "mtcars"   "chickwts" "quakes"

sheet.1 <- read_excel(file)                    # Importiere erstes Arbeitsblatt
head(sheet.1, 4)

## # A tibble: 4 × 5
##   Sepal.Length Sepal.Width Petal.Length Petal.Width Species
##          <dbl>       <dbl>        <dbl>       <dbl> <chr>
## 1          5.1         3.5          1.4         0.2 setosa
## 2          4.9         3            1.4         0.2 setosa
## 3          4.7         3.2          1.3         0.2 setosa
## 4          4.6         3.1          1.5         0.2 setosa

sheet.car1 <- read_excel(file, sheet="mtcars") # Importiere Arbeitsblatt
head(sheet.car1, 4)

## # A tibble: 4 × 11
```

```
##     mpg   cyl  disp    hp  drat    wt  qsec    vs    am  gear  carb
##   <dbl> <dbl> <dbl> <dbl> <dbl> <dbl> <dbl> <dbl> <dbl> <dbl> <dbl>
## 1  21       6   160   110  3.9   2.62  16.5     0     1     4     4
## 2  21       6   160   110  3.9   2.88  17.0     0     1     4     4
## 3  22.8     4   108    93  3.85  2.32  18.6     1     1     4     1
## 4  21.4     6   258   110  3.08  3.22  19.4     1     0     3     1
sheet.car2 <- read_excel(file, sheet="mtcars", col_names= FALSE)# Ohne Spaltenname
## New names:
## • `` -> `...1`
## • `` -> `...2`
## • `` -> `...3`
## • `` -> `...4`
## • `` -> `...5`
## • `` -> `...6`
## • `` -> `...7`
## • `` -> `...8`
## • `` -> `...9`
## • `` -> `...10`
## • `` -> `...11`
head(sheet.car2, 4)
## # A tibble: 4 × 11
##   ...1  ...2  ...3  ...4  ...5  ...6  ...7  ...8  ...9  ...10 ...11
##   <chr> <chr> <chr> <chr> <chr> <chr> <chr> <chr> <chr> <chr> <chr>
## 1 mpg   cyl   disp  hp    drat  wt    qsec  vs    am    gear  carb
## 2 21    6     160   110   3.9   2.62  16.46 0     1     4     4
## 3 21    6     160   110   3.9   2.875 17.02 0     1     4     4
## 4 22.8  4     108   93    3.85  2.32  18.61 1     1     4     1
sheet.car3 <- read_excel(file, sheet="mtcars", range="C1:F6")   # Range "C1:F6"
head(sheet.car3, 4)
## # A tibble: 4 × 4
##    disp    hp  drat    wt
##   <dbl> <dbl> <dbl> <dbl>
## 1   160   110  3.9   2.62
## 2   160   110  3.9   2.88
## 3   108    93  3.85  2.32
## 4   258   110  3.08  3.22
#- Import Tabelle von Webseite------------------------------------------------
url <- "https://www.bundesliga.com/de/bundesliga/tabelle"
content<-xml2::read_html(url)                            # Lese Webseite
list<-rvest::html_table(content,header=TRUE)[[1]]        # Tabelle Bundesliga
tabelle<-as.data.frame(list)                             # Konvertierung in Dataframe
tabelle<-tabelle[,c(-1,-3,-4,-6)]                        # Leere Spalten entfernen
colnames(tabelle)<-c("Pos","Verein","Spiele","Punkte",# Spaltenbezeichnungen
                     "S","U","N","Tore","Diff")
tabelle[,2:9]                                            # Ausgabe
##                                       Verein Spiele Punkte S  U     N Tore Diff
## 1              FCB BayernFC Bayern München     15     10 4  1 49:13   36   34
## 2           SCF FreiburgSport-Club Freiburg     15      9 3  3 25:17    8   30
## 3                    RBL LeipzigRB Leipzig     15      8 4  3 30:21    9   28
## 4       SGE FrankfurtEintracht Frankfurt     15      8 3  4 32:24    8   27
## 5         FCU Union Berlin1. FC Union Berlin     15      8 3  4 24:20    4   27
## 6            BVB DortmundBorussia Dortmund     15      8 1  6 25:21    4   25
## 7                WOB WolfsburgVfL Wolfsburg     15      6 5  4 24:20    4   23
## 8   BMG M'gladbachBorussia Mönchengladbach     15      6 4  5 28:24    4   22
## 9                 SVW BremenSV Werder Bremen     15      6 3  6 25:27   -2   21
```

```
## 10                M05 Mainz1. FSV Mainz 05    15     5 4  6 19:24   -5   19
## 11          TSG HoffenheimTSG Hoffenheim    15     5 3  7 22:22    0   18
## 12      B04 LeverkusenBayer 04 Leverkusen    15     5 3  7 25:26   -1   18
## 13                    KOE Köln1. FC Köln    15     4 5  6 21:29   -8   17
## 14              FCA AugsburgFC Augsburg    15     4 3  8 18:26   -8   15
## 15                  BSC BerlinHertha BSC    15     3 5  7 19:22   -3   14
## 16           VFB StuttgartVfB Stuttgart    15     3 5  7 18:27   -9   14
## 17            BOC BochumVfL Bochum 1848    15     4 1 10 14:36  -22   13
## 18              S04 SchalkeFC Schalke 04    15     2 3 10 13:32  -19    9

#- Export von Dateien----------------------------------------------------------
write_csv(kfz, "kfz1000.csv")                          # Exportiere CSV-Datei
write.xlsx(formel1,"formel1-kopie.xlsx")               # Exportiere XLSX-Datei
#- Verzeichnisse und Dateien---------------------------------------------------
list.files()                                           # Dateien anzeigen

##  [1] "autos.csv"              "Basis"                  "datasets.xlsx"
##  [4] "flugdatenmodell.RData" "formel1-kopie.xlsx"     "formel1.xlsx"
##  [7] "kfz1000.csv"            "ml.funktionen.RData"    "ml.tunepar.RData"
## [10] "Wizard_Of_Oz.txt"
```

6.2 Preprocessing

Especially after data has been imported, it is often necessary to improve the quality. This includes, among other things, suitable naming of variables, the conversion of data formats for better evaluation, the deletion of unnecessary data as well as dealing with empty or faulty data sets.

6.2.1 Selection

After objects are present in the memory as a result of data import and other activities, objects that are no longer needed should be deleted. To get an overview of the existing objects in the workspace or memory, the function `ls()` or `objects()` is suitable. Individual objects can be deleted with the function `rm()` if they are no longer needed. This creates space in the memory for the necessary data. Alternatively, in RStudio it is also possible in the dialog to switch to the grid view in the Global Environment, then select the desired objects and delete them via the brush symbol.

Objects that are still needed for processing, but contain data sets or observations (rows of a dataframe) or variables (columns of a dataframe) that are not necessary for processing, should be deleted or deselected, as described in Sect. 3.4 as best practices for selection.

Simple selection operations were already introduced in Sect. 5.3.3. In addition, the package `tidyverse` offers the functions `select()` and `filter()` to select columns or rows of a dataframe.

Columns that are no longer needed can be removed with the function `select()`. To select only the columns with the variables of the teams and driver names from the dataframe formel1, for example, the command `select(formel1,Team,Fahrer)` can be applied. The selection can also be made via the column numbers. To select columns 4 to 7, the command `select(formel1,4:7)` is appropriate. Instead of specifying the columns that should remain, it is also possible to specify the columns that should be removed, e.g. `select(formel1,-Platzierung,-Punkte)`. A minus sign is prefixed to the unwanted columns.

If you intend to select certain rows from a dataframe, the function `filter()` can be used, e.g. `filter(formel1, Team=="Ferrari")`, to select only the records of the Ferrari team. Any number of conditions can be specified, separated by commas, e.g. `filter(formel1,Team=="Ferrari",Ort== "Spain")` returns all rows that concern races in Spain and the Ferrari team. The conditions passed to the `filter()` function can contain any logical operators.

Selection

```
## Selektion -------------------------------------------------------------------
rm(list=ls())                                         # Arbeitsspeicher leeren
#- Vektorvariablen--------------------------------------------------------------
x1 <- c("Tobi","Silke","Klaus","Tobi","Jay","Rob")    # Vektor x1 (Freunde)
x2 <- c(42,356,123,200,0,23)                          # Vektor X2 (Facebookfreunde)
x3 <- c(53,17,31,NA,82,71)                            # Vektor X3 (Alter)
x4 <- c("blau","grün","grün","blau",                  # Vektor X4 (Augenfarbe)
        "gemischt","blau")
df <- data.frame(x1,x2,x3,x4)                         # Dataframe
#- Objekte anzeigen und löschen-------------------------------------------------
ls()                                                  # Objekte anzeigen

## [1] "df" "x1" "x2" "x3" "x4"

rm(x2,x3)                                             # Objekte löschen
ls()                                                  # Objekte anzeigen

## [1] "df" "x1" "x4"

#- Objekte anzeigen und löschen-------------------------------------------------
head(formel1, 7)                                      # Anzeige erste 7 Zeilen

## # A tibble: 7 × 9
##   Rennen Ort    Startposition Rennwagen Fahrer Team  Platzierung Punkte WRPunkte
##    <dbl> <chr>          <dbl>     <dbl> <chr>  <chr>       <dbl>  <dbl>    <dbl>
## 1      1 Austr…             1        44 Lewis… Merc…           2     18     50
## 2      1 Austr…             2         7 Kimi … Ferr…           3     15     33.3
## 3      1 Austr…             3         5 Sebas… Ferr…           1     25    100
## 4      1 Austr…             4        33 Max V… Red …           6      8     16.7
## 5      1 Austr…             5        20 Kevin… Haas…          NA      0     NA
## 6      1 Austr…             6         8 Romai… Haas…          NA      0     NA
## 7      1 Austr…             7        27 Nico … Rena…           7      6     14.3

str(formel1)                                          # Df 420 Zeilen x 9 Spalten

## tibble [420 × 9] (S3: tbl_df/tbl/data.frame)
##  $ Rennen       : num [1:420] 1 1 1 1 1 1 1 1 1 1 ...
##  $ Ort          : chr [1:420] "Australia" "Australia" "Australia" "Australia" ...
##  $ Startposition: num [1:420] 1 2 3 4 5 6 7 8 9 10 ...
##  $ Rennwagen    : num [1:420] 44 7 5 33 20 8 27 3 55 14 ...
##  $ Fahrer       : chr [1:420] "Lewis Hamilton HAM" "Kimi Räikkönen RAI" "Sebastia
n Vettel VET" "Max Verstappen VER" ...
##  $ Team         : chr [1:420] "Mercedes" "Ferrari" "Ferrari" "Red Bull Racing TAG
 Heuer" ...
##  $ Platzierung  : num [1:420] 2 3 1 6 NA NA 7 4 10 5 ...
##  $ Punkte       : num [1:420] 18 15 25 8 0 0 6 12 1 10 ...
##  $ WRPunkte     : num [1:420] 50 33.3 100 16.7 NA ...

formel1Extrakt1<-select(formel1,Team,Fahrer)          # Spalten selektieren
str(formel1Extrakt1)                                  # Df 420 Zeilen x 2 Spalten

## tibble [420 × 2] (S3: tbl_df/tbl/data.frame)
##  $ Team  : chr [1:420] "Mercedes" "Ferrari" "Ferrari" "Red Bull Racing TAG Heuer"
 ...
##  $ Fahrer: chr [1:420] "Lewis Hamilton HAM" "Kimi Räikkönen RAI" "Sebastian Vette
l VET" "Max Verstappen VER" ...

formel1Extrakt2<-select(formel1,4:7)                  # Spalten selektieren
str(formel1Extrakt2)                                  # Df 420 Zeilen x 4 Spalten

## tibble [420 × 4] (S3: tbl_df/tbl/data.frame)
##  $ Rennwagen  : num [1:420] 44 7 5 33 20 8 27 3 55 14 ...
```

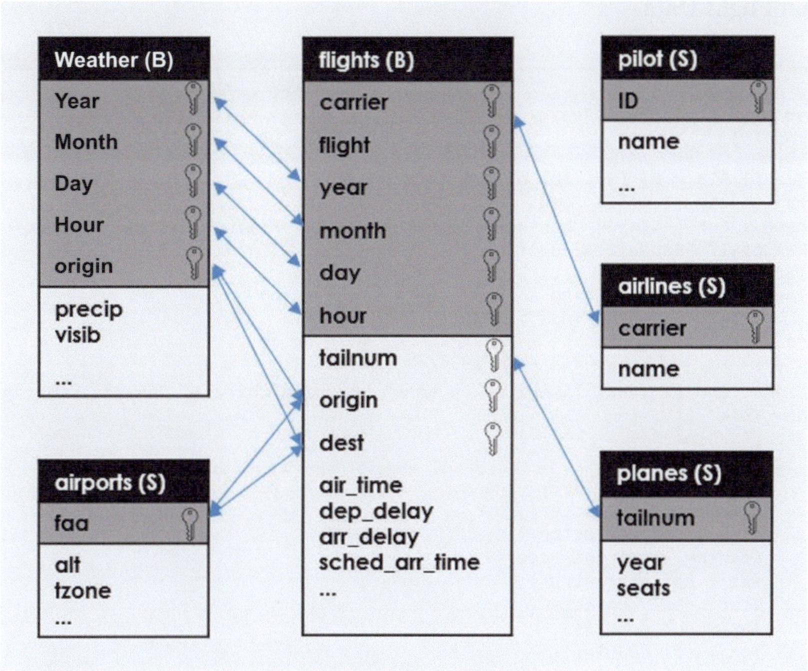

Fig. 6.2 Tables Flight Data (S=Master Data, B=Movement Data)

The following code uses tables with data on flight data from New York airports from the nycflights13 package (see Fig. 6.2):

- Weather(weather): ID(year,month,day,hour,origin), precipitation(precip), timezone(tzone) …
- Airport(airports): ID(faa), altitude above sea level(alt), timezone (tzone) …
- Flight(flights): ID(flight,year,month,day,hour), airline(carrier), aircraft(tailnum), departure airport(origin), destination airport(dest), flight duration(air_time), departure delay(dep_delay), arrival delay(arr_delay) …
- Pilot(pilot): ID(ID), Name(name) …
- Airline(airlines): ID(carrier), Name(name)
- Aircraft(planes): ID(tailnum), year of construction(year), number of seats(seats) …

If the tables also contain variables that are no longer needed (described in the figure with …), these can be removed from the tables. To determine the extent of the memory savings, the function `object.size()` can be used, which displays the memory requirement of an object.

As shown in the following code, the selection significantly reduces the memory requirement, e.g. for airports from 292 KB to 125 KB, for aircraft from 384 KB to 240 KB, for weather from 2.7 MB to 1 MB and for flights from 40 MB to 28 MB. Thus, a larger number of data records can be made available in memory for analysis and the performance of the analyses can be improved.

Selection Flight Data

```
## Selektion -----------------------------------------------------------------
library(nycflights13)                              # Paket erforderlich
#- Stammdaten Flughäfen--------------------------------------------------------
str(airports)                                      # Typ und Anzahl

## tibble [1,458 × 8] (S3: tbl_df/tbl/data.frame)
##  $ faa  : chr [1:1458] "04G" "06A" "06C" "06N" ...
##  $ name : chr [1:1458] "Lansdowne Airport" "Moton Field Municipal Airport" "Schau
mburg Regional" "Randall Airport" ...
##  $ lat  : num [1:1458] 41.1 32.5 42 41.4 31.1 ...
##  $ lon  : num [1:1458] -80.6 -85.7 -88.1 -74.4 -81.4 ...
##  $ alt  : num [1:1458] 1044 264 801 523 11 ...
##  $ tz   : num [1:1458] -5 -6 -6 -5 -5 -5 -5 -5 -5 -8 ...
##  $ dst  : chr [1:1458] "A" "A" "A" "A" ...
##  $ tzone: chr [1:1458] "America/New_York" "America/Chicago" "America/Chicago" "Am
erica/New_York" ...
##  - attr(*, "spec")=
##   .. cols(
##   ..   id = col_double(),
##   ..   name = col_character(),
##   ..   city = col_character(),
##   ..   country = col_character(),
##   ..   faa = col_character(),
##   ..   icao = col_character(),
##   ..   lat = col_double(),
##   ..   lon = col_double(),
##   ..   alt = col_double(),
##   ..   tz = col_double(),
##   ..   dst = col_character(),
##   ..   tzone = col_character()
##   .. )
dim(airports)                                      # 1458 Zeilen x 8 Spalten

## [1] 1458    8

head(airports,5)                                   # 5 Datensätze anzeigen

## # A tibble: 5 × 8
##   faa   name                              lat   lon   alt    tz dst   tzone
##   <chr> <chr>                           <dbl> <dbl> <dbl> <dbl> <chr> <chr>
## 1 04G   Lansdowne Airport                41.1 -80.6  1044    -5 A     America/New…
## 2 06A   Moton Field Municipal Airport    32.5 -85.7   264    -6 A     America/Chi…
## 3 06C   Schaumburg Regional              42.0 -88.1   801    -6 A     America/Chi…
## 4 06N   Randall Airport                  41.4 -74.4   523    -5 A     America/New…
## 5 09J   Jekyll Island Airport            31.1 -81.4    11    -5 A     America/New…

object.size(airports)                              # Größe des Dataframes in Bytes
## 292968 bytes

airports.orig<-airports                            # Kopie mit allen Merkmalen
# In der Realität sollte kein zusätzliches Objekt, z.B. airports.orig angelegt
# werden. Hier wird dies nur getan, um später aufzuzeigen, wie der Speicherbedarf
# der durch Joins der Tabellen entstehenden Basistabelle erheblich höher wäre
airports<-select(airports,                         # Selektion relevanter Variablen
                 c("faa","alt","tzone"))
dim(airports)                                      # 1458 Zeilen x 3 Spalten

## [1] 1458    3

head(airports,5)                                   # 5 Datensätze anzeigen

## # A tibble: 5 × 3
```

```
##   faa     alt tzone
##   <chr> <dbl> <chr>
## 1 04G    1044 America/New_York
## 2 06A     264 America/Chicago
## 3 06C     801 America/Chicago
## 4 06N     523 America/New_York
## 5 09J      11 America/New_York

object.size(airports)                          # Größe des Dataframes in Bytes

## 125160 bytes

#- Stammdaten Fluglinien---------------------------------------------------------
str(airlines)                                  # Typ und Anzahl

## tibble [16 × 2] (S3: tbl_df/tbl/data.frame)
##  $ carrier: chr [1:16] "9E" "AA" "AS" "B6" ...
##  $ name   : chr [1:16] "Endeavor Air Inc." "American Airlines Inc." "Alaska Airli
nes Inc." "JetBlue Airways" ...

dim(airlines)                                  # 16 Zeilen x 2 Spalten

## [1] 16  2

head(airlines,5)                               # 5 Datensätze anzeigen

## # A tibble: 5 × 2
##   carrier name
##   <chr>   <chr>
## 1 9E      Endeavor Air Inc.
## 2 AA      American Airlines Inc.
## 3 AS      Alaska Airlines Inc.
## 4 B6      JetBlue Airways
## 5 DL      Delta Air Lines Inc.

object.size(airlines)                          # Größe des Dataframes in Bytes

## 3352 bytes

airlines.orig<-airlines                        # Kopie mit allen Merkmalen
# In der Realität sollte kein zusätzliches Objekt, z.B. airlines.orig angelegt
# werden. Hier wird dies nur getan, um später aufzuzeigen, wie der Speicherbedarf
# der durch Joins der Tabellen entstehenden Basistabelle erheblich höher wäre
airlines<-select(airlines,c("carrier","name")) # Selektion der Merkmale
#- Stammdaten Flugzeuge----------------------------------------------------------
str(planes)                                    # Typ und Anzahl

## tibble [3,322 × 9] (S3: tbl_df/tbl/data.frame)
##  $ tailnum     : chr [1:3322] "N10156" "N102UW" "N103US" "N104UW" ...
##  $ year        : int [1:3322] 2004 1998 1999 1999 2002 1999 1999 1999 1999 1999 .
..
##  $ type        : chr [1:3322] "Fixed wing multi engine" "Fixed wing multi engine"
 "Fixed wing multi engine" "Fixed wing multi engine" ...
##  $ manufacturer: chr [1:3322] "EMBRAER" "AIRBUS INDUSTRIE" "AIRBUS INDUSTRIE" "AI
RBUS INDUSTRIE" ...
##  $ model       : chr [1:3322] "EMB-145XR" "A320-214" "A320-214" "A320-214" ...
##  $ engines     : int [1:3322] 2 2 2 2 2 2 2 2 2 2 ...
##  $ seats       : int [1:3322] 55 182 182 182 55 182 182 182 182 182 ...
##  $ speed       : int [1:3322] NA NA NA NA NA NA NA NA NA NA ...
##  $ engine      : chr [1:3322] "Turbo-fan" "Turbo-fan" "Turbo-fan" "Turbo-fan" ...

dim(planes)                                    # 3299 Zeilen x 9 Spalten

## [1] 3322    9
```

```
head(planes,5)                                    # 5 Datensätze anzeigen

## # A tibble: 5 × 9
##   tailnum  year type                    manuf…¹ model engines seats speed engine
##   <chr>   <int> <chr>                   <chr>   <chr>   <int> <int> <int> <chr>
## 1 N10156   2004 Fixed wing multi engine EMBRAER EMB-…       2    55    NA Turbo…
## 2 N102UW   1998 Fixed wing multi engine AIRBUS… A320…       2   182    NA Turbo…
## 3 N103US   1999 Fixed wing multi engine AIRBUS… A320…       2   182    NA Turbo…
## 4 N104UW   1999 Fixed wing multi engine AIRBUS… A320…       2   182    NA Turbo…
## 5 N10575   2002 Fixed wing multi engine EMBRAER EMB-…       2    55    NA Turbo…
## # … with abbreviated variable name ¹manufacturer

object.size(planes)                               # Größe des Dataframes in Bytes

## 384432 bytes

planes.orig<-planes                               # Kopie mit allen Merkmalen
# In der Realität sollte kein zusätzliches Objekt, z.B. planes.orig angelegt
# werden. Hier wird dies nur getan, um später aufzuzeigen, wie der Speicherbedarf
# der durch Joins der Tabellen entstehenden Basistabelle erheblich höher wäre
planes<-select(planes,c("tailnum","year","seats"))  # Selektion der Merkmale
dim(planes)                                       # 3299 Zeilen x 3 Spalten

## [1] 3322    3

object.size(planes)                               # Größe des Dataframes in Bytes

## 240304 bytes

#- Bewegungsdaten Wetter---------------------------------------------------------
str(weather)                                      # Typ und Anzahl

## tibble [26,115 × 15] (S3: tbl_df/tbl/data.frame)
##  $ origin    : chr [1:26115] "EWR" "EWR" "EWR" "EWR" ...
##  $ year      : int [1:26115] 2013 2013 2013 2013 2013 2013 2013 2013 2013 2013 ..
.
##  $ month     : int [1:26115] 1 1 1 1 1 1 1 1 1 1 ...
##  $ day       : int [1:26115] 1 1 1 1 1 1 1 1 1 1 ...
##  $ hour      : int [1:26115] 1 2 3 4 5 6 7 8 9 10 ...
##  $ temp      : num [1:26115] 39 39 39 39.9 39 ...
##  $ dewp      : num [1:26115] 26.1 27 28 28 28 ...
##  $ humid     : num [1:26115] 59.4 61.6 64.4 62.2 64.4 ...
##  $ wind_dir  : num [1:26115] 270 250 240 250 260 240 240 250 260 260 ...
##  $ wind_speed: num [1:26115] 10.36 8.06 11.51 12.66 12.66 ...
##  $ wind_gust : num [1:26115] NA NA NA NA NA NA NA NA NA NA ...
##  $ precip    : num [1:26115] 0 0 0 0 0 0 0 0 0 0 ...
##  $ pressure  : num [1:26115] 1012 1012 1012 1012 1012 ...
##  $ visib     : num [1:26115] 10 10 10 10 10 10 10 10 10 10 ...
##  $ time_hour : POSIXct[1:26115], format: "2013-01-01 01:00:00" "2013-01-01 02:00:
00" ...

dim(weather)                                      # 26115 Zeilen x 15 Spalten

## [1] 26115    15

head(weather,5)                                   # 5 Datensätze

## # A tibble: 5 × 15
##   origin  year month   day  hour  temp  dewp humid wind_dir wind_speed wind_gust
##   <chr>  <int> <int> <int> <int> <dbl> <dbl> <dbl>    <dbl>      <dbl>     <dbl>
## 1 EWR     2013     1     1     1  39.0  26.1  59.4      270      10.4         NA
## 2 EWR     2013     1     1     2  39.0  27.0  61.6      250       8.06        NA
## 3 EWR     2013     1     1     3  39.0  28.0  64.4      240      11.5         NA
## 4 EWR     2013     1     1     4  39.9  28.0  62.2      250      12.7         NA
```

```
## 5 EWR     2013     1     1     5  39.0  28.0  64.4      260      12.7         NA
## # … with 4 more variables: precip <dbl>, pressure <dbl>, visib <dbl>,
## #   time_hour <dttm>

object.size(weather)                                # Größe in Bytes

## 2719280 bytes

weather.orig<-weather                               # Kopie mit allen Merkmalen
# In der Realität sollte kein zusätzliches Objekt, z.B. weather.orig angelegt
# werden. Hier wird dies nur getan, um später aufzuzeigen, wie der Speicherbedarf
# der durch Joins der Tabellen entstehenden Basistabelle erheblich höher wäre
spalten<-c("year","month","day","hour","origin","precip","visib")
weather<-select(weather,all_of(spalten))            # Selektion der Merkmale
dim(weather)                                        # 26115 Zeilen x 7 Spalten

## [1] 26115     7

object.size(weather)                                # Größe in Bytes

## 1046384 bytes

#- Bewegungsdaten Flüge--------------------------------------------------------
str(flights)                                        # Typ und Anzahl

## tibble [336,776 × 19] (S3: tbl_df/tbl/data.frame)
##  $ year          : int [1:336776] 2013 2013 2013 2013 2013 2013 2013 2013 2013 20
13 ...
##  $ month         : int [1:336776] 1 1 1 1 1 1 1 1 1 1 ...
##  $ day           : int [1:336776] 1 1 1 1 1 1 1 1 1 1 ...
##  $ dep_time      : int [1:336776] 517 533 542 544 554 554 555 557 557 558 ...
##  $ sched_dep_time: int [1:336776] 515 529 540 545 600 558 600 600 600 600 ...
##  $ dep_delay     : num [1:336776] 2 4 2 -1 -6 -4 -5 -3 -3 -2 ...
##  $ arr_time      : int [1:336776] 830 850 923 1004 812 740 913 709 838 753 ...
##  $ sched_arr_time: int [1:336776] 819 830 850 1022 837 728 854 723 846 745 ...
##  $ arr_delay     : num [1:336776] 11 20 33 -18 -25 12 19 -14 -8 8 ...
##  $ carrier       : chr [1:336776] "UA" "UA" "AA" "B6" ...
##  $ flight        : int [1:336776] 1545 1714 1141 725 461 1696 507 5708 79 301 ...
##  $ tailnum       : chr [1:336776] "N14228" "N24211" "N619AA" "N804JB" ...
##  $ origin        : chr [1:336776] "EWR" "LGA" "JFK" "JFK" ...
##  $ dest          : chr [1:336776] "IAH" "IAH" "MIA" "BQN" ...
##  $ air_time      : num [1:336776] 227 227 160 183 116 150 158 53 140 138 ...
##  $ distance      : num [1:336776] 1400 1416 1089 1576 762 ...
##  $ hour          : num [1:336776] 5 5 5 5 6 5 6 6 6 6 ...
##  $ minute        : num [1:336776] 15 29 40 45 0 58 0 0 0 0 ...
##  $ time_hour     : POSIXct[1:336776], format: "2013-01-01 05:00:00" "2013-01-01 0
5:00:00" ...

dim(flights)                                        # 336776 Zeilen x 19 Spalten

## [1] 336776     19

head(flights,5)                                     # 5 Datensätze anzeigen

## # A tibble: 5 × 19
##    year month   day dep_time sched_dep…¹ dep_d…² arr_t…³ sched…⁴ arr_d…⁵ carrier
##   <int> <int> <int>    <int>       <int>   <dbl>   <int>   <int>   <dbl> <chr>
## 1  2013     1     1      517         515       2     830     819      11 UA
## 2  2013     1     1      533         529       4     850     830      20 UA
## 3  2013     1     1      542         540       2     923     850      33 AA
## 4  2013     1     1      544         545      -1    1004    1022     -18 B6
## 5  2013     1     1      554         600      -6     812     837     -25 DL
## # … with 9 more variables: flight <int>, tailnum <chr>, origin <chr>,
## #   dest <chr>, air_time <dbl>, distance <dbl>, hour <dbl>, minute <dbl>,
```

```
## #   time_hour <dttm>, and abbreviated variable names ¹sched_dep_time,
## #   ²dep_delay, ³arr_time, ⁴sched_arr_time, ⁵arr_delay

object.size(flights)                             # Größe des Dataframes in Bytes

## 40650104 bytes

flights.orig<-flights                            # Kopie mit allen Merkmalen
# In der Realität sollte kein zusätzliches Objekt, z.B. flights.orig angelegt
# werden. Hier wird dies nur getan, um später aufzuzeigen, wie der Speicherbedarf
# der durch Joins der Tabellen entstehenden Basistabelle erheblich höher wäre
flights<-select(flights,                         # Selektion relevanter Variablen
                c("carrier","flight","year","month","day","hour",
                  "tailnum","origin","dest","air_time","dep_delay","arr_delay",
                  "sched_arr_time"))
dim(flights)                                     # 336776 Zeilen x 13 Spalten

## [1] 336776     13

head(flights,5)                                  # 5 Datensätze anzeigen

## # A tibble: 5 × 13
##    carrier flight  year month   day  hour tailnum origin dest  air_time dep_delay
##    <chr>    <int> <int> <int> <int> <dbl> <chr>   <chr>  <chr>    <dbl>     <dbl>
## 1 UA        1545  2013     1     1     5 N14228  EWR    IAH        227         2
## 2 UA        1714  2013     1     1     5 N24211  LGA    IAH        227         4
## 3 AA        1141  2013     1     1     5 N619AA  JFK    MIA        160         2
## 4 B6         725  2013     1     1     5 N804JB  JFK    BQN        183        -1
## 5 DL         461  2013     1     1     6 N668DN  LGA    ATL        116        -6
## # … with 2 more variables: arr_delay <dbl>, sched_arr_time <int>

object.size(flights)                             # Größe des Dataframes in Bytes

## 28524920 bytes
```

6.2.2 Validation

For objective and reliable analyses, data must be accurate and meaningful. Often, encodings for products, cost centers, customers, suppliers, locations, currencies, etc. are used in movement data. These encodings (foreign keys to master data) should be used uniformly in all datasets used and used to validate the movement data. Through this process of validation, for example, data records with sales data can be filtered out that refer to an invalid product number, an invalid country, or an invalid currency. This process of validation is also referred to as **data integration**, within which the data from various data sources are integrated into a uniform, high-quality database.

The prerequisite for such validation of data and data integration is metadata, i.e., data about the data. The **metadata** describe for each table its content, which columns (variables) exist and what meaning they have (name or explanation) and also the data type. The metadata should also note which variables are key fields, i.e., suitable for identifying a data record (row) in the table and whether this key is a unique key (Unique Key; no second entry with the same key, e.g., airline "LH" for Lufthansa) or a simple key

(Non-Unique Key; multiple entries with the same key allowed, e.g., names of the pilots "Hans Müller"). The validation can then check whether, for example, a unique key in the table really only occurs once. For each characteristic, the metadata should also describe whether there is a limited set of valid values, e.g., for a number for the months the value set from 1 to 12, for the age of the pilots no negative numbers etc. In addition to a fixed lower and upper limit as in the aforementioned examples, the value set can also be done via a foreign key, which means that there must be a data record with this key in another table (S=master data), e.g., an airline with the ID in the table of airlines, which is in the table flights (B=movement data) in the feature carrier or the ID of an airport in the table of airports (S=master data), which is in the table flights in the feature origin and dest (see Fig. 6.2). This master data such as airline, airport, customer, product, currency and the like must be uniformly coded in all data. The existence of this metadata and master data forms the basis for quality assurance and validation of the data.

If invalid data are found in the process of validation, e.g., flights with invalid airport, flights with invalid airline, sales data with invalid product number or invalid currency, then these faulty data records need to be checked. If the check can clarify how an error can be eliminated, then this should be done, e.g., if a data entry error has led to LK for Lufthansa being listed in the list of airlines instead of LH for Lufthansa, then this can be corrected. However, if the age of a pilot was entered as −33, then the age should be temporarily marked with NA (not available) until the correct age of the pilot is known.

Missing data can be determined with the function `is.na()`. The function returns TRUE for the parameter in brackets if the element is NA and FALSE if the element is NOT NA. The function returns a vector of the same length as the parameter vector and shows for each element the position (index) with TRUE where NA is in a vector.

Faulty or inconsistent data are the data for which a feature specification is missing (NA) or an invalid entry is present. Missing or faulty data sets should ideally be corrected. If this is not possible, then different approaches can be taken with these data sets:

1. Leave data unchanged, but do not consider the data sets with missing values in analyses: Many functions support this through the parameter `na.rm = TRUE`.
2. Delete the data sets with missing values: If you want to delete all data sets with missing values, this can be achieved with the function `na.omit()` or the function `drop_na()` from the package `tidyr`, which is included in the package `tidyverse`.
3. Replace the missing values with the mean, median, or mode: The advantage of this variant is that not an entire data set falls away for the analysis. This can be achieved with the help of the logical check `is.na()` in conjunction with an IF condition like `ifelse()`.

Instead of using the function `is.na()`, it is advisable to combine it with the function `colSums()`, e.g., `colSums(is.na())`, to display for a dataframe per column how many variable contents have the entry NA (NA = not available), i.e., are missing. In the master data table of the airports, the time zone information is missing for three airports. This can be corrected in the master data table. There are no missing variable values in the data sets for the airline master data and the movement data for the weather. In the aircraft master data, the information on the year in which the aircraft was manufactured is missing for 70 aircraft. This will limit the analysis possibilities if, for example, it is to be investigated whether there is a correlation between delay and age of an aircraft. The use of the parameter na.rm=TRUE, however, prevents the absence of variable values from leading to a termination of the calculation. In the table of movement data for flights, 2512 entries for the aircraft (tailnum), 9430 for the flight time (airtime), 8255 for departure delay (dep_delay), and 9430 for arrival delay (arr_delay) are missing. This limits the quality of the evaluation, although evaluations are still possible.

In addition to checking for missing data, the quality of the variable values should also be validated. With the flight data, the validation of the month information of the flights for the interval from 1 to 12 shows that all values are within the valid interval. Regarding the correctness of the information on the airlines, there is also a pleasing result. The comparison of the information on the airline of all 336,776 data sets via the foreign key shows that all values are valid airlines. The destination airports (dest) of the flights are compared with the airport table via the foreign key. This identifies 7602 faulty data sets. The error correction is carried out for the two airports BQN and SJU by adding the airports to the master data table airports. After this correction, only 887 faulty data sets for flights remain. Assuming that the airports PSE and STT do not exist, these 887 data sets in the flight movement data (flights) are modified by storing NA (not available) in the variable dest. With this, all correctable errors are corrected and the validation is completed. In reality, the validation of the other variables, including departure airport, time… would have to be performed just like for the here validated variables month, airline, and destination airport.

Validation

```
## Validierung -----------------------------------------------------------------
#- Stammdaten Flughäfen-------------------------------------------------------------
colSums(is.na(airports))                              # Anzahl NAs je Spalte

##   faa   alt tzone
##     0     0     3

err.airports<-filter(airports,is.na(airports$tzone))      # Keine TimeZone-Angabe
head(err.airports,5)

## # A tibble: 3 × 3
##   faa     alt tzone
##   <chr> <dbl> <chr>
## 1 EEN     149 <NA>
## 2 LRO      12 <NA>
## 3 YAK      33 <NA>

airports$tzone[airports$faa=="EEN"]<-"America/New_York"  # Korrektur
airports$tzone[airports$faa=="LRO"]<-"America/New_York"  # Korrektur
airports$tzone[airports$faa=="YAK"]<-"America/Anchorage" # Korrektur
colSums(is.na(airports))                              # Alle fehlenden Werte korrigiert

##   faa   alt tzone
##     0     0     0

#- Stammdaten Fluglinien-------------------------------------------------------------
dim(airlines)                                         # Größe der Tabelle Zeilen/Spalten

## [1] 16  2

colSums(is.na(airlines))                              # Anzahl NAs je Spalte

## carrier    name
##       0       0

#- Stammdaten Flugzeuge---------------------------------------------------------------
dim(planes)                                           # Größe der Tabelle Zeilen/Spalten

## [1] 3322    3

colSums(is.na(planes))                                # Anzahl NAs je Spalte

## tailnum    year   seats
##       0      70       0

err.planes<-filter(planes,is.na(planes$year))  # Flugzeuge ohne Angabe year
head(err.planes,5)
## # A TIBBLE: 5 × 3
##   TAILNUM  YEAR SEATS
##   <CHR>   <INT> <INT>
## 1 N14558     NA    55
## 2 N15555     NA    55
## 3 N15574     NA    55
## 4 N174US     NA   199
## 5 N177US     NA   199

# Fehlende Angaben bei year können bei der Analyse berücksichtigt durch na.rm=TRUE
planes$age<-2013-planes$year                          # Variable Alter neu(Daten aus 2013)
planes<-select(planes,-"year")                        # Variable Year entfernen
#- Bewegungsdaten Wetter--------------------------------------------------------------
dim(weather)                                          # Größe der Tabelle Zeilen/Spalten

## [1] 26115     7

colSums(is.na(weather))                               # Anzahl NAs je Spalte
```

```
##    year  month    day   hour origin precip  visib
##      0      0      0      0      0      0      0

#- Bewegungsdaten Flüge-------------------------------------------------------------
dim(flights)                                  # Größe der Tabelle Zeilen/Spalten

## [1] 336776     13

colSums(is.na(flights))                       # Anzahl NAs je Spalte

##        carrier         flight           year          month            day
##              0              0              0              0              0
##           hour        tailnum         origin           dest       air_time
##              0           2512              0              0           9430
##      dep_delay      arr_delay sched_arr_time
##           8255           9430              0

#- Bewegungsdaten Flüge: Monat 1-12------------------------------------------------
err.monat<-filter(flights,!(month %in% 1:12))  # Validierung Monat
cat("Fehler zu Monat:",dim(err.monat)[1],"\n")

## Fehler zu Monat: 0

#- Bewegungsdaten Flüge: Fluggesellschaft in Stammdatentabelle? -----------------
err.airlines<-filter(flights,!(carrier %in% airlines$carrier))
cat("Fehler zu Fluggesellschaft:",dim(err.airlines)[1],"\n")

## Fehler zu Fluggesellschaft: 0

#- Bewegungsdaten Flüge: Flughafen in Stammdatentabelle? ------------------------
err.airports<-filter(flights,!(dest %in% airports$faa))
cat("Fehler zu Flughafen:",dim(err.airports)[1],"\n")

## Fehler zu Flughafen: 7602

table(err.airports$dest)                      # 7602 Fehler zu vier Flughäfen

##
##  BQN  PSE  SJU  STT
##  896  365 5819  522

#- Bewegungsdaten Flüge: Korrektur von 2 Fehlern in Flughafen-Stammdaten ---------
airports<-rbind(airports,c("BQN",237,"America/Chicago")) # Rafael Hernandez Airp.
airports<-rbind(airports,c("SJU",9,"America/Chicago"))   # Luis Munoz Marin Airp.
err.airports<-filter(flights,!(dest %in% airports$faa))
table(err.airports$dest)                     # Fehler von 7602->887 reduziert

##
## PSE STT
## 365 522

cat("Fehler zu Flughafen:",dim(err.airports)[1],"\n")

## Fehler zu Flughafen: 887

#- Bewegungsdaten Flüge: Korrektur von 2 Fehlern zu NA in Flug-Bewegungsdaten ----
flights$dest[flights$dest %in% c("PSE","STT")] <- NA # NA für ungültige Flughäfen
err.airports<-filter(flights,!(dest %in% airports$faa) & !is.na(dest))
cat("Fehler zu Flughafen:",dim(err.airports)[1],"\n")

## Fehler zu Flughafen: 0
```

6.2.3 Merging

After validating the data in the individual tables, they are often integrated with each other in a form effective for data analysis, so-called **base tables**. This is done especially when not all relevant information is present in one table or even deliberately stored in separate tables in a redundancy-free, normalized form effective for transaction data processing according to the Entity-Relationship Model.

The fusion of tables x and y can be done in R with the function `merge()`. The parameters `by.x` and `by.y` specify the column names over which the join or fusion should occur. The result of an inner join is a table that combines the contents of table x and the contents of table y when there is an identical entry for the key specified with `by.x` in table y with the key `by.y`. If the parameter `all.x=TRUE` is used and there is no entry for the key `by.x` in `by.y`, then the columns otherwise filled with the values of table y are filled with NA (Outer Join, Partial Match). If the parameter `all.x=TRUE` is not used and there is no entry for the key `by.x` in `by.y`, the record with `by.x` is not included in the result table (Inner Join, Full Match), as shown in Fig. 6.3.

In the following code, there are initially two dataframes, one with the director names and their nationality and another dataframe with the film names and their director. To transfer all this information as a basis for evaluations into a base table, it is necessary to merge the information on film name, director name, and nationality of the director. The columns each represent a variable and the rows each represent an observation (record). The characteristics or values of the variables are in the cells of the dataframes fusion1 to fusion5.

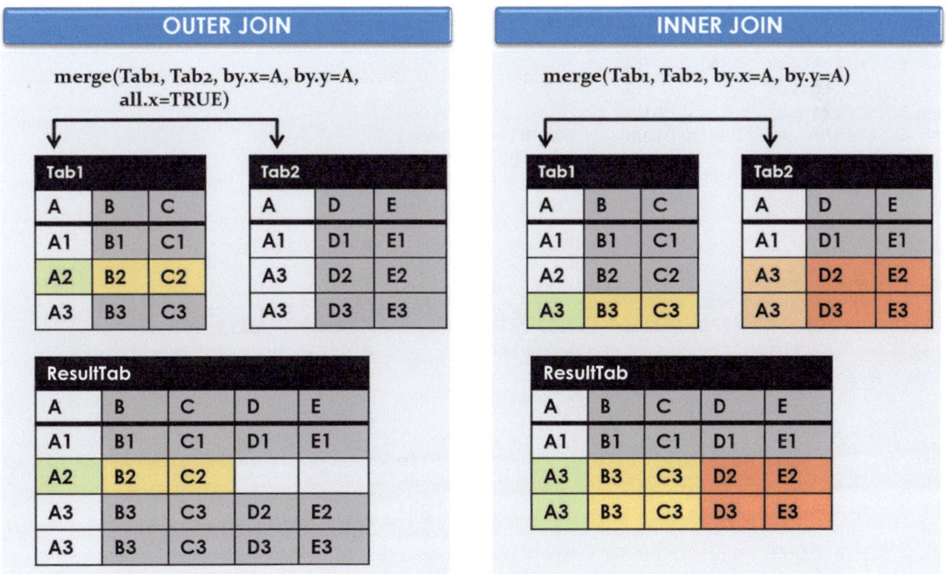

Fig. 6.3 Table Joins

Merge

```
## Fusion -------------------------------------------------------------------
regisseur <- data.frame(                                     # Regisseure
    Name = c("Spielberg","Scorsese","Hitchcock","Schlöndorff","Polanski"),
    Nationalität = c("USA","USA","GB","Deutschland","Polen"),
    stringsAsFactors=FALSE)
regisseur

##          Name Nationalität
## 1   Spielberg          USA
## 2    Scorsese          USA
## 3   Hitchcock           GB
## 4 Schlöndorff  Deutschland
## 5    Polanski        Polen

film <- data.frame(                                          # Filme
    Film = c("Der Weiße Hai","The Irishman","Psycho","Die Vögel",
             "Catch Me If You Can","Die Blechtrommel","Chinatown"),
    Regisseur = c("Spielberg","Scorsese","Hitchcock","Hitchcock",
             "Spielberg","Schlöndorff","Polanski"),
    stringsAsFactors=FALSE)
film

##                  Film   Regisseur
## 1       Der Weiße Hai   Spielberg
## 2        The Irishman    Scorsese
## 3              Psycho   Hitchcock
## 4           Die Vögel   Hitchcock
## 5 Catch Me If You Can   Spielberg
## 6    Die Blechtrommel Schlöndorff
## 7           Chinatown    Polanski

fusion1 <- merge(regisseur,film,                             # Inner Join
                 by.x="Name",by.y="Regisseur")
fusion1                                                      # ->Join Regisseur mit Film

##          Name Nationalität                Film
## 1   Hitchcock           GB              Psycho
## 2   Hitchcock           GB           Die Vögel
## 3    Polanski        Polen           Chinatown
## 4 Schlöndorff  Deutschland    Die Blechtrommel
## 5    Scorsese          USA        The Irishman
## 6   Spielberg          USA       Der Weiße Hai
## 7   Spielberg          USA Catch Me If You Can

dim(fusion1)                                                 # Df 7 Zeilen x 3 Spalten

## [1] 7 3

regisseur.neu <-  c("deJenlis","Frankreich")                 # Neuer Regisseur
regisseur <- rbind(regisseur,regisseur.neu)                  # Regisseure zusammenfassen
fusion2 <- merge(regisseur,film,                             # Inner Join
                 by.x="Name",by.y="Regisseur")
fusion2                                                      # ->Keine Zeile neuer Regisseur

##          Name Nationalität                Film
## 1   Hitchcock           GB              Psycho
## 2   Hitchcock           GB           Die Vögel
## 3    Polanski        Polen           Chinatown
## 4 Schlöndorff  Deutschland    Die Blechtrommel
## 5    Scorsese          USA        The Irishman
## 6   Spielberg          USA       Der Weiße Hai
## 7   Spielberg          USA Catch Me If You Can
```

```
dim(fusion2)                                    # Df 7 Zeilen x 3 Spalten

## [1] 7 3

fusion3 <- merge(regisseur,film,                # Outer Join
                by.x="Name",by.y="Regisseur",
                all.x = TRUE)
fusion3                                         # ->Zeile Regisseur ohne Film

##          Name Nationalität              Film
## 1    deJenlis   Frankreich              <NA>
## 2   Hitchcock           GB            Psycho
## 3   Hitchcock           GB         Die Vögel
## 4    Polanski        Polen         Chinatown
## 5 Schlöndorff  Deutschland   Die Blechtrommel
## 6    Scorsese          USA      The Irishman
## 7   Spielberg          USA      Der Weiße Hai
## 8   Spielberg          USA Catch Me If You Can

dim(fusion3)                                    # Df 8 Zeilen x 3 Spalten

## [1] 8 3

regisseur.neu <-  c("Polanski","Frankreich")    # Neuer Regisseur - Name doppelt
regisseur <- rbind(regisseur,regisseur.neu)     # Regisseure zusammenfassen
fusion4 <- merge(regisseur,film,                # Inner Join
                by.x="Name",by.y="Regisseur")
fusion4                                         # ->Zwei Zeilen Polanski-Film

##          Name Nationalität              Film
## 1   Hitchcock           GB            Psycho
## 2   Hitchcock           GB         Die Vögel
## 3    Polanski        Polen         Chinatown
## 4    Polanski   Frankreich         Chinatown
## 5 Schlöndorff  Deutschland   Die Blechtrommel
## 6    Scorsese          USA      The Irishman
## 7   Spielberg          USA      Der Weiße Hai
## 8   Spielberg          USA Catch Me If You Can

dim(fusion4)                                    # Df 8 Zeilen x 3 Spalten

## [1] 8 3

fusion5 <- merge(regisseur,film,                # Outer Join
                by.x="Name",by.y="Regisseur",
                all.x = TRUE)
fusion5                                         # ->Zeile Regisseur ohne Film

##          Name Nationalität              Film
## 1    deJenlis   Frankreich              <NA>
## 2   Hitchcock           GB            Psycho
## 3   Hitchcock           GB         Die Vögel
## 4    Polanski        Polen         Chinatown
## 5    Polanski   Frankreich         Chinatown
## 6 Schlöndorff  Deutschland   Die Blechtrommel
## 7    Scorsese          USA      The Irishman
## 8   Spielberg          USA      Der Weiße Hai
## 9   Spielberg          USA Catch Me If You Can
dim(fusion5)                                    # Df 9 Zeilen x 3 Spalten

## [1] 9 3
```

A fusion for the flight data tables into a base table is also helpful, for example, to facilitate evaluations of flight delays. With the flight data, the dataframes weather, airports, flights, airlines, and planes initially exist as independent tables. To transfer all the information desired for the analysis into a base table, it is necessary to merge the information via the associated key fields. After successful fusion of the tables, the data is then available in the dataframe flight in the form of a base table. In Fig. 6.4, the key fields are represented by the symbol of a key. Variables that do not originally come from the dataframe flights, but are added to the base table flight via join from other dataframes, are marked with an *. The fusion of the tables creates a base table with 336,776 records and 27 variables, which take up 62 MB in memory. If the tables had been merged into a base table without selection, a base table with 336,776 records and 56 variables would have been created, which would have taken up 139 MB in memory.

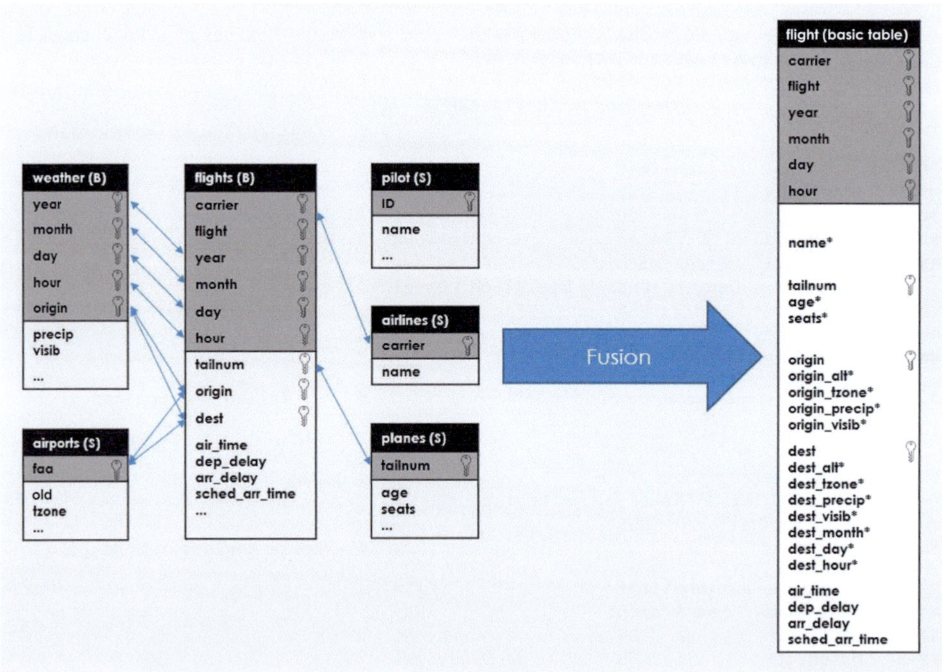

Fig. 6.4 Merging Flight Data into Base Table or DataCube

Merging Flight Data

```
## Fusion der Tabellen, um Faktentabelle (DataCube) zu erstellen -----------------
## Join zum Hinzufügen von origin_alt + origin_tzone ----------------------------
dim(flights)                                        # 336776 Zeilen x 13 Spalten

## [1] 336776     13

flug<-merge(flights,airports,                       # Inner Join über origin=faa
            by.x="origin",by.y="faa")
dim(flug)                                           # 336776 Zeilen x 15 Spalten

## [1] 336776     15

flug<-rename(flug,"origin_alt"="alt","origin_tzone"="tzone")# Spalten umbenennen
## Join zum Hinzufügen von dest_alt + dest_tzone ---------------------------------
colSums(is.na(flug))                                # Anzahl NAs je Spalte

##          origin        carrier          flight            year           month
##               0              0               0               0               0
##             day           hour         tailnum            dest        air_time
##               0              0            2512             887            9430
##       dep_delay      arr_delay  sched_arr_time      origin_alt    origin_tzone
##            8255           9430               0               0               0

# 887 Datensätze haben keine Angabe zu dest
flug.innerjoin.dest<-merge(flug,airports,           # Inner Join über dest=faa
            by.x="dest",by.y="faa")
dim(flug.innerjoin.dest)                            # 335889 Zeilen x 17 Spalten

## [1] 335889     17

# 887 Datensätze gehen bei dem Innter Join verloren, da kein Eintrag zu
# dem Zielflufhafen dest in der Tabelle flug existiert. Die Qualität der
# Stammdatentabelle der Flughäfen sollte unbedingt erhöht werden, indem fehlende
# Daten ergänzt werden.
# Damit die 887 Datensätze mit den anderen Informationen für die Analyse nicht
# verloren gehen, wird statt des Inner jetzt ein Outer Join durchgeführt. Für die
# 887 Datensätze fehlt dann natürlich trotzdem die Information zur Höhe über dem
# Meeresspiegel (alt) und der Zeitzone (tzone) für die Analysen.
rm(flug.innerjoin.dest)
flug<-merge(flug,airports,all.x=TRUE,               # Outer Join über dest=faa
            by.x="dest",by.y="faa")
dim(flug )                                          # 336776 Zeilen x 17 Spalten

## [1] 336776     17

flug<-rename(flug,"dest_alt"="alt","dest_tzone"="tzone")     # Spalten umbenennen
colSums(is.na(flug))                                # Anzahl NAs je Spalte

##            dest         origin         carrier          flight            year
##             887              0               0               0               0
##           month            day            hour         tailnum        air_time
##               0              0               0            2512            9430
##       dep_delay      arr_delay  sched_arr_time      origin_alt    origin_tzone
##            8255           9430               0               0               0
##        dest_alt     dest_tzone
##             887            887

## Join zum Hinzufügen von name ----------------------------------------------------
flug<-merge(flug,airlines,                          # Inner Join über carrier=carrier
            by.x="carrier",by.y="carrier")
dim(flug )                                          # 336776 Zeilen x 18 Spalten

## [1] 336776     18

## Join zum Hinzufügen von age + seats -------------------------------------------
```

```
colSums(is.na(flug))                           # Anzahl NAs je Spalte

##         carrier            dest          origin          flight            year
##               0             887               0               0               0
##           month             day            hour         tailnum        air_time
##               0               0               0            2512            9430
##       dep_delay       arr_delay sched_arr_time      origin_alt    origin_tzone
##            8255            9430               0               0               0
##        dest_alt      dest_tzone            name
##             887             887               0

# 2512 Datensätze haben keine Angabe zu tailnum
flug.innerjoin.tailnum<-merge(flug,planes,      # Inner Join über tailnum=tailnum
            by.x="tailnum",by.y="tailnum")
dim(flug.innerjoin.tailnum)                    # 284170 Zeilen x 20 Spalten

## [1] 284170     20

# 52606 Datensätze gehen bei dem Innter Join verloren, da entweder die Angabe
# tailnum in der Tabelle flug fehlt (2512 Datensätze) oder aber kein Eintrag zu
# dem Flugzeug mit der tailnum in der Tabelle planes existiert (50094 Datensätze).
# Die Qualität der Stammdatentabelle der Flugzeuge sollte unbedingt erhöht werden,
# indem fehlende Daten ergänzt werden.
# Damit die 52606 Datensätze mit den anderen Informationen für die Analyse nicht
# verloren gehen, wird statt des Inner jetzt ein Outer Join durchgeführt. Für die
# 52606 Datensätze fehlt dann natürlich trotzdem die Information zum Alter
# (age) und der Sitzplatzanzahl (seats) des Flugzeugs für die Analysen. Die
# Angabe zum Alter fehlt sogar bei 57912 Datensätzen, denn zu 70 Flugzeugen in
# der Tabelle planes fehlt die Angabe age und beim Join fehlt dies dann bei allen
# Flügen dieser Maschinen.
rm(flug.innerjoin.tailnum)
flug<-merge(flug,planes,all.x=TRUE,            # Outer Join über tailnum=tailnum
            by.x="tailnum",by.y="tailnum")
dim(flug)                                      # 336776 Zeilen x 20 Spalten

## [1] 336776     20

colSums(is.na(planes))                         # Anzahl NAs je Spalte

## tailnum   seats     age
##       0       0      70

colSums(is.na(flug))                           # Anzahl NAs je Spalte

##         tailnum         carrier            dest          origin          flight
##            2512               0             887               0               0
##            year           month             day            hour        air_time
##               0               0               0               0            9430
##       dep_delay       arr_delay sched_arr_time      origin_alt    origin_tzone
##            8255            9430               0               0               0
##        dest_alt      dest_tzone            name           seats             age
##             887             887               0           52606           57912

fehlend<-filter(flug,is.na(flug$age))          # Flüge ohne Alter von Flugzeug
dim(table(fehlend$tailnum))                    # 791 Flugstammdaten ohne Alter

## [1] 791

head(sort(table(fehlend$tailnum),decreasing=T),5)  # 5 erste Flugzeuge ohne Alter

##
## N725MQ N722MQ N723MQ N713MQ N735MQ
##    575    513    507    483    396

# Sobald die 791 Altersangaben zu den Flugzeugen hinzugefügt wurden, kann auch
```

```
# ein Inner Join ohne Datenverlust erfolgen.
## Join zum Hinzufügen von origin_precip + origin_visib --------------------------
flug.innerjoin.origin<-merge(flug,weather,     # Inner Join über Abflughafen+Date
            by.x=c("origin","year","month","day","hour"),
            by.y=c("origin","year","month","day","hour"))
dim(flug.innerjoin.origin)                      # 335220 Zeilen x 22 Spalten

## [1] 335220     22

# Für 1556 Flugdaten fehlt die Wetterinformation und daher werden diese
# beim Inner Join gelöscht. Um dies zu vermeiden erfolgt ein Outer Join
rm(flug.innerjoin.origin)
flug<-merge(flug,weather,all.x=TRUE,            # Outer Join über Abflughafen+Date

            by.y=c("origin","year","month","day","hour"))
dim(flug)                                       # 336776 Zeilen x 22 Spalten

## [1] 336776     22

flug<-rename(flug,"origin_precip"="precip","origin_visib"="visib") # Spalten um.
## Join zum Hinzufügen von dest_precip + dest_visib ------------------------------
# Bevor der Join der Wetterdaten am Zielflughafen zur Zeit des Landevorgangs
# erfolgen kann, muss die Landezeit berechnet werden auf Basis der Angabe zu
# sched_arr_time.
flug$dest_hour<-as.integer(flug$sched_arr_time/100)
flug$dest_day<-flug$day
flug$dest_month<-flug$month
require(lubridate)
for(i in 1:nrow(flug)){                        # Fälle, wo Tag/Monat anders
  if(flug$dest_hour[i]<flug$hour[i]) {         # als Abflugtag/Monat
    tag<-ymd(paste(flug$year[i],flug$month[i],flug$day[i]))+1
    flug$dest_day[i]<-day(tag)
    flug$dest_month[i]<-month(tag)
  }
}
flug.innerjoin.dest<-merge(flug,weather,        # Inner Join Zielfughafen+Date
            by.x=c("dest","year","dest_month","dest_day","dest_hour"),
            by.y=c("origin","year","month","day","hour"))
dim(flug.innerjoin.dest)                        # 1 Zeile x 27 Spalten

## [1]  1 27

# Nur für einen der Zielflughäfen (LGA) liegt Wetterinformation in der Tabelle
# weather vor. Daher erzeugt der Inner Join nur einen Datensatz. Da diese
# Information zum Wetter am Zielflughafen fehlt, kann eine Analyse, ob die
# Verspätung von den Wetterverhältnissen am Zielflughafen abhängt nicht
# durchgeführt werden.
# Um alle anderen Informationen auswerten zu können, erfolgt ein Outer Join
rm(flug.innerjoin.dest)
flug<-merge(flug,weather,all.x=TRUE,            # Outer Join Zielfughafen+Date
            by.x=c("dest","year","dest_month","dest_day","dest_hour"),
            by.y=c("origin","year","month","day","hour"))
dim(flug)                                       # 336776 Zeilen x 27 Spalten

## [1] 336776     27

flug<-rename(flug,"dest_precip"="precip","dest_visib"="visib") # Spalten umben.
colSums(is.na(flug))                            # Anzahl NAs je Spalte

##         dest         year   dest_month     dest_day    dest_hour
##          887            0            0            0            0
##       origin        month          day         hour      tailnum
##            0            0            0            0         2512
```

```
##        carrier         flight        air_time       dep_delay       arr_delay
##              0              0            9430            8255            9430
## sched_arr_time     origin_alt    origin_tzone        dest_alt      dest_tzone
##              0              0               0             887             887
##           name          seats             age   origin_precip    origin_visib
##              0          52606           57912            1556            1556
##    dest_precip     dest_visib
##         336775         336775

object.size(flug)                               # Größe in Bytes

## 62211176 bytes

## Join der ORIGINALEN Tabellen (ohne Selektion), um Basistabelle zu erstellen ---
flug.orig<-merge(flights,airports.orig,                # Inner origin=faa
            by.x="origin",by.y="faa")
flug.orig<-rename(flug.orig,"origin_alt"="alt","origin_tzone"="tzone")
flug.orig<-merge(flug.orig,airports.orig,all.x=TRUE,# Outer dest=faa
                 by.x="dest",by.y="faa")
flug.orig<-rename(flug.orig,"dest_alt"="alt","dest_tzone"="tzone")
flug.orig<-merge(flug.orig,airlines.orig,              # Inner carrier=carrier
            by.x="carrier",by.y="carrier")
flug.orig<-merge(flug.orig,planes.orig,all.x=TRUE,  # Outer tailnum=tailnum
            by.x="tailnum",by.y="tailnum")
flug.orig<-rename(flug.orig,"year"="year.x","year_plane"="year.y")
flug.orig<-merge(flug.orig,weather.orig,all.x=TRUE, # Outer Abflughafen Wetter
            by.x=c("origin","year","month","day","hour"),
            by.y=c("origin","year","month","day","hour"))
flug.orig<-rename(flug.orig,"origin_precip"="precip","origin_visib"="visib")
flug.orig<-merge(flug.orig,weather.orig,all.x=TRUE, # Outer Zielflughafen Wetter
            by.x=c("dest","year","month","day","hour"),
            by.y=c("origin","year","month","day","hour"))
flug.orig<-rename(flug.orig,"dest_precip"="precip","dest_visib"="visib")
dim(flug.orig)                                  # 336776 Zeilen x 56 Spalten

## [1] 336776     56

object.size(flug.orig)                          # Größe in Bytes

## 139013728 bytes
```

6.2.4 Transformation

After the data has been selected, merged, and validated, it needs to be transformed for effective analysis. This involves implementing the best practices for naming, coding, and structural conventions presented in Sect. 3.4.

- **Naming conventions**:
  - Object names should be as short as possible.
  - Object names should consist only of lowercase letters if the programming language distinguishes between upper and lower case, as is the case with R. This helps to avoid unnecessary errors. Delimiters such as "_" or "." are suitable for

indicating details such as the year, e.g. “sales_2021” or “sales.2021”. Variables (columns) in a table (dataframe, tibble, matrix, array) can be defined either uniformly in lowercase or starting with a capital letter. It should be considered that the column names will later also be displayed with their name in the tabular or graphical output.

- **Coding conventions**:
  - Date information
    If only a year is given, it should be coded as an integer.
    If a year and month are given, the day should be indicated with 01 and this should be coded as a date variable.
    If a year, month, and day are given, this should be coded as a date variable.
    Nominal variables should be coded as factor variables. In the case of binary variables that have Yes/No as an option, coding as a logical variable (TRUE/FALSE) is also possible.
    Ordinal variables should be coded as factor variables with ordered=TRUE.
    Metric variables should be coded either as integer or numeric.
- **Structural conventions**:
  - Base tables (dataframe, tibble, matrix, array), which form the basis for evaluations, should consist of columns, each representing a variable (key field or descriptive feature), and rows, each containing an observation (data record). The manifestations or values of the variables are in the cells of the table. This format is also referred to as “long format” or “narrow format” in R. The so-called “wide format”, in which different manifestations of values of a variable are in different columns, making their evaluation more difficult, should be avoided for base tables.
  - Key fields in tables, which serve to identify data records, e.g. customer number or order number, should always be in the first columns (left in the table) as they are of particular importance.

When implementing the naming conventions, the names of the objects should be chosen to be as self-explanatory as possible. The renaming of an object can be done by copying and then deleting the original object. A special feature of R is that the copy of an object does not consume additional memory as long as the object remains unchanged, as only a pointer to the object is created. The function `colnames()` is suitable for renaming columns (variables) in dataframes or tibbles, and for (rows) the function `rownames()` (note: `rownames()` does not work with tibbles).

The coding conventions can be implemented with the help of a conversion of the data types. For example, the conversion of variables from the character type to factor variables or from numeric variables to integer variables reduces the memory requirement. The functions `as.numeric()`, `as.logical()`, `as.character()` and `as.Date()` are suitable for converting data types. The conversion into a date variable is possible for any date format with the parameter `format`, e.g. `as.Date("06.10.2022",format="%d.%m.%Y")`.

Ordinal variables of the character data type can be converted into an ordinal factor variable with level designation and order using the parameter `order=TRUE`, e.g. `grades.ordinal<-factor(grades, order=TRUE,levels=c("very good","good","satisfactory","sufficient","insufficient"))`. With the parameter `exclude=NULL`, manifestations with NA can also be considered if necessary. To improve the display of results, it is sometimes also desired that levels that do not occur are eliminated, which is possible with the argument `drop`, e.g. `grades.ordinal<-grades.ordinal[drop=T]`.

Prior to evaluations, it is often necessary to recode metric variables, e.g. the uniform conversion of weight specifications in pounds, kilograms, and tons into one unit of measurement, e.g. kg or of amounts of money into a uniform currency. In this case, simple arithmetic operations are usually used.

If all variables are correctly named and correspond to the desired data type, it may still be necessary to store the data in a data structure optimal for evaluation, according to the SDMX data model presented in Sect. 3.1 in a base table. A **base table** represents a multidimensional data cube, OLAP-Cube, DataCube or Datamart, in which all key fields (navigation attributes or dimensions) and features (KPIs and attributes) should be included that are necessary for the analysis. In Fig. 6.4 the features added to the base table by the fusion are marked with *. This process step of creating a base table, which is subsequently supplemented by the addition and summarization of data, is also referred to as loading (ETL = Extraction, Transformation, Loading) into the data management level of the Business Analytics Framework (see Fig. 3.2).

Transformation

```
## Umformung ------------------------------------------------------------------
#- Objekte anlegen -------------------------------------------------------------
v1 <- c("Tobi","Silke","Klaus","Behi","Jay","Rob")# Variable v1 (Freunde)
v2 <- c(42,356,123,200,0,23)                      # Variable v2 (Freunde facebook)
v3 <- c("53","17","31",NA,"82","71")              # Variable v3 (Alter)
v4 <- c("107000","87000",NA,NA,NA,"44000")        # Variable v4 (Einkommen)
v5 <- c("Euro","Dollar",NA,NA,NA,"Euro")          # Variable v5 (Währung)
v6 <- c("blau","grün","grün","blau",              # Variable v6 (Augenfarbe)
        "gemischt","blau")
df <- data.frame(v1,v2,v3,v4,v5,v6)               # Dataframe
### Umformung von Objekt: dataframe df ###########################################
## Namenskonventionen-----------------------------------------------------------
freunde<-df                                       # Objekt mit neuem Namen
rm(df)                                            # Altes Objekt löschen
str(freunde)                                      # Ausgabe Typ und Anzahl

## 'data.frame':    6 obs. of  6 variables:
##  $ v1: chr  "Tobi" "Silke" "Klaus" "Behi" ...
##  $ v2: num  42 356 123 200 0 23
##  $ v3: chr  "53" "17" "31" NA ...
##  $ v4: chr  "107000" "87000" NA NA ...
##  $ v5: chr  "Euro" "Dollar" NA NA ...
##  $ v6: chr  "blau" "grün" "grün" "blau" ...

colnames(freunde)<- c("Freund","Facebook-Freunde",# Umbenennen Variablen in Df
                      "Alter","Einkommen","Waehrung","Augenfarbe")
str(freunde)                                      # Ausgabe Typ und Anzahl

## 'data.frame':    6 obs. of  6 variables:
##  $ Freund          : chr  "Tobi" "Silke" "Klaus" "Behi" ...
##  $ Facebook-Freunde: num  42 356 123 200 0 23
##  $ Alter           : chr  "53" "17" "31" NA ...
##  $ Einkommen       : chr  "107000" "87000" NA NA ...
##  $ Waehrung        : chr  "Euro" "Dollar" NA NA ...
##  $ Augenfarbe      : chr  "blau" "grün" "grün" "blau" ...

## Codierungskonventionen-------------------------------------------------------
#- Umcodierung Nominale Character-Variable in Faktorvariable---------------------
str(freunde$Freund)                               # Ausgabe Typ und Anzahl

##  chr [1:6] "Tobi" "Silke" "Klaus" "Behi" "Jay" "Rob"
freunde$Freund<- as.factor(freunde$Freund)        # Levels ohne Reihenfolge
str(freunde$Freund)                               # Ausgabe Typ und Anzahl

##  Factor w/ 6 levels "Behi","Jay","Klaus",..: 6 5 3 1 2 4

#- Umcodierung Character-Variable in Metrische Variable--------------------------
str(freunde$Alter)                                # Ausgabe Typ und Anzahl

##  chr [1:6] "53" "17" "31" NA "82" "71"

freunde$Alter<- as.numeric(freunde$Alter)         # Konvertierung in numerisch
str(freunde$Alter)                                # Ausgabe Typ und Anzahl

##  num [1:6] 53 17 31 NA 82 71

#- Umcodierung Character-Variable in Metrische Variable mit einheitlicher Währung-
str(freunde$Einkommen)                            # Einkommen in versch. Währung

##  chr [1:6] "107000" "87000" NA NA NA "44000"

freunde$Einkommen <- as.numeric(freunde$Einkommen)# Konvertierung in numerisch
str(freunde$Waehrung)                             # Währung
```

```
## chr [1:6] "Euro" "Dollar" NA NA NA "Euro"

wechselkurs.dollar <- 0.97                          # Wechselkurs Dollar->Euro
for (s in 1:ncol(freunde)){
  if(!is.na(freunde$Einkommen[s])) {
    if (freunde$Waehrung[s] == "Dollar"){
      freunde$Einkommen[s]<-freunde$Einkommen[s] * wechselkurs.dollar
      freunde$Waehrung[s] <- "Euro"
      print(freunde$Einkommen[s])
    }
  }
}

## [1] 84390

str(freunde$Einkommen)                              # Beräge in €

## num [1:6] 107000 84390 NA NA NA ...

str(freunde$Waehrung)                               # Einheitliche Währung €

## chr [1:6] "Euro" "Euro" NA NA NA "Euro"

#- Umcodierung Nominale Character-Variable in Faktorvariable---------------------
freunde$Waehrung<- as.factor(freunde$Waehrung)     # Levels ohne Reihenfolge
str(freunde$Waehrung)                               # Ausgabe Typ und Anzahl

## Factor w/ 1 level "Euro": 1 1 NA NA NA 1

#- Umcodierung Nominale Character-Variable in Faktorvariable---------------------
str(freunde$Augenfarbe)                             # Ausgabe Typ und Anzahl

## chr [1:6] "blau" "grün" "grün" "blau" "gemischt" "blau"

table(freunde$Augenfarbe)                           # Häufigkeit alphabetisch

##
##     blau gemischt     grün
##        3        1        2

freunde$Augenfarbe <- factor(freunde$Augenfarbe,   # Levels mit Reihenfolge
                       levels=c("blau","grün","braun","gemischt"))
str(freunde$Augenfarbe)                             # Ausgabe Typ und Anzahl
## Factor w/ 4 levels "blau","grün",..: 1 2 2 1 4 1

table(freunde$Augenfarbe)                           # Häufigkeit nach Levels

##
##     blau     grün    braun gemischt
##        3        2        0        1

## Objekt nach Anwendung der Konventionen-----------------------------------------
str(freunde)                                        # Ausgabe Typ und Anzahl

## 'data.frame':     6 obs. of  6 variables:
##  $ Freund          : Factor w/ 6 levels "Behi","Jay","Klaus",..: 6 5 3 1 2 4
##  $ Facebook-Freunde: num  42 356 123 200 0 23
##  $ Alter           : num  53 17 31 NA 82 71
##  $ Einkommen       : num  107000 84390 NA NA NA ...
##  $ Waehrung        : Factor w/ 1 level "Euro": 1 1 NA NA NA 1
##  $ Augenfarbe      : Factor w/ 4 levels "blau","grün",..: 1 2 2 1 4 1
```

The transformation of the base table flight with 336,776 data records and 27 variables can reduce the memory requirement once again from 62 MB by 21 MB to 41 MB, i.e. by more than 30%.

Flight Data Transformation

```
## Umformung -------------------------------------------------------------------
### Umformung von Objekt: dataframe flug ########################################
## Namenskonventionen-----------------------------------------------------------
str(flug)                                       # Keine Umbenennung erforderlich

## 'data.frame':    336776 obs. of  27 variables:
##  $ dest          : chr  "ABQ" "ABQ" "ABQ" "ABQ" ...
##  $ year          : int  2013 2013 2013 2013 2013 2013 2013 2013 2013 2013 ...
##  $ dest_month    : num  4 4 4 4 4 4 4 4 4 5 ...
##  $ dest_day      : int  22 23 24 25 26 27 28 29 30 1 ...
##  $ dest_hour     : int  19 23 23 23 23 23 23 23 23 23 ...
##  $ origin        : chr  "JFK" "JFK" "JFK" "JFK" ...
##  $ month         : int  4 4 4 4 4 4 4 4 4 5 ...
##  $ day           : int  22 23 24 25 26 27 28 29 30 1 ...
##  $ hour          : num  16 20 20 20 20 20 20 20 20 20 ...
##  $ tailnum       : chr  "N821JB" "N547JB" "N655JB" "N655JB" ...
##  $ carrier       : chr  "B6" "B6" "B6" "B6" ...
##  $ flight        : int  1505 1505 1505 1505 1505 1505 1505 1505 1505 1505 ...
##  $ air_time      : num  256 274 278 274 258 246 243 239 236 246 ...
##  $ dep_delay     : num  42 -3 12 -2 -5 -5 -8 -7 0 -4 ...
##  $ arr_delay     : num  31 15 70 35 14 -19 -17 -14 -5 -23 ...
##  $ sched_arr_time: int  1915 2304 2304 2304 2304 2304 2304 2304 2304 2308 ...
##  $ origin_alt    : chr  "13" "13" "13" "13" ...
##  $ origin_tzone  : chr  "America/New_York" "America/New_York" "America/New_York"
"America/New_York" ...
##  $ dest_alt      : chr  "5355" "5355" "5355" "5355" ...
##  $ dest_tzone    : chr  "America/Denver" "America/Denver" "America/Denver" "Ameri
ca/Denver" ...
##  $ name          : chr  "JetBlue Airways" "JetBlue Airways" "JetBlue Airways" "Je
tBlue Airways" ...
##  $ seats         : int  200 200 200 200 200 200 200 200 200 200 ...
##  $ age           : num  1 11 6 6 10 7 9 8 10 14 ...
##  $ origin_precip : num  0 0 0 0 0 0 0 0 0 0 ...
##  $ origin_visib  : num  10 10 0.5 10 10 10 10 10 10 10 ...
##  $ dest_precip   : num  NA NA NA NA NA NA NA NA NA NA ...
##  $ dest_visib    : num  NA NA NA NA NA NA NA NA NA NA ...

size.vorumformung<-object.size(flug)           # Größe in Bytes
## Codierungskonventionen-------------------------------------------------------
flug$dest<-as.factor(flug$dest)                 # Konvertierung zu Faktor
flug$origin<-as.factor(flug$origin)
flug$tailnum<-as.factor(flug$tailnum)
flug$carrier<-as.factor(flug$carrier)
flug$flight<-as.factor(flug$flight)
flug$origin_tzone<-as.factor(flug$origin_tzone)
flug$dest_tzone<-as.factor(flug$dest_tzone)
flug$name<-as.factor(flug$name)
flug$dest_month<-as.integer(flug$dest_month)   # Konvertierung zu Integer
flug$air_time<-as.integer(flug$air_time)
flug$dep_delay<-as.integer(flug$dep_delay)
flug$arr_delay<-as.integer(flug$arr_delay)
flug$origin_alt<-as.integer(flug$origin_alt)
flug$dest_alt<-as.integer(flug$dest_alt)
flug$age<-as.integer(flug$age)
flug$origin_precip<-as.integer(flug$origin_precip)
flug$origin_visib<-as.integer(flug$origin_visib)
## Objekt nach Anwendung der Konventionen---------------------------------------
str(flug)                                       # Ausgabe Typ und Anzahl

## 'data.frame':    336776 obs. of  27 variables:
##  $ dest          : Factor w/ 103 levels "ABQ","ACK","ALB",..: 1 1 1 1 1 1 1 1 1 1
```

```
...
## $ year          : int  2013 2013 2013 2013 2013 2013 2013 2013 2013 2013 ...
## $ dest_month    : int  4 4 4 4 4 4 4 4 4 5 ...
## $ dest_day      : int  22 23 24 25 26 27 28 29 30 1 ...
## $ dest_hour     : int  19 23 23 23 23 23 23 23 23 23 ...
## $ origin        : Factor w/ 3 levels "EWR","JFK","LGA": 2 2 2 2 2 2 2 2 2 2 ...
## $ month         : int  4 4 4 4 4 4 4 4 4 5 ...
## $ day           : int  22 23 24 25 26 27 28 29 30 1 ...
## $ hour          : num  16 20 20 20 20 20 20 20 20 20 ...
## $ tailnum       : Factor w/ 4043 levels "D942DN","N0EGMQ",..: 3262 2002 2592 259
2 2065 2472 2176 2398 2086 1756 ...
## $ carrier       : Factor w/ 16 levels "9E","AA","AS",..: 4 4 4 4 4 4 4 4 4 4 ...
## $ flight        : Factor w/ 3844 levels "1","2","3","4",..: 1346 1346 1346 1346
1346 1346 1346 1346 1346 1346 ...
## $ air_time      : int  256 274 278 274 258 246 243 239 236 246 ...
## $ dep_delay     : int  42 -3 12 -2 -5 -5 -8 -7 0 -4 ...
## $ arr_delay     : int  31 15 70 35 14 -19 -17 -14 -5 -23 ...
## $ sched_arr_time: int  1915 2304 2304 2304 2304 2304 2304 2304 2304 2308 ...
## $ origin_alt    : int  13 13 13 13 13 13 13 13 13 13 ...
## $ origin_tzone  : Factor w/ 1 level "America/New_York": 1 1 1 1 1 1 1 1 1 1 ...
## $ dest_alt      : int  5355 5355 5355 5355 5355 5355 5355 5355 5355 5355 ...
## $ dest_tzone    : Factor w/ 7 levels "America/Anchorage",..: 3 3 3 3 3 3 3 3 3 3
 ...
## $ name          : Factor w/ 16 levels "AirTran Airways Corporation",..: 10 10 10
 10 10 10 10 10 10 10 ...
## $ seats         : int  200 200 200 200 200 200 200 200 200 200 ...
## $ age           : int  1 11 6 6 10 7 9 8 10 14 ...
## $ origin_precip : int  0 0 0 0 0 0 0 0 0 0 ...
## $ origin_visib  : int  10 10 0 10 10 10 10 10 10 10 ...
## $ dest_precip   : num  NA NA NA NA NA NA NA NA NA NA ...
## $ dest_visib    : num  NA NA NA NA NA NA NA NA NA NA ...

size.nachumformung<-object.size(flug)          # Größe in Bytes
cat("Reduktion von:",round((size.vorumformung-size.nachumformung)/1000000,0),
    "MegaBytes")

## Reduktion von: 21 MegaBytes

cat("Einsparung von:",(1-round(size.nachumformung/size.vorumformung,2))*100,"Prozent
")

## Einsparung von: 34 Prozent
```

6.2.5 Addition

Additions to base tables are made when the evaluations are to be based on characteristics that are not yet present in the base table.

For example, for the base table of friends, there is an indication of age. However, evaluations should also be based on age groups. To optimize the performance of these evaluations when they are performed multiple times, the one-time calculation of the age group makes sense. Such coding of the ordinal variable age group based on the numerical variable age can be achieved with the following coding: `age_group<-factor(levels=c("Children", "Adults", "Retirees"),ordered=T)`. Subsequently, the assignments of the data records of the dataframe are achieved via logical conditions, e.g. `age_group[df$age<18]<- "Children"; age_group[df$age>=18 & df$age<=67] <- "Adults"` and `age_group[df$age>67] <- "Retirees"`.

Addition

```
## Ergänzung --------------------------------------------------------------
#- Neue Ordinale Variable ------------------------------------------------
str(freunde)                                      # Ausgabe

## 'data.frame':    6 obs. of  6 variables:
##  $ Freund          : Factor w/ 6 levels "Behi","Jay","Klaus",..: 6 5 3 1 2 4
##  $ Facebook-Freunde: num  42 356 123 200 0 23
##  $ Alter           : num  53 17 31 NA 82 71
##  $ Einkommen       : num  107000 84390 NA NA NA ...
##  $ Waehrung        : Factor w/ 1 level "Euro": 1 1 NA NA NA 1
##  $ Augenfarbe      : Factor w/ 4 levels "blau","grün",..: 1 2 2 1 4 1

freunde$Altersgruppe <-                           # Zusätzliche Variable
  case_when(freunde$Alter < 18      ~ "Kinder",
    between(freunde$Alter, 18, 67) ~ "Erwachsene",
            freunde$Alter >= 67    ~ "Rentner")
freunde$Altersgruppe <- factor(freunde$Altersgruppe, ordered=TRUE,
                        levels=c("Kinder","Erwachsene","Rentner"))
str(freunde)                                      # Ausgabe

## 'data.frame':    6 obs. of  7 variables:
##  $ Freund          : Factor w/ 6 levels "Behi","Jay","Klaus",..: 6 5 3 1 2 4
##  $ Facebook-Freunde: num  42 356 123 200 0 23
##  $ Alter           : num  53 17 31 NA 82 71
##  $ Einkommen       : num  107000 84390 NA NA NA ...
##  $ Waehrung        : Factor w/ 1 level "Euro": 1 1 NA NA NA 1
##  $ Augenfarbe      : Factor w/ 4 levels "blau","grün",..: 1 2 2 1 4 1
##  $ Altersgruppe    : Ord.factor w/ 3 levels "Kinder"<"Erwachsene"<..: 2 1 2 NA 3
3

require(vtree)                                    # Paket Zuordnungsbaum Codierung

vtree(freunde,"Alter Altersgruppe",sameline=T)  # Zuordnungsbaum Codierung
```

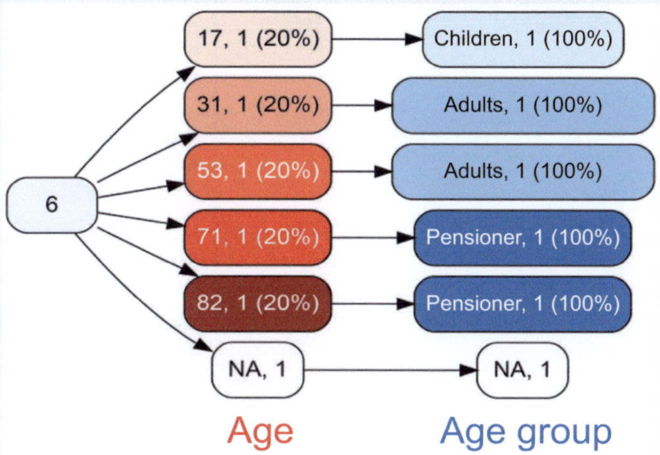


```
table(freunde$Altersgruppe)                       # Häufigkeit

##
##     Kinder Erwachsene    Rentner
##          1          2          2
```

Similarly, the characteristic age group can also be added to the base table flight to identify age groups of the aircraft. If evaluations are not only to refer to one day, one month or one year, but possibly also to a quarter, this characteristic can also be constructed and added based on the month. In Fig. 6.5, the added variables are marked with **.

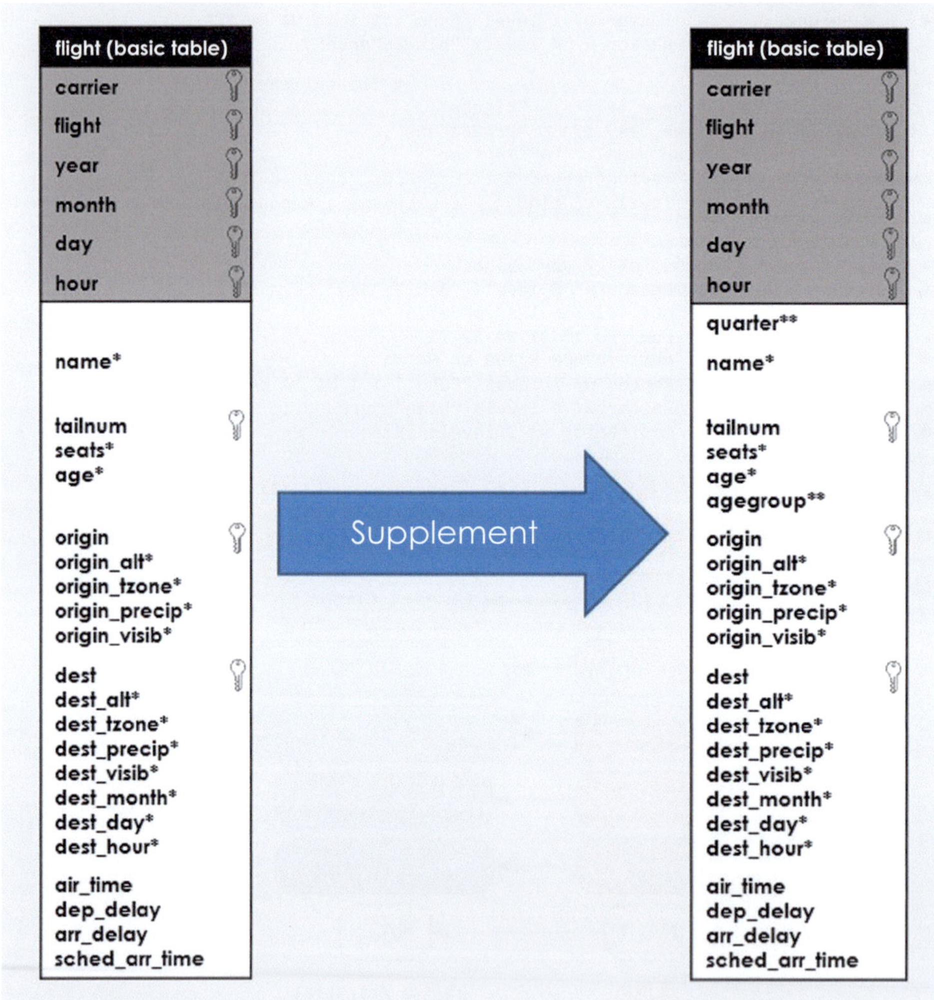

Fig. 6.5 Addition Flight Data

Addition Flight Data

```
## Ergänzung ----------------------------------------------------------------
#- Neue Ordinale Variable --------------------------------------------------
str(flug)                                                    # Ausgabe
## 'data.frame':    336776 obs. of  27 variables:
##  $ dest          : Factor w/ 103 levels "ABQ","ACK","ALB",..: 1 1 1 1 1 1 1 1 1 1
 ...
##  $ year          : int  2013 2013 2013 2013 2013 2013 2013 2013 2013 2013 ...
##  $ dest_month    : int  4 4 4 4 4 4 4 4 4 5 ...
##  $ dest_day      : int  22 23 24 25 26 27 28 29 30 1 ...
##  $ dest_hour     : int  19 23 23 23 23 23 23 23 23 23 ...
##  $ origin        : Factor w/ 3 levels "EWR","JFK","LGA": 2 2 2 2 2 2 2 2 2 2 ...
##  $ month         : int  4 4 4 4 4 4 4 4 4 5 ...
##  $ day           : int  22 23 24 25 26 27 28 29 30 1 ...
##  $ hour          : num  16 20 20 20 20 20 20 20 20 20 ...
##  $ tailnum       : Factor w/ 4043 levels "D942DN","N0EGMQ",..: 3262 2002 2592 259
2 2065 2472 2176 2398 2086 1756 ...
##  $ carrier       : Factor w/ 16 levels "9E","AA","AS",..: 4 4 4 4 4 4 4 4 4 4 ...
##  $ flight        : Factor w/ 3844 levels "1","2","3","4",..: 1346 1346 1346 1346
1346 1346 1346 1346 1346 1346 ...
##  $ air_time      : int  256 274 278 274 258 246 243 239 236 246 ...
##  $ dep_delay     : int  42 -3 12 -2 -5 -5 -8 -7 0 -4 ...
##  $ arr_delay     : int  31 15 70 35 14 -19 -17 -14 -5 -23 ...
##  $ sched_arr_time: int  1915 2304 2304 2304 2304 2304 2304 2304 2304 2308 ...
##  $ origin_alt    : int  13 13 13 13 13 13 13 13 13 13 ...
##  $ origin_tzone  : Factor w/ 1 level "America/New_York": 1 1 1 1 1 1 1 1 1 1 ...
##  $ dest_alt      : int  5355 5355 5355 5355 5355 5355 5355 5355 5355 5355 ...
##  $ dest_tzone    : Factor w/ 7 levels "America/Anchorage",..: 3 3 3 3 3 3 3 3 3 3
 ...
##  $ name          : Factor w/ 16 levels "AirTran Airways Corporation",..: 10 10 10
 10 10 10 10 10 10 10 ...
##  $ seats         : int  200 200 200 200 200 200 200 200 200 200 ...
##  $ age           : int  1 11 6 6 10 7 9 8 10 14 ...
##  $ origin_precip : int  0 0 0 0 0 0 0 0 0 0 ...
##  $ origin_visib  : int  10 10 0 10 10 10 10 10 10 10 ...
##  $ dest_precip   : num  NA NA NA NA NA NA NA NA NA NA ...
##  $ dest_visib    : num  NA NA NA NA NA NA NA NA NA NA ...
flug$agegroup <-                                             # Zusätzliche Variable
  case_when(flug$age < 10      ~ "new",
    between(flug$age, 10, 20) ~ "middle-aged",
            flug$age >= 21     ~ "old")
flug$agegroup <- factor(flug$agegroup, ordered=TRUE,
                        levels=c("new","middle-aged","old"))
str(flug)                                                    # Ausgabe
## 'data.frame':    336776 obs. of  28 variables:
##  $ dest          : Factor w/ 103 levels "ABQ","ACK","ALB",..: 1 1 1 1 1 1 1 1 1 1
 ...
##  $ year          : int  2013 2013 2013 2013 2013 2013 2013 2013 2013 2013 ...
##  $ dest_month    : int  4 4 4 4 4 4 4 4 4 5 ...
##  $ dest_day      : int  22 23 24 25 26 27 28 29 30 1 ...
##  $ dest_hour     : int  19 23 23 23 23 23 23 23 23 23 ...
##  $ origin        : Factor w/ 3 levels "EWR","JFK","LGA": 2 2 2 2 2 2 2 2 2 2 ...
##  $ month         : int  4 4 4 4 4 4 4 4 4 5 ...
##  $ day           : int  22 23 24 25 26 27 28 29 30 1 ...
##  $ hour          : num  16 20 20 20 20 20 20 20 20 20 ...
##  $ tailnum       : Factor w/ 4043 levels "D942DN","N0EGMQ",..: 3262 2002 2592 259
2 2065 2472 2176 2398 2086 1756 ...
##  $ carrier       : Factor w/ 16 levels "9E","AA","AS",..: 4 4 4 4 4 4 4 4 4 4 ...
##  $ flight        : Factor w/ 3844 levels "1","2","3","4",..: 1346 1346 1346 1346
1346 1346 1346 1346 1346 1346 ...
```

```
## $ air_time      : int  256 274 278 274 258 246 243 239 236 246 ...
## $ dep_delay     : int  42 -3 12 -2 -5 -5 -8 -7 0 -4 ...
## $ arr_delay     : int  31 15 70 35 14 -19 -17 -14 -5 -23 ...
## $ sched_arr_time: int  1915 2304 2304 2304 2304 2304 2304 2304 2304 2308 ...
## $ origin_alt    : int  13 13 13 13 13 13 13 13 13 13 ...
## $ origin_tzone  : Factor w/ 1 level "America/New_York": 1 1 1 1 1 1 1 1 1 1 ...
## $ dest_alt      : int  5355 5355 5355 5355 5355 5355 5355 5355 5355 5355 ...
## $ dest_tzone    : Factor w/ 7 levels "America/Anchorage",..: 3 3 3 3 3 3 3 3 3 3
 ...
## $ name          : Factor w/ 16 levels "AirTran Airways Corporation",..: 10 10 10
 10 10 10 10 10 10 10 ...
## $ seats         : int  200 200 200 200 200 200 200 200 200 200 ...
## $ age           : int  1 11 6 6 10 7 9 8 10 14 ...
## $ origin_precip : int  0 0 0 0 0 0 0 0 0 0 ...
## $ origin_visib  : int  10 10 0 10 10 10 10 10 10 10 ...
## $ dest_precip   : num  NA NA NA NA NA NA NA NA NA NA ...
## $ dest_visib    : num  NA NA NA NA NA NA NA NA NA NA ...
## $ agegroup      : Ord.factor w/ 3 levels "new"<"middle-aged"<..: 1 2 1 1 2 1 1 1
 2 2 ...

table(flug$agegroup)                                       # Häufigkeit

##
##          new middle-aged         old
##       104813      139894       34157

#- Neue Ordinale Variable -----------------------------------------------------
flug$quarter <-                                            # Zusätzliche Variable
  case_when(flug$month < 4     ~ "Q1",
    between(flug$month, 4, 6) ~ "Q2",
    between(flug$month, 7, 9) ~ "Q3",
            flug$month >= 10   ~ "Q4")
flug$quarter <- factor(flug$quarter, ordered=TRUE,
                       levels=c("Q1","Q2","Q3","Q4"))
str(flug)                                                  # Ausgabe

## 'data.frame':    336776 obs. of  29 variables:
## $ dest          : Factor w/ 103 levels "ABQ","ACK","ALB",..: 1 1 1 1 1 1 1 1 1 1
 ...
## $ year          : int  2013 2013 2013 2013 2013 2013 2013 2013 2013 2013 ...
## $ dest_month    : int  4 4 4 4 4 4 4 4 4 5 ...
## $ dest_day      : int  22 23 24 25 26 27 28 29 30 1 ...
## $ dest_hour     : int  19 23 23 23 23 23 23 23 23 23 ...
## $ origin        : Factor w/ 3 levels "EWR","JFK","LGA": 2 2 2 2 2 2 2 2 2 2 ...
## $ month         : int  4 4 4 4 4 4 4 4 4 5 ...
## $ day           : int  22 23 24 25 26 27 28 29 30 1 ...
## $ hour          : num  16 20 20 20 20 20 20 20 20 20 ...
## $ tailnum       : Factor w/ 4043 levels "D942DN","N0EGMQ",..: 3262 2002 2592 259
2 2065 2472 2176 2398 2086 1756 ...
## $ carrier       : Factor w/ 16 levels "9E","AA","AS",..: 4 4 4 4 4 4 4 4 4 4 ...
## $ flight        : Factor w/ 3844 levels "1","2","3","4",..: 1346 1346 1346 1346
1346 1346 1346 1346 1346 1346 ...
## $ air_time      : int  256 274 278 274 258 246 243 239 236 246 ...
## $ dep_delay     : int  42 -3 12 -2 -5 -5 -8 -7 0 -4 ...
## $ arr_delay     : int  31 15 70 35 14 -19 -17 -14 -5 -23 ...
## $ sched_arr_time: int  1915 2304 2304 2304 2304 2304 2304 2304 2304 2308 ...
## $ origin_alt    : int  13 13 13 13 13 13 13 13 13 13 ...
## $ origin_tzone  : Factor w/ 1 level "America/New_York": 1 1 1 1 1 1 1 1 1 1 ...
## $ dest_alt      : int  5355 5355 5355 5355 5355 5355 5355 5355 5355 5355 ...
```

```
## $ dest_tzone    : Factor w/ 7 levels "America/Anchorage",..: 3 3 3 3 3 3 3 3 3 3
 ...
## $ name          : Factor w/ 16 levels "AirTran Airways Corporation",..: 10 10 10
 10 10 10 10 10 10 10 ...
## $ seats         : int  200 200 200 200 200 200 200 200 200 200 ...
## $ age           : int  1 11 6 6 10 7 9 8 10 14 ...
## $ origin_precip : int  0 0 0 0 0 0 0 0 0 0 ...
## $ origin_visib  : int  10 10 0 10 10 10 10 10 10 10 ...
## $ dest_precip   : num  NA NA NA NA NA NA NA NA NA NA ...
## $ dest_visib    : num  NA NA NA NA NA NA NA NA NA NA ...
## $ agegroup      : Ord.factor w/ 3 levels "new"<"middle-aged"<..: 1 2 1 1 2 1 1 1
 2 2 ...
## $ quarter       : Ord.factor w/ 4 levels "Q1"<"Q2"<"Q3"<..: 2 2 2 2 2 2 2 2 2 2
...

require(vtree)                                     # Paket Zuordnungsbaum
vtree(flug,"quarter month",sameline=TRUE)          # Zuordnungsbaum Codierung
```

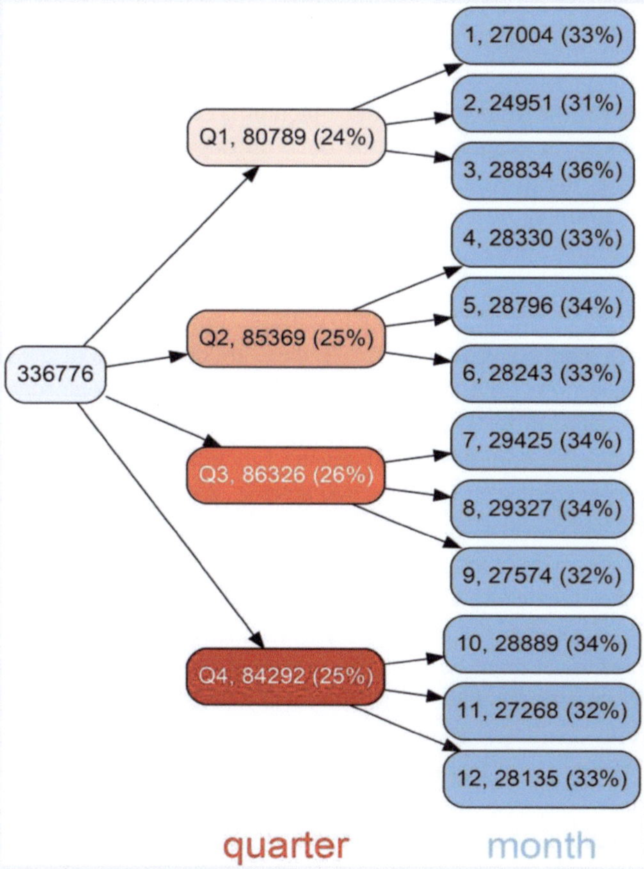


```
table(flug$quarter)                                # Häufigkeit
##
##    Q1    Q2    Q3    Q4
## 80789 85369 86326 84292
```

6.2.6 Summary

Once all the variables required for the analysis are available in a base table, the base table can be used for evaluations. Analyses often refer to data from many data sets and are therefore complex in terms of computer or CPU time.

An analysis of the average delay overall, per airline, per departure airport, per quarter, per month etc. must process all 336,776 records of the base table flight and this cannot be avoided. However, if one of these analyses is performed more than once, then the question arises whether it is not more effective to perform the calculation only once and store the results in the base table, so that in a repetition of the analysis simply the already calculated results from the base table are read.

For example, if the delays per airport for the four airlines with the most flights are to be analyzed, this affects 215,583 records. The result of the calculation can be stored in the base table in the variable avg_delay. In Fig. 6.6 the added variable is marked with ***.

Grouping by variables of a dataframe using the function `group_by()` facilitates the evaluation. For example, if an evaluation of the delay at departure per airport and per airline is to be made, then the dataframe can be grouped by the command `group_by(origin,carrier`. Grouping works particularly well in conjunction with the function `summarize()`, which applies a function to all records of a dataframe, e.g. `summarize(avg_delay=round(mean(dep_delay, na.rm=TRUE),0)`, to determine the average delay of the departure time. With the functions `summarize_all()`, `summarize_if()` and `summarize_at()` there are derivatives of the function `summarize()`. The addition _all allows the execution of more than one function. The addition _if allows a condition to be attached to the execution and the addition _at allows the selection of specific columns on which the function or functions should be applied, e.g. `summarize_at(.vars=vars(dep_delay,air_time), .funs=list(Min=min,Max=max),na.rm=TRUE)`.

To measure the factor by which the memory requirement for the object flight increases, the function `object.size()` is used. To measure the factor by which the computing time (CPU time) improves, the function `system.time()` is used. The calculation of the CPU time becomes more accurate by executing the function a thousand times and then considering the average value of all thousand repetitions. It turns out that the memory requirement only increases minimally by a factor of 1.1, while the performance of the evaluation improves by a factor of 33. This means that the CPU time for 2 evaluations without summary consumes more CPU time than 30 evaluations with summary, the CPU time for 100 evaluations without summary requires more CPU time than 3000 evaluations with summary and the CPU time for 1000 evaluations without summary requires more CPU time than 30,000 evaluations with summary. The conclusion of this investigation is that an analysis that is performed more than once, a summary is recommended if you want to minimize the execution time. This also applies to machine learning.

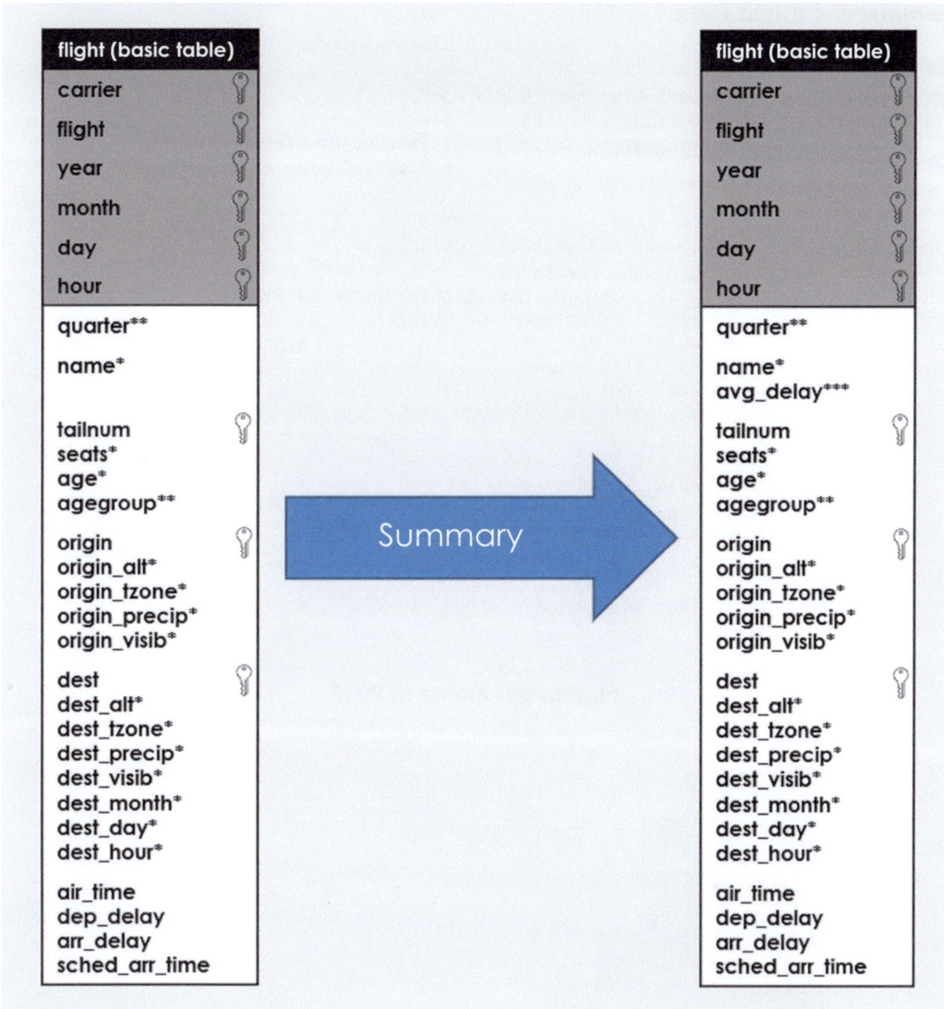

Fig. 6.6 Summary of flight data

Summary of Flight Data

```
## Zusammenfassung ------------------------------------------------------------
#- Flugdaten in Faktorvariablen nach deren Häufigkeit ordnen
flug$carrier<-fct_infreq(flug$carrier)      # Faktorvariable nach Häufikeit ordnen
flug$name<-fct_infreq(flug$name)            # Faktorvariable nach Häufikeit ordnen
dim(flug)                                   # 336776 Zeilen x 29 Spalten

## [1] 336776     29

## Exemplarische Analyse der Basistabelle (DataCube)----------------------------
#-Analysefunktion für einfache Charts
chart <- function(variable, name, xachse, yachse) {       # Funktion für Charts
  if (is.numeric(variable) | is.integer(variable)){
    barplot(table(variable), space = 0,                   # Space zwischen Balken
            xlab = xachse, main = name,
            border="black", col="grey",las=2)
    boxplot(variable,data=variable,main=name,xlab="",ylab=xachse)
  } else {
    par(oma=c(2,0,0,0))                                   # Platz für Labels
    barplot(table(variable), space = 0,                   # Space zwischen Balken
            xlab = xachse, main = name,
            border="black", col="grey",las=2)
  }
}
chart(flug$carrier, "Flüge je Fluggesellschaft in 2013", # Plot Flüge/Carrier
      "Fluggesellschaft","")
```

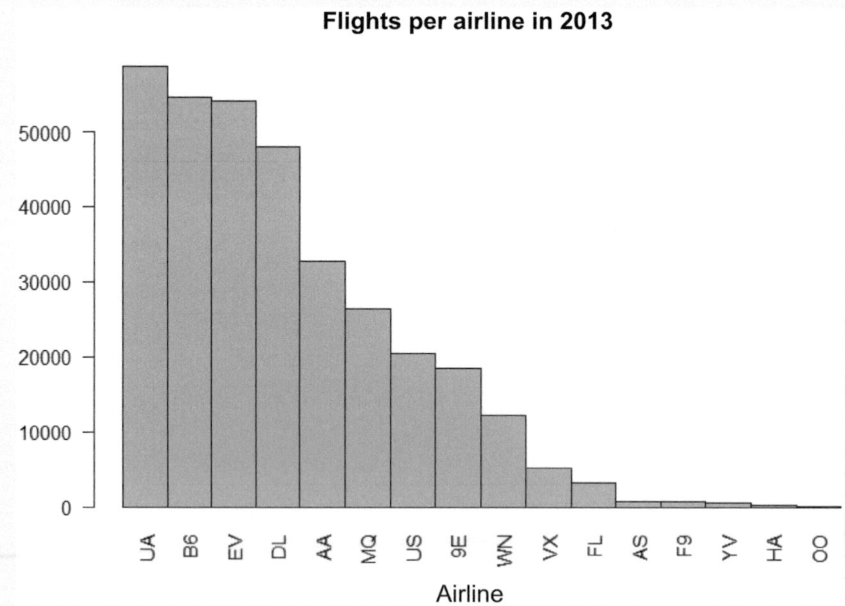


```
chart(flug$age, "Alter der Flugzeuge","Jahre","")         # Plot Flugzeugalter
```

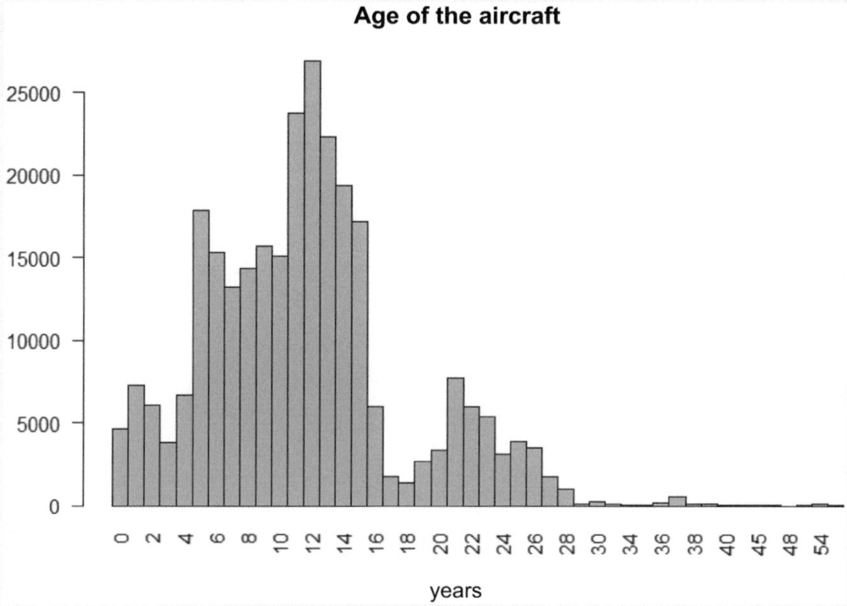

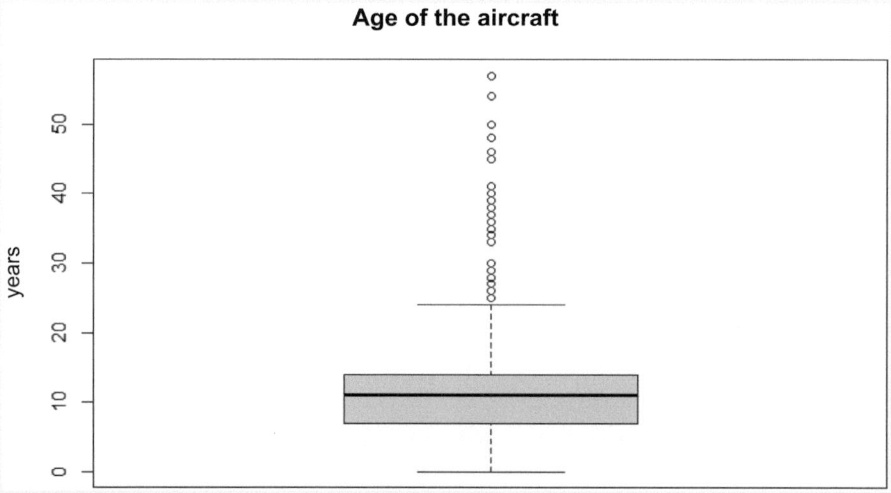


```
#-Gruppiertes Chart---------------------------------------------------------------
carrier.top4<-levels(flug$carrier)[1:4]                      # Top4 Fluggesellschaften
carrier.top4

## [1] "UA" "B6" "EV" "DL"

dim(flug[flug$carrier %in% carrier.top4,])                   # Top4 Anzahl Flüge

## [1] 215583     29

delay<-flug[flug$carrier %in% carrier.top4,] %>%
         group_by(origin,carrier) %>%
```

```
        summarize(avg_delay=round(mean(dep_delay,na.rm=TRUE),0),.groups="keep")
delay<-merge(delay,airlines,                                  # Inner Join
        by.x="carrier",by.y="carrier")                        # um name hinzuzufügen
library(viridis)                                              # Paket für Farbenblinde

## Loading required package: viridisLite

##
## Attaching package: 'viridis'

## The following object is masked from 'package:scales':
##
##    viridis_pal

ggplot(delay,                                                 # Barchart gruppiert
  aes(x=origin,y=avg_delay,group=name,fill=name))+
  geom_bar(stat="identity",position = "dodge")+
  scale_color_viridis(discrete = TRUE, guide = "none")+
  ggtitle("Durchschnittliche Verspätung beim Abflug")+
  labs(x="Abflughafen",y="Verspätung in Minuten")+
  guides(fill=guide_legend(title="Fluggesellschaft"))
```

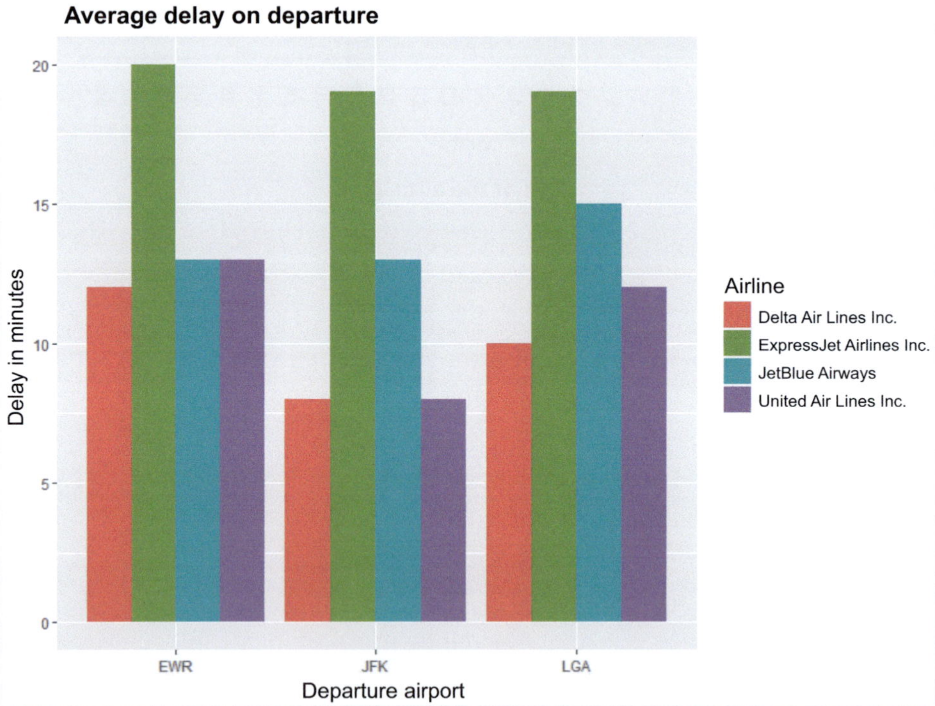


```
## Untersuchung der Performanz bei Einfügen von Zusammenfassung------------------
#- Neue Numerische Variable mit Aggregation der durchschnittlichen Verspätung ----
n<-1000                                                       # 1000 Wiederholungen
ohnezus.cpu.sec<-system.time(replicate(n,{                    # CPU-Sekunden ohne Zus.
  delay<-flug[flug$carrier %in% carrier.top4,] %>%
          group_by(origin,carrier) %>%
          summarize(avg_delay=round(mean(dep_delay,na.rm=TRUE),0),.groups="keep")
  delay<-merge(delay,airlines,                                # Inner Join
          by.x="carrier",by.y="carrier")                      # um name hinzuzufügen
```

```
}))[1]/n*100
cat("CPU-Sekunden bei Zugriff ohne Aggregation:",round(ohnezus.cpu.sec,2))

## CPU-Sekunden bei Zugriff ohne Aggregation: 5.27

ohnezus.memory<-object.size(flug)                    # Speicherbedarf ohne Zus.
flug$avg_delay<-NA                                   # Neue Variable in Basist.
for (i in 1:nrow(delay)){                            # Ergebnisse in Basist.
    flug<-ungroup(flug) %>%                          # speichern
      add_row(year=2013,
              origin=delay[i,]$origin,
              carrier=delay[i,]$carrier,
              name=delay[i,]$name,
              avg_delay=delay[i,]$avg_delay)
}
zus.memory<-object.size(flug)                        # Speicherbedarf mit Zus.
rm(delay)                                            # Vektor löschen
zus.cpu.sec<-system.time(replicate(n,{               # CPU Zeit mit Zus.
  delay<-flug[!is.na(flug$avg_delay),
              c("year","origin","carrier","name","avg_delay")]
}))[1]/n*100
cat("CPU-Sekunden bei Zugriff auf Aggregation:",round(zus.cpu.sec,2))

## CPU-Sekunden bei Zugriff auf Aggregation: 0.19

cat("Faktor des zusätzlichen Memory-Bedarfs:",       # Fazit
    round(zus.memory/ohnezus.memory,2))

## Faktor des zusätzlichen Memory-Bedarfs: 1.12

cat("Faktor der CPU-Zeitersparnis:",
    round(ohnezus.cpu.sec/zus.cpu.sec,2))

## Faktor der CPU-Zeitersparnis: 27.9

#- Vergleich der CPU-Zeit---------------------------------------------------------
analyse.ohne<-2
analyse.mit<-round(1+(analyse.ohne-1)*ohnezus.cpu.sec/zus.cpu.sec,0)
cat("CPU-Zeit:",analyse.ohne,"Analysen ohne Zus.=",analyse.mit,"Analysen mit Zus.")

## CPU-Zeit: 2 Analysen ohne Zus.= 29 Analysen mit Zus.

analyse.ohne<-100
analyse.mit<-round(1+(analyse.ohne-1)*ohnezus.cpu.sec/zus.cpu.sec,0)
cat("CPU-Zeit:",analyse.ohne,"Analysen ohne Zus.=",analyse.mit,"Analysen mit Zus.")

## CPU-Zeit: 100 Analysen ohne Zus.= 2764 Analysen mit Zus.

analyse.ohne<-1000
analyse.mit<-round(1+(analyse.ohne-1)*ohnezus.cpu.sec/zus.cpu.sec,0)
cat("CPU-Zeit:",analyse.ohne,"Analysen ohne Zus.=",analyse.mit,"Analysen mit Zus.")

## CPU-Zeit: 1000 Analysen ohne Zus.= 27878 Analysen mit Zus.

#- Export von Dateien mit Zusammenfassung----------------------------------------
save(flug,                                           # Exportieren Flugdaten
     airlines,
     airports,
     planes,
     weather,
     file="Flug.RData")
#load("Flug.RData")                                  # Laden Flugdaten
```

6.2.7 Best Practices Case Study

The application of best practices for data transformation will be presented once again using the example of data on alcohol consumption. The fivethirtyeight package includes, among other things, the dataframe drinks, which contains data on the alcohol consumption of various countries in 2010, as reported by the World Health Organization. The information provides the consumption of pure alcohol per person in liters (total_litres_of_pure_alcohol) and the consumption of beer, wine, and spirits per person in terms of the number of servings. The following will present how the best practices (see Sect. 3.4) can be applied:

- Application of **selection conventions**: If the information on pure alcohol per person in liters is not needed for the evaluations, then this column can be deleted. If only the countries Germany, USA, Sweden, and Italy are to be considered for the evaluation, all data records of the other countries can be deleted. In this way, the table with 193 rows and 5 columns is reduced to 4 rows and 4 columns. In the master data table of the countries, only the information for translating the country name into German and English next to the country code should be used. Furthermore, the subsequent analyses should only include the countries of Europe, so all other data records should be deleted. In this way, the table with 288 rows and 621 columns is significantly reduced to 51 rows and 3 columns.
- Application of **structure conventions**: The information on the consumption of the three types of drinks beer, wine, and spirits is stored in three different columns (wide format), although the type of drink represents the variable and beer, wine, and spirits only represent different characteristics or values of the variable. The function `pivot_longer()` from the package `tidyr`, which is included in the package `tidyverse`, supports the transformation into the long format, which is desired for a base table. After executing the function, the characteristics of the variable type are stored in the column type.
- Application of **coding conventions**: Since the variable type and country are nominal variables and are coded as character type variables, a transformation into a factor type variable should be carried out. This is possible with the function `factor()`.

In addition, errors are discovered during validation because, for example, there is no country code for the Aland Islands from Eurostat. The Aland Islands belong to Finland and can therefore be removed from the master data table of countries. During the validation of the country information in the base table, it is noticed that there is a data record for the USA, although the evaluations should only be limited to Europe. Therefore, this data record can also be removed.

In the course of merging the base table with the master data table of the countries, the variables for the German-speaking country name and the country coding according to Eurostat are added.

The transformation includes, among other things, the renaming of the variables to the German designations. The base table is supplemented to also add a statement on the consumed volume of alcoholic beverages per type. In order to quickly query the total consumption of all alcoholic beverages per country in liters, this variable is also inserted into the base table during the summary.

After implementing the best practices, the base table is in a format suitable for evaluations.

Best Practices Case Study for Data Provision

```
## Best Practices Fallstudie ------------------------------------------------------
library(fivethirtyeight)                                # Paket mit Daten

## Some larger datasets need to be installed separately, like senators and
## house_district_forecast. To install these, we recommend you install the
## fivethirtyeightdata package by running:
## install.packages('fivethirtyeightdata', repos =
## 'https://fivethirtyeightdata.github.io/drat/', type = 'source')

library(countrycode)
## Selektion ---------------------------------------------------------------------
#- Stammdaten Länder-------------------------------------------------------------
land<-codelist                                         # Tabelle mit Ländern
dim(land)                                              # 288 Zeilen x 621 Spalten

## [1] 288 621

land<-land[,c("country.name.de","iso.name.en",         # Spalten selektieren
              "eurostat")] %>%
  filter(land$continent == "Europe")                   # Zeilen selektieren
str(land)                                              # 51 Zeilen x 3 Spalten
```

```
## tibble [51 × 3] (S3: tbl_df/tbl/data.frame)
##  $ country.name.de: chr [1:51] "Ålandinseln" "Albanien" "Andorra" "Österreich" ..
.
##  $ iso.name.en    : chr [1:51] "Åland Islands" "Albania" "Andorra" "Austria" ...
##  $ eurostat       : chr [1:51] NA "AL" "AD" "AT" ...

#- Bewegungsdaten Alkoholkonsum-----------------------------------------------
alk.konsum<-drinks                                      # Tabelle mit Alkoholkonsum
str(alk.konsum)                                         # 193 Zeilen x 5 Spalten

## tibble [193 × 5] (S3: tbl_df/tbl/data.frame)
##  $ country                     : chr [1:193] "Afghanistan" "Albania" "Algeria" "A
ndorra" ...
##  $ beer_servings               : int [1:193] 0 89 25 245 217 102 193 21 261 279 .
..
##  $ spirit_servings             : int [1:193] 0 132 0 138 57 128 25 179 72 75 ...
##  $ wine_servings               : int [1:193] 0 54 14 312 45 45 221 11 212 191 ...
##  $ total_litres_of_pure_alcohol: num [1:193] 0 4.9 0.7 12.4 5.9 4.9 8.3 3.8 10.4
9.7 ...
##  - attr(*, "spec")=
##   .. cols(
##   ..   country = col_character(),
##   ..   beer_servings = col_integer(),
##   ..   spirit_servings = col_integer(),
##   ..   wine_servings = col_integer(),
##   ..   total_litres_of_pure_alcohol = col_double()
##   .. )

alk.konsum<- alk.konsum %>%
  select(-5) %>%                                        # Spalten selektieren
  filter(country %in% c("Germany","USA",                # Zeilen selektieren
                        "Sweden","Italy"))
str(alk.konsum)                                         # 4 Zeilen x 4 Spalten

## tibble [4 × 4] (S3: tbl_df/tbl/data.frame)
##  $ country        : chr [1:4] "Germany" "Italy" "Sweden" "USA"
##  $ beer_servings  : int [1:4] 346 85 152 249
##  $ spirit_servings: int [1:4] 117 42 60 158
##  $ wine_servings  : int [1:4] 175 237 186 84
##  - attr(*, "spec")=
##   .. cols(
```

```
##    ..   country = col_character(),
##    ..   beer_servings = col_integer(),
##    ..   spirit_servings = col_integer(),
##    ..   wine_servings = col_integer(),
##    ..   total_litres_of_pure_alcohol = col_double()
##    .. )
## Validierung----------------------------------------------------------------
#- Stammdaten Länder---------------------------------------------------------
colSums(is.na(land))                                  # Anzahl NAs je Spalte

## country.name.de      iso.name.en         eurostat
##               0                0                1

err.land<-filter(land,is.na(land$eurostat))           # Keine Angabe zu Eurostat
cat("Fehler zu Land:",dim(err.land)[1],"\n")

## Fehler zu Land: 1

head(err.land,5)                                      # Alandinseln Teil Finland

## # A tibble: 1 × 3
##   country.name.de iso.name.en   eurostat
##   <chr>           <chr>         <chr>
## 1 Ålandinseln     Åland Islands <NA>

dim(land)                                             # 51 Zeilen x 3 Spalten

## [1] 51  3

land<-filter(land,!is.na(land$eurostat))              # Korrektur (Entfernung)
dim(land)                                             # 50 Zeilen x 3 Spalten

## [1] 50  3

colSums(is.na(land))                                  # Anzahl NAs je Spalte

## country.name.de      iso.name.en         eurostat
##               0                0                0

#- Bewegungsdaten Alkoholkonsum----------------------------------------------
colSums(is.na(alk.konsum))                            # Anzahl NAs je Spalte

##         country   beer_servings spirit_servings   wine_servings
##               0               0               0               0

#- Bewegungsdaten Alkoholkonsum: country in Stammdatentabelle? --------------
err.alk.konsum<-filter(alk.konsum,!(country %in% land$iso.name.en))
cat("Fehler zu Land:",dim(err.alk.konsum)[1],"\n")

## Fehler zu Land: 1

head(err.alk.konsum,5)                                # USA nicht in Europa

## # A tibble: 1 × 4
##   country beer_servings spirit_servings wine_servings
##   <chr>           <int>           <int>         <int>
## 1 USA               249             158            84

dim(alk.konsum)                                       # 4 Zeilen x 4 Spalten

## [1] 4 4

alk.konsum<-filter(alk.konsum,country %in% land$iso.name.en)# Korrektur
dim(alk.konsum)                                       # 3 Zeilen x 4 Spalten

## [1] 3 4
```

```
colSums(is.na(Land))                                   # Anzahl NAs je Spalte
## country.name.de     iso.name.en        eurostat
##               0               0               0

## Fusion ------------------------------------------------------------------
## Join zum Hinzufügen von country.name.de und eurostat-------------------------
dim(alk.konsum)                                         # 3 Zeilen x 4 Spalten

## [1] 3 4

alk.konsum<-merge(alk.konsum,Land,                      # Inner Join über country
         by.x="country",by.y="iso.name.en")
dim(alk.konsum)                                         # 3 Zeilen x 6 Spalten

## [1] 3 6

## Umformung ----------------------------------------------------------------
#- Umformung Namenskonventionen----------------------------------------------
alk.konsum<- rename(alk.konsum,                         # Spalten umbenennen
               Land.en=country,
               Bier=beer_servings,
               Wein=wine_servings,
               Spirituosen=spirit_servings,
               Land=country.name.de,
               Landescode=eurostat)
#- Umformung Codierungskonventionen -----------------------------------------
#- Umcodierung Nominale Character-Variablen in Faktorvariablen--------------------
str(alk.konsum)                                         # Ausgabe Typ und Anzahl

## 'data.frame':    3 obs. of  6 variables:
##  $ Land.en    : chr  "Germany" "Italy" "Sweden"
##  $ Bier       : int  346 85 152
##  $ Spirituosen: int  117 42 60
##  $ Wein       : int  175 237 186
##  $ Land       : chr  "Deutschland" "Italien" "Schweden"
##  $ Landescode : chr  "DE" "IT" "SE"

alk.konsum$Land.en<-factor(alk.konsum$Land.en)          # Transformation Codierung
alk.konsum$Land<-factor(alk.konsum$Land)                # Transformation Codierung
alk.konsum$Landescode<-factor(alk.konsum$Landescode)    # Transformation Codierung
str(alk.konsum)                                         # 3 Zeilen x 6 Spalten

## 'data.frame':    3 obs. of  6 variables:
##  $ Land.en    : Factor w/ 3 levels "Germany","Italy",..: 1 2 3
##  $ Bier       : int  346 85 152
##  $ Spirituosen: int  117 42 60
##  $ Wein       : int  175 237 186
##  $ Land       : Factor w/ 3 levels "Deutschland",..: 1 2 3
##  $ Landescode : Factor w/ 3 levels "DE","IT","SE": 1 2 3

alk.konsum                                              # Datensätze

##   Land.en Bier Spirituosen Wein        Land Landescode
## 1 Germany  346         117  175 Deutschland         DE
## 2   Italy   85          42  237     Italien         IT
## 3  Sweden  152          60  186    Schweden         SE

#- Umformung Strukturkonventionen -------------------------------------------
alk.konsum<-pivot_longer(alk.konsum,                    # Transformation in Basistab.
                    cols=-c("Land","Land.en","Landescode"),
                    names_to="Typ",
                    values_to="Portionen")
alk.konsum$Typ<-factor(alk.konsum$Typ,                  # Transformation Codierung
```

```
                    levels=c("Bier","Wein","Spirituosen"))
str(alk.konsum)                                            # 9 Zeilen x 5 Spalten

## tibble [9 × 5] (S3: tbl_df/tbl/data.frame)
##  $ Land.en   : Factor w/ 3 levels "Germany","Italy",..: 1 1 1 2 2 2 3 3 3
##  $ Land      : Factor w/ 3 levels "Deutschland",..: 1 1 1 2 2 2 3 3 3
##  $ Landescode: Factor w/ 3 levels "DE","IT","SE": 1 1 1 2 2 2 3 3 3
##  $ Typ       : Factor w/ 3 levels "Bier","Wein",..: 1 3 2 1 3 2 1 3 2
##  $ Portionen : int [1:9] 346 117 175 85 42 237 152 60 186

alk.konsum                                                 # Datensätze

## # A tibble: 9 × 5
##   Land.en Land        Landescode Typ         Portionen
##   <fct>   <fct>       <fct>      <fct>           <int>
## 1 Germany Deutschland DE         Bier              346
## 2 Germany Deutschland DE         Spirituosen       117
## 3 Germany Deutschland DE         Wein              175
## 4 Italy   Italien     IT         Bier               85
## 5 Italy   Italien     IT         Spirituosen        42
## 6 Italy   Italien     IT         Wein              237
## 7 Sweden  Schweden    SE         Bier              152
## 8 Sweden  Schweden    SE         Spirituosen        60
## 9 Sweden  Schweden    SE         Wein              186

## Ergänzung ------------------------------------------------------------------
#- Ergänzung der Literangabe: je Portion Bier=354ml, Wein=147ml, Spirit.=44ml-----
alk.konsum$Liter<-NA
for(i in 1:nrow(alk.konsum)){                              # Liter berechnen
  if(alk.konsum$Typ[i]=="Bier") {
    alk.konsum$Liter[i]<-alk.konsum$Portionen[i]*0.354
} else if(alk.konsum$Typ[i]=="Wein")  {
    alk.konsum$Liter[i]<-alk.konsum$Portionen[i]*0.147
} else if(alk.konsum$Typ[i]=="Spirituosen")  {
    alk.konsum$Liter[i]<-alk.konsum$Portionen[i]*0.044
} else {
    alk.konsum$Liter[i]<-NA
}}
str(alk.konsum)                                            # 9 Zeilen x 6 Spalten

## tibble [9 × 6] (S3: tbl_df/tbl/data.frame)
##  $ Land.en   : Factor w/ 3 levels "Germany","Italy",..: 1 1 1 2 2 2 3 3 3
##  $ Land      : Factor w/ 3 levels "Deutschland",..: 1 1 1 2 2 2 3 3 3
##  $ Landescode: Factor w/ 3 levels "DE","IT","SE": 1 1 1 2 2 2 3 3 3
##  $ Typ       : Factor w/ 3 levels "Bier","Wein",..: 1 3 2 1 3 2 1 3 2
##  $ Portionen : int [1:9] 346 117 175 85 42 237 152 60 186
##  $ Liter     : num [1:9] 122.48 5.15 25.72 30.09 1.85 ...

alk.konsum                                                 # Datensätze

## # A tibble: 9 × 6
##   Land.en Land        Landescode Typ         Portionen  Liter
##   <fct>   <fct>       <fct>      <fct>           <int>  <dbl>
## 1 Germany Deutschland DE         Bier              346 122.
## 2 Germany Deutschland DE         Spirituosen       117   5.15
## 3 Germany Deutschland DE         Wein              175  25.7
## 4 Italy   Italien     IT         Bier               85  30.1
## 5 Italy   Italien     IT         Spirituosen        42   1.85
## 6 Italy   Italien     IT         Wein              237  34.8
## 7 Sweden  Schweden    SE         Bier              152  53.8
## 8 Sweden  Schweden    SE         Spirituosen        60   2.64
## 9 Sweden  Schweden    SE         Wein              186  27.3
```

```
## Zusammenfassung ------------------------------------------------------------
Liter.Summe<-alk.konsum %>%
           group_by(Land,Land.en,Landescode) %>%
           summarize(Liter=round(sum(Liter,na.rm=TRUE),0),.groups="keep")

alk.konsum$Liter.Summe<-NA                              # Neue Variable in Basist.
for (i in 1:nrow(Liter.Summe)){                         # Ergebnisse in Basist.
    alk.konsum<-ungroup(alk.konsum) %>%                 # speichern
      add_row(Land=Liter.Summe[i,]$Land,
              Land.en=Liter.Summe[i,]$Land.en,
              Landescode=Liter.Summe[i,]$Landescode,
              Liter.Summe=Liter.Summe[i,]$Liter)
}
str(alk.konsum)                                         # 9 Zeilen x 6 Spalten

## tibble [12 × 7] (S3: tbl_df/tbl/data.frame)
##  $ Land.en    : Factor w/ 3 levels "Germany","Italy",..: 1 1 1 2 2 2 3 3 3 1 ...
##  $ Land       : Factor w/ 3 levels "Deutschland",..: 1 1 1 2 2 2 3 3 3 1 ...
##  $ Landescode : Factor w/ 3 levels "DE","IT","SE": 1 1 1 2 2 2 3 3 3 1 ...
##  $ Typ        : Factor w/ 3 levels "Bier","Wein",..: 1 3 2 1 3 2 1 3 2 NA ...
##  $ Portionen  : int [1:12] 346 117 175 85 42 237 152 60 186 NA ...
##  $ Liter      : num [1:12] 122.48 5.15 25.72 30.09 1.85 ...
##  $ Liter.Summe: num [1:12] NA NA NA NA NA NA NA NA NA 153 ...

alk.konsum                                              # Datensätze

## # A tibble: 12 × 7
##    Land.en Land        Landescode Typ         Portionen  Liter Liter.Summe
##    <fct>   <fct>       <fct>      <fct>           <int>  <dbl>       <dbl>
##  1 Germany Deutschland DE         Bier              346 122.            NA
##  2 Germany Deutschland DE         Spirituosen       117   5.15          NA
##  3 Germany Deutschland DE         Wein              175  25.7           NA
##  4 Italy   Italien     IT         Bier               85  30.1           NA
##  5 Italy   Italien     IT         Spirituosen        42   1.85          NA
##  6 Italy   Italien     IT         Wein              237  34.8           NA
##  7 Sweden  Schweden    SE         Bier              152  53.8           NA
##  8 Sweden  Schweden    SE         Spirituosen        60   2.64          NA
##  9 Sweden  Schweden    SE         Wein              186  27.3           NA
## 10 Germany Deutschland DE         <NA>               NA  NA            153
## 11 Italy   Italien     IT         <NA>               NA  NA             67
## 12 Sweden  Schweden    SE         <NA>               NA  NA             84

## Auswertung-------------------------------------------------------------------
library(viridis)                                        # Paket für Farbenblinde

ggplot(alk.konsum[!is.na(alk.konsum$Portionen),],       # Barchart gruppiert Portion
  aes(x=Land,y=Portionen,group=Typ,fill=Typ))+
  geom_bar(stat="identity",position = "dodge")+
  scale_color_viridis(discrete = TRUE, guide = "none")+
  ggtitle("Alkoholkonsum")+
  labs(x="Land",y="Portionen")+
  guides(fill=guide_legend(title="Getränk"))
```

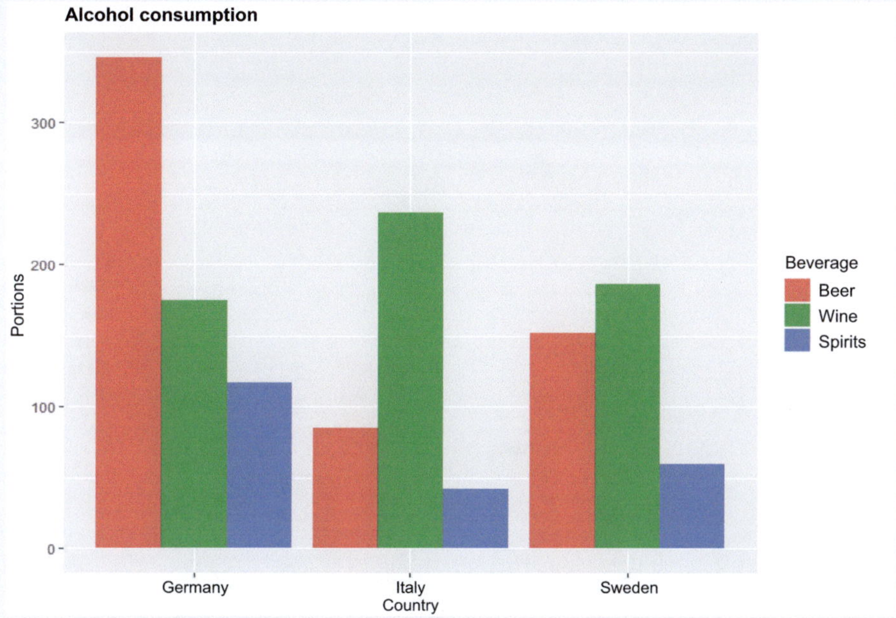


```
ggplot(alk.konsum[!is.na(alk.konsum$Liter),],          # Barchart gruppiert Liter
  aes(x=Land,y=Liter,group=Typ,fill=Typ))+
  geom_bar(stat="identity",position = "dodge")+
  scale_color_viridis(discrete = TRUE, guide = "none")+
  ggtitle("Alkoholkonsum")+
  labs(x="Land",y="Liter")+
  guides(fill=guide_legend(title="Getränk"))
```

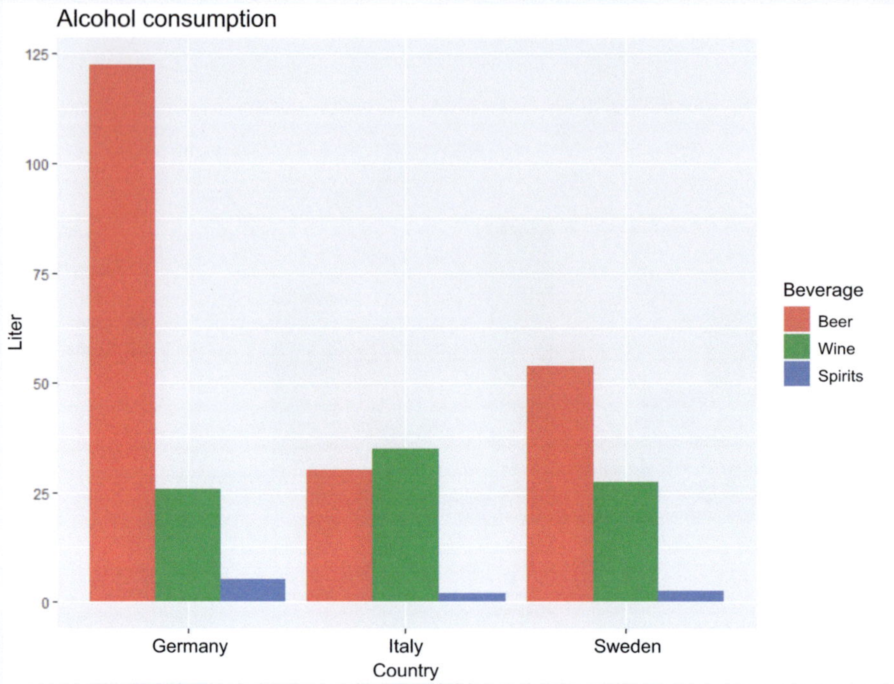


```
ggplot(alk.konsum[!is.na(alk.konsum$Liter.Summe),],     # Barchart gruppiert Gesamt
  aes(x=Land,y=Liter.Summe,group=Land,fill=Land))+
  geom_bar(stat="identity",position = "dodge")+
  scale_color_viridis(discrete = TRUE, guide = "none")+
  ggtitle("Alkoholkonsum insgesamt")+
  labs(x="Land",y="Liter alkoholischer Getränke")+
  guides(fill=guide_legend(title="Land"))
```

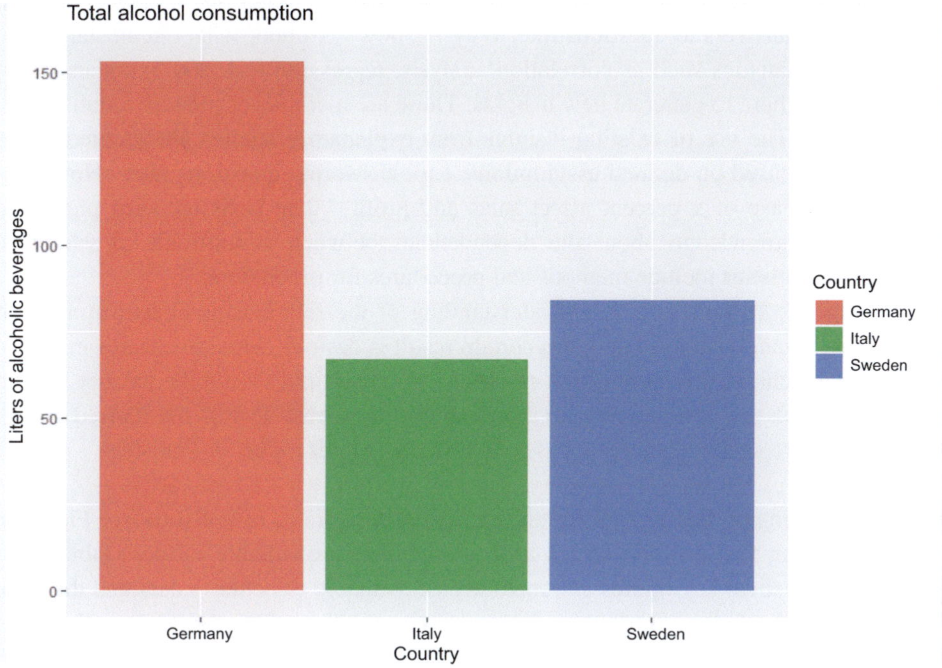

6.3 Exploratory Data Analysis

Once data has been cleaned and transformed, it is often helpful to gain an initial overview of it before applying machine learning models. Understanding data, also known as **Exploratory Data Analysis (EDA)** in English, serves to answer practical questions such as: "What is …", "How is …", "Why is …" and "How should …" and thus to gain insights that can be assigned to one or more of the following categories:

- **Exploratory**: This approach is best suited for topics that are not well understood and have not yet been extensively researched. Initial ideas have been developed and relevant variables or questions have already been uncovered. This usually takes place before the application of statistics, by considering which data should be collected to answer the questions or to identify suitable data that has already been collected. Simple descriptive statistics can be applied at this stage to get a first impression of the data and data quality.
- **Descriptive**: The aim is to accurately describe relevant variables and answer questions of the following type: What …, when …, where …, who …. Descriptive statistics are often used to describe the relevant facts.

- **Explanatory**: Based on the analysis of descriptive insights, explanatory research tries to find answers to questions like: Why …, how …, under what circumstances … Advanced statistics such as correlation analysis, regression test, and hypothesis testing are used here to generate new insights. These are methods of inductive statistics.
- **Predictive**: The use of existing insights from explanatory studies allows predictions to be made based on defined assumptions, e.g., answering questions like: "How does a price increase of x percent affect sales and profit?" The basis for such predictive research is knowledge about the dependencies between a multitude of variables. Inductive statistics include methods and procedures for predictions.
- **Prescriptive**: If there is a solid understanding of the relationships between a set of variables, it can be determined, if a certain result is desired, which measures might be suitable to achieve this. A question could be: "At what time should the journey start in order to reach the destination with a probability of at least 95% at the desired time?" The answer to such questions is possible with the help of inductive statistics.

The understanding of data is significantly supported by methods of statistics (see Fig. 6.7).

The following first presents which analysis methods are suitable for base tables, and then which statistical key figures and methods are available to better understand the data. Since visualizations also contribute to a better understanding of the data, this will also be presented.

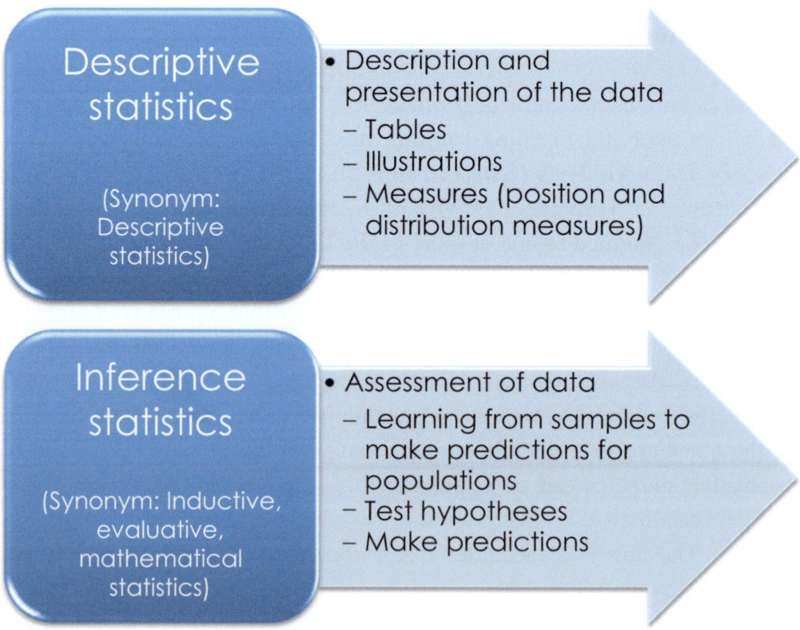

Fig. 6.7 Types of Statistics

6.3.1 Analysis Methods for Base Tables

A data model with many tables is not suitable for complex analyses and should therefore be transferred into a so-called base table (DataCube) beforehand, which contains all the key fields and features required for the analysis. In Sect. 6.2, the flight data of the New York airports EWR (Newark Liberty Intl.), JFK (John F Kennedy Intl.) and LGA (La Guardia) to destinations within the USA, to Puerto Rico and the American Virgin Islands from the year 2013 were transformed into such a base table in order to examine the delays of the flights more closely and to find out which factors have an influence on the delays.

According to the model of the Business Analytic Framework presented in Chap. 3 (see Fig. 6.8), data is divided into **master data**, code lists or dimensions (data that is constant, e.g. products, customers, airplanes, airports) and **transaction data** or facts (data that is constantly generated over time, e.g. order data, booking data, transaction data, flight data, weather data). All master data should have a unique key (ID = identification key) so that an ID can be clearly assigned to a product, customer, airplane, airport, etc.

In addition to the transaction data table with the flight connections (flights), the flight data model also provided transaction data with weather information (weather). The master data tables for validating and integrating the data are the tables with airports (airports), airplanes (planes), and airlines (airlines). In Fig. 6.9, the master data tables are marked with an S behind the table name and the transaction data with a B.

The evaluations of data usually refer to the transaction data, as these change over time and are therefore particularly "interesting". By analyzing the transaction data, questions can then be answered, e.g. how much of which product was sold in a period or which airline has the least delays and does this vary depending on the weather or the month.

Assuming the goal of the evaluation is the analysis of flight delays and the following influencing factors are to be examined:

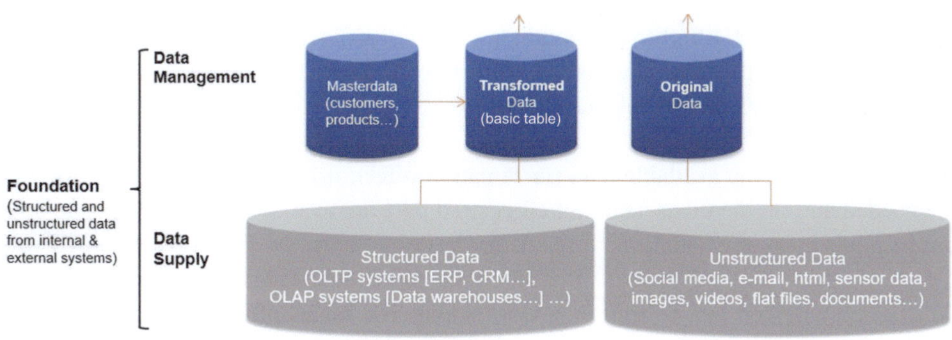

Fig. 6.8 Loading data into Data-Management-Layer

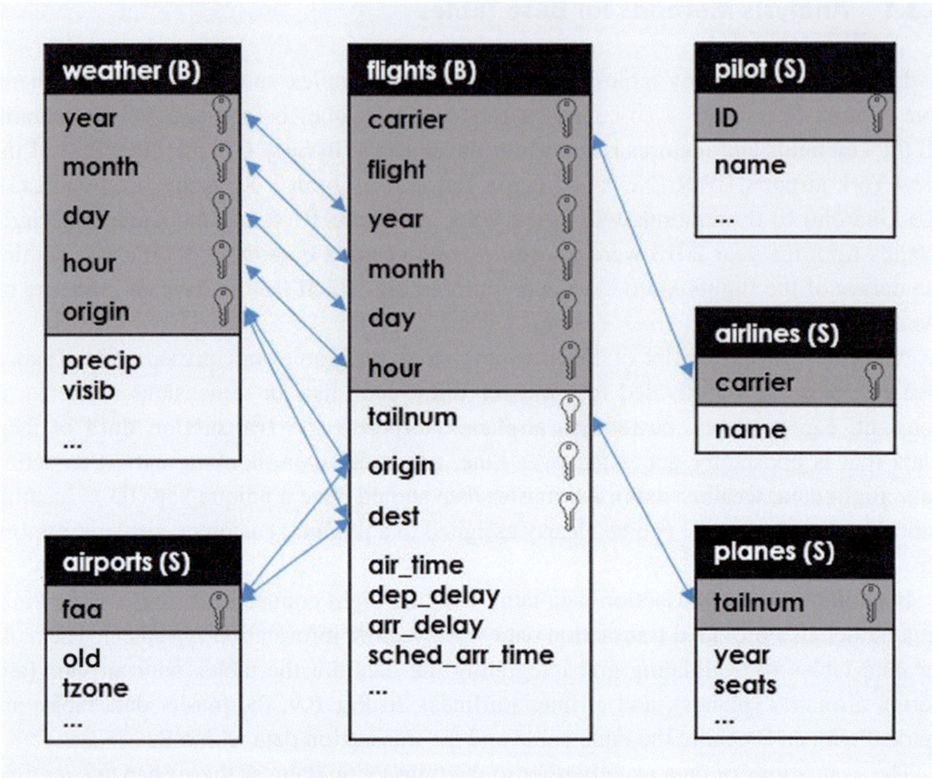

Fig. 6.9 Flight data model (S=Master data, B=Transaction data)

- Airport(airports): ID(faa), altitude above sea level(alt), time zone (tzone)
- Aircraft(planes): ID(tailnum), year of construction(year), number of seats(seats)
- Airline(airlines): ID(carrier), name(name)
- Weather(weather): ID(origin,year,month,day,hour), precipitation(precip), visibility(visib)
- Staff: Pilot
- Flight(flights): ID(flight,year,month,day,hour), airline(carrier), aircraft(tailnum), departure airport(origin), destination airport(dest), flight duration(air_time), departure delay(dep_delay), arrival delay(arr_delay)

In the analysis of the available data (information offer), it is noticeable that not all data are needed and on the other hand not all desired data are available. The unavailable data on the pilots prevent an analysis of the dependence of delays on flight personnel. To correct this in the future, this information should be added. If the data do not exist, it should be checked whether data collection is worthwhile. The tables of the data model contain significantly more variables than are necessary for the above analysis. The unnecessary features are marked with .... The tables have different field types:

- Key field (gray background with key symbol)
- Foreign key feature (white background with key symbol)
- Simple feature or descriptive variable (white background).

In order to be able to make efficient evaluations, it is necessary to select which data are relevant for the analysis (selection) after loading the data and before the analyses, and to ensure sufficient data quality for these selected data (validation). A conversion of the data types can also be useful. Another part of the preprocessing of the data can be a coding, e.g. 1=female, 2=male, as well as the transformation of units, e.g. the uniform conversion of weight specifications from pounds and tons to kg or of amounts of money into a uniform currency. If additional features are desired for the analyses, these can be constructed, such as the feature quarter (quarter) based on the month. Data can also be summarized and made available as aggregations for a performant analysis.

Since ad hoc analyses often have to satisfy very different information needs, it is necessary to construct a base table specifically tailored to each one. However, if analyses repeatedly refer back to the same base table, it makes sense to save this base table as a file or in a database. In this way, the complex process of data transformation or preprocessing (selection, validation, fusion, transformation, addition, summary) does not have to be repeated over and over again. This helps to increase the performance of the evaluations.

Base tables (DataCubes) consist of navigation attributes (dimensions) and key figures (Key Performance Indicators, KPI). The navigation attributes serve to select relevant feature expressions, e.g. specific airlines or airports in a selected time window. The DataCube of the base table for the flight data can be represented three-dimensionally for the navigation attributes time, airport and airline in Fig. 6.10. In fact, there are more dimensions (navigation attributes) in the DataCube and the dimension time alone contains a five-level hierarchy, which allows the granularity of the display to be changed from year to quarter, month, day and hour. Only to explain the concept of a DataCube, the representation here is limited to three dimensions for visualization reasons.

The advantage of transforming the data into base tables is that the following analytical operations can be easily realized on base tables (see Fig. 6.11):

- Slicing: Apply a filter so that a slice is extracted from the DataCube.
- Dicing: Apply a filter to more than one dimension to extract a subset, a sub-cube.
- Drill-Across: Rotate the axes of the DataCube so that a different dimension is selected on the first/front level.
- Drill-Down or Roll-Up: Choose a finer granularity of a navigation attribute and thus select more detailed information or choose a less fine granularity and display the information in a more aggregated form.

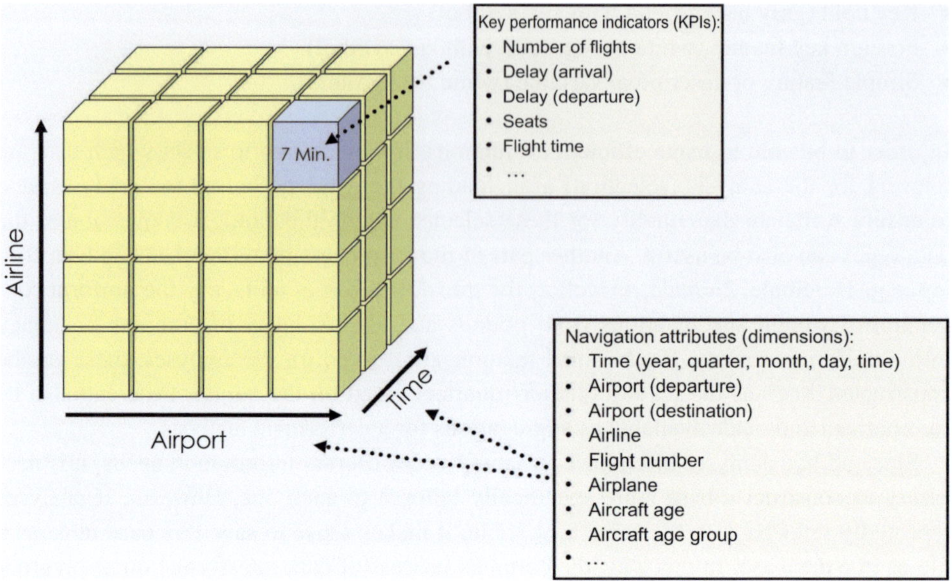

Fig. 6.10 Base table or DataCube flight data

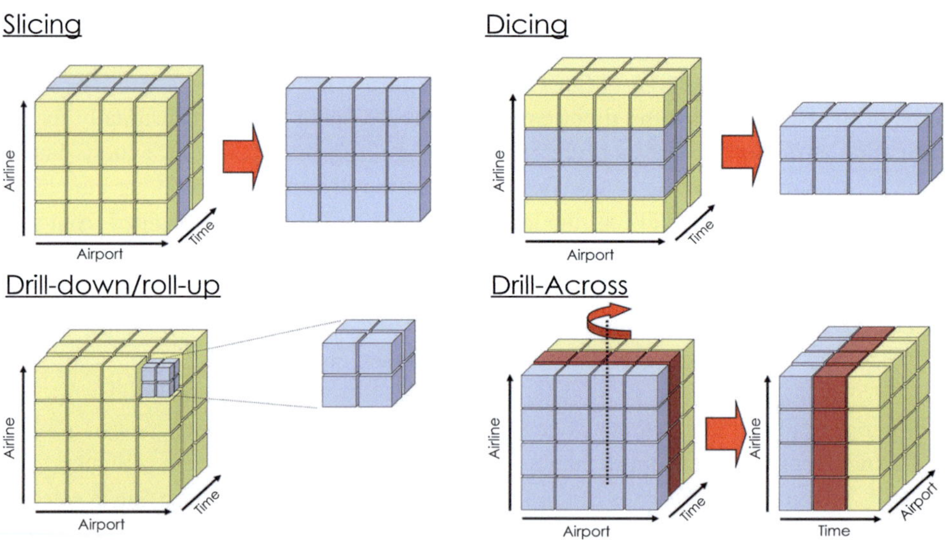

Fig. 6.11 Analysis methods for base tables or DataCubes

Analysis Methods for Basic Tables

```
#- DataCube-----------------------------------------------------------------------
carrier.top4<-levels(flug$carrier)[1:4]                    # Top4 Fluggesellschaften
#- Slicing: nur 1.Quartal---------------------------------------------------------
delay<-filter(flug,flug$carrier %in% carrier.top4 & flug$quarter==”Q1”) %>%
         group_by(origin,carrier,quarter) %>%
         summarize(avg_delay=round(mean(dep_delay,na.rm=TRUE),0),.groups=”keep”)
ggplot(delay,                                                # Barchart 1. Quartal
  aes(x=origin,y=avg_delay,group=carrier,fill=carrier))+
  geom_bar(stat=”identity”,position = “dodge”)+
  scale_color_viridis(discrete = TRUE, guide = “none”)+
  ggtitle(“1.Quartal”)+
  labs(x=”Abflughafen”,y=”Verspätung in Minuten”)+
  guides(fill=guide_legend(title=”Fluggesellschaft”))
```

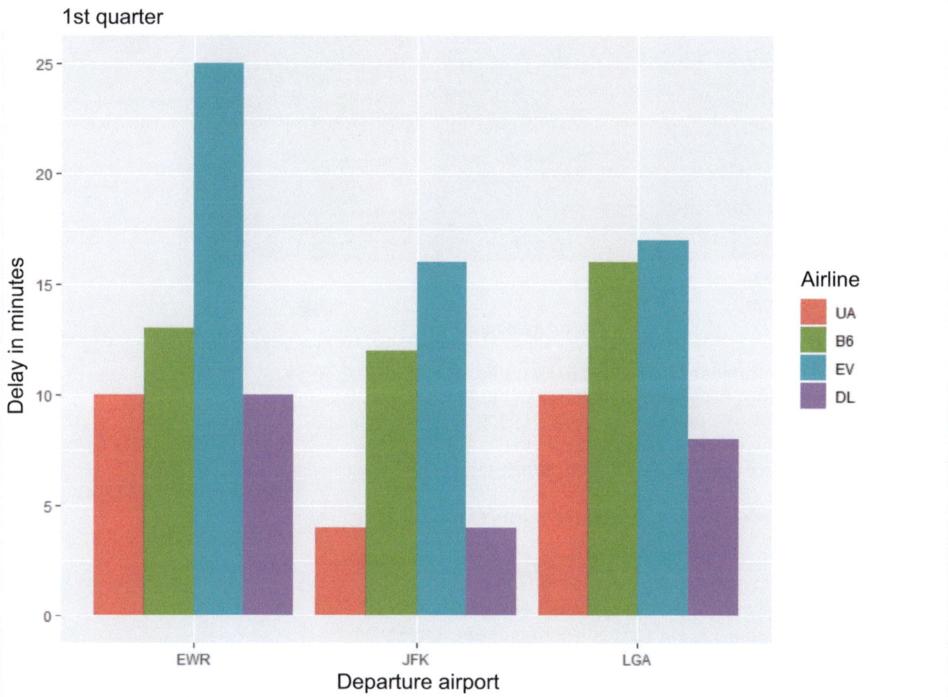


```
#- Slicing: nur 2.Quartal---------------------------------------------------------
delay<-filter(flug,flug$carrier %in% carrier.top4 &
                  flug$quarter==”Q2”) %>%
         group_by(origin,carrier,quarter) %>%
         summarize(avg_delay=round(mean(dep_delay,na.rm=TRUE),0),.groups=”keep”)
ggplot(delay,                                                # Barchart 2. Quartal
  aes(x=origin,y=avg_delay,group=carrier,fill=carrier))+
  geom_bar(stat=”identity”,position = “dodge”)+
  scale_color_viridis(discrete = TRUE, guide = “none”)+
  ggtitle(“2. Quartal“)+
  labs(x=“Abflughafen“,y=“Verspätung in Minuten“)+
  guides(fill=guide_legend(title=“Fluggesellschaft“))
```

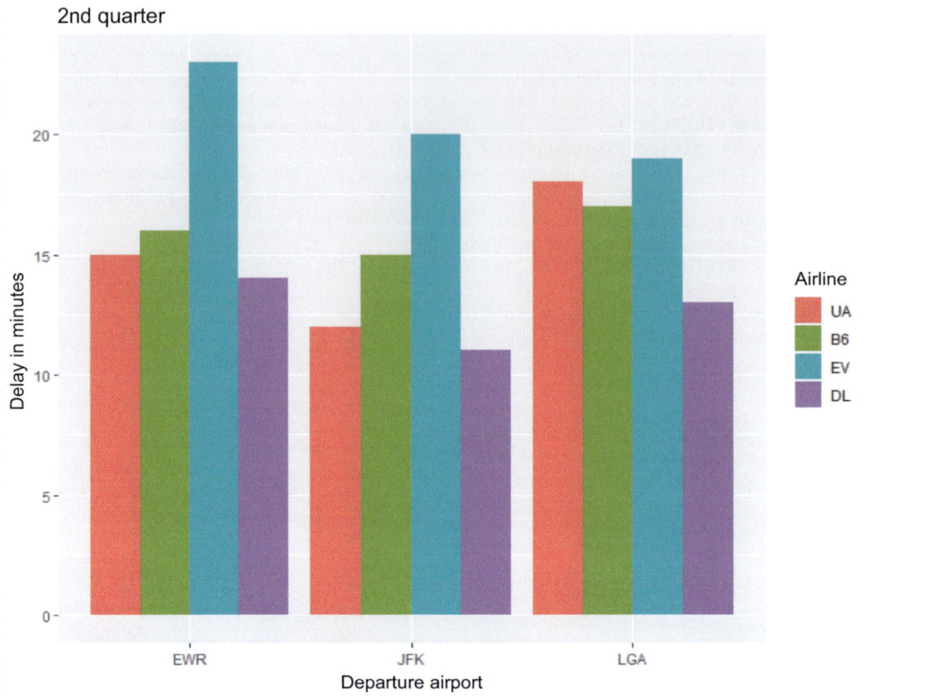


```
#- Dicing: nur Fluggesellschaft(B6+EV) und nur 2.+3. Quartal--------------------
delay<-filter(flug,flug$carrier %in% c("B6","EV") &
                  flug$quarter%in% c("Q2","Q3")) %>%
        group_by(origin,carrier,quarter) %>%
        summarize(avg_delay=round(mean(dep_delay,na.rm=TRUE),0),.groups="keep")
ggplot(delay,                                         # Barchart 2. Quartal
  aes(x=origin,y=avg_delay,group=carrier,fill=carrier))+
  geom_bar(stat="identity",position = "dodge")+
  scale_color_viridis(discrete = TRUE, guide = "none")+
  ggtitle("2.+3. Quartal")+
  labs(x="Abflughafen",y="Verspätung in Minuten")+
  guides(fill=guide_legend(title="Fluggesellschaft"))
```

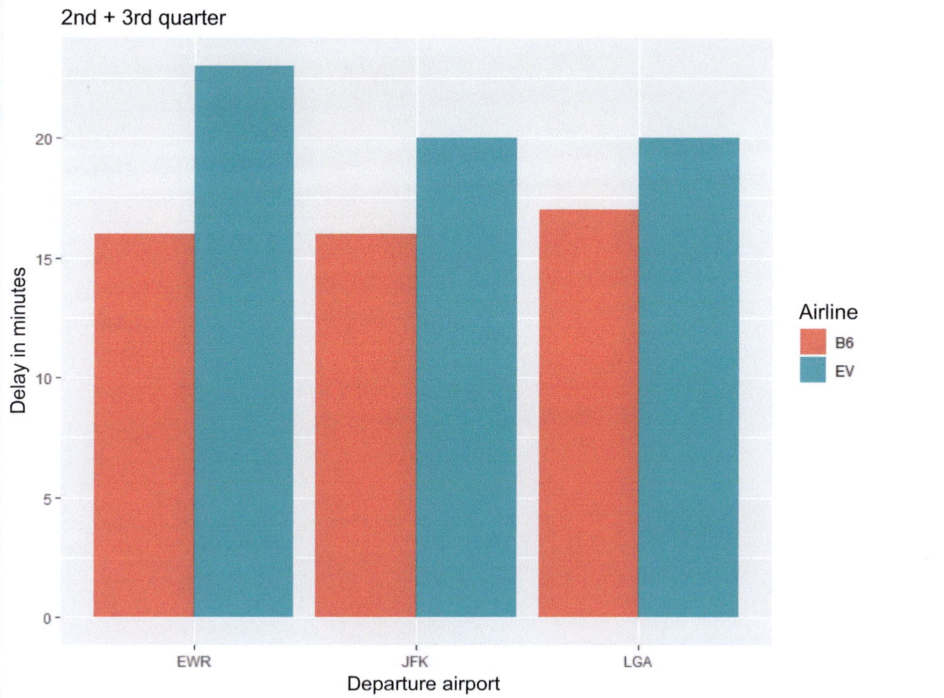


```
#- Drill-Down: nur LGA, nur Delta Airlines, 1. Quartal, Drill-Down je Monat------
delay<-filter(flug,flug$origin=="LGA" &
                   flug$carrier=="DL" &
                   flug$quarter=="Q1") %>%
         group_by(origin,carrier,month) %>%
         summarize(avg_delay=round(mean(dep_delay,na.rm=TRUE),0),.groups="keep")
ggplot(delay,                                            # Barchart LGA, Delta
  aes(x=month,y=avg_delay,group=carrier,fill=carrier))+  # 1. Quartal
  geom_bar(stat="identity",position = "dodge")+
  scale_color_viridis(discrete = TRUE, guide = "none")+
  ggtitle("1.Quartal LaGuardia")+
  labs(x="Monat",y="Verspätung in Minuten")+
  guides(fill=guide_legend(title="Fluggesellschaft"))
```

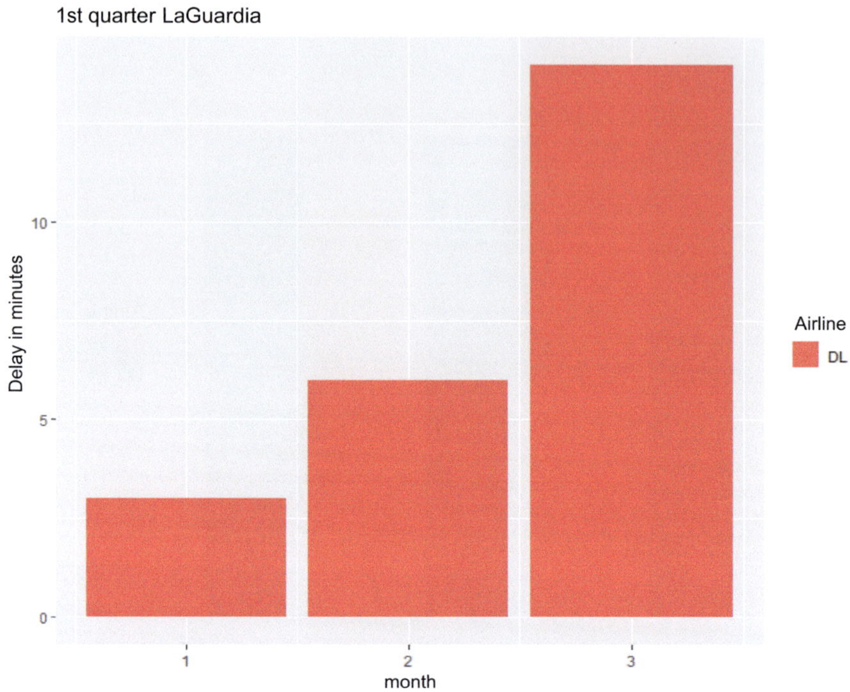


```
#- Drill-Across: Basis um Achse drehen x=Flughafen,z=Quartal, x z tauschen--------
delay<-filter(flug,flug$carrier %in% carrier.top4 &
                   !is.na(flug$quarter)) %>%
         group_by(origin,carrier,quarter) %>%
         summarize(avg_delay=round(mean(dep_delay,na.rm=TRUE),0),.groups="keep")
delay.q1<-delay[delay$quarter=="Q1",]                          # 1. Quartal
head(delay.q1)
```

```
## # A tibble: 6 × 4
## # Groups:   origin, carrier, quarter [6]
##   origin carrier quarter avg_delay
##   <fct>  <fct>   <ord>       <dbl>
## 1 EWR    UA      Q1             10
## 2 EWR    B6      Q1             13
## 3 EWR    EV      Q1             25
## 4 EWR    DL      Q1             10
## 5 JFK    UA      Q1              4
## 6 JFK    B6      Q1             12
```

```
p1<-ggplot(delay.q1,                                           # Barchart 1. Quartal
  aes(x=origin,y=avg_delay,group=carrier,fill=carrier))+
  geom_bar(stat="identity",position = "dodge")+
  scale_color_viridis(discrete = TRUE, guide = "none")+
  ggtitle("1.Quartal")+
  labs(x="Flughafen",y="Verspätung in Minuten")+
  guides(fill=guide_legend(title="Fluglinie"))
delay.ua<-delay[delay$carrier=="UA",]                          # United Airlines
head(delay.ua)
```

```
## # A tibble: 6 × 4
## # Groups:   origin, carrier, quarter [6]
##   origin carrier quarter avg_delay
##   <fct>  <fct>   <ord>       <dbl>
## 1 EWR    UA      Q1             10
## 2 EWR    UA      Q2             15
## 3 EWR    UA      Q3             14
## 4 EWR    UA      Q4             11
## 5 JFK    UA      Q1              4
## 6 JFK    UA      Q2             12

p2<-ggplot(delay.ua,                                          # Barchart United Airl.
      aes(x=quarter,y=avg_delay,group=origin,fill=origin))+
      geom_bar(stat="identity",position = "dodge")+
      scale_color_viridis(discrete = TRUE, guide = "none")+
      ggtitle("United Airlines")+
      labs(x="Quartal",y="Verspätung in Minuten")+
      guides(fill=guide_legend(title="Flughafen"))
delay.ewr<-delay[delay$origin=="EWR",]                        # EWR
head(delay.ewr)

## # A tibble: 6 × 4
## # Groups:   origin, carrier, quarter [6]
##   origin carrier quarter avg_delay
##   <fct>  <fct>   <ord>       <dbl>
## 1 EWR    UA      Q1             10
## 2 EWR    UA      Q2             15
## 3 EWR    UA      Q3             14
## 4 EWR    UA      Q4             11
## 5 EWR    B6      Q1             13
## 6 EWR    B6      Q2             16

p3<-ggplot(delay.ewr,                                         # Barchart EWR
      aes(x=carrier,y=avg_delay,group=quarter,fill=quarter))+
      geom_bar(stat="identity",position = "dodge")+
      scale_color_viridis(discrete = TRUE, guide = "none")+
      ggtitle("EWR, Newark Liberty")+
      labs(x="Fluglinie",y="Verspätung in Minuten")+
      guides(fill=guide_legend(title="Quartal"))
library(gridExtra)                                            # Paket erforderlich
grid.arrange(p1,p2,p3,nrow=2,ncol=2)                          # Grafiken ausgeben
```

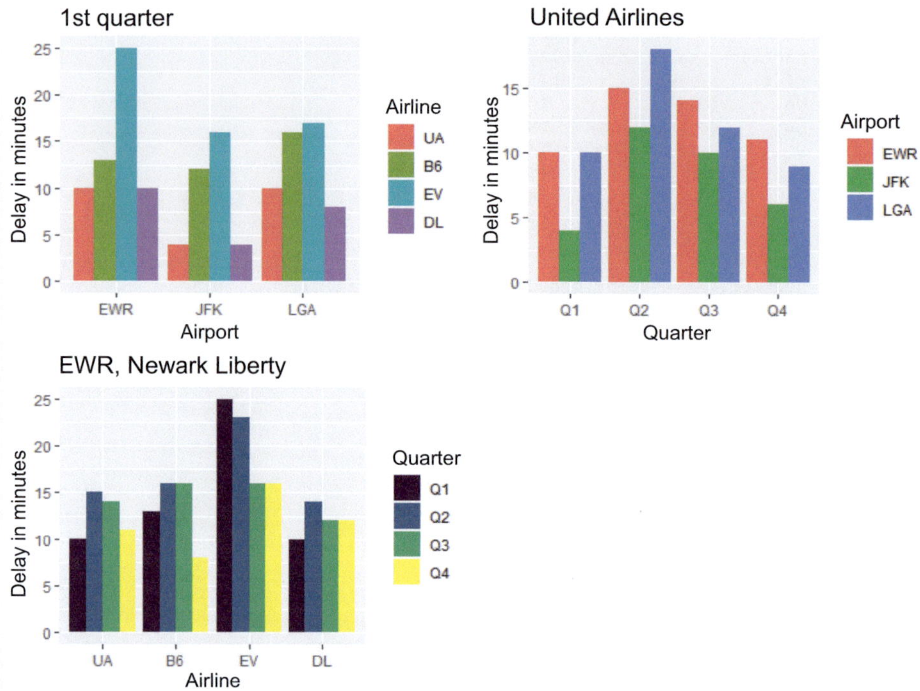


```
#-Slicing 4x je Quartal-----------------------------------------------------------
#-Analyse Abflugverspätungen je Flughafen, Fluggesellschaft und Quartal-----------
delay<-filter(flug,flug$carrier %in% carrier.top4 & !is.na(flug$quarter)) %>%
         group_by(origin,name,quarter) %>%
         summarize(avg_delay=round(mean(dep_delay,na.rm=TRUE),0),.groups=”keep”)
delay.q1<-delay[delay$quarter==”Q1”,]                              # 1. Quartal
delay.q2<-delay[delay$quarter==”Q2”,]                              # 2. Quartal
delay.q3<-delay[delay$quarter==”Q3”,]                              # 3. Quartal
delay.q4<-delay[delay$quarter==”Q4”,]                              # 4. Quartal
head(delay.q1)
```

```
## # A tibble: 6 × 4
## # Groups:   origin, name, quarter [6]
##   origin name                     quarter avg_delay
##   <fct>  <fct>                    <ord>       <dbl>
## 1 EWR    United Air Lines Inc.    Q1             10
## 2 EWR    JetBlue Airways          Q1             13
## 3 EWR    ExpressJet Airlines Inc. Q1             25
## 4 EWR    Delta Air Lines Inc.     Q1             10
## 5 JFK    United Air Lines Inc.    Q1              4
## 6 JFK    JetBlue Airways          Q1             12
```

```
p1<-ggplot(delay.q1,                                               # Barchart 1. Quartal
  aes(x=origin,y=avg_delay,group=name,fill=name))+
  geom_bar(stat=”identity”,position = “dodge”)+
  scale_color_viridis(discrete = TRUE, guide = “none”)+
  ggtitle(“1.Quartal”)+
  labs(x=”Flughafen”,y=”Verspätung in Minuten”)+
  guides(fill=guide_legend(title=”Fluglinie”))
```

```
p2<-ggplot(delay.q2,                                    # Barchart 2. Quartal
  aes(x=origin,y=avg_delay,group=name,fill=name))+
  geom_bar(stat="identity",position = "dodge")+
  scale_color_viridis(discrete = TRUE, guide = "none")+
  ggtitle("2.Quartal")+
  labs(x="Flughafen",y="Verspätung in Minuten")+
  guides(fill=guide_legend(title="Fluglinie"))
p3<-ggplot(delay.q3,                                    # Barchart 3. Quartal
  aes(x=origin,y=avg_delay,group=name,fill=name))+
  geom_bar(stat="identity",position = "dodge")+
  scale_color_viridis(discrete = TRUE, guide = "none")+
  ggtitle("3.Quartal")+
  labs(x="Flughafen",y="Verspätung in Minuten")+
  guides(fill=guide_legend(title="Fluglinie"))
p4<-ggplot(delay.q4,                                    # Barchart 4. Quartal
  aes(x=origin,y=avg_delay,group=name,fill=name))+
  geom_bar(stat="identity",position = "dodge")+
  scale_color_viridis(discrete = TRUE, guide = "none")+
  ggtitle("4.Quartal")+
  labs(x="Flughafen",y="Verspätung in Minuten")+
  guides(fill=guide_legend(title="Fluglinie"))
grid.arrange(p1,p2,p3,p4,nrow=2,ncol=2)                 # Grafiken ausgeben
```

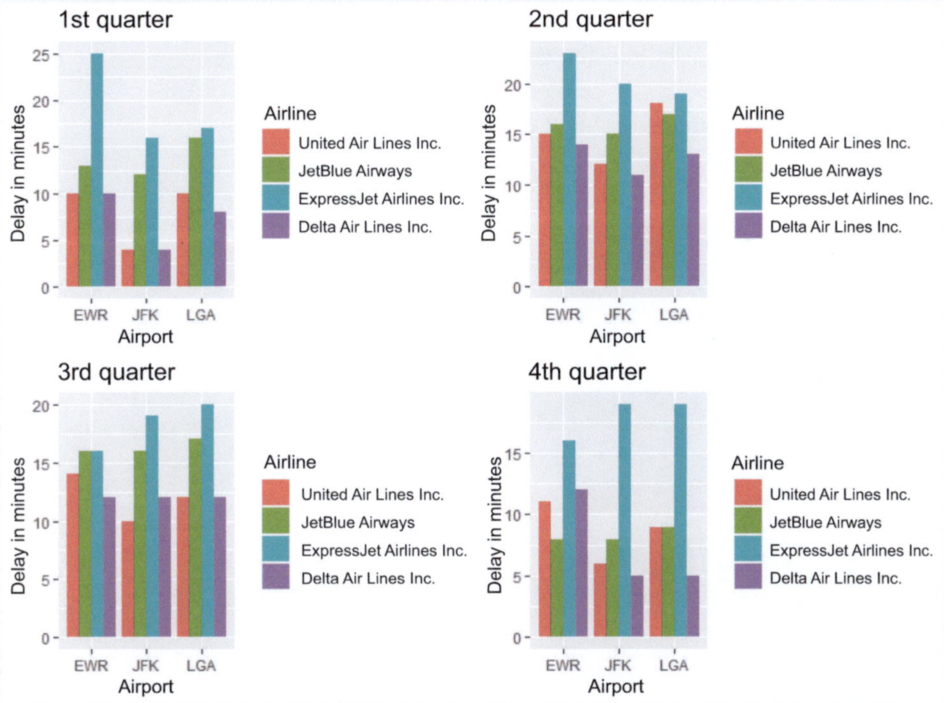

6.3.2 Statistical Analysis Methods

In addition to the simple analysis of basic tables, the consideration of statistical key figures is also suitable for gaining a better understanding of the data. Descriptive statistics distinguish between so-called measures of central tendency and measures of dispersion (see Fig. 6.12). Measures of central tendency include the mode, median, and mean. Measures of dispersion include the range, deviation, variance, and standard deviation.

Not all measures of central tendency and dispersion can be determined for each variable. Therefore, it is important to first differentiate the different types of variables: categorical or qualitative variables and cardinal, metric, numerical, or quantitative variables (see Fig. 6.13).

Qualitative variables can be divided into nominal variables without ranking (binary variable: special case with two characteristics) and ordinal variables with ranking. Examples of qualitative variables are:

- **Nominal**
  - Family status: Married, divorced, single
  - Payment mode: Cash, transfer, credit card
  - Logic: True, false
  - Gender: Male, female

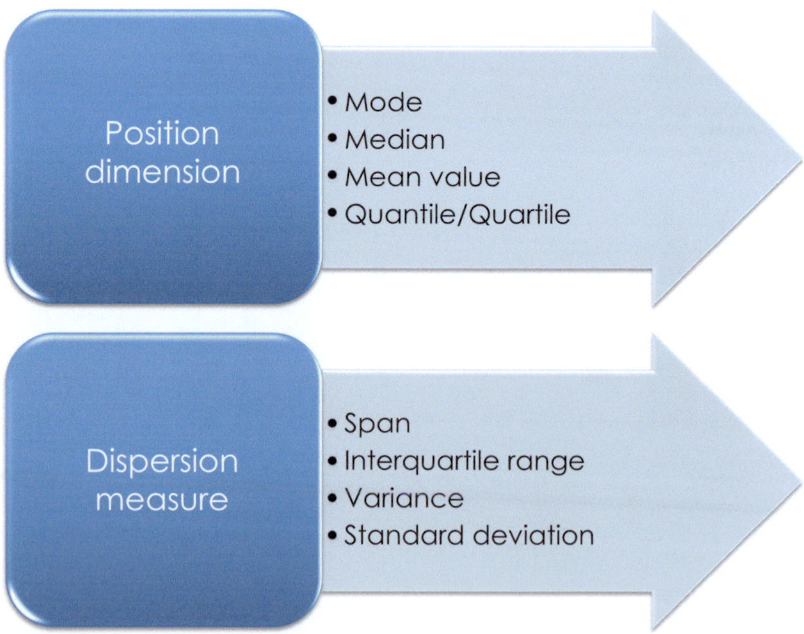

Fig. 6.12 Key figures of descriptive statistics

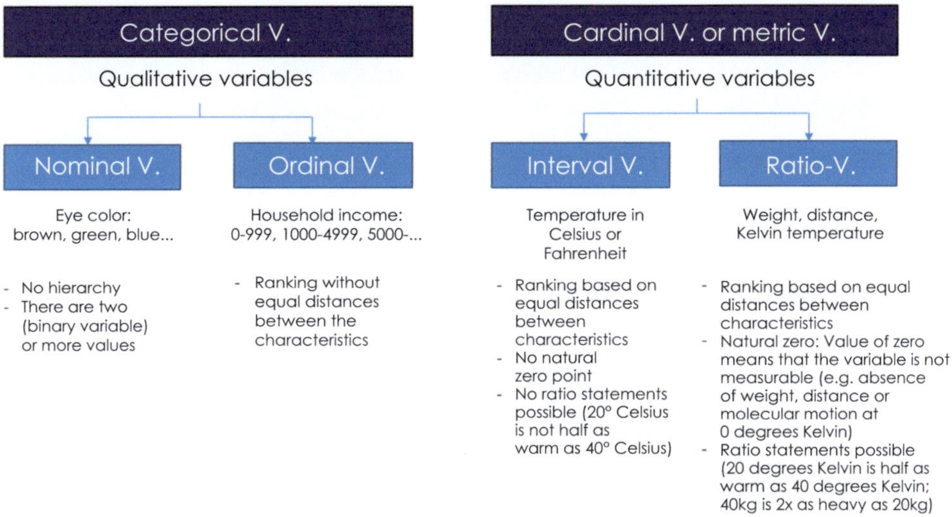

Fig. 6.13 Types of variables

- **Ordinal**
  - Income: Low, medium, high
  - Satisfaction: Very good, good, medium, bad, very bad
  - Degree of agreement: Strongly agree, agree, neutral, disagree, strongly disagree
  - Perception: Very cold, cold, normal, hot, very hot

Quantitative variables can be divided into interval variables, which do not have a natural zero point and therefore do not allow ratio statements, and ratio variables, which have a natural zero point and therefore allow ratio statements. Examples of quantitative variables are:

- **Interval**
  - Temperature in degrees Celsius
  - Year
- **Ratio**
  - Price
  - Age
  - Weight

The following evaluations of descriptive statistics can be applied to the different types of variables (see Fig. 6.14).

	Nominal variable	Ordinal variable	Interval variable	Ratio variable
Frequency distribution	x	x	x	x
Mode	x	x	x	x
Median		x	x	x
Mean value			x	x
Quantile/Quartile			x	x
Variance			x	x
Standard deviation			x	x
Span			x	x
Addition, subtraction			x	x
Multiplication, division				x

Fig. 6.14 Descriptive statistics per variable type

6.3.2.1 Measures of location and dispersion

The arithmetic mean, also known as the average, is the most important measure of central tendency. To calculate this value, you first sum all individual values and then divide them by the number of individual values (see Fig. 6.15). The letter μ (pronounced "my" or "mu") is used for the mean of populations, while the mean of samples is described with $\bar{x}$. The mean is a valuable measure, but it can be strongly influenced by so-called outliers, extremely low or high values. In R, the function `mean()` can be used to calculate the mean. Extreme individual values can significantly influence the mean. Therefore, in addition to the mean, the median is used as another measure, which is not influenced by extreme values. The median is determined by first sorting all individual values in ascending order. For an odd number of values, the median is the individual value in the middle. For an even number of values, the median is calculated from the mean of the two individual values in the middle. In R, the function `median()` can be used to determine the median.

Another measure of location is the mode, the value that occurs most frequently. The mode can be determined by counting how often each of the values occurs. The value that

Fig. 6.15 Formulas for calculating the mean

$$\mu = \frac{\sum_{i=1}^{N} x_i}{N} \qquad \bar{x} = \frac{\sum_{i=1}^{n} x_i}{n}$$

occurs most frequently is the mode. A disadvantage of the mode as a measure of central tendency is that it does not have to lie in the middle of the values. Therefore, the mean, median, and mode are usually considered together to get an idea of the distribution of a variable.

In R, there is no specific function for calculating the mode. However, the function `table()` can be helpful as it indicates the frequency of occurrence of the values in tabular form. The mode for the variable x can be determined using the following command: `as.character(filter(as.data.frame(table(x)), as.data.frame(table(x))$Freq==max(as.data.frame (table(x))$Freq))$Var1`.

While measures of location describe the average values of a data set, measures of dispersion describe how far the values are spread out or scattered and how far they are from the average values. Looking at measures of location alone is often not sufficient, as samples with the same mean or median can have very different dispersion, as shown in Fig. 6.16. Looking at measures of dispersion alone is also not sufficient to gain an understanding of the data.

To better understand dispersion, the measures of range and interquartile range are used, among others. The range describes the distance between the largest and smallest value. In the example from Fig. 6.17, the range is $351-12=339$. In R, there is no stand-

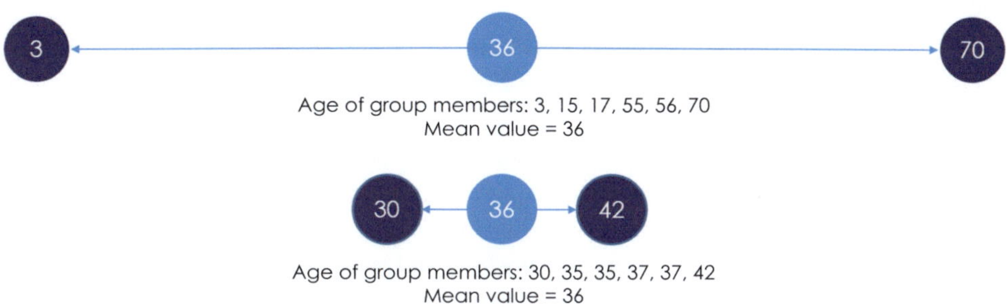

Fig. 6.16 Mean unsuitable as a measure of dispersion

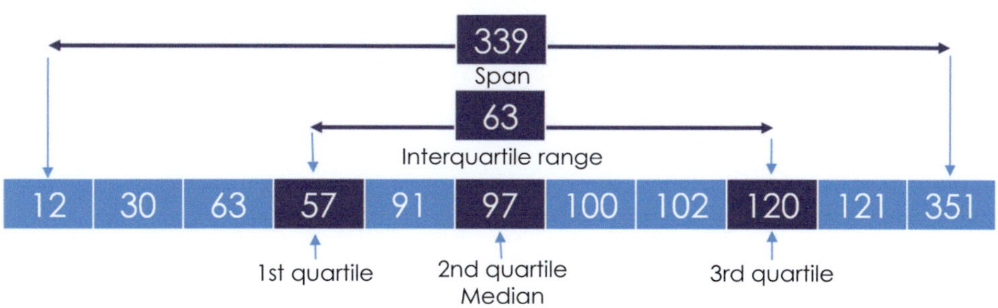

Fig. 6.17 Quantiles, quartiles and interquartile range

ard function for calculating the range. However, the function `range()` is helpful as it displays the smallest and largest value. The range for the variable x can then be determined with the command `max(range(x, na.rm=TRUE))-min(range(x,na.rm=TRUE))`. Even more convenient is the command `max(x)-min(x)`.

The median represents the value that lies in the middle. The median is also called the 2nd quartile or the 50% quantile. Quantile values are measures of location and indicate the value for which a certain percentage of all values is lower. The 50% quantile indicates the value for which 50% of all values are smaller. Thus, the 0% quantile represents the smallest value and the 100% quantile the largest value. The 25% quantile is also referred to as the 1st quartile (quartile, as the set is divided into four sections each with 25%), the 50% quantile as the 2nd quartile or median, and the 75% quantile as the 3rd quartile. The function `quantile()` in R calculates the quartiles and the minimum and maximum.

The distance between the 1st and 3rd quartile is also known as the interquartile range (English: Interquartile range, IQR) and indicates the dispersion of the middle 50% of all values. To determine the IQR, the function `IQR()` can be used in R.

Suppose there is a vector named anzahl that contains the following values 351, 120, 12, 30, 121, 63, 57, 91, 97, 100, 102. To find out how large the number in the vector anzahl must be if 70% of all values are to be smaller or the number is to belong to the top 30%, the following command `quantile(anzahl,probs=0.70)` can be used and results in the value 109.2.

A summary of essential measures is also provided by the function `summary()`. It outputs the minimum, maximum, 1st to 3rd quartile, mean and possibly the number of missing values.

Another significant measure of dispersion is the variance. Mathematically, the variance is calculated as the average of the squared deviations (square of the deviations of all values from the mean, see Fig. 6.18). The simple sum of all deviations from the mean is insufficient, as the deviations can take positive and negative values, which can cancel each other out, even if the deviations are large. A small value of this sum would therefore not be meaningful. Also, the sum of the squares of the deviations (Sum of Squared Errors, SS) would not be a good measure of deviation, as the measure would increase with an increasing number of values, even though the average deviation is not higher. For a population, the variance σ^2 (Sigma squared) is therefore calculated as the sum of the

$$\sigma^2 = \frac{SS}{N} = \frac{\Sigma(x_i - \mu)}{N} \qquad s^2 = \frac{SS}{n-1} = \frac{\Sigma(x_i - \bar{x})}{n-1}$$

Fig. 6.18 Formulas for calculating variance

squared deviations (SS) divided by the number of observations (N = size of the population, if the entire population was observed). The sample variance s^2, on the other hand, results from the sum of the squared deviations (SS) divided by the number of degrees of freedom (degree of freedom = n – 1, n = number of observations of the sample).

In R, the function `var()` is used to calculate the variance, and by default, it calculates the sample variance s^2, thus dividing by n-1. To calculate the population variance σ^2 (sigma squared), the following function can be defined if necessary: `sigmaquadrat<-function(x) {n=length(x);var(x, na.rm=T) * (n-1)/n}`.

The disadvantage of variance is that it is based on squared deviations and is therefore somewhat inflated. To correct this inaccuracy, the square root of the variance is taken to determine the standard deviation of the population (sigma) or the standard deviation of the sample s. In R, the function `sd()` (Standard Deviation) can be used to calculate the standard deviation of the sample s. To calculate the standard deviation of the population (sigma), the following function can be used if necessary: `sigma<-function(x) {n=length(x);sqrt(var(x, na.rm=T) * (n-1)/n)}`. The standard deviation is the most commonly used measure of dispersion.

To display the most important measures of location and dispersion for a variable, the function `ml.summary()` from the machinelearning package can be used, which also displays a histogram of the distribution.

Measures of Location and Dispersion

```
#- Lage- und Streuungsmaße anzeigen------------------------------------------
# Verhältnisvariablen
summary(flug$dep_delay)

##    Min. 1st Qu.  Median    Mean 3rd Qu.    Max.    NA's
##  -43.00   -5.00   -2.00   12.64   11.00 1301.00    8267

sum<-ml.summary(flug$dep_delay,"Verspätung Abflug","Minuten",rel=T)
```

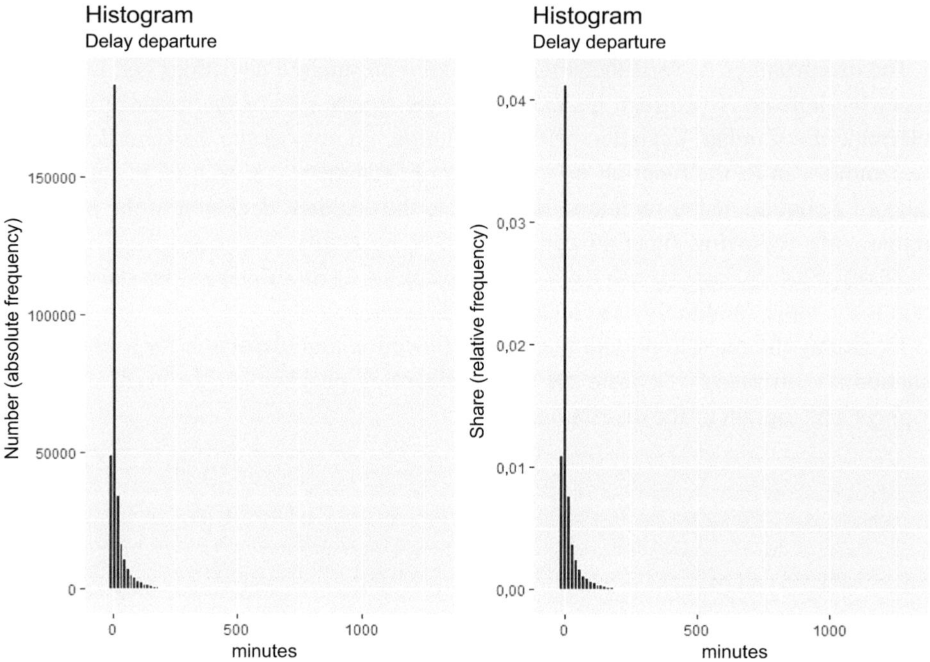


```
sum

##              25%
## anzahl       "336788"
## anzahl.na    "8267"
## modus        "-5"
## median       "-2"
## mean         "12.64"
## min          "-43"
## max          "1301"
## sd           "40.21"
## q1           "-5"
## q3           "11"
## iqr          "16"
## whisker.min  "-29"
## whisker.max  "35"
## skewness     "4.8"
## skewness.txt "Rechte Schiefe: 4.8 > 0, positive Schiefe,linkssteil,rechtsschief"
## kurtosis     "43.95"
## kurtosis.txt "Steilgipflig mit Exzess Kurtosis 43.95 > 0"
```

```
summary(flug$hour)

##    Min. 1st Qu.  Median    Mean 3rd Qu.    Max.    NA's
##    1.00    9.00   13.00   13.18   17.00   23.00      12

sum<-ml.summary(flug$hour,"Volle Stunde des Abflugs","Stunde",rel=T)
```

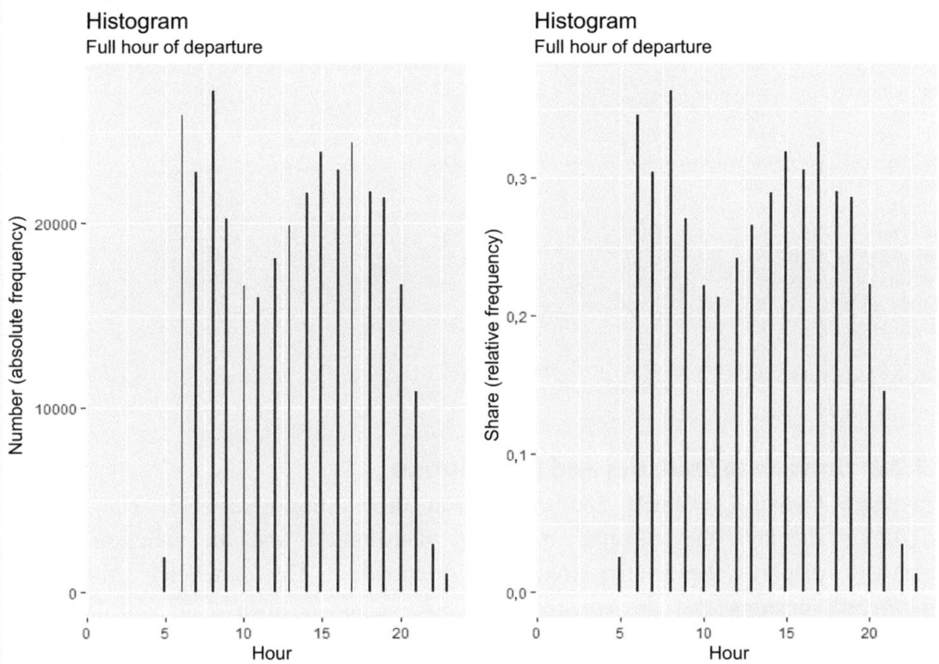


```
sum

##                25%
## anzahl         "336788"
## anzahl.na      "12"
## modus          "8"
## median         "13"
## mean           "13.18"
## min            "1"
## max            "23"
## sd             "4.66"
## q1             "9"
## q3             "17"
## iqr            "8"
## whisker.min    "-3"
## whisker.max    "29"
## skewness       "0"
## skewness.txt "Näherungsweise Normalverteilt. Schiefe: 0"
## kurtosis       "-1.21"
## kurtosis.txt "Flachgipflig mit Exzess Kurtosis -1.21 < 0"

# Nominale Variablen
summary(airports$tzone)

##    Length      Class      Mode
```

```
##      1460 character character

sum<-ml.summary(airports$tzone,"Zeitzone Flughafen")
sum

##              [,1]
## anzahl       „1460"
## anzahl.na    „0"
## modus        „America/New_York"
## median       NA
## mean         NA
## min          NA
## max          NA
## sd           NA
## q1           NA
## q3           NA
## iqr          NA
## whisker.min  NA
## whisker.max  NA
## skewness     NA
## skewness.txt NA
## kurtosis     NA
## kurtosis.txt NA
```

6.3.2.2 Univariate Statistics and Distribution

Univariate statistics examine the distribution of a single variable. The function `table()` displays the absolute frequency distribution and the function `prop.table()` displays the relative frequency distribution. The parameter `useNA = "always"` ensures that the frequency of NA values is always displayed, even if no NA values are present, e.g., `table(flug$month, useNA="always")`. The function `sort()` can be used to display the result sorted by frequency, e.g., `sort(abs,decreasing=T)`. The function `cumsum()` allows the relative frequencies to be accumulated.

Univariate Statistics and Distribution

```
#- Univariate Statistik und Verteilung------------------------------------------
abs<-table(flug$month, useNA="always")
abs

##
##     1     2     3     4     5     6     7     8     9    10    11    12  <NA>
## 26483 23690 27973 27662 28233 27234 28485 28841 27122 28653 27035 27110     0

rel<-round(100*prop.table(abs),1)
rel

##
##    1    2    3    4    5    6    7    8    9   10   11   12 <NA>
##  8.1  7.2  8.5  8.4  8.6  8.3  8.7  8.8  8.3  8.7  8.2  8.3  0.0

cum<-round(cumsum(rel),0)
cum

##    1    2    3    4    5    6    7    8    9   10   11   12 <NA>
##    8   15   24   32   41   49   58   67   75   84   92  100  100
```

For some statistical applications, a normal frequency distribution is assumed. Since distributions of values can also be asymmetric, the skewness and kurtosis of a distribution are measured.

A frequency distribution is usually represented in the form of a histogram, in which all observed values of a variable are plotted on the horizontal axis and the height of the bars indicates how often a value occurs. The following properties of the distribution are differentiated:

- Skewness indicates the type and strength of asymmetry of a distribution (see Figs. 6.19 and 6.20)
  - Positive skewness, left-steep or right-skewed distribution
  - Negative skewness, left-skewed or right-steep distribution
- Kurtosis is a measure of the steepness or "peakedness" of a distribution (see Fig. 6.21)
  - Steep-peaked, supergaussian or leptokurtic
  - Flat-peaked, subgaussian or platykurtic

Whether a distribution is skewed can be determined based on the measures of location mode, median, and mean. The function `skewness()` from the e1071 package calculates the skewness in R.

The value for kurtosis is 3 in the case of a normal distribution. If the value of kurtosis is above 3, it is a narrow-peaked, pointed distribution, and for values less than 3, it is a broad-peaked, flat distribution with observations close to the mean.

To better assess the extent of kurtosis, the kurtosis of a distribution is compared with the kurtosis of a normal distribution. For this purpose, the excess is calculated as kurtosis minus 3 of the kurtosis of a normal distribution. If the value of the excess is zero, it

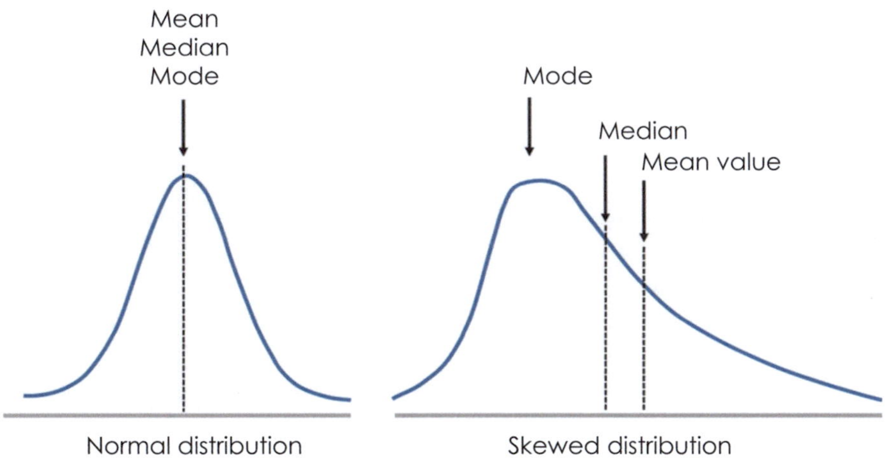

Fig. 6.19 Normal distribution versus skewness

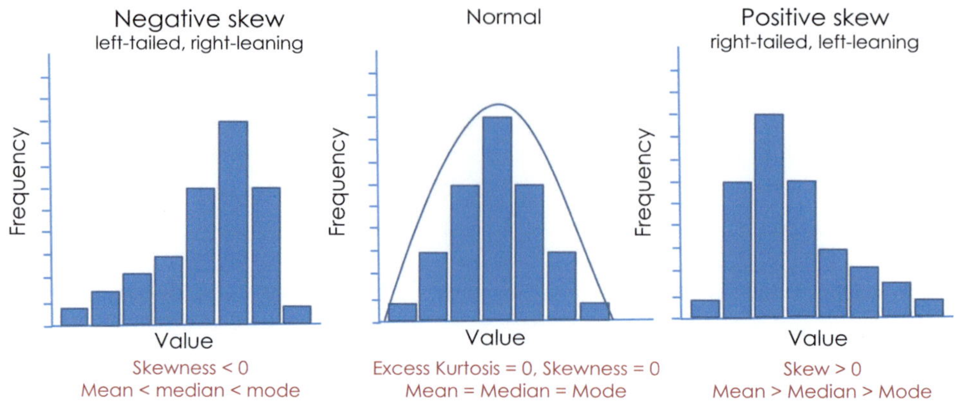

Fig. 6.20 Distributions without and with skewness

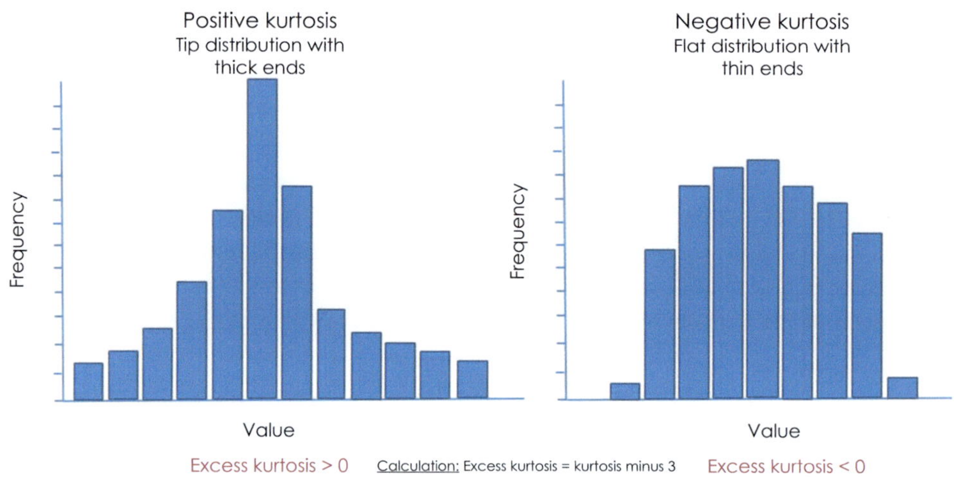

Fig. 6.21 Kurtosis

is approximately a normal distribution. If the value is above zero, it is a narrow-peaked, pointed distribution, and for values less than zero, it is a broad-peaked, flat distribution with observations close to the mean.

In R, the kurtosis can be calculated with the function `kurtosis()` from the e1071 package.

To calculate the skewness and kurtosis for a numerical variable, the function `ml.summary()` from the machinelearning package can also be used, which also outputs a histogram and box plot at the same time.

Distribution

```
#- Verteilung-----------------------------------------------------------------------
sum<-ml.summary(autos$PS,"PS Gebrauchtwagen","PS",box=T)          # Verhältnisvar.
```

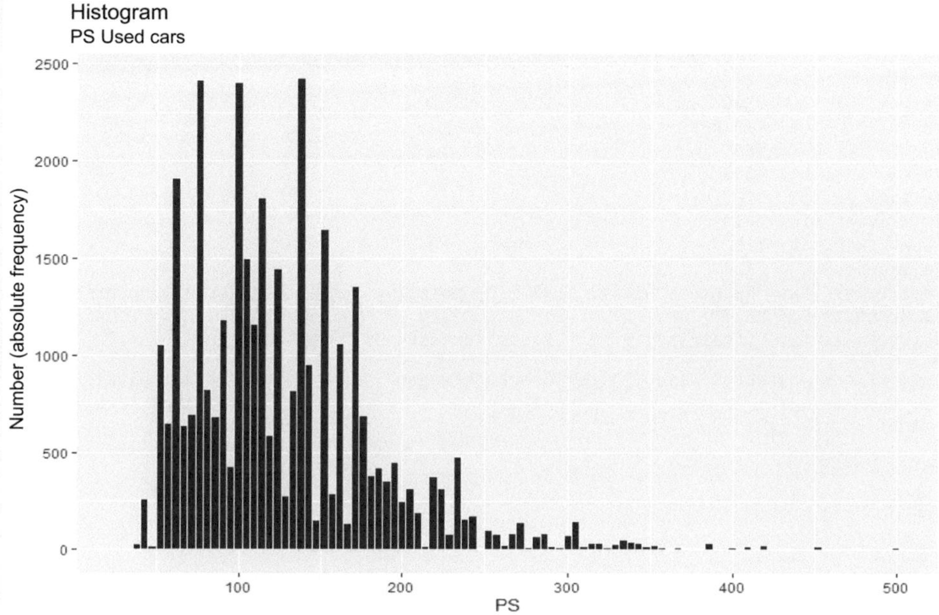

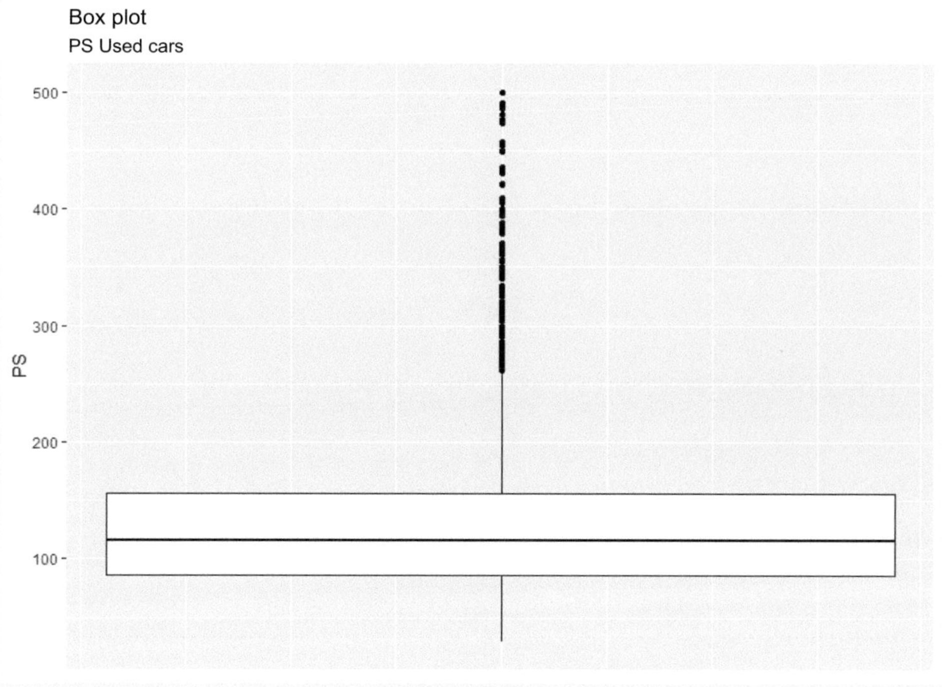

```
sum
##                25%
## anzahl         "34864"
## anzahl.na      "0"
## modus          "75"
## median         "116"
## mean           "128.48"
## min            "29"
## max            "500"
## sd             "59.43"
## q1             "86"
## q3             "156"
## iqr            "70"
## whisker.min    "-19"
## whisker.max    "261"
## skewness       "1.47"
## skewness.txt "Rechte Schiefe: 1.47 > 0, positive Schiefe,linkssteil,rechtsschief"
## kurtosis       "3.7"
## kurtosis.txt "Steilgipflig mit Exzess Kurtosis 3.7 > 0"

sum<-ml.summary(autos$Alter,"Alter Gebrauchtwagen","Jahre",box=T)# Verhältnisvar.
```

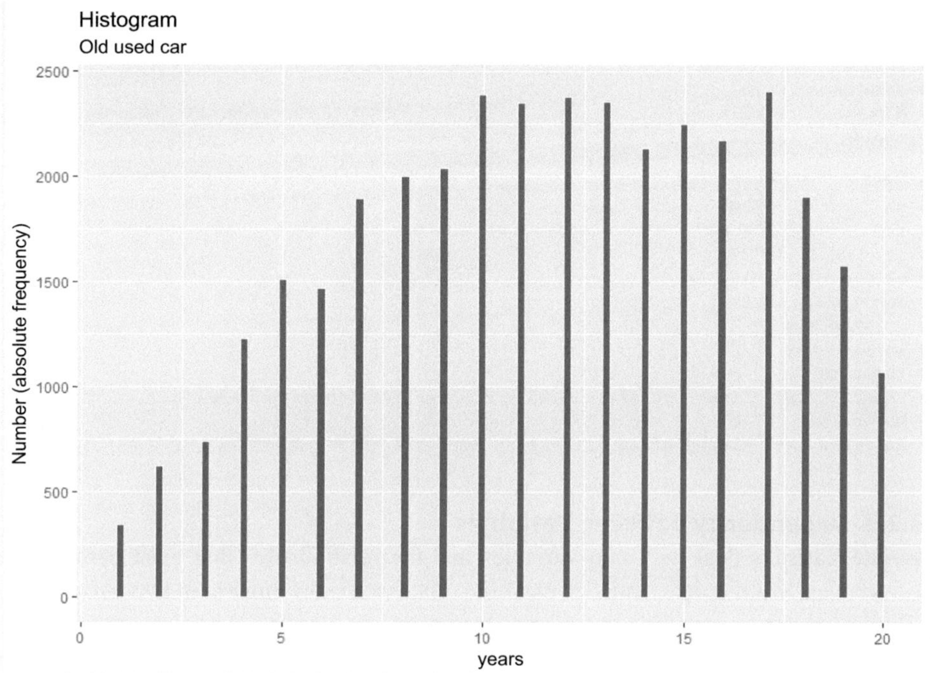
Histogram
Old used car
2500
2000
1500
1000
500
0
Number (absolute frequency)
0
5
10
15
20
years

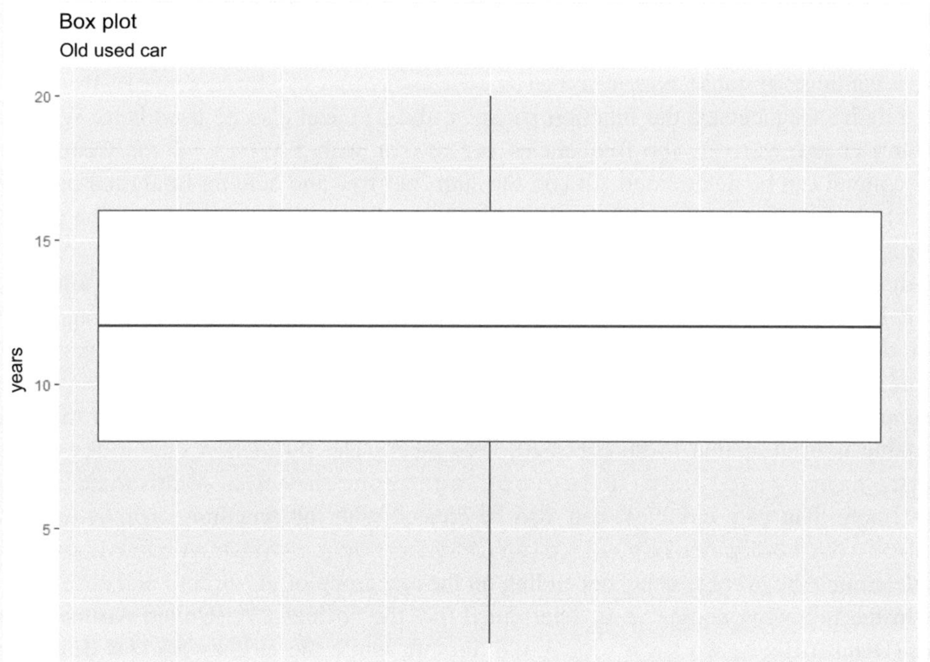
Box plot
Old used car
20
15
10
5
years

```
sum

##                25%
## anzahl         "34864"
## anzahl.na      "0"
## modus          "17"
## median         "12"
## mean           "11.67"
## min            "1"
## max            "20"
## sd             "4.84"
## q1             "8"
## q3             "16"
## iqr            "8"
## whisker.min    "-4"
## whisker.max    "28"
## skewness       "-0.18"
## skewness.txt "Linke Schiefe: -0.18 < 0, negative Schiefe,linksschief,rechtssteil"
## kurtosis       "-0.9"
## kurtosis.txt "Flachgipflig mit Exzess Kurtosis -0.9 < 0"
```

6.3.2.3 Dependency between Variables

Bivariate statistics deal with two variables and the relationships that exist between the two variables. Multivariate statistics examine relationships of more than two variables.

While univariate statistics consider a frequency distribution, bivariate statistics use contingency tables to display the conditional frequencies for each value combination of the variables. It is common to display the independent variable in the columns and the dependent variable in the rows. In R, the function `table()` is suitable for creating a contingency table, e.g., `table(sp1$agegroup,sp1$carrier)`. To display the relative frequencies, the function `prop.table()` can also be used here. With the parameter `margin=1`, the frequencies per row or with `margin=2` the frequencies per column can be determined. To add the sums per row and column (marginal distributions), the function `addmargins()` can be applied to a contingency table. The parameter `margin=1` adds the sums only for rows and with `margin=2` only for columns. To display only the marginal distributions, the function `margin.table()` is suitable.

It is often also interesting to apply a function to groupings, e.g., the average body size depending on gender. The function `tapply()` is suitable for this. The first argument in the function call is the dependent variable, here the delay at departure, the second argument is the grouping variable, here the age group of the aircraft, and the third variable is the function to be executed, here mean. The instruction then looks as follows: `tapply(sp1$dep_delay,sp1$agegroup,mean)`. Multivariate statistics (more than two variables) can also be created with the function `tapply()`, e.g., `tapply(sp1$dep_delay,list(sp1$agegroup, sp1$carrier),mean)`, to determine the average delay depending on the age group of the aircraft and airline.

In the following coding, it is determined that the airlines UA (United Airlines) and DL (Delta Airlines) work with many old aircraft, while the airlines B6 (Jet Blue) and EV (`ExpressJet Airlines`) almost exclusively use new aircraft. It turns out that the delay of the old aircraft with an average of 11 minutes compared to middle-aged and new aircraft with 13 minutes even looks better. The analyses contribute to a better understanding of the data.

Contingency Tables

```
#- Kontingenztabellen-----------------------------------------------------------
carrier.top4<-levels(flug$carrier)[1:4]         # Top4 Fluggesellschaften
sp1<-flug[flug$carrier %in% carrier.top4 &      # Stichprobe 1 (sp1)
          !is.na(flug$dep_delay),]
sp1$carrier<-droplevels(sp1$carrier)            # Levels in Faktorvar. Löschen
abs<-table(sp1$agegroup,sp1$carrier)            # Absolute Häufigkeit
abs

##
##                UA    B6    EV    DL
##   new         11721 42751 10677  1377
##   middle-aged 37269  9751 39159 29871
##   old          7127     0     0 15882

rel<-100*prop.table(abs)                        # Relative Häufigkeit in %
rel

##
##                       UA         B6         EV         DL
##   new          5.7012914 20.7948051  5.1934723  0.6697959
##   middle-aged 18.1282681  4.7430503 19.0475959 14.5297565
##   old          3.4666926  0.0000000  0.0000000  7.7252718

round(100*prop.table(abs,margin=1),2)           # Relative Häufigkeit je Zeile

##
##                UA    B6    EV    DL
##   new         17.62 64.26 16.05  2.07
##   middle-aged 32.11  8.40 33.74 25.74
##   old         30.97  0.00  0.00 69.03

round(100*prop.table(abs,margin=2),2)           # Relative Häufigkeit je Spalte

##
##                UA    B6    EV    DL
##   new         20.89 81.43 21.42  2.92
##   middle-aged 66.41 18.57 78.58 63.38
##   old         12.70  0.00  0.00 33.70

addmargins(abs)                                 # Randverteilungen hinzufügen

##
##                  UA     B6     EV     DL    Sum
##   new         11721  42751  10677   1377  66526
##   middle-aged 37269   9751  39159  29871 116050
##   old          7127      0      0  15882  23009
##   Sum         56117  52502  49836  47130 205585

addmargins(rel)

##
##                        UA          B6          EV          DL         Sum
##   new           5.7012914  20.7948051   5.1934723   0.6697959  32.3593647
##   middle-aged  18.1282681   4.7430503  19.0475959  14.5297565  56.4486709
##   old           3.4666926   0.0000000   0.0000000   7.7252718  11.1919644
##   Sum          27.2962522  25.5378554  24.2410682  22.9248243 100.0000000

addmargins(abs)

##
##                  UA     B6     EV     DL    Sum
##   new         11721  42751  10677   1377  66526
##   middle-aged 37269   9751  39159  29871 116050
##   old          7127      0      0  15882  23009
```

```
##   Sum           56117  52502  49836  47130 205585

addmargins(abs,margin=1)                            # Randv. Hinzufügen für Zeilen

##
##                    UA    B6    EV    DL
##   new           11721 42751 10677  1377
##   middle-aged 37269  9751 39159 29871
##   old            7127     0     0 15882
##   Sum           56117 52502 49836 47130

addmargins(abs,margin=2)                            # Randv. Hinzufügen für Spalten

##
##                     UA     B6     EV     DL    Sum
##   new            11721  42751  10677   1377  66526
##   middle-aged  37269   9751  39159  29871 116050
##   old             7127      0      0  15882  23009

margin.table(abs)                                   # Nur Randverteilung

## [1] 205585

margin.table(abs,1)                                 # Randv. Nur Zeilen

##
##         new middle-aged         old
##       66526      116050       23009

margin.table(abs,2)                                 # Randv. Nur Spalten

##
##    UA    B6    EV    DL
## 56117 52502 49836 47130

tapply(sp1$dep_delay,sp1$agegroup,mean)             # Gruppenauswertung 1 unabh.V.

##         new middle-aged         old
##    13.73328    13.99421    11.16481

tapply(sp1$dep_delay,sp1$agegroup,summary)          # Gruppenauswertung 1 unabh.V.


## $new
##     Min. 1st Qu.  Median    Mean 3rd Qu.    Max.
##   -43.00   -5.00   -1.00   13.73   13.00  548.00
##
## $`middle-aged`
##     Min. 1st Qu.  Median    Mean 3rd Qu.    Max.
##   -32.00   -4.00   -1.00   13.99   13.00  899.00
##
## $old
##     Min. 1st Qu.  Median    Mean 3rd Qu.    Max.
##   -33.00   -5.00   -2.00   11.16    7.00  960.00
tapply(sp1$dep_delay,sp1$name,mean)                 # Gruppenauswertung 1 unabh.V.

##      Delta Air Lines Inc. ExpressJet Airlines Inc.          JetBlue Airways
##                  9.264505                19.955390                13.022522
##    United Air Lines Inc.
##                 12.106073

tapply(sp1$dep_delay,sp1$name,summary)              # Gruppenauswertung 1 unabh.V.

## $`Delta Air Lines Inc.`
```

```
##     Min. 1st Qu.  Median    Mean 3rd Qu.    Max.
## -33.000  -5.000  -2.000   9.264   5.000 960.000
##
## $`ExpressJet Airlines Inc.`
##     Min. 1st Qu.  Median    Mean 3rd Qu.    Max.
##  -32.00   -5.00   -1.00   19.96   25.00  548.00
##
## $`JetBlue Airways`
##     Min. 1st Qu.  Median    Mean 3rd Qu.    Max.
##  -43.00   -5.00   -1.00   13.02   12.00  502.00
##
## $`United Air Lines Inc.`
##     Min. 1st Qu.  Median    Mean 3rd Qu.    Max.
##  -20.00   -4.00    0.00   12.11   11.00  483.00

tapply(sp1$dep_delay,                                  # Gruppenauswertung 2 unabh.V.
       list(sp1$agegroup,sp1$carrier),mean)

##                   UA       B6       EV        DL
## new         12.33862 12.99619 19.24642  5.740741
## middle-aged 12.23473 13.14983 20.12350  8.429982
## old         11.37519       NA       NA 11.070394
```

To further investigate the relationship of variables, the calculation of correlation is also suitable. A correlation indicates the direction (positive or negative) and strength of a linear dependency between two variables (the slope of the line in the figures corresponds to the correlation coefficient). A positive correlation exists when high values of variable A coincide with high values of variable B (see Fig. 6.22). A negative correlation exists when high values of variable A coincide with low values of variable B (see Fig. 6.23). The strength of a correlation is expressed by the correlation coefficient, which lies between -1 (strong negative correlation) and +1 (strong positive correlation). A correlation is stronger when the slope of the correlation function approaches -1 or $+1$.

A value of zero means that there is no correlation and the further a value deviates from zero upwards or downwards, the stronger the correlation (see Fig. 6.24).

A correlation coefficient should not be considered in isolation, as outliers or even the distribution can greatly influence the calculation. It is therefore recommended to always consider the correlation coefficient together with the associated plot, as can be seen from Fig. 6.25. In the plot at the top left, the variables appear to be normally distributed. The plot at the top right shows a relationship between the variables, but not a linear one. The plot at the bottom left shows a perfect linear correlation between the two variables, but the calculation is greatly influenced by the outlier. The plot at the bottom right shows how even a single extreme outlier can result in a correlation coefficient that seems unsuitable. To avoid misinterpretations, the correlation coefficient should therefore always be considered together with the associated plot.

A prerequisite for the significance of a Pearson correlation coefficient of numerical variables is the fulfillment of the following conditions:

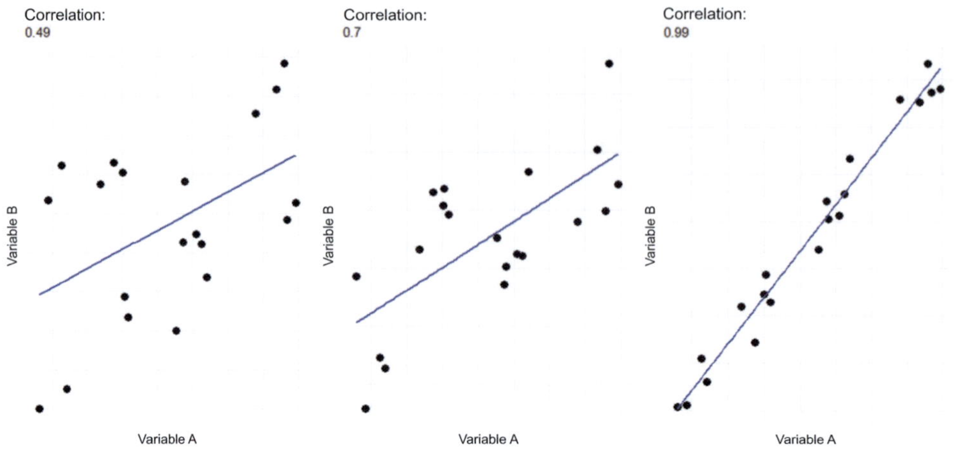

Fig. 6.22 Positive Correlation

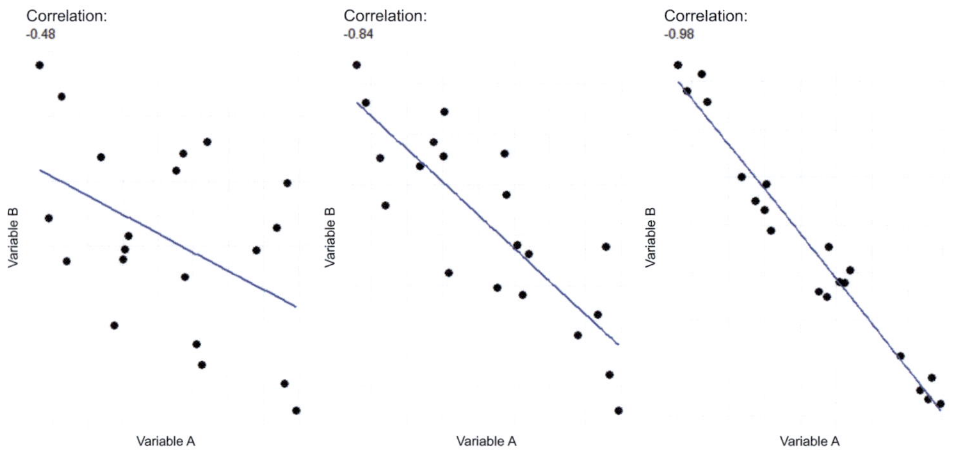

Fig. 6.23 Negative Correlation

- Bivariate normal distribution: The bivariate normal distribution (also two-dimensional normal distribution) describes a normal distribution of one variable for each value of the other variable.
- Homoscedasticity (uniform dispersion of both variables)
- No outliers
- No clusters

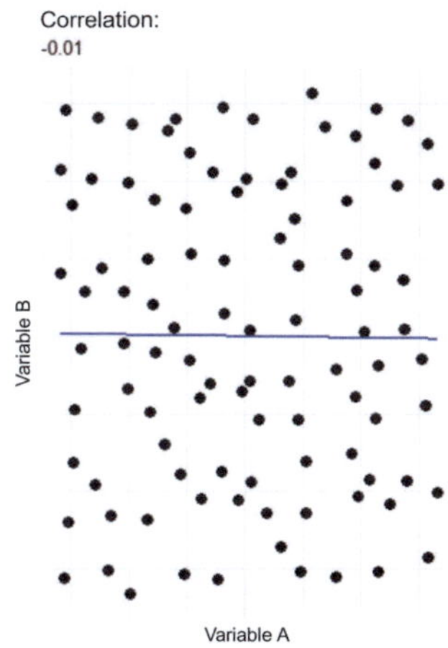

Fig. 6.24 No Correlation

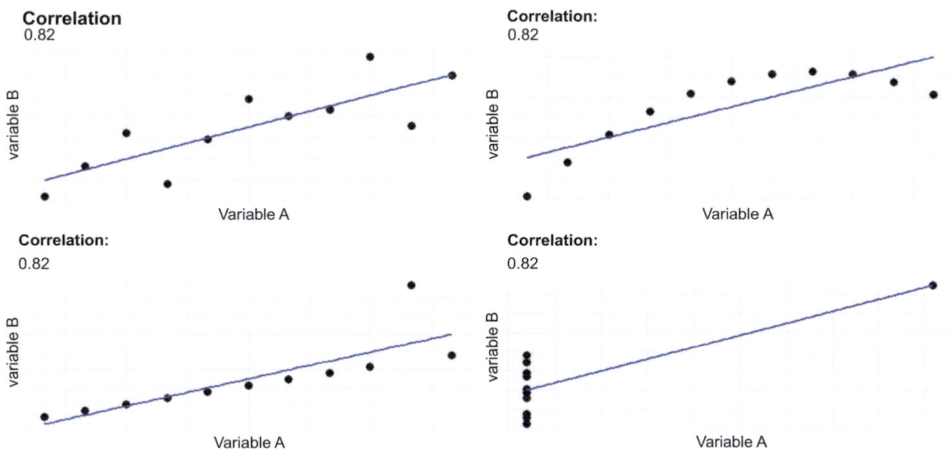

Fig. 6.25 Anscombe Quartet

The Fig. 6.26 shows the calculation of the correlation coefficients based on information from two hundred used cars. A negative correlation between age and price expresses that the price decreases with increasing age. A positive correlation between horsepower and price indicates that used cars with more horsepower are more expensive.

However, even a statistically significant correlation is never evidence of causality, i.e., a cause-and-effect relationship. A statistically significant correlation merely confirms the dependence of the variables. Recognizing such a dependence can be very valuable in gaining a better understanding of the data, e.g., what influence the price has on demand.

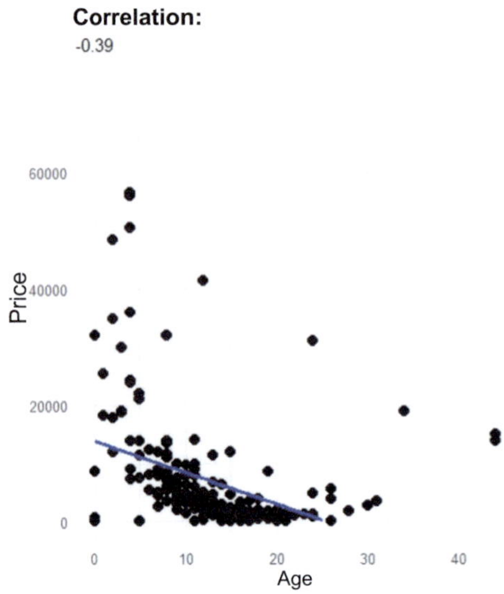

Fig. 6.26 Correlation coefficients

In R, the Pearson correlation coefficient r for interval variables can be calculated using the function `cor()`. However, a more clever calculation and output is achieved by the function `correlate()` from the corrr package. For ordinal-scaled variables, either the Spearman rank correlation coefficient r_s (Spearman's Rho) or the Kendall rank correlation coefficient T (Kendall's Tau) is calculated. The correlations according to Spearman and Kendall can also be calculated with the function `cor()` or `correlate()` if the parameter `method="spearman"` or `method="kendall"` is used.

The function `correlate()` can also be applied to a large number of variables that are presented in the form of a matrix to create a correlation matrix that determines the correlation coefficient between all variables. In addition to the function `correlate()`, the results of the correlation matrix can be arranged with the function `rearrange()` so that similarly correlated variables are shown in clusters on. Since a correlation matrix shows all coefficients in mirror image twice, the function `shave()` is suitable for displaying the mirrored, redundant values with NA. If the function `fashion()` is then applied to the result of the function `shave()`, the NAs are completely removed. The graphical output of the correlation matrix can be done using the function `rplot()`.

Another way to analyze the relationship between variables is a plot matrix, which in addition to the correlation coefficient for at least one numerical variable, shows two other plots depending on the variable type, e.g., a scatter plot and a distribution function. This is supported by the function `ggpairs()` from the Ggally package. The creation of a correlation matrix and a correlation plot is presented in detail in Sect. 6.3.3.2.7.

Earlier, the function `cor()` was introduced. If you want to determine not only the correlation but also its significance, you can use the function `cor.test()`, which calculates both the correlation and its significance. To get a first impression of the dependency of two variables, a plot of the variables under consideration is often helpful. In Fig. 6.26, the variables age and price and PS and price are shown. If you calculate the correlation coefficient according to Pearson, the result is -0.39 as a value and a significance of $p = 1.17\text{e-}08$. Suppose the hypothesis (alternative hypothesis) is that a correlation of -0.39 exists and the corresponding null hypothesis is that there is no relationship between the variables. Since the p-value, which indicates the degree of error, is less than the usual rejection limit of 0.05 (5%), the null hypothesis is rejected. Consequently, the alternative hypothesis of a correlation between age and price is accepted. Also, the null hypothesis that there is no relationship between PS and price can be rejected, as the p-value is again significantly less than 5%. Thus, with at least 95% probability, the statement of the alternative hypothesis is correct, that a correlation between the two variables exists.

The effect size (correlation factor) of a correlation is considered weakly relevant from 0.1, moderately strong from 0.3, and strong from 0.5.

Descriptive Statistics: Correlations

```
#- Korrelation 2 Variablen----------------------------------------------------
library(corrr)                                                # Paket erforderlich
cor(autos[,c(„Kilometer“,“Preis“)],use=“na.or.complete“)      # Pearson

##               Kilometer      Preis
## Kilometer  1.0000000 -0.4827629
## Preis     -0.4827629  1.0000000

correlate(autos[,c(“Kilometer”,”Preis”)])                     # Pearson

##
## Correlation method: ‘pearson’
## Missing treated using: ‘pairwise.complete.obs’

## # A tibble: 2 × 3
##   term       Kilometer  Preis
##   <chr>          <dbl>  <dbl>
## 1 Kilometer     NA      -0.483
## 2 Preis         -0.483 NA

correlate(autos[,c(“PS”,”Kategorie”)],method=”spearman”)     # Spearman: Fehler!

## Error in stats::cor(x = x, y = y, use = use, method = method): ‘x’ must be numeri
c

autos$Kategorie.n <- as.numeric(autos$Kategorie)              # Faktor->Numerisch
correlate(autos[,c(„PS“,“Kategorie.n“)],method=“spearman“) # Spearman: Erfolgreich

##
## Correlation method: ‘spearman’
## Missing treated using: ‘pairwise.complete.obs’

## # A tibble: 2 × 3
##   term              PS Kategorie.n
##   <chr>          <dbl>       <dbl>
## 1 PS            NA          -0.199
## 2 Kategorie.n -0.199        NA

#- Korrelation n Variablen----------------------------------------------------
library(MASS)                                         # Paket erforderlich

data(mtcars)                                          # Daten importieren aus MASS
detach(“package:MASS”, unload = TRUE)                 # Paket nicht mehr erforderlich

mtcars$l100km <-round(235.215/mtcars$mpg,1)           # Verbrauch mpg -> l/100km
mtcars$disp <- round(16.387*mtcars$disp,0)            # Hubraum Kubik-Inch -> ccm
mtcars$wt <- round(0.453592*mtcars$wt*1000,0)         # Gewicht Pfund -> kg
mtcars<-dplyr::select(mtcars,-mpg,-drat,-vs,-carb) # Spalten entfernen
mtcars[14:21,]                                        # Anzeige

##                     cyl disp  hp   wt  qsec am gear l100km
## Merc 450SLC           8 4520 180 1715 18.00  0    3   15.5
## Cadillac Fleetwood    8 7735 205 2381 17.98  0    3   22.6
## Lincoln Continental   8 7538 215 2460 17.82  0    3   22.6
## Chrysler Imperial     8 7210 230 2424 17.42  0    3   16.0
## Fiat 128              4 1290  66  998 19.47  1    4    7.3
## Honda Civic           4 1240  52  733 18.52  1    4    7.7
## Toyota Corolla        4 1165  65  832 19.90  1    4    6.9
## Toyota Corona         4 1968  97 1118 20.01  0    3   10.9

cor(mtcars)                                           # Korrelationsmatrix cor()
```

```
##              cyl       disp         hp         wt       qsec         am
## cyl     1.0000000  0.9020411  0.8324475  0.7825300 -0.5912421 -0.5226070
## disp    0.9020411  1.0000000  0.7909211  0.8880114 -0.4336736 -0.5912239
## hp      0.8324475  0.7909211  1.0000000  0.6587227 -0.7082234 -0.2432043
## wt      0.7825300  0.8880114  0.6587227  1.0000000 -0.1746919 -0.6925583
## qsec   -0.5912421 -0.4336736 -0.7082234 -0.1746919  1.0000000 -0.2298609
## am     -0.5226070 -0.5912239 -0.2432043 -0.6925583 -0.2298609  1.0000000
## gear   -0.4926866 -0.5556111 -0.1257043 -0.5833703 -0.2126822  0.7940588
## l100km  0.8156956  0.8809961  0.7636462  0.8906788 -0.3863496 -0.5406040
##              gear     l100km
## cyl    -0.4926866  0.8156956
## disp   -0.5556111  0.8809961
## hp     -0.1257043  0.7636462
## wt     -0.5833703  0.8906788
## qsec   -0.2126822 -0.3863496
## am      0.7940588 -0.5406040
## gear    1.0000000 -0.4803364
## l100km -0.4803364  1.0000000
```

```
correlate(mtcars)                                 # Korrelationsmatrix correlate()
```

```
##
## Correlation method: 'pearson'
## Missing treated using: 'pairwise.complete.obs'
```

```
## # A tibble: 8 × 9
##   term      cyl   disp     hp     wt   qsec     am   gear l100km
##   <chr>   <dbl>  <dbl>  <dbl>  <dbl>  <dbl>  <dbl>  <dbl>  <dbl>
## 1 cyl    NA      0.902  0.832  0.783 -0.591 -0.523 -0.493  0.816
## 2 disp    0.902 NA      0.791  0.888 -0.434 -0.591 -0.556  0.881
## 3 hp      0.832  0.791 NA      0.659 -0.708 -0.243 -0.126  0.764
## 4 wt      0.783  0.888  0.659 NA     -0.175 -0.693 -0.583  0.891
## 5 qsec   -0.591 -0.434 -0.708 -0.175 NA     -0.230 -0.213 -0.386
## 6 am     -0.523 -0.591 -0.243 -0.693 -0.230 NA      0.794 -0.541
## 7 gear   -0.493 -0.556 -0.126 -0.583 -0.213  0.794 NA     -0.480
## 8 l100km  0.816  0.881  0.764  0.891 -0.386 -0.541 -0.480 NA
```

```
rearrange(correlate(mtcars))                      # Clustern ähnlicher Werte
```

```
##
## Correlation method: 'pearson'
## Missing treated using: 'pairwise.complete.obs'
```

```
## # A tibble: 8 × 9
##   term     disp    cyl     wt l100km     hp   qsec   gear     am
##   <chr>   <dbl>  <dbl>  <dbl>  <dbl>  <dbl>  <dbl>  <dbl>  <dbl>
## 1 disp   NA      0.902  0.888  0.881  0.791 -0.434 -0.556 -0.591
## 2 cyl     0.902 NA      0.783  0.816  0.832 -0.591 -0.493 -0.523
## 3 wt      0.888  0.783 NA      0.891  0.659 -0.175 -0.583 -0.693
## 4 l100km  0.881  0.816  0.891 NA      0.764 -0.386 -0.480 -0.541
## 5 hp      0.791  0.832  0.659  0.764 NA     -0.708 -0.126 -0.243
## 6 qsec   -0.434 -0.591 -0.175 -0.386 -0.708 NA     -0.213 -0.230
## 7 gear   -0.556 -0.493 -0.583 -0.480 -0.126 -0.213 NA      0.794
## 8 am     -0.591 -0.523 -0.693 -0.541 -0.243 -0.230  0.794 NA
```

```
r<-shave(rearrange(correlate(mtcars)))            # Auf Dreieck reduzieren
```

```
##
## Correlation method: 'pearson'
## Missing treated using: 'pairwise.complete.obs'
```

```
r                                                 # Ausgabe
```

```
## # A tibble: 8 × 9
##   term      disp    cyl     wt l100km     hp   qsec   gear     am
##   <chr>    <dbl>  <dbl>  <dbl>  <dbl>  <dbl>  <dbl>  <dbl>  <dbl>
## 1 disp    NA     NA     NA     NA     NA     NA     NA        NA
## 2 cyl      0.902 NA     NA     NA     NA     NA     NA        NA
## 3 wt       0.888  0.783 NA     NA     NA     NA     NA        NA
## 4 l100km   0.881  0.816  0.891 NA     NA     NA     NA        NA
## 5 hp       0.791  0.832  0.659  0.764 NA     NA     NA        NA
## 6 qsec    -0.434 -0.591 -0.175 -0.386 -0.708 NA     NA        NA
## 7 gear    -0.556 -0.493 -0.583 -0.480 -0.126 -0.213 NA        NA
## 8 am      -0.591 -0.523 -0.693 -0.541 -0.243 -0.230  0.794    NA
fashion(shave(rearrange(correlate(mtcars))))       # Anzeige 2 Nachkommastellen
##
## Correlation method: 'pearson'
## Missing treated using: 'pairwise.complete.obs'

##     term disp  cyl   wt l100km   hp qsec gear am
## 1   disp
## 2    cyl  .90
## 3     wt  .89  .78
## 4 l100km  .88  .82  .89
## 5     hp  .79  .83  .66    .76
## 6   qsec -.43 -.59 -.17   -.39 -.71
## 7   gear -.56 -.49 -.58   -.48 -.13 -.21
## 8     am -.59 -.52 -.69   -.54 -.24 -.23  .79
rplot(r,print_cor = TRUE)                          # Ausgabe als Grafik
```

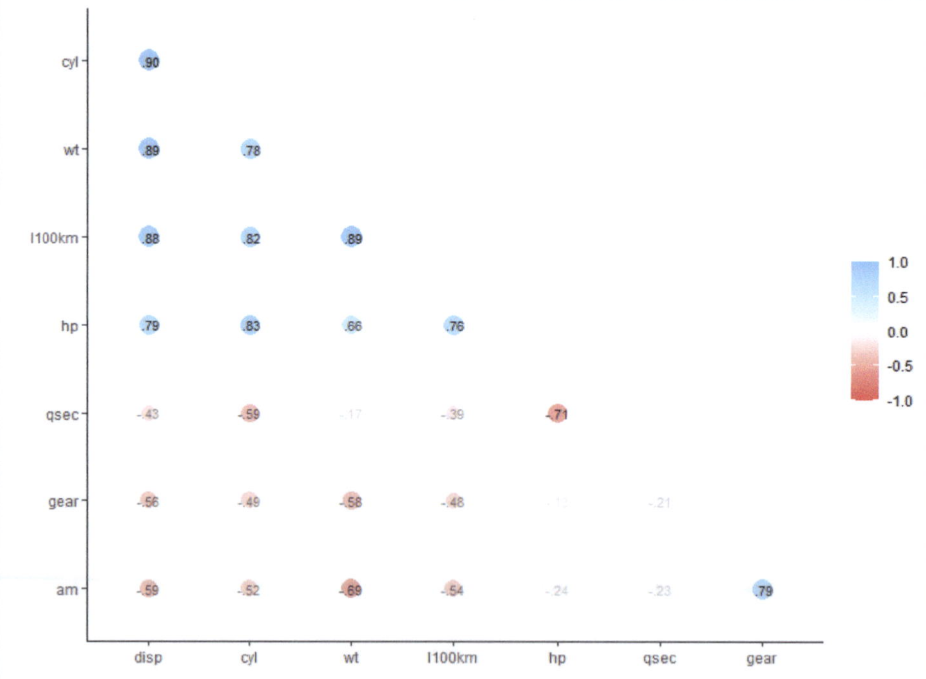


```
#- Korrelation und Plots n Variablen-------------------------------------------
library(Ggally)                                    # Paket erforderlich
ggpairs(autos,columns=c("Alter","PS","Preis"))     # Cor für numerische Variablen
```

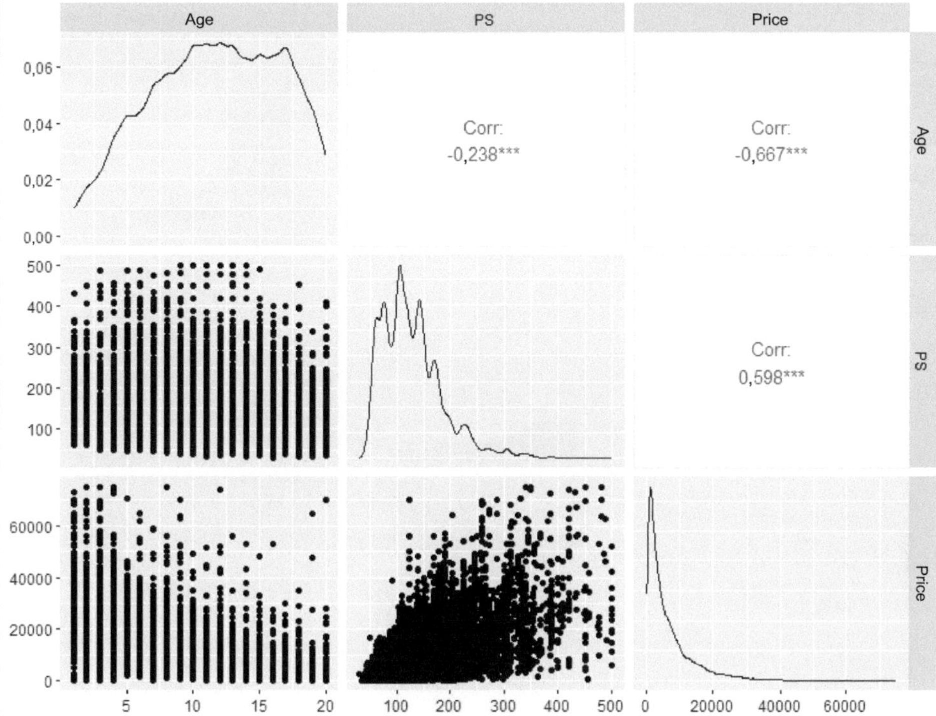


```
autos <- autos[1:50,]                                    # Reduktion auf 50 Datensätze
ggpairs(autos, columns = c(„Preis“,“Kilometer“,“Kategorie“), aes(fill=Kategorie))
```

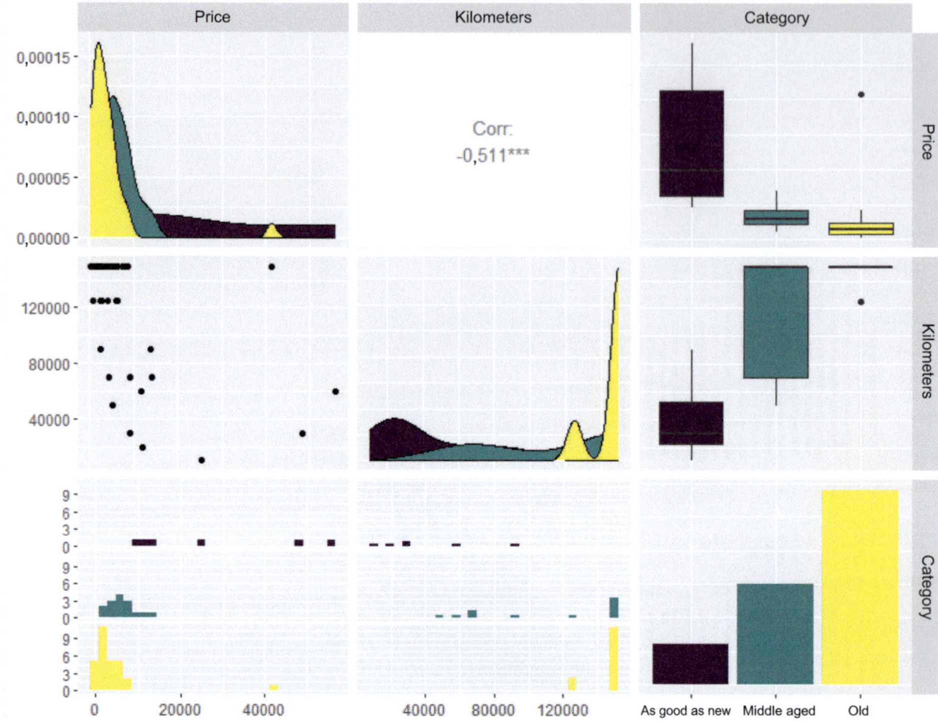


```
#- Korrelation mit Signifikanztest---------------------------------------------
plot(autos$Alter,autos$Preis)                        # Plot
```

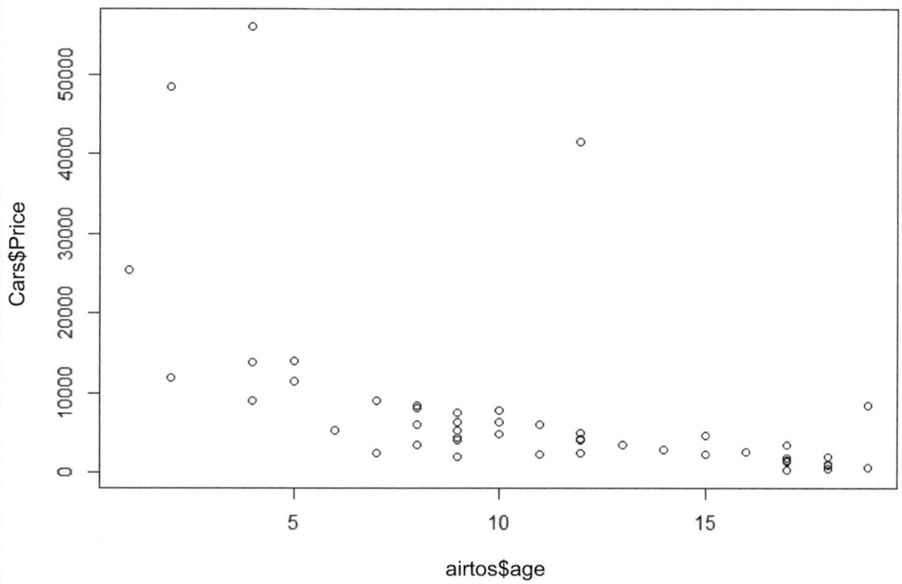


```
cor(autos$Alter,autos$Preis)                          # Pearson

## [1] -0.5330323

cor.test(autos$Alter,autos$Preis)                     # Pearson mit Signifikanz

##
##  Pearson's product-moment correlation
##
## data:  autos$Alter and autos$Preis
## t = -4.3647, df = 48, p-value = 6.741e-05
## alternative hypothesis: true correlation is not equal to 0
## 95 percent confidence interval:
##  -0.7065504 -0.2990549
## sample estimates:
##        cor
## -0.5330323

cor.test(autos$PS,autos$Preis)                        # Pearson mit Signifikanz

##
##  Pearson's product-moment correlation
##
## data:  autos$PS and autos$Preis
## t = 8.5795, df = 48, p-value = 2.974e-11
## alternative hypothesis: true correlation is not equal to 0
## 95 percent confidence interval:
##  0.6377652 0.8683121
## sample estimates:
##       cor
## 0.7780017
```

6.3.3 Data Visualization

Visualization greatly facilitates the understanding of data. The function `ggplot()` from the package `ggplot2` (gg stands for grammar of graphics) is a powerful function for creating graphics in R. The package `ggplot2` is part of the package `tidyverse` and therefore does not need to be installed and activated separately, provided the package `tidyverse` is already installed and activated. The figures are output in the designated graphics window, unless otherwise specified. Alternatively, the created figures can also be saved as a graphic file using the function `ggsave()`. The size of the generated file can be specified using the parameters `units` (cm, in, mm) in conjunction with the specification `height=` and `width`.

The function `ggplot()` is passed the data required for the figure as parameters. In addition, the so-called aesthetic mappings such as the assignment of the x- and y-axis or the color are specified with the function `aes()`. The form of the figure is defined by the function `geom_*()`, e.g., `geom_line()` for a line or `geom_point()` for a scatterplot. Each function call of the functions `ggplot()`, `aes()` and `geom_*` creates its own so-called layer, a figure level. Several figure levels can then be displayed on top of each other in a figure. To add another layer to an already existing layer, the + is used, e.g., `ggplot(data=mpg) + geom_point(mapping=aes(x=displacement,y=consumption))`. The basic structure of the ggplot function follows the syntax `ggplot(data=<DATA>) + <GEOM_FUNCTION>(mapping=aes(<MAPPINGS>))`, where DATA stands for a dataframe, GEOM_FUNCTION for a type of figure, and MAPPINGS for the assignment of the variables.

Instead of displaying the output of ggplot() directly on the screen, it is also possible to store the output in a variable, so `p1<-ggplot(data=mpg) + geom_point(mapping=aes(x=displacement,y=consumption)` causes the output to be stored in the variable p1. With the command `p1`, the output can then be displayed on the screen at any time.

The function `grid.arrange()` allows more than one figure to be displayed on a page. The parameters ncol and nrow allow to specify how many columns and rows the page should be divided into. In each segment, a previously stored figure can then be output, e.g. `grid.arrange(p1,p2,ncol=2)`, to output the figures p1 and p2 created with ggplot side by side.

Visualization with ggplot()

```
#- Dataframe mpg und mpg.teilmenge erstellen--------------------------------------
mpg<-ggplot2::mpg                                 # Daten aus Paket ggplot2 laden
mpg<-rename(mpg,"Hubraum"="displ","Klasse"="class")# Spalten umbenennen
mpg$Verbrauch <-round(235.215/mpg$cty,1)          # Verbrauch mpg -> l/100km
mpg$Getriebe<-                                    # Zusätzliche Variable
  case_when(str_detect(mpg$trans,"auto") ~ „automatik",
            str_detect(mpg$trans,"man")  ~ „manuell")
mpg.teilmenge <- filter(mpg,Klasse %in%  c(„subcompact","2seater","pickup"))
#- ggplot-------------------------------------------------------------------------
ggplot(data=mpg) +                                # Ausgabe von Scatter-Plot
  geom_point(mapping=aes(x=Verbrauch,y=Hubraum))
```

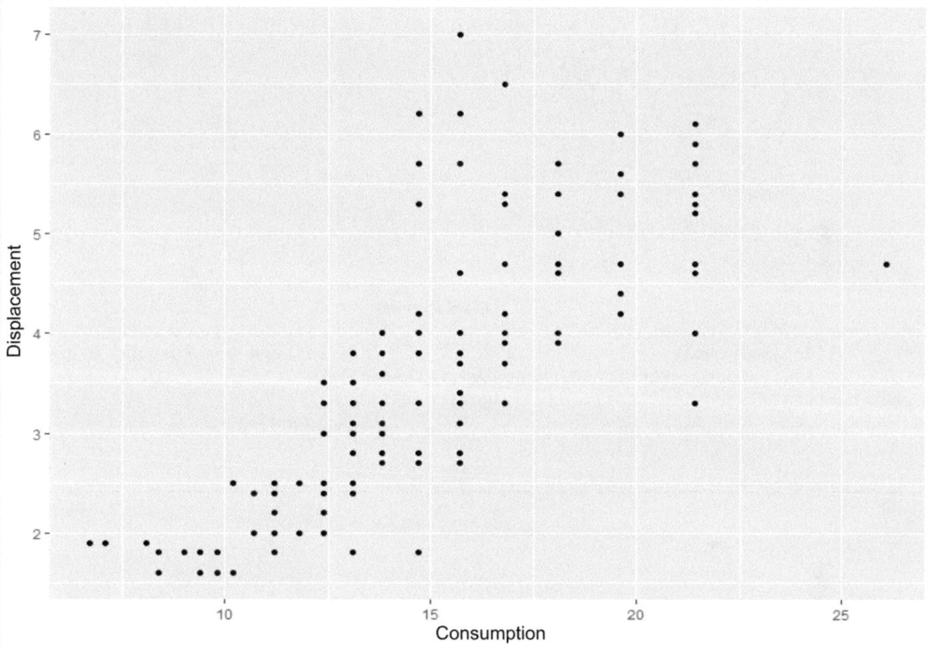


```
p1<-ggplot(data=mpg)+                             # Speichern von Ausgabe in p1
  geom_point(mapping=aes(x=Hubraum,y=Verbrauch))
p1                                                # Ausgabe p1
```

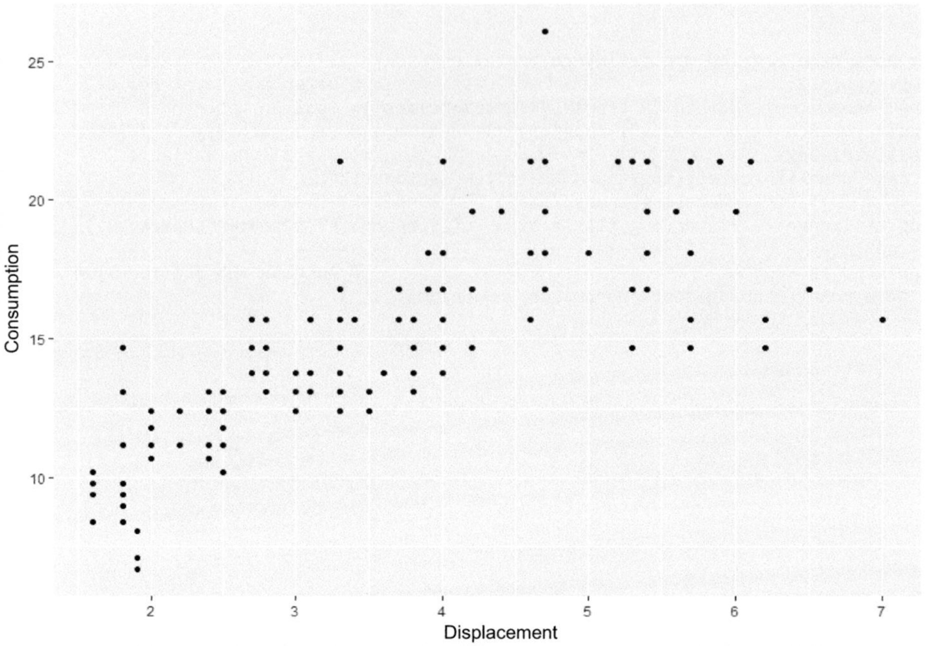


```
p2<-ggplot(data=alkohol,                           # Speichern von Ausgabe in p2
          aes(x=Land,y=Portionen,group=Typ,fill=Typ))+
  geom_bar(stat="identity",position="dodge")
p2                                                 # Ausgabe p2
```

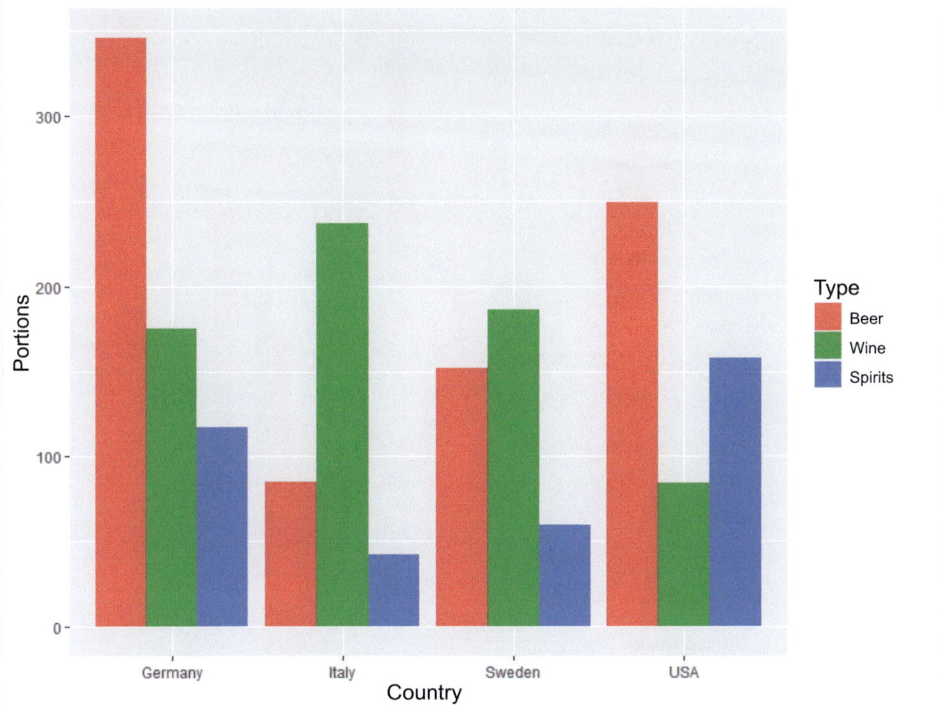


```
grid.arrange(p1,p2,nrow=1,ncol=2)                  # Ausgabe p1+p2 in 2 Spalten
```

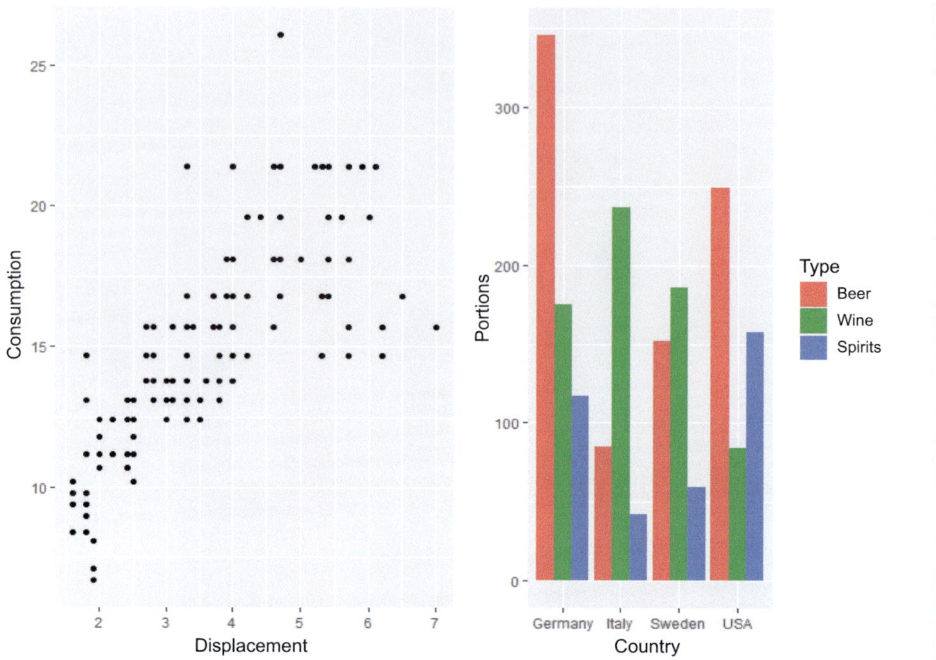

6.3.3.1 Optical Design

In Sect. 3.2 best practices for visualization were introduced. In visualization, it is important, among other things, to use a uniformity of form, font, and color to improve the readability of visualizations. In R, this is supported by the specification of labels, scales, themes, and color palettes.

6.3.3.1.1 Labeling and Scaling

For all figures, it is valuable to use appropriate titles, subtitles, axis labels, source information, and possibly figure numbers.

Titles, subtitles, axis labels, source information, and figure numbers can be set using the function `labs()` with the parameters `title`, `subtitle`, `x`, `y`, `caption` and `tag`. The caption is displayed below the figure on the right and is suitable for source information. The tag is useful, among other things, for specifying a figure number. If you want to remove one of the previously made labels, this can be achieved by assigning the value NULL to the respective parameter, e.g., `subtitle=NULL` to hide the subtitle. To change the heading, the elements of the legend, and their colors, the function `scale_fill_manual()` can be used. Its parameter `name` allows you to determine the legend title, `labels` for the elements, and `values` for the colors. Alternatively, the function `guides()` can also be used.

The scaling of the axes can be set with the functions `xlim()` and `ylim()`, e.g., `xlim(0,10)+ylim(0,30)`. To avoid under- or over-emphasis in visualizations, both axes should generally start at 0 for ratio variables. The axis function `axis()` allows further options for formatting the axes. If an axis label should exceptionally not be displayed, the functions `xlab(NULL)` and `ylab(NULL)` can be used. If both the axis scaling and the axis label should be removed, the function `guides(x="none",y="none")` is suitable. The scaling of the axes can also be modified with the functions `scale_x_continuous()` and `scale_y_continuous()` using the parameters name, breaks, labels, limits, and trans, e.g., a logarithmic axis transformation with `scale_y_continuous(trans="log10")`.

Visual Design with Labels and Scales

```
ggplot(data=mpg) +
  geom_point(mapping=aes(x=Hubraum, y=Verbrauch))+
  labs(title="Scatter-Plot",                          # Titel,Untertitel
       subtitle="Spritverbrauch",
       x="Hubraum in Litern",                         # x-Achse Verhältnis-Variable
       y="Verbrauch in Litern je 100 km",             # y-Achse Verhältnis-Variable
       caption="Quelle:(Heesen,2023)",tag="Abb-1") # Quelle, Tag
```

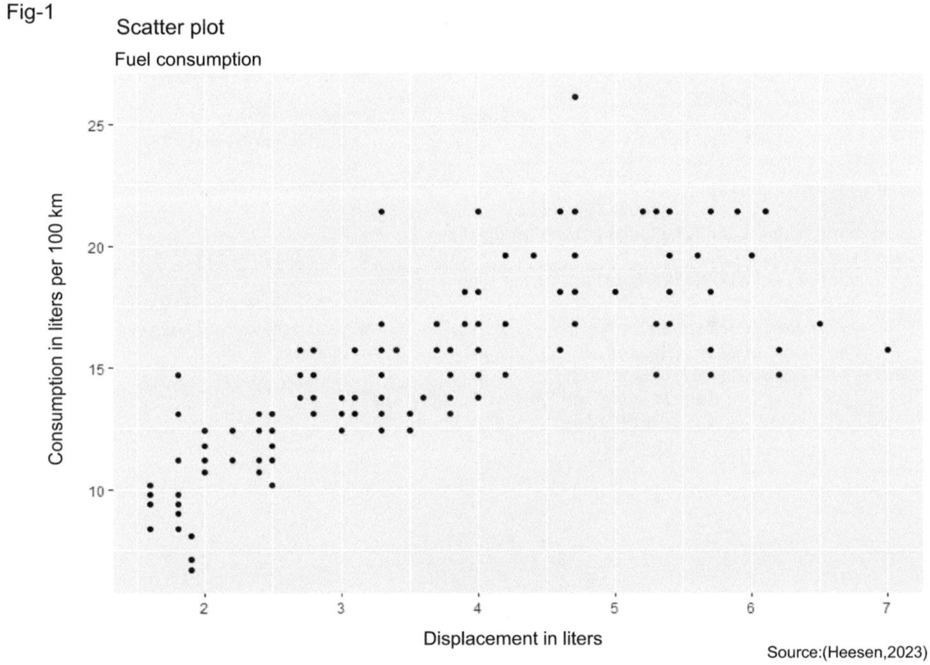


```
ggplot(data=alkohol,aes(x=Land,y=Portionen,
                        group=Typ,fill=Typ))+
  geom_bar(stat="identity",position="dodge")+
  labs(title="Bar-Chart",                             # Titel,Untertitel
       subtitle="Alkoholkonsum",
       x="Land",                                      # x-Achse Nominal-Variable
       y="Anzahl Portionen",                          # y-Achse Verhältnis-Variable
       caption="Quelle:(Heesen,2023)",tag="Abb-2")+# Quelle, Tag
  guides(fill=guide_legend(title="Getränk"))
```

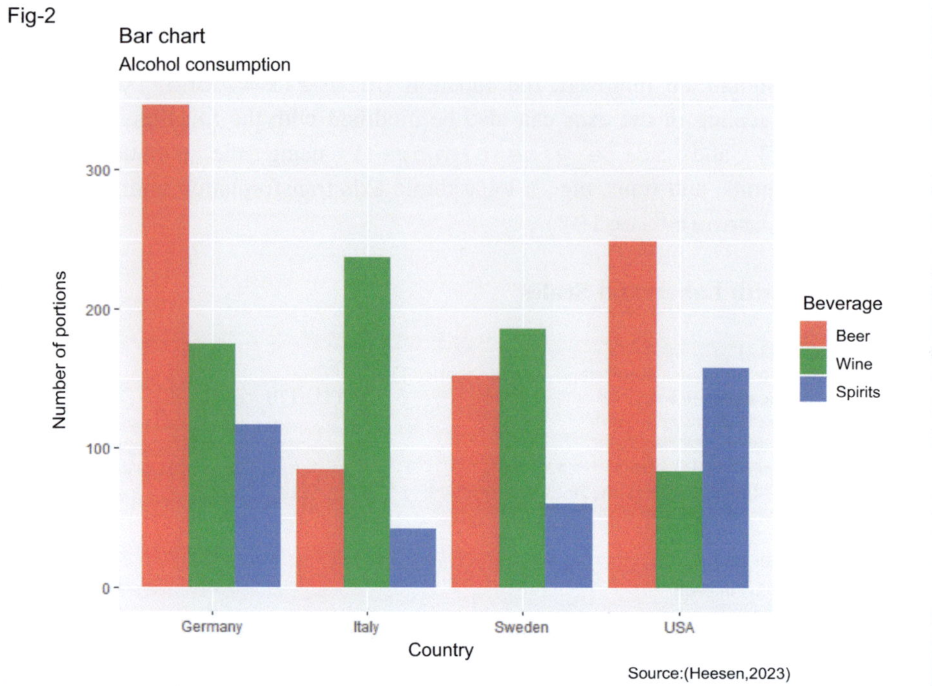


```
ggplot(data=alkohol,aes(x=Land,y=Portionen,
                      group=Typ,fill=Typ))+
  geom_bar(stat="identity",position="dodge")+
  labs(title="Bar-Chart",                          # Titel,Untertitel
       subtitle="Alkoholkonsum",
       x="Land",                                   # x-Achse Nominal-Variable
       y="Anzahl Portionen",                       # y-Achse Verhältnis-Variable
       caption="Quelle:(Heesen,2023)",tag="Abb-3")+# Quelle, Tag
  scale_fill_manual(name="Getränk:",
                    labels=c("Bier","Wein","Schnaps"),
                    values=c("blue","red","green"))
```

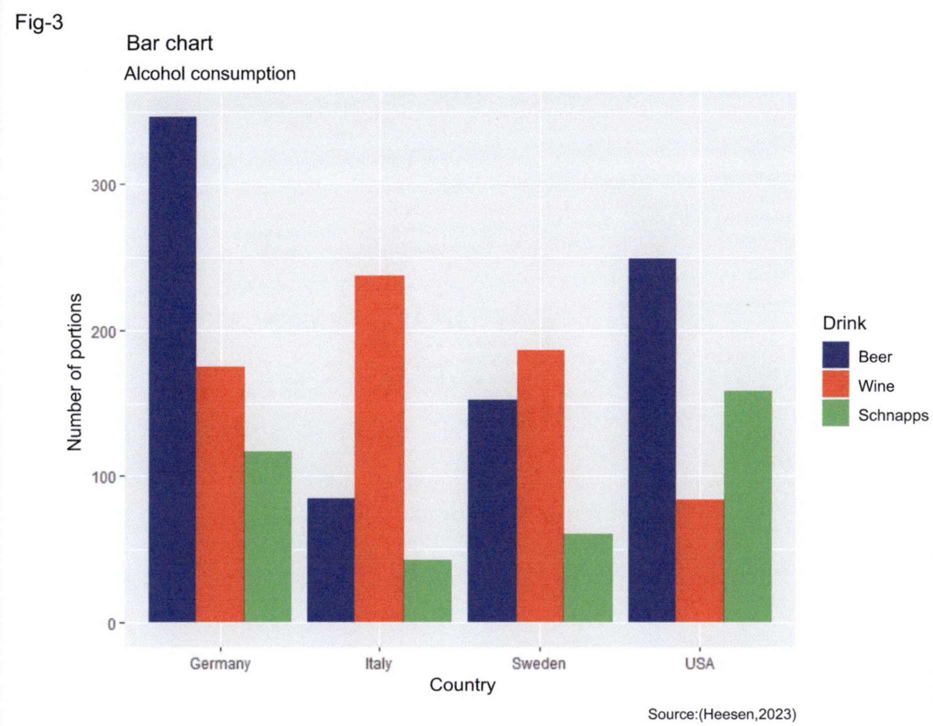


```
#- Skalen-------------------------------------------------------------------------
# Skala festlegen
ggplot(data=mpg) +
  geom_point(mapping=aes(x=Hubraum, y=Verbrauch))+
  labs(title="Scatter-Plot",                           # Titel,Untertitel
       subtitle="Spritverbrauch",
       x="Hubraum in Litern",                          # x-Achse Verhältnis-Variable
       y="Verbrauch in Litern je 100 km",              # y-Achse Verhältnis-Variable
       tag="Abb-1")+                                   # Tag
   xlim(0,10)+ylim(0,30)                               # x von 0-10, y von 0-30
```

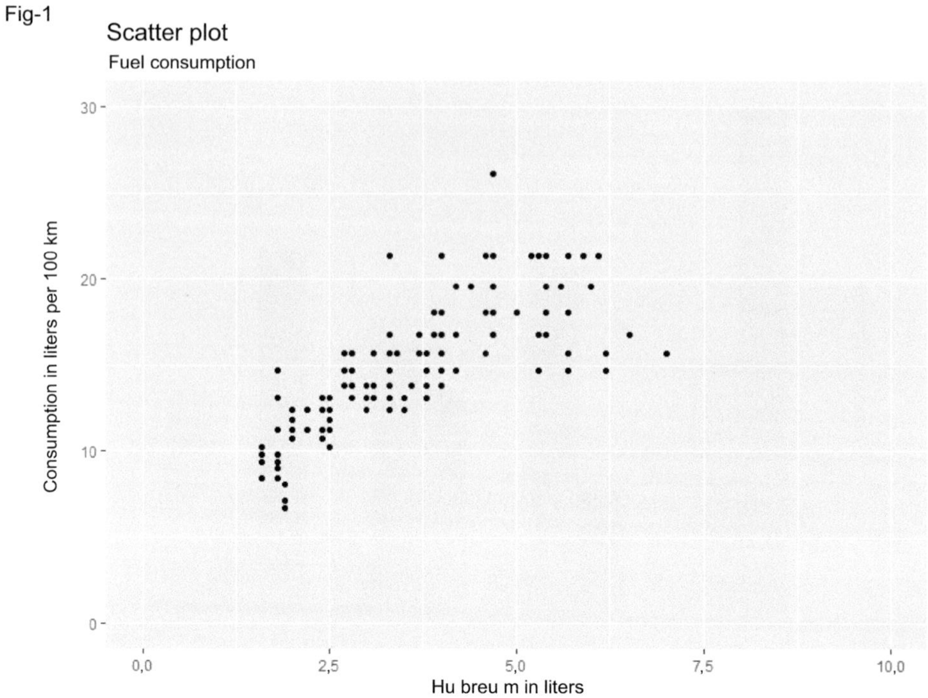


```
# Skalenbeschriftung ausblenden
p1<-ggplot(data=mpg) +
  geom_point(mapping=aes(x=Hubraum, y=Verbrauch))+
  labs(title="Scatter-Plot",          # Titel,Untertitel
       subtitle="Ohne X-Beschriftung",
       x="Hubraum in Litern",                       # x-Achse Verhältnis-Variable
       y="Verbrauch in Litern je 100 km",           # y-Achse Verhältnis-Variable
       tag="Abb-1")+                                # Tag
   xlab(NULL)                                       # x-Achsenbeschriftung ausbl.
# Skala ausblenden
p2<-ggplot(data=mpg) +
  geom_point(mapping=aes(x=Hubraum, y=Verbrauch))+
  labs(title="Scatter-Plot",                        # Titel,Untertitel
       subtitle="Ohne X-Skala",
       x="Hubraum in Litern",                       # x-Achse Verhältnis-Variable
       y="Verbrauch in Litern je 100 km",           # y-Achse Verhältnis-Variable
       tag="Abb-2")+                                # Tag
   guides(x="none")                                 # x-Achse komplett ausblenden
# Skala + Skalenbeschriftung ausblenden
p3<-ggplot(data=mpg) +
  geom_point(mapping=aes(x=Hubraum, y=Verbrauch))+
  labs(title="Scatter-Plot",                        # Titel,Untertitel
       subtitle="Ohne X-Details",
       x="Hubraum in Litern",                       # x-Achse Verhältnis-Variable
       y="Verbrauch in Litern je 100 km",           # y-Achse Verhältnis-Variable
       tag="Abb-3")+                                # Tag
   xlab(NULL)+guides(x="none")                      # x-Achse komplett ausblenden
grid.arrange(p1,p2,p3,ncol=3)                       # Ausgabe p1+p2+p3 in 3 Spalten
```

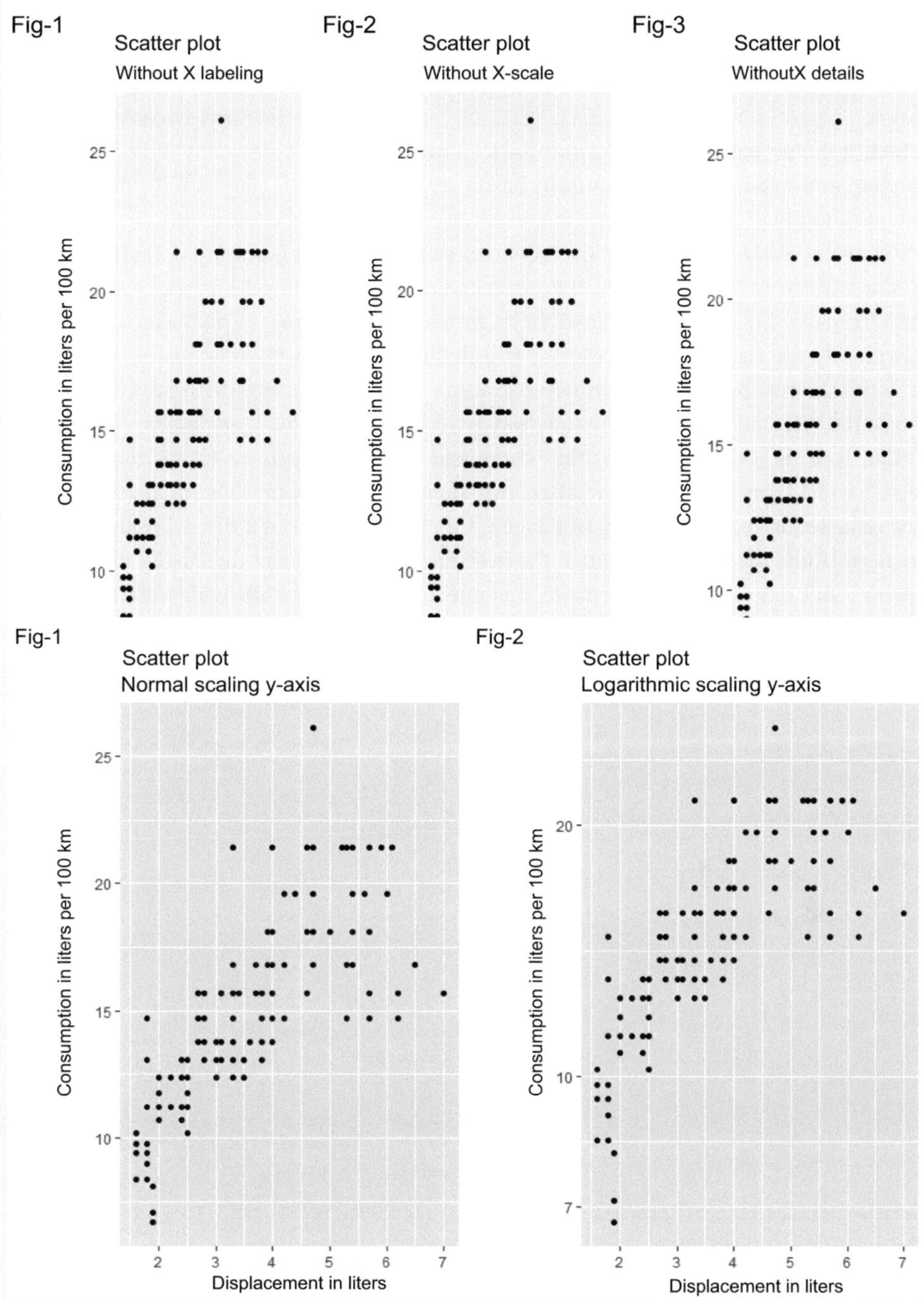
Fig-1
Scatter plot
Without X labeling
Consumption in liters per 100 km
25
20
15
10
Fig-2
Scatter plot
Without X-scale
Consumption in liters per 100 km
25
20
15
10
Fig-3
Scatter plot
WithoutX details
Consumption in liters per 100 km
25
20
15
10
Fig-1
Scatter plot
Normal scaling y-axis
Consumption in liters per 100 km
25
20
15
10
2 3 4 5 6 7
Displacement in liters
Fig-2
Scatter plot
Logarithmic scaling y-axis
Consumption in liters per 100 km
20
10
7
2 3 4 5 6 7
Displacement in liters

6.3.3.1.2 Topics

Further design options for visualizations can be selected via the `theme_*()` functions, e.g. `theme_classic()` for a white background without grid, `theme_grey()`, `theme_dark()` and `theme_minimal()` with grid and different background color. In fact, a theme includes many more basic settings such as colors, text alignment, font, margins, and much more. The current basic settings can be displayed with the `theme_get()` function.

Regarding fonts, you can choose between sans, serif, and mono by default, by applying the `theme()` function with the `text` parameter, e.g. `theme(text=element_text(size=20,family="serif",face="bold"))`. This can be added individually for each figure. However, if a default font is to be defined for the entire R session, this can be achieved with the `theme_set()` function to load a specific set or via the `theme_update()` function to change individual attributes. The change of the font is done e.g. via the command `theme_update(text=element_text(family="serif"))`. After the standards for your own organization have been adjusted in a theme, e.g. theme.eigen, this can be saved in a file using the `save()` function. If this file is then loaded at the beginning of all R script files using the `load()` function and thus loads the theme.eigen object, the organization-wide standard can be applied using the `theme_set()` function. All visualizations will then implement these visualization standards.

Optical Design with Themes

```
#- Themes und Schriftarten---------------------------------------------------
p1<-ggplot(data=mpg) +                              # Ausgabe von Scatter-Plot
  geom_point(mapping=aes(x=Hubraum,y=Verbrauch))+   # in Variable p1
  labs(title="Scatter-Plot",subtitle="Normal")      # Titel,Untertitel
p1a<-p1+theme_dark()+labs(subtitle="theme_dark")    # Themes
p1b<-p1+theme_minimal()+labs(subtitle="theme_minimal")
p1c<-p1+theme_classic()+labs(subtitle="theme_classic")
p1d<-p1+theme_grey()+labs(subtitle="theme_grey")
grid.arrange(p1a,p1b,p1c,p1d,nrow=2,ncol=2)         # Ausgabe in 2 Zeilen+2 Spalten
```

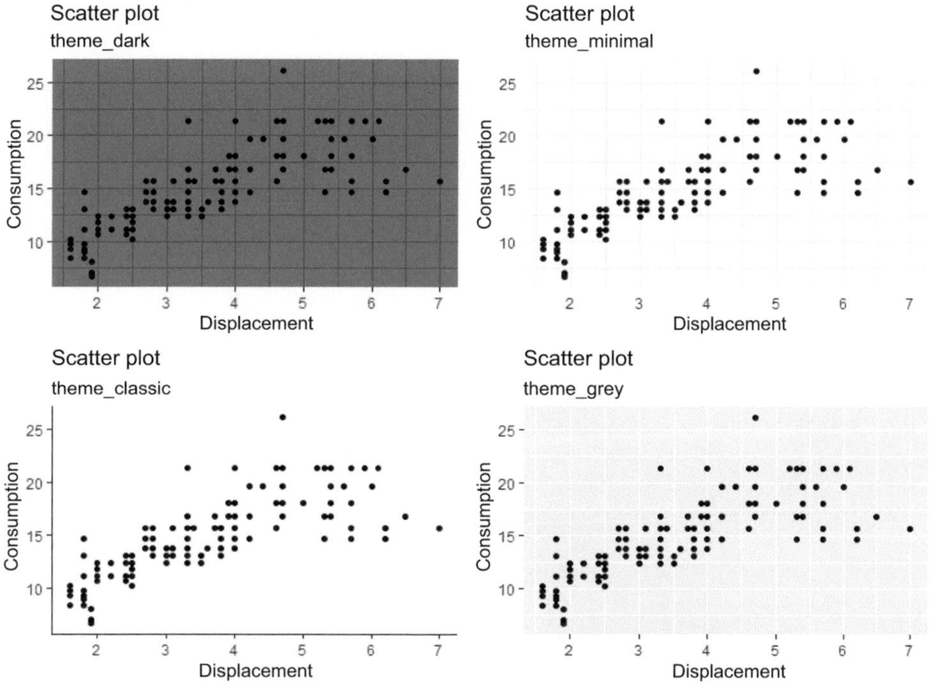


```
p1e<-p1+theme(text=element_text(family="serif"))+
        labs(subtitle="font=serif")
p1f<-p1+theme(text=element_text(size=14,family="sans",face="bold"))+
        labs(subtitle="font=sans,bold,14")
grid.arrange(p1,p1e,p1f,nrow=2,ncol=2)              # Ausgabe in 2 Zeilen+2 Spalten
```

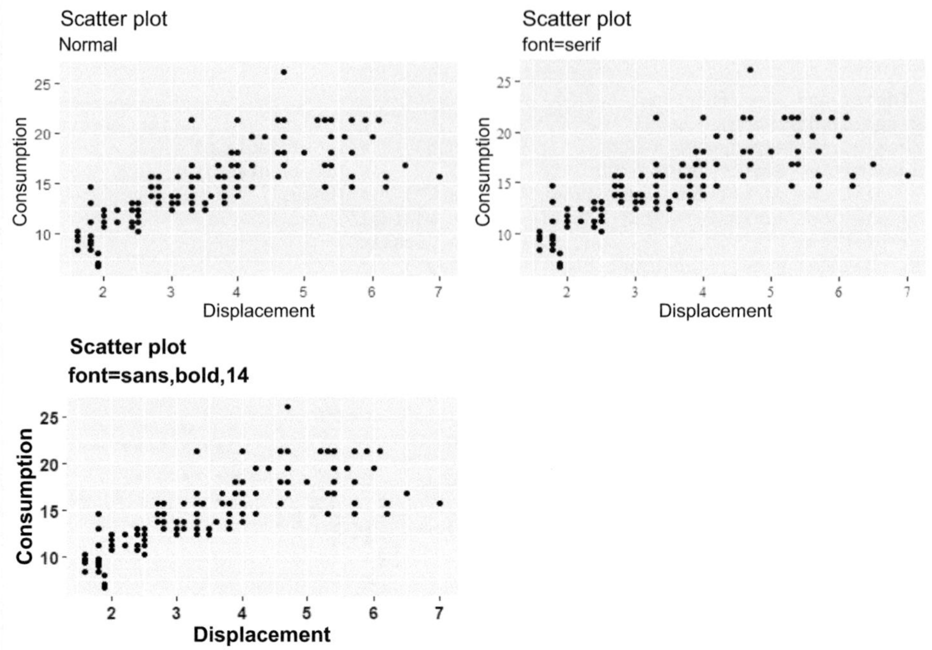


```
theme.norm <- theme_get()                        # Voreinstellungen->theme.norm
ggplot(mpg,aes(x=Hubraum,y=Verbrauch))+          # Normalschrift
  geom_point()
```

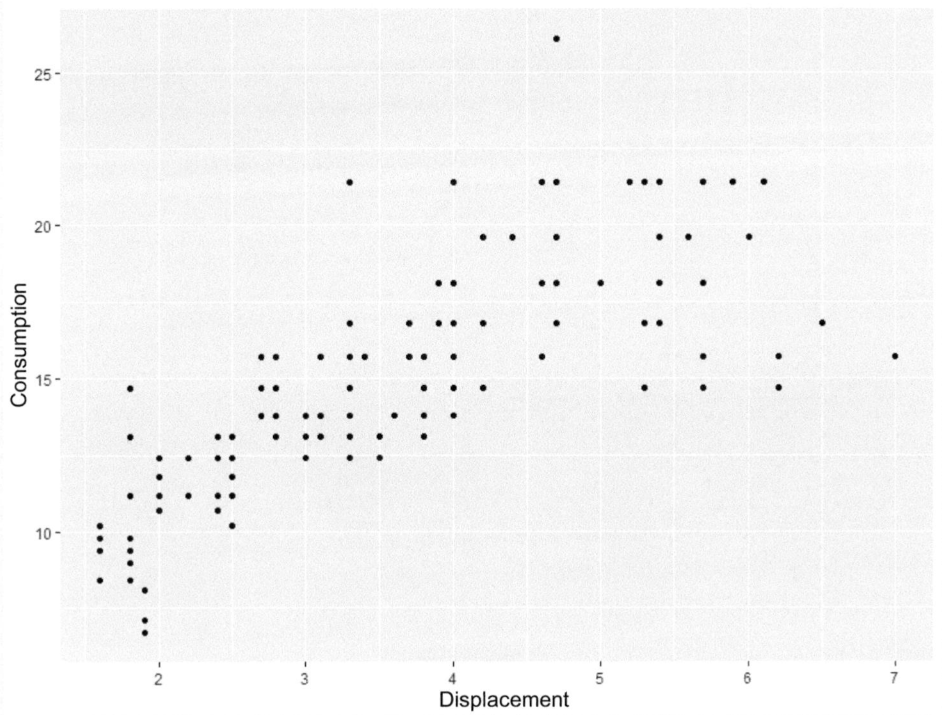


```
#- Modifikation von Theme und Speicherung in Variable theme.eigen-----------------
theme_update(text=element_text(family="serif",size=18,colour="red"))
theme.own<-theme_get()
ggplot(mpg,aes(x=Hubraum,y=Verbrauch))+           # Modifizierte Normalschrift
  geom_point()
```

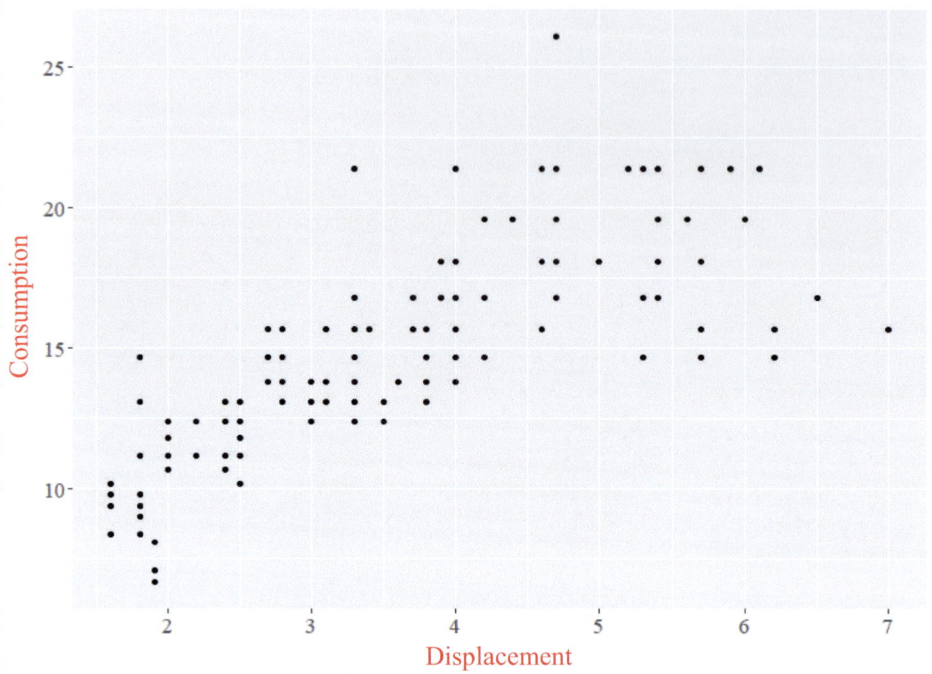


```
#- Speichern von theme.eigen in Datei zur allgemeinen Verwendung ----------------
#save(theme.own,file="visualisierungsstandard.RData")
rm(theme.own)                                    # Löschen von theme.eigen
#- Laden von theme.eigen aus Datei in allen R-Skripts zu Beginn------------------
theme_set(theme.norm)                              # Zurück zu Voreinstellungen
ggplot(mpg,aes(x=Hubraum,y=Verbrauch))+            # Normalschrift
  geom_point()
```

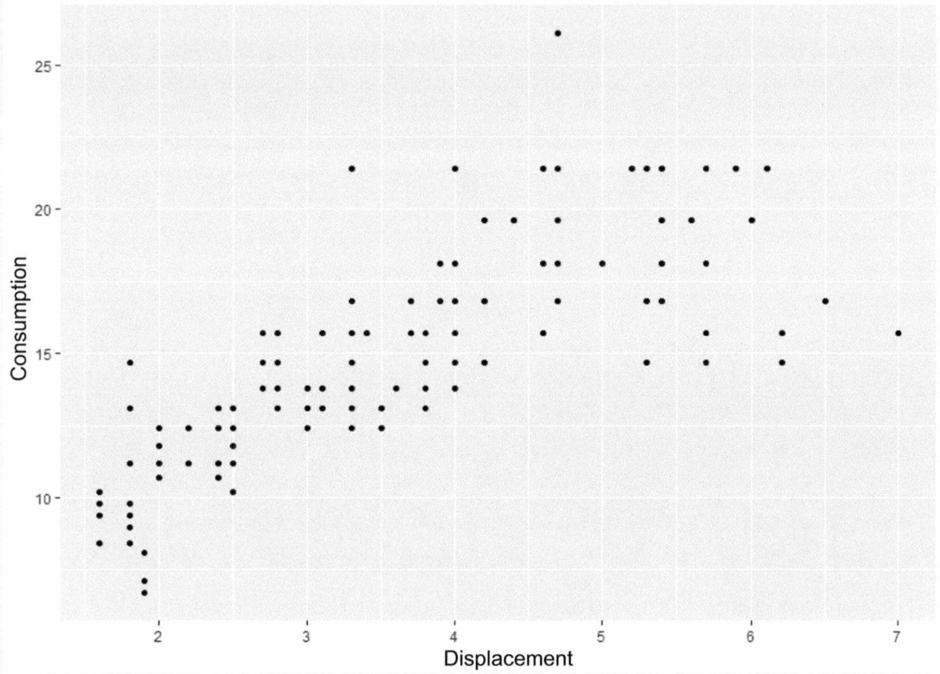


```
load("visualisierungsstandard.RData")          # Laden von theme.eigen
theme_set(theme.own)
ggplot(mpg,aes(x=Hubraum,y=Verbrauch))+        # Modifizierte Normalschrift
  geom_point()
```

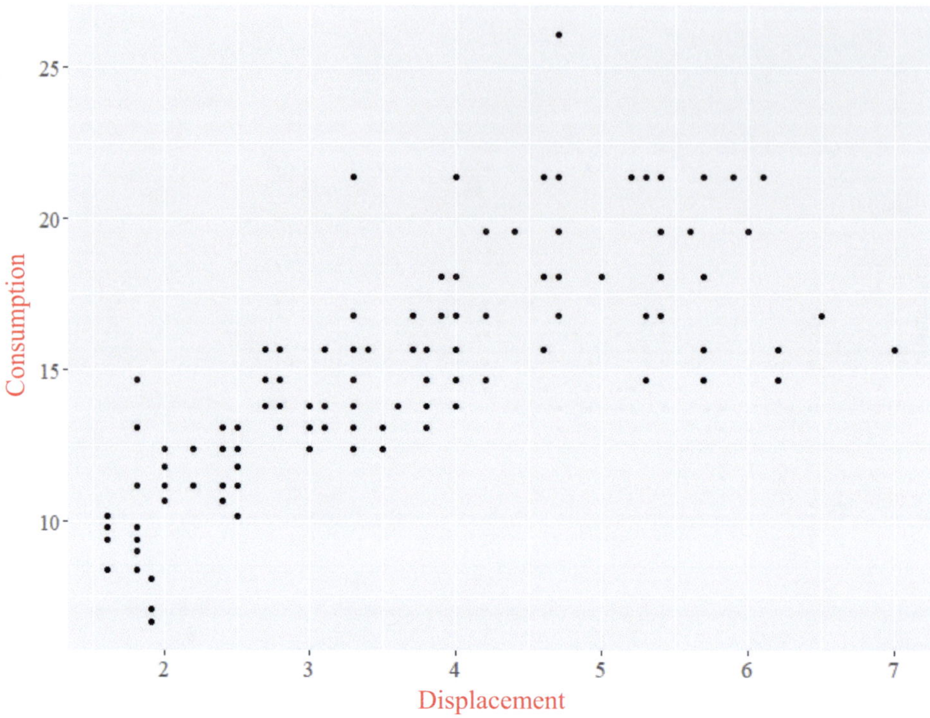


```
theme_set(theme.norm)                                    # Zurück zu Voreinstellungen
```

6.3.3.1.3 Colors

Colors also play a major role in figures. Colors can be defined using the `rgb()` function according to the RGBA model. In the RGBA model, RGB stands for Red, Green, Blue (colors are formed by mixing the three primary colors) and A for the alpha channel, which indicates transparency. Alternatively to the RGBA model, colors can also be defined via the `hsv()` function using the Hue-Saturation-Value or `hcl()` through the Hue-Chroma-Luminance. All colors are ultimately stored in hexadecimal form, where e.g. red is stored as (255,0,0) in decimal and as "#FF0000" in hexadecimal, which means 100% red component and 0% blue and green component.

Popular color palettes facilitate the selection of colors. Established among others are the color palettes of Cynthia Brewer and Mark Harrower, which are included in the ColorBrewer package and can be copied into your own palette via the `brewer.pal()` function, e.g. `meinepalette1<-brewer.pal(11,"Set2")`. The number in the bracket indicates how many colors should be used. In conjunction with the `ggplot()` function, the functions `scale_fill_brewer()` and `scale_color_brewer()` with the `palette` parameter are useful, e.g. `scale_color_brewer(palette="Dark2")`.

Brewer's color palettes are divided into three categories. The first category is for **sequential** color gradients from light to dark, which are particularly suitable for ordinal and metric variables. The second category with strongly distinguishing colors is particularly suitable for **nominal** variables. The third category is suitable for **spectral** representations, in which the center is neutral, such as when using a Likert scale. In Fig. 6.27, the color palettes are shown, which are also suitable for color-blind people.

Color palettes specifically developed for color-blind people are also provided in the viridis package (see Fig. 6.28). Color blindness along with color vision deficiency (especially red-green deficiency) affects about five percent of the world's population, predominantly men. Therefore, it seems advisable to use palettes in color selection that also allow color-blind and color-deficient people to distinguish based on grayscale. In conjunction with the function `ggplot()`, the functions `scale_fill_viridis()` and `scale_color_viridis()` with the parameter `option` are available, e.g. `scale_color_viridis(option="D")`, where the options A stand for "magma", B for "inferno", C for "plasma" and D for "viridis". If no parameter `option` is specified, viridis is chosen as the standard.

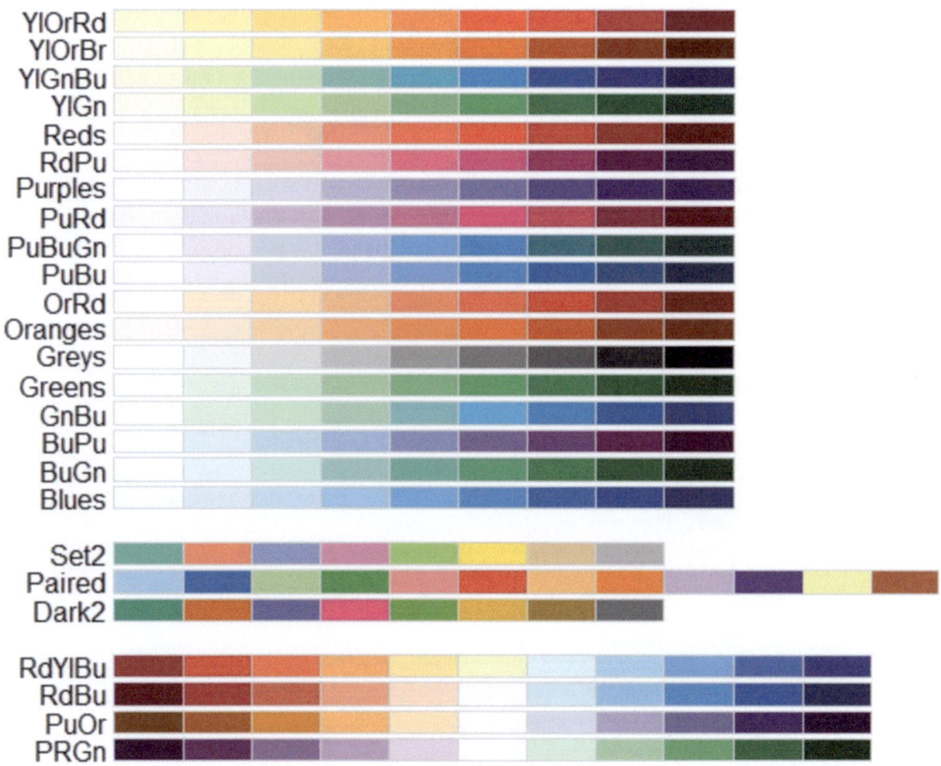

Fig. 6.27 Brewer color palettes for color-blind people

Fig. 6.28 Viridis color palettes for color-blind people

In the previous figures, colors were already used to differentiate categories. For this purpose, the parameters `color` and `fill` of the function `aes()` were used. The assignment of colors is done with one color, for example, with the specification `color="red"`. For a multitude of colors, the function `scale_colour_manual()` with the parameter `values` is suitable, which can be specified either via the function `c()` with `values= ("red","white","green")` or via color palettes with `values=nameofPalette`.

In addition to the color palettes already mentioned, there is also the possibility to work only with grayscale. For this purpose, the functions `scale_fill_grey()` and `scale_color_grey()` can be used. Very individual color transitions can be created with the function `colorRampPalette()`. To create a list with seven colors from blue to white to red, the following instruction can be used: `colors<-colorRampPalette(c("blue","white","red"))(7)`. These generated colors can then be used in a figure with the function `scale_fill_manual()`, e.g. `scale_fill_manual(values=colors)`. The function `scale_fill_gradient2()` also supports color gradients. The midpoint parameter additionally determines at which data value the middle of the three colors in the following instruction should be used: `scale_fill_gradient2(midpoint=0,low="red",mid="white",high="green")`.

For the implementation of organization-wide standards in visualization, it is also recommended to use specially created color palettes. These can be compiled with the functions mentioned above, then saved in a file and subsequently read in all R scripts for initialization.

Optical Design with Colors

```
#  Farbpaletten laden-----------------------------------------------------------
library("RColorBrewer")                                # Farbpaletten von Brewer
library(viridis)                                       # Farbpaletten viridis
# Anwendung der Farbpaletten für nominale Variablen-----------------------------
p0<-ggplot(alkohol,aes(x=Land,y=Portionen,group=Typ,fill=Typ))+
  geom_bar(stat="identity",position="dodge")+
  labs(title="Standard",subtitle="Alkoholkonsum",x="Land",y="Portionen")+
  guides(fill=guide_legend(title="Getränk"))
p1<-p0+scale_fill_brewer(palette="Dark2")+
    labs(title="Nominale Farbpalette",
         subtitle="Brewer: Dark2")
p2<-p0+scale_color_viridis(discrete=TRUE,guide="none")+
    labs(title="Nominale Farbpalette",
         subtitle="Viridis")
p3<-p0+scale_fill_grey()+
    labs(title="Nominale Farbpalette",
         subtitle="Graustufen")
grid.arrange(p1,p2,p3,nrow=2,ncol=2)                   # Ausgabe
```

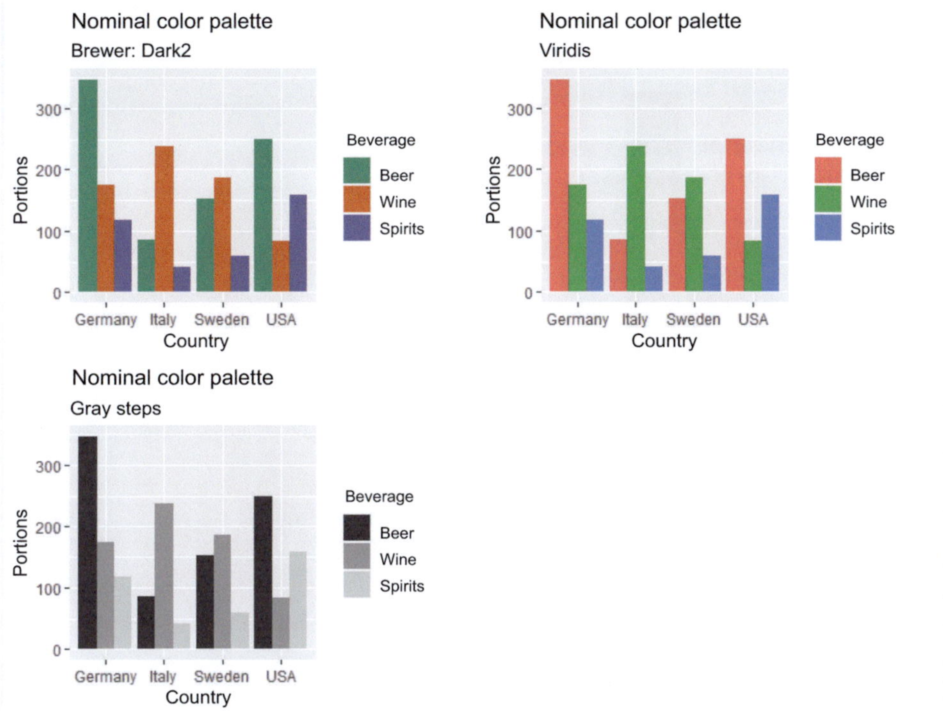


```
# Anwendung der Farbpaletten für sequentielle Variablen--------------------------
set.seed(126)
siege=data.frame(Spieler=1:12,                        # 12 Spieler
                 Siege=sample(1:9,12,replace=T))      # 12 Zufallszahlen von 1-9
siege$Ampel<-                                         # Zusätzliche Variable
  case_when(siege$Siege < 4  ~ "rot",
    between(siege$Siege,4,6) ~ "gelb",
            siege$Siege >= 7 ~ "grün")
siege$Ampel<-factor(siege$Ampel,
                   levels=c("grün","gelb","rot"), ordered = T)
ggplot(siege)+
    geom_col()+
    aes(x=Spieler,y=Siege,fill=factor(Siege))+
    scale_x_continuous(breaks=c(1:12))+
    scale_y_continuous(breaks=c(1:9))+
    labs(title="Sequentielle Farbpalette",
       subtitle="Brewer: OrRd")+
    scale_fill_brewer(name="Siege",palette="OrRd")+
    guides(fill=guide_legend(reverse=TRUE))           # Legende hohe Werte oben
```

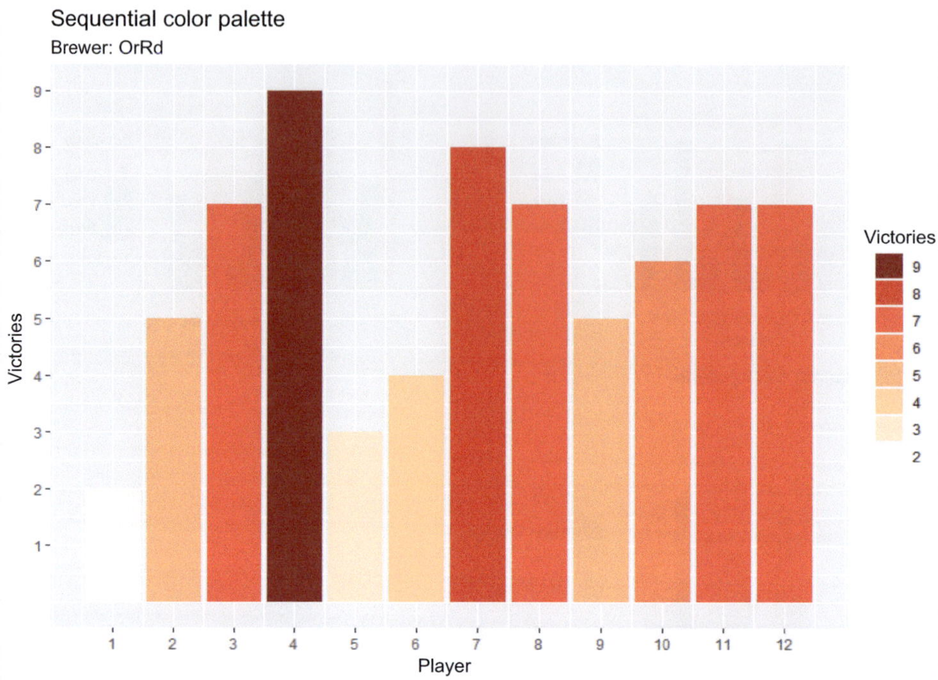


```
# Anwendung der Farbpaletten für spektrale Variablen(z.B. Likert-Skala)----------
likert <- factor(c("sehr zufrieden","zufrieden",      # Ordinale Likert-Variable
                   "neutral","unzufrieden",
                   "sehr unzufrieden"),
  order=TRUE,
  levels=c("sehr zufrieden","zufrieden","neutral",
           "unzufrieden","sehr unzufrieden"))
str(likert)

##  Ord.factor w/ 5 levels "sehr zufrieden"<..: 1 2 3 4 5

set.seed(126)
zimmer=sample(likert,15,prob=c(2,5,3,0,1),replace=T) # Zimmerbewertungen
set.seed(126)
service=sample(likert,15,prob=c(5,2,1,1,0),replace=T)# Servicebewertungen
set.seed(126)
essen=sample(likert,15,prob=c(0,1,3,4,1),replace=T)  # Essensbewertungen
umfrage<-data.frame(Gast=1:15,                        # Dataframe mit Bewertungen
                    Frage=c(rep("1",15),rep("2",15),rep("3",15)),
                    Bewertung=c(zimmer,service,essen))
farben<-colorRampPalette(c("green","white","red"))(5)# 5 Farben von grün->rot
ggplot(umfrage) +
  geom_bar(aes(x=Frage,fill=Bewertung))+              # Stacked Bar-Chart
  scale_fill_manual(values=farben)+                   # Farben aus Vektor
  labs(title="Gästezufriedenheit",                    # Beschriftungen
       subtitle="Likert-Skala",y="Anzahl")+
  scale_x_discrete("Bewertung",
                   labels=c("Zimmer","Service","Essen"))
```

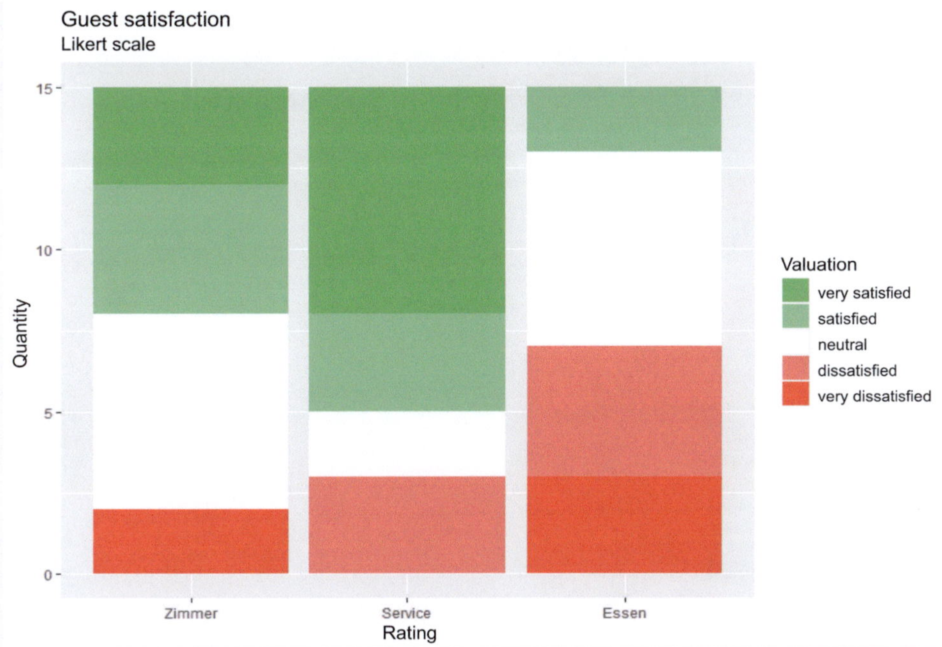


```
# Anwendung der Farbpaletten für spektrale Variablen (Beispiel 2)-----------------
leipzig$Farbstufen<-cut(leipzig$wert,
                   c(1,0.4,0.2,0,-0.1,-0.2,-0.4,-1))  # Farbstufe in 7 Gruppen
levels(leipzig$Farbstufen)<-c("-1 bis -0,2",          # Gruppenbezeichnungen
                  "-0,2 bis -0,1","-0,1 bis 0","0",
                  "0 bis 0,1","0,1 bis 0,2","0,2 bis 1")
farben<-colorRampPalette(c("red","white","green"))(7)  # 7 Farben von grün->rot
ggplot(leipzig) +
  aes(x=reorder(token,anzahl),                          # Häufigste Worte oben
      y=anzahl,group=Farbstufen,fill=Farbstufen)+       # Farbe nach Farbstufe
  scale_fill_manual(values=farben)+                     # Farben aus Vektor
  geom_col()+                                           # Balkendiagramm
  labs(title="Emotionale Worte im Text (basierend auf Leipzig-Wörterbuch)",
       subtitle="negative Werte=negative Emotion, positive Werte=positive Emotion",
       y="Anzahl",x="Worte") +
  coord_flip()                                          # Balkendiagramm gedreht
```

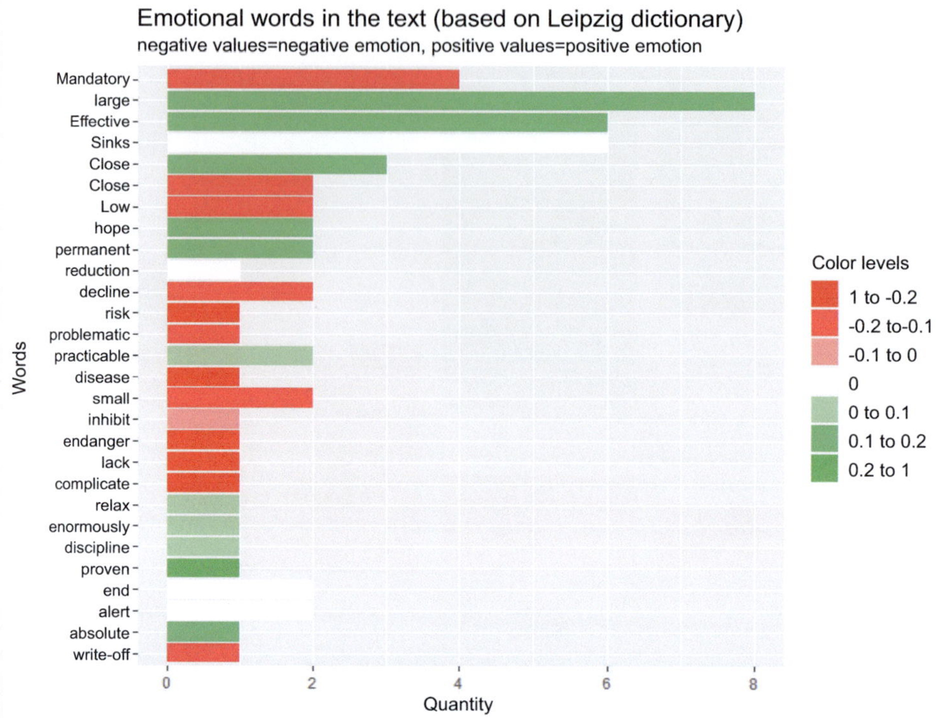


```
p1<-ggplot(leipzig) +
  aes(x=token,
      y=wert,fill=wert) +                          # Farbe nach Wert
  scale_fill_gradient2(midpoint=0,low="red",       # Farben von rot->grün
                       mid="white",high="green") + # weiß=Wert von 0
  geom_col()+                                      # Balkendiagramm
  labs(title="Emotionale Worte im Text",           # Beschriftungen
       subtitle="Leipzig-Wörterbuch (weiß=0)",
       y="Anzahl",x="Worte")+
  coord_flip()
p2<-ggplot(leipzig) +
  aes(x=token,
      y=wert,fill=wert) +                          # Farbe nach Wert
  scale_fill_gradient2(midpoint=-0.2,low="red",    # Farben von rot->grün
                       mid="white",high="green") + # weiß=Wert von -0.2
  geom_col()+                                      # Balkendiagramm
  labs(title="Emotionale Worte im Text",           # Beschriftungen
       subtitle="Leipzig-Wörterbuch (weiß=-0,2)",
       y="Anzahl",x="Worte")+
  coord_flip()
grid.arrange(p1,p2,ncol=2)                         # Ausgabe
```

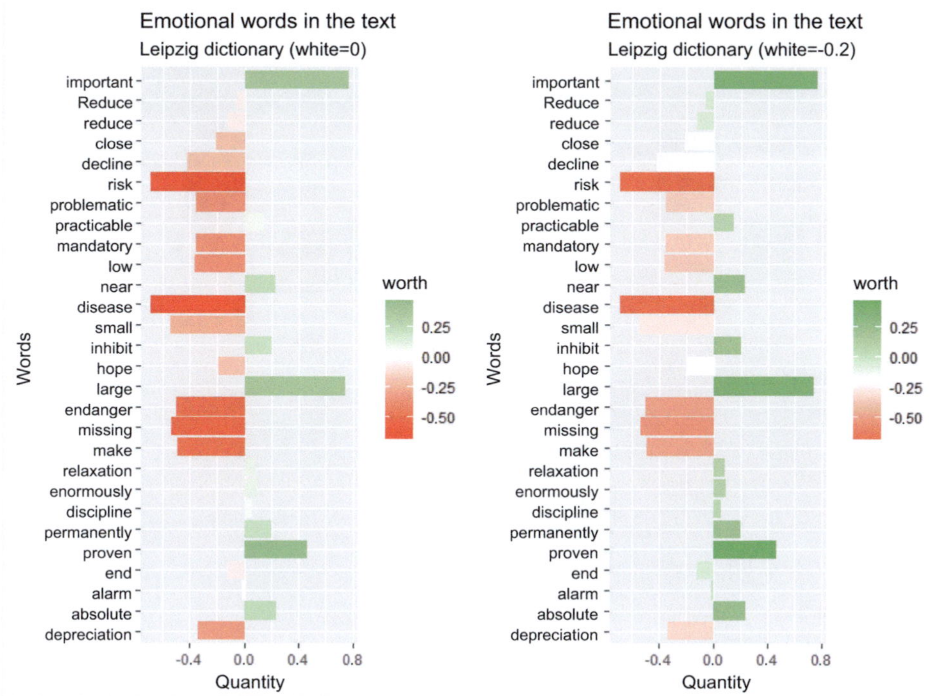


```
leipzig.teil<-filter(leipzig,Farbstufen %in%                    # Teilmenge bilden
                     c("-0,2 bis -0,1","0",
                       "0,1 bis 0,2"))
p1<-ggplot(leipzig.teil,aes(x=Farbstufen,y=anzahl,
                        fill=Farbstufen))+                    # Farbe nach Farbstufe
    geom_boxplot()+                                             # Boxplot
    labs(y="Anzahl",x="Wert")+                                  # Beschriftungen
    labs(title="Farbzuweisung",                                 # Beschriftungen
         subtitle="via Spalte 'Farbzuweisung' des Dataframes")
p2<-p1 + scale_fill_manual(values=c("#999999","#E69F00",# Manuelle Farbzuweisung
         "#56B4E9"))+                                           # via Hexadezimalcode
    labs(title="Farbzuweisung",                                 # Beschriftungen
         subtitle="via Hexadezimalcode")
p3<-p1 + scale_fill_manual(values=farben[3:5])+                # Manuelle Farbzuweisung
                                                                # aus Vektor "farben"
    labs(title="Farbzuweisung",                                 # Beschriftungen
         subtitle="aus Vektor")
grid.arrange(p1,p2,p3,ncol=2)                                   # Ausgabe
```

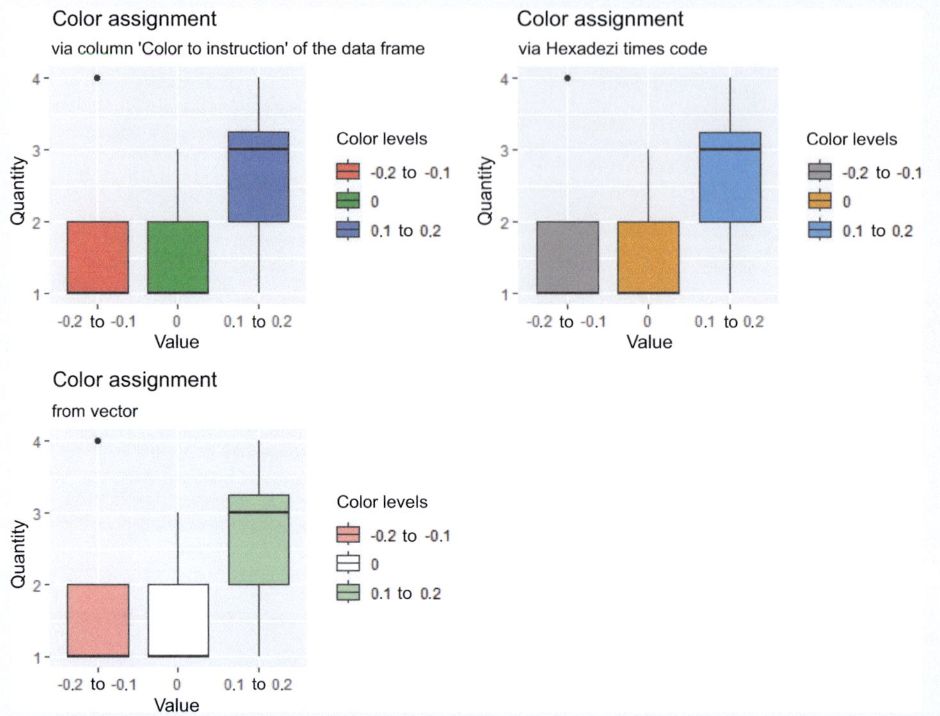


```
# Organisationsspezifische Farbpaletten definieren--------------------------------
colour.own.ampel<-c("#53D335","#E0CA28","#DD382B")       # Definieren
colour.own.nomin.1<-brewer.pal(11,"Paired")
colour.own.nomin.2<-brewer.pal(8,"Dark2")
colour.own.seque.1<-brewer.pal(9,"Blues")
colour.own.seque.2<-rev(viridis(11,option = "D"))
colour.own.spect.1<-colorRampPalette(c("green","white","red"))(11)
colour.own.spect.2<-rev(brewer.pal(11,"RdYlBu"))
# Organisationsspezifische Farbpaletten speichern---------------------------------
save(theme.own, theme.standard,                          # Speichern
     colour.own.ampel,
     colour.own.seque.1,colour.own.seque.2,
     colour.own.nomin.1,colour.own.nomin.2,
     colour.own.spect.1,colour.own.spect.2,
     file="visualisierungsstandard.RData")
# Nominale organisationsspezifische Farbpaletten verwenden-----------------------
p1<-ggplot(alkohol,aes(x=Land,y=Portionen,group=Typ,fill=Typ))+
  geom_bar(stat="identity",position="dodge")+
  labs(x="Land",y="Portionen")+
  labs(title="Organisationsspezifische Farben",
       subtitle="Nominale Farbpalette: colour.own.nomin.1")+
  guides(fill=guide_legend(title="Getränk"))+
  scale_fill_manual(values=colour.own.nomin.1)
p2<-ggplot(alkohol,aes(x=Land,y=Portionen,group=Typ,fill=Typ))+
  geom_bar(stat="identity",position="dodge")+
  labs(x="Land",y="Portionen")+
  labs(title="Organisationsspezifische Farben",
```

```
        subtitle="Nominale Farbpalette: colour.own.nomin.2")+
  guides(fill=guide_legend(title="Getränk"))+
  scale_fill_manual(values=colour.own.nomin.2)
grid.arrange(p1,p2,ncol=2)                                  # Ausgabe
```

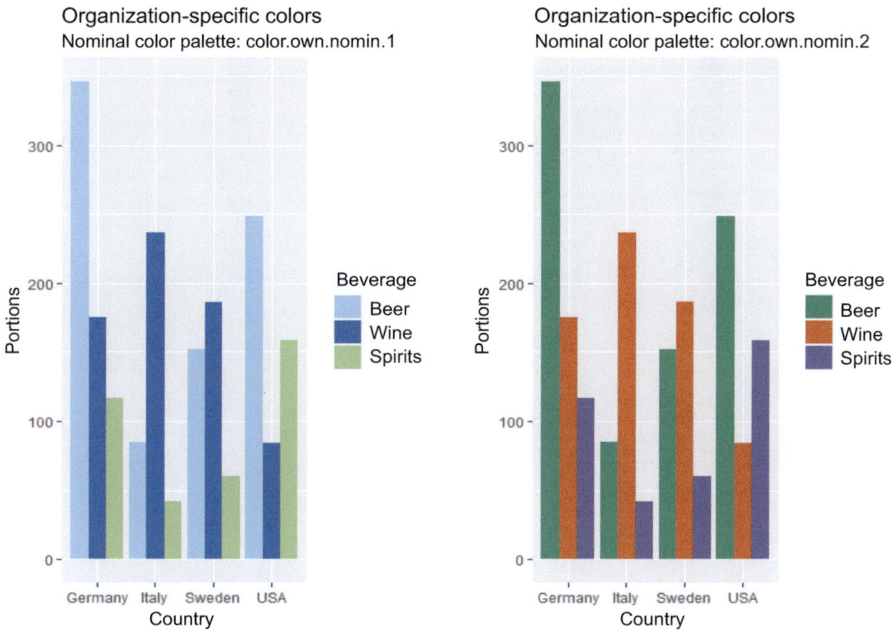


```
# Sequentielle organisationsspezifische Farbpaletten verwenden--------------------
p1<-ggplot(siege)+
    geom_col()+
    aes(x=Spieler,y=Siege,fill=factor(Siege))+
    scale_x_continuous(breaks=c(1:12))+
    scale_y_continuous(breaks=c(1:9))+
    labs(title="Organisationsspezifische Farben",
       subtitle="Sequentielle Farbpalette: colour.own.seque.1")+
    scale_fill_manual(name="Siege",values=colour.own.seque.1)+
    guides(fill=guide_legend(reverse=TRUE))
p2<-ggplot(siege)+
    geom_col()+
    aes(x=Spieler,y=Siege,fill=factor(Siege))+
    scale_x_continuous(breaks=c(1:12))+
    scale_y_continuous(breaks=c(1:9))+
    labs(title="Organisationsspezifische Farben",
       subtitle="Sequentielle Farbpalette: colour.own.seque.2")+
    scale_fill_manual(name="Siege",values=colour.own.seque.2)+
    guides(fill=guide_legend(reverse=TRUE))
grid.arrange(p1,p2,ncol=2)                                  # Ausgabe
```

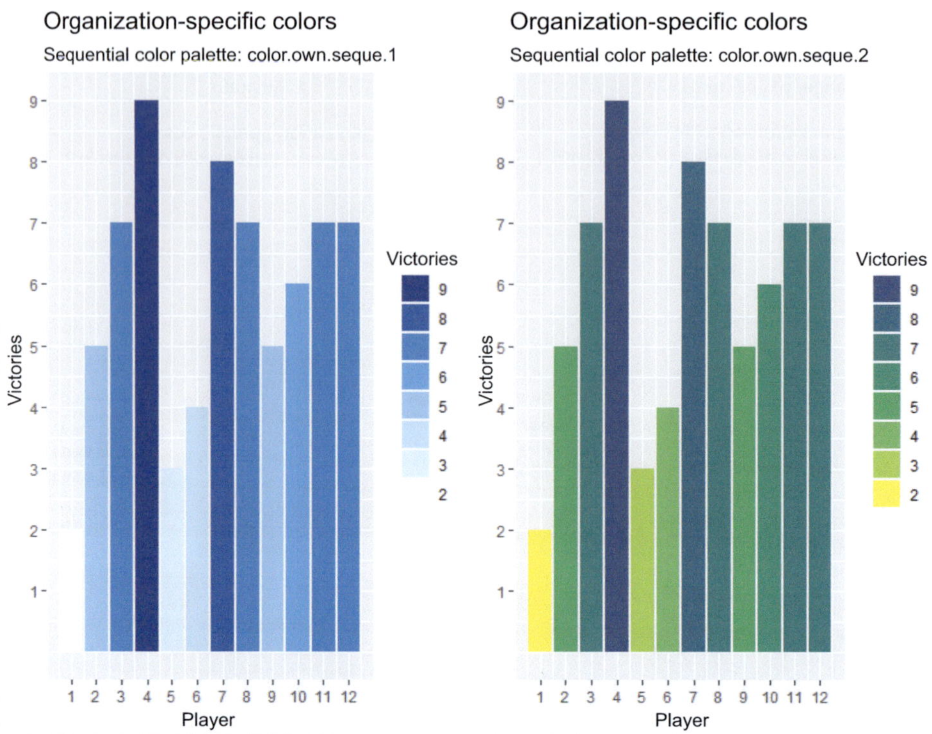


```
# Spektrale organisationsspezifische Farbpaletten verwenden----------------------
p1<-ggplot(umfrage) +
  geom_bar(aes(x=Frage,fill=Bewertung))+                # Stacked Bar-Chart
  scale_fill_manual(values=colour.own.spect.1[c(1,4,6,8,11)])+
  labs(title="Organisationsspezifische Farben",
       subtitle="Spektrale Farbpalette: colour.own.spect.1")+
  scale_x_discrete("Bewertung",
                   labels=c("Zimmer","Service","Essen"))
p2<-ggplot(umfrage) +
  geom_bar(aes(x=Frage,fill=Bewertung))+                # Stacked Bar-Chart
  scale_fill_manual(values=colour.own.spect.2[c(1,4,6,8,11)])+
  labs(title="Organisationsspezifische Farben",
       subtitle="Spektrale Farbpalette: colour.own.spect.2")+
  scale_x_discrete("Bewertung",
                   labels=c("Zimmer","Service","Essen"))
grid.arrange(p1,p2,ncol=2)                              # Ausgabe
```

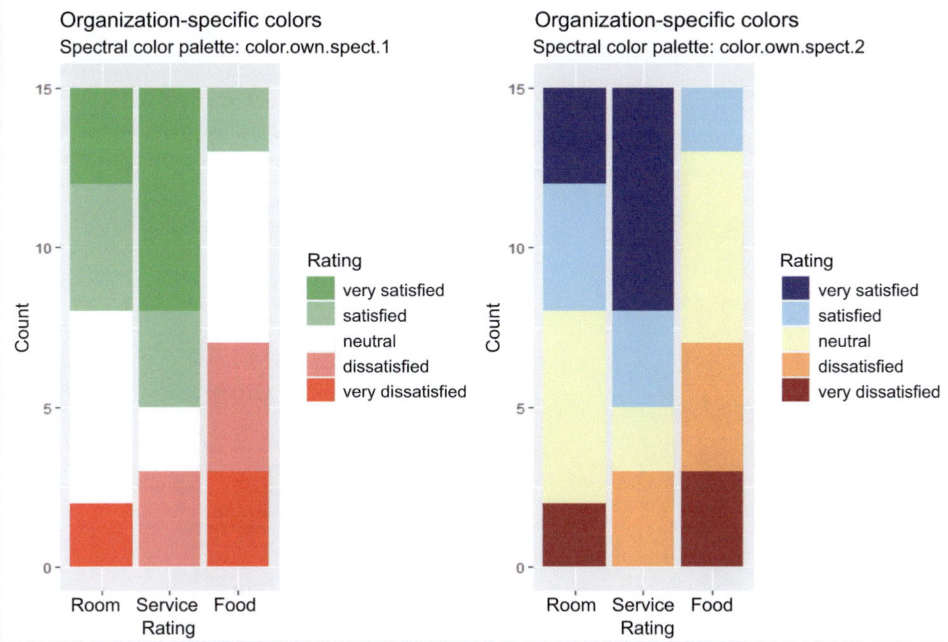


```
# Ampel als organisationsspezifische Farbpalette verwenden------------------------
ggplot(siege)+
  geom_col()+
  aes(x=Spieler,y=Siege,fill=Ampel)+
  scale_x_continuous(breaks=c(1:12))+
  scale_y_continuous(breaks=c(1:9))+
  labs(title="Organisationsspezifische Farben",
     subtitle="Ampel Farbpalette: colour.own.ampel")+
  scale_fill_manual(name="Performanz",values=colour.own.ampel)
```

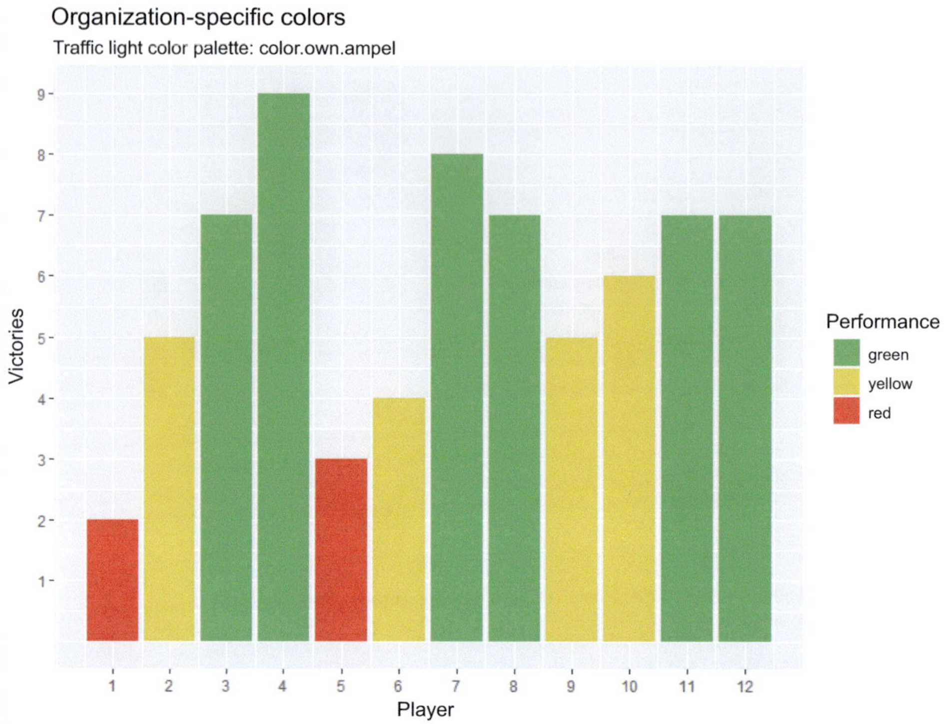

6.3.3.2 Types of Illustrations

The choice of the appropriate type of illustration also depends on the type of variable and the number of variables. For a metric variable, for example, the function `geom_histogram()` is suitable for displaying the absolute frequency of values. The parameter `bins` allows the specification of the number of bars in this function. To display the relative frequency of values in percent, the y-axis can be specified with `y=after_stat(density)`. To achieve a distinction depending on a cardinal variable, e.g. gender, this can be defined in the function `aes()` with the parameter `fill=gender`. By default, the variable values specified with the fill parameter are displayed stacked. The parameter `position`with the characteristics `identity`in connection with transparency via the parameter `alpha` allows the variable values to be displayed one after the other. If the parameter `position` is specified with the value `dodge`, the variable values are displayed side by side.

For a categorical variable, the function `geom_bar()`, which represents a bar chart, is suitable for displaying frequencies. If a different coloring of the categories is desired, this can be achieved with the parameter `fill` in the function `aes()`. As with the function `geom_histogram()`, the parameter `position` can be used to define whether the bars should be displayed on top of each other, one after the other, or side by side.

When one wants to represent two metric variables and their relationship to each other, the function `geom_point()` is suitable. The parameters `size` and `alpha` can also be used here, and to avoid overplotting, the function `geom_jitter()` may be useful. To incorporate a third variable, especially a categorical variable, into the figure, the parameter `shape` or `color` in the function `aes()` is suitable. The function `geom_line()` is also suitable for two metric variables. The parameters `linetype` (blank, solid, dashed, dotted, longdash, twodash) and `size` allow design options. If a categorical grouping variable is to be used in addition to represent multiple lines or points, this can be achieved with the parameter `group` in the function `aes()`.

Especially when multiple grouping variables are used, which are represented in figures with the help of colors and shapes, a figure can become confusing. Here, the function `facet_wrap()` or `facet_grid()` can be helpful. The function `facet_wrap()` creates a separate figure for each category of a grouping variable, and the function `facet_grid()` does the same for each combination of all categories of two grouping variables.

Another helpful function is the function `grid.arrange()`, with the help of which several figures can be displayed together on one page. The parameters `ncol` and `nrow` allow to specify how many columns and rows the page should be divided into. In each segment, a previously saved figure can then be output, e.g., `grid.arrange(p1,p2,ncol=2)`, to output the figures p1 and p2 created with ggplot side by side.

6.3.3.2.1 Histogram

The distribution of a variable can be well represented in a histogram. Histograms can either be normally distributed (not skewed) or have a left or right skew if the data tend to be more towards high or low values. The skewness of a distribution describes the type and strength of asymmetry. It indicates whether and how strongly the distribution is inclined to the right (right-steep, left-skewed, negative skewness) or to the left (left-steep, right-skewed, positive skewness).

In the case of negative skewness (skewness < 0), one speaks of a left-skewed or right-steep distribution; it typically falls off more gently on the left side than on the right. The majority of the values are then to the right of the mean. Therefore, with negative skewness, the mean is smaller (further to the left) than the median.

In the case of positive skewness (skewness > 0), one speaks of a right-skewed or left-steep distribution; it falls off more gently on the right side than on the left. The majority of the values are then to the left of the mean. Therefore, with positive skewness, the mean is larger (further to the right) than the median.

A distribution without skewness is referred to as approximately normally distributed, as the values are distributed approximately symmetrically on both sides. In this case, the mean and median are approximately identical.

In R, the Geom `geom_histogram()` in the function `ggplot()` is used for the representation of histograms.

```
#- Histogramm-----------------------------------------------------------------
p1<-ggplot(mpg.teilmenge)+
    aes(x=Hubraum)+                                # Absolute Häufigkeit
    geom_histogram(bins=20,co“="whi“e")+          # Histogramm 20 Balken
    labs(titl“="Histogramm",
         subtitl“="Absolute Häufigke“t",
         “="Hubraum in Lite“n", “="Anza“l")
p2<-p1+aes(y=after_stat(density))+                # Relative Häufikeit
    labs(subtitl“="Relative Häufigke“t",
         “="Ante“l")
p3<-p1+aes(fill=Klasse)+                          # Abs.Häufikeit je Klasse
    scale_fill_manual(values=colour.own.nomin.2)+
    labs(subtitl“="Absolute Häufigkeit je Klas“e",
         “="Anza“l")
grid.arrange(p1,p2,p3,nrow=2,ncol=2)              # Ausgabe
```

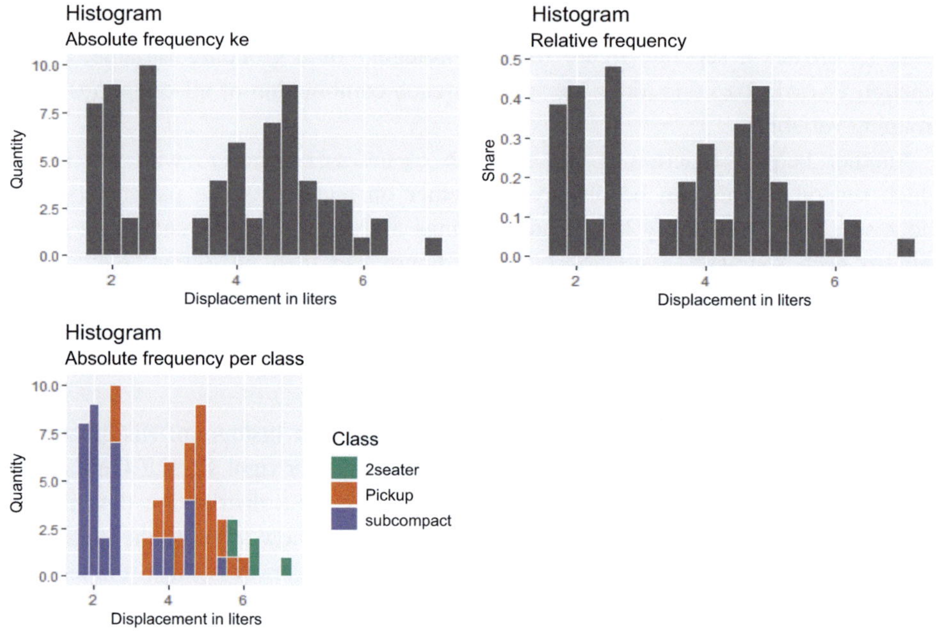

6.3.3.2.2 Bar Chart

Another type of representation is the bar chart, in which all data records are depicted one after the other on the x-axis, while the y-axis shows the value of the variable.

In R, the Geom `geom_bar()` in the function `ggplot()` is used for the representation of bar charts.

Bar Chart

```
#- Bar-Chart 1 Metrische Variable + 1 Nominale Variable--------------------------
p1<-ggplot(mpg.teilmenge)+
    aes(x=Hubraum)+                                  # x-Achse: Metrische Variable
    geom_bar()+                                      # Bar-Chart
    scale_fill_manual(values=colour.own.nomin.2)+
    labs(title="Bar-Chart",
         subtitle="Absolute Häufigkeit",
         x="Hubraum in Litern", y="Anzahl")
p2<-p1+aes(fill=Klasse)+                             # Farbe je Klasse
    labs(subtitle="Absolute Häufigkeit je Klasse")
grid.arrange(p1,p2,ncol=2)                           # Ausgabe
```

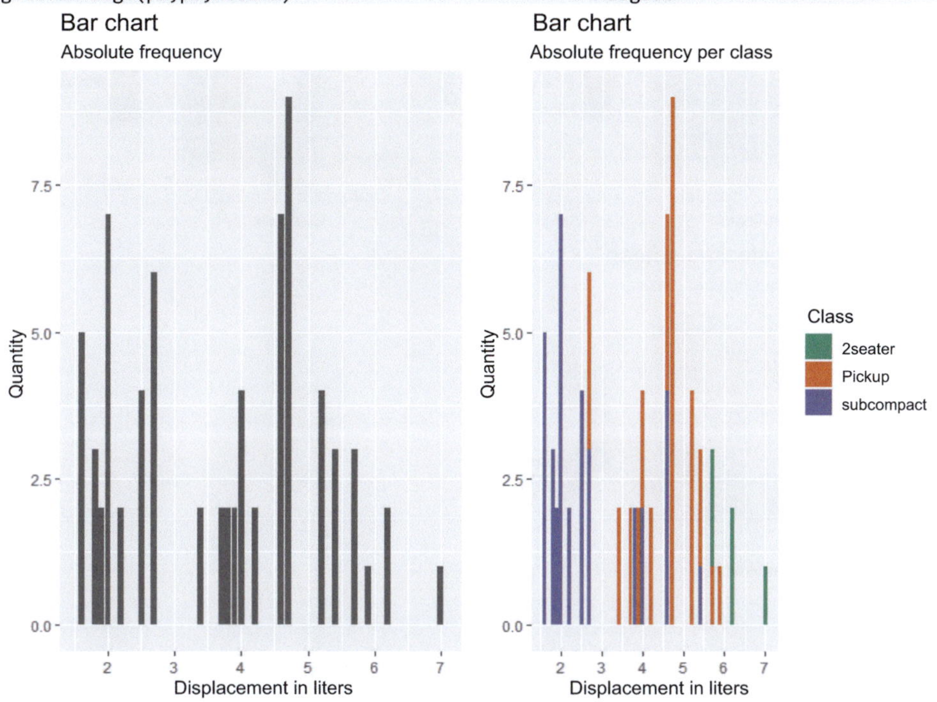


```
#- Bar-Chart 2 Kategoriale Variablen-------------------------------------------
mpg.tab<-as.data.frame(table(mpg.teilmenge$Getriebe,mpg.teilmenge$Klasse))
colnames(mpg.tab)<-c("Getriebe","Klasse","Anzahl")
p1<-ggplot(mpg.tab)+
    aes(x=Getriebe,y=Anzahl,fill=Klasse)+            # x-Achse: Nominale Variable
    geom_bar(position="stack",stat="identity")+      # Bar-Chart stacked
    labs(title="Bar-Chart stacked",
         subtitle="Absolute Häufigkeit je Klasse",
         x="Getriebe", y="Anzahl")+
    scale_fill_manual(values=colour.own.nomin.2)
p2<-ggplot(mpg.tab)+
    aes(x=Getriebe,y=Anzahl,fill=Klasse)+
    geom_bar(position="dodge",stat="identity")+      # Bar-Chart nebeneinander
    labs(title="Bar-Chart nebeneinander",
         subtitle="Absolute Häufigkeit je Klasse",
         x="Getriebe", y="Anzahl")+
    scale_fill_manual(values=colour.own.nomin.2)
```

```
p3<-ggplot(mpg.tab)+
    aes(x=Getriebe,y=Anzahl,fill=Klasse)+
    geom_bar(position="fill",stat="identity")+     # Bar-Chart stacked zu 100%
    labs(title="Bar-Chart stacked Anteile",
         subtitle="Relative Häufigkeit je Klasse",
         x="Getriebe", y="Anteil")+
    scale_fill_manual(values=colour.own.nomin.2)
grid.arrange(p1,p2,p3,ncol=2)                       # Ausgabe
```

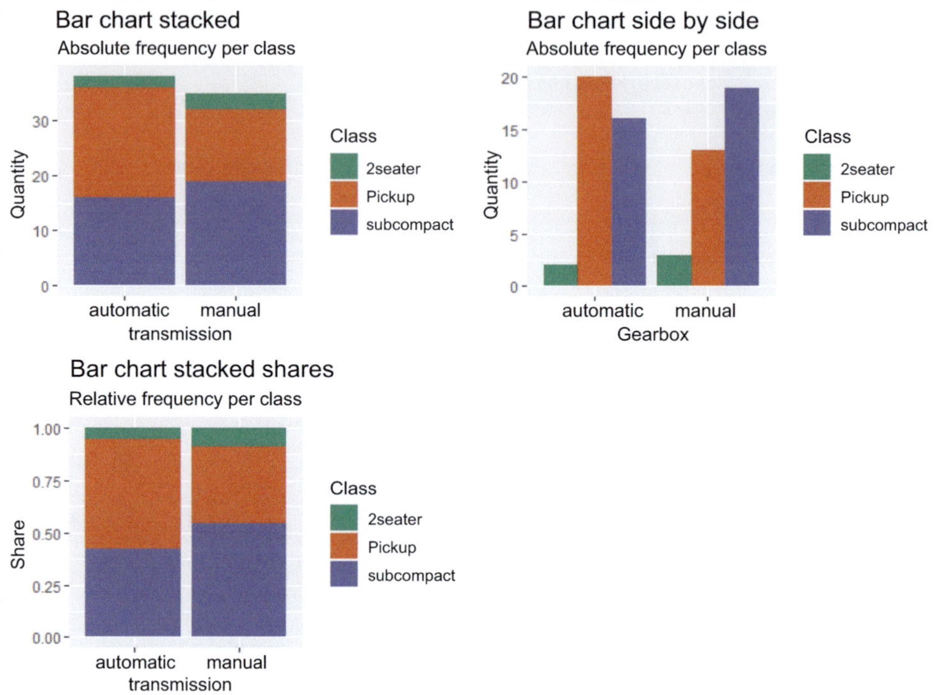

6.3.3.2.3 Line-Chart

A line chart is suitable for displaying the temporal progression of one or more variables. In R, the geom `geom_line()` is used in the function `ggplot()` for the representation of line charts.

Line-Chart

```
#- Line-Chart -----------------------------------------------------------------
str(zeitreihe)

## 'data.frame':    300 obs. of  3 variables:
##  $ Zeitreihe: int  1 2 3 4 5 6 7 8 9 10 ...
##  $ Variable : Factor w/ 3 levels "X1","X2","X3": 1 1 1 1 1 1 1 1 1 1 ...
##  $ Wert     : num  3.67 13.13 12.59 11.56 9.89 ...

ggplot(zeitreihe, aes(x = Zeitreihe, y = Wert, color = Variable)) +
  geom_line()
```

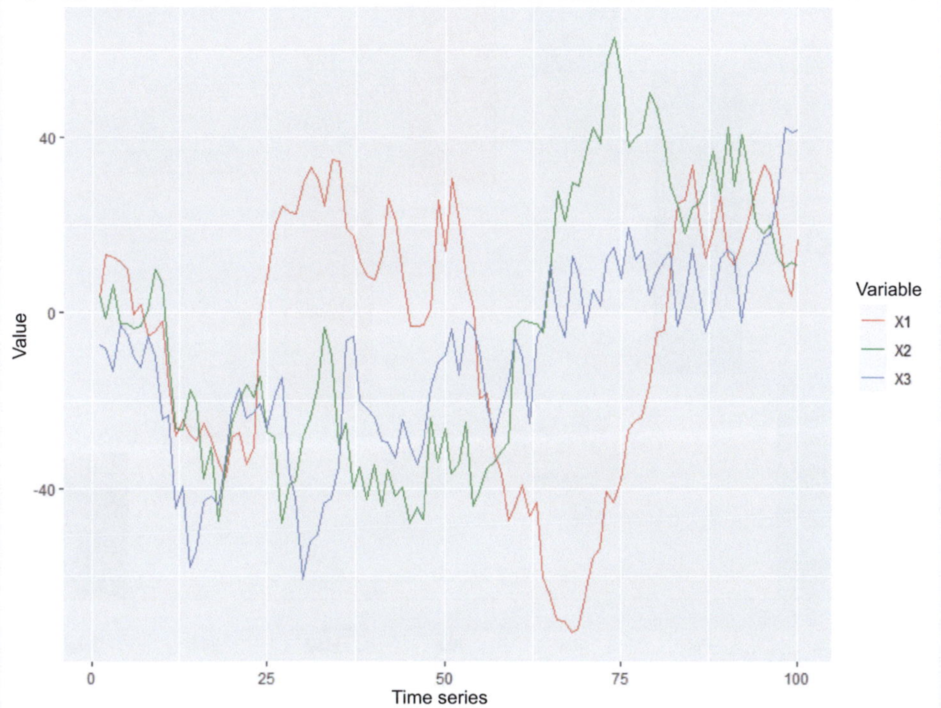

6.3.3.2.4 Waterfall-Chart

A waterfall chart is suitable for showing how an initial value is increased and/or decreased by a series of further values, leading to a final value.

In R, the function `waterfall()` from the package `waterfalls` is used for the representation of waterfall charts. The parameter `calc_total=True` determines whether the difference of the final value should also be displayed.

Waterfall-Chart

```
#- Waterfall-Chart -----------------------------------------------------------
---
Monat <- c("Jan","Feb","Mär","Apr","Mai","Jun")
Wert <- c(3000, 4000, 1000,-2500, -3000, -500)
df <- data.frame(x = Monat, y = Wert)
library(waterfalls)                                    # Paket erforderlich
waterfall(df,calc_total = TRUE)
```

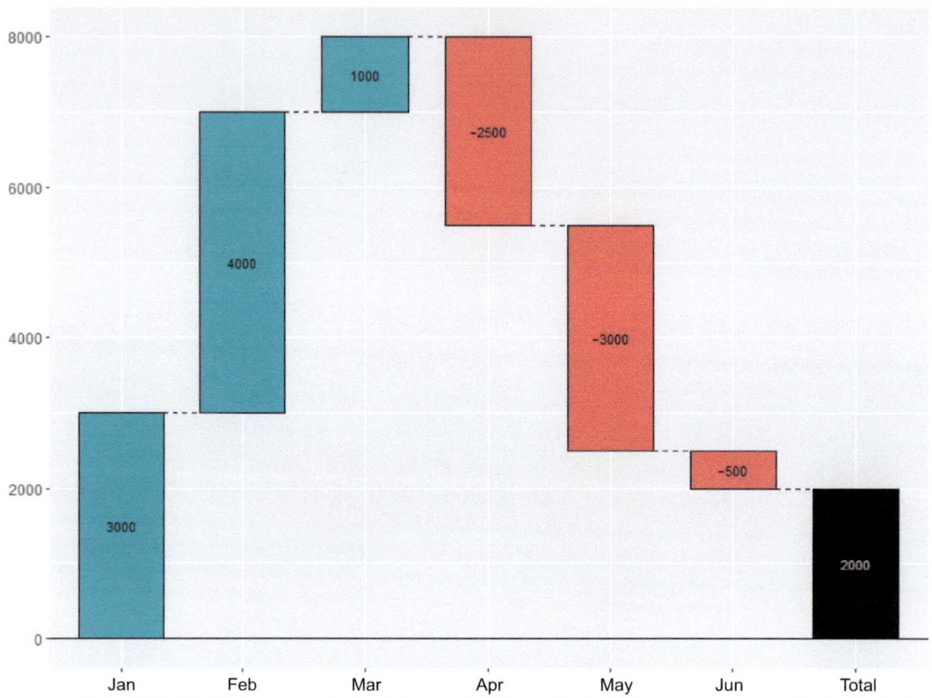

6.3.3.2.5 Box-Plot

A box plot, or box-whisker plot, is a diagram used for the graphical representation of the distribution of a metric variable. It summarizes various location and dispersion measures in one representation and quickly gives an impression of the range in which the data lies and how it is distributed over this range. The box is described by three lines. The upper limit of the box describes the 0.75 quantile (3rd quartile), the line in the middle of the box describes the 0.5 quantile (2nd quartile, median), and the lower limit of the box describes the 0.25 quantile (1st quartile). This shows the range in which 50% of the middle values are distributed. The box is supplemented by antennas (whiskers), which usually do not extend further up or down than 1.5 times the IQR (interquartile range = 3rd quartile – 1st quartile) from the edge of the box. The antennas are displayed so that they reach up to the last real data point, which is not further than 1.5 times the IQR from the upper or lower edge of the box. Data points that are even further away are represented as circles and describe outliers, i.e., extreme values. If you do not want outliers to be displayed and instead want the antennas to be displayed from Q1 to minimum and from Q3 to maximum, this is possible with the parameter `range=0`.

In R, the geom `geom_boxplot()` is used in the function `ggplot()` for the representation of box plots. The engine size of the cars is shown in a box plot. Before the output of the box plot, the quartiles, the IQR, and the maximum endpoints of the whiskers are calculated using the functions `mean()` and `quantile()`, which can also be visually read from the figure. A second box plot shows that it is also possible to display several box plots separated by grouping, here separated by class.

Box-Plot

```
#- Box-Plot--------------------------------------------------------------------
# Boxplot Autos: Berechnung der Kennzahlen
boxpl<-data.frame(Name=as.character(),Wert=as.numeric())
boxpl[1,]<-c("1. Quartil",round(quantile(mpg$Hubraum,0.25),1))
boxpl[2,]<-c(" Median, 2. Quartil",round(mean(mpg$Hubraum),1))
boxpl[3,]<-c("3. Quartil",round(quantile(mpg$Hubraum,0.75),1))
boxpl[4,]<-c("IQR",as.numeric(boxpl[3,2])-as.numeric(boxpl[1,2]))
boxpl[5,]<-c("Whisker Min",as.numeric(boxpl[1,2])-1.5*as.numeric(boxpl[4,2]))
boxpl[6,]<-c("Whisker Max",as.numeric(boxpl[3,2])+1.5*as.numeric(boxpl[4,2]))
boxpl                                                                # Anzeige

##                    Name Wert
## 1           1. Quartil  2.4
## 2  Median, 2. Quartil  3.5
## 3           3. Quartil  4.6
## 4                  IQR  2.2
## 5          Whisker Min -0.9
## 6          Whisker Max  7.9

# Boxplot mpg
ggplot(mpg,                                                          # Dataframe
  mapping = aes(y=Hubraum)) +                                        # y-Achse
  labs(title="Box-Plot",
       subtitle="Motorgröße der Autos",y="Hubraum in Litern") +
  geom_boxplot() +                                                   # Box-Plot
  theme(axis.title.x=element_blank(),                                # x-Achse ohne
        axis.text.x=element_blank(),                                 # Beschriftung
        axis.ticks.x=element_blank())
```

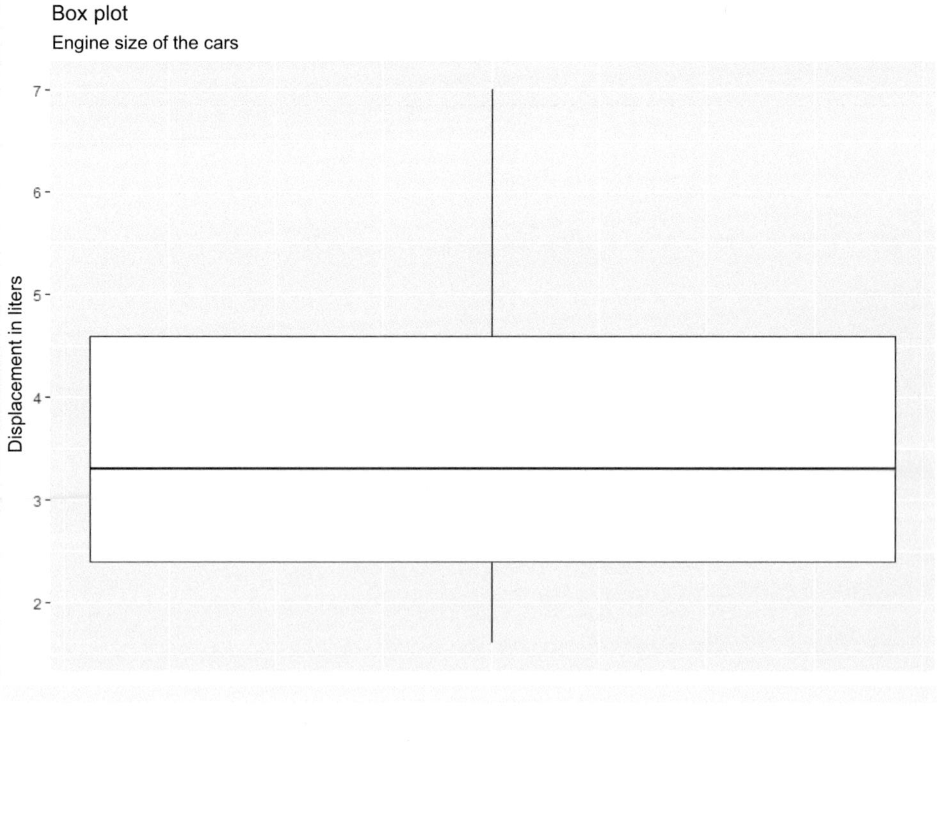

```
# Boxplot mpg getrennt nach Klasse
ggplot(mpg,                                                         # Dataframe
  mapping = aes(x=Klasse,y=Hubraum,color=Klasse)) +                 # x+y-Achsen
  labs(title="Box-Plot",
       subtitle="Motorgröße der Autos nach Klasse",
       x="Klasse", y="Hubraum in Litern")+
  geom_boxplot()+                                                   # Box-Plot
  scale_colour_manual(values=colour.own.nomin.2)
```

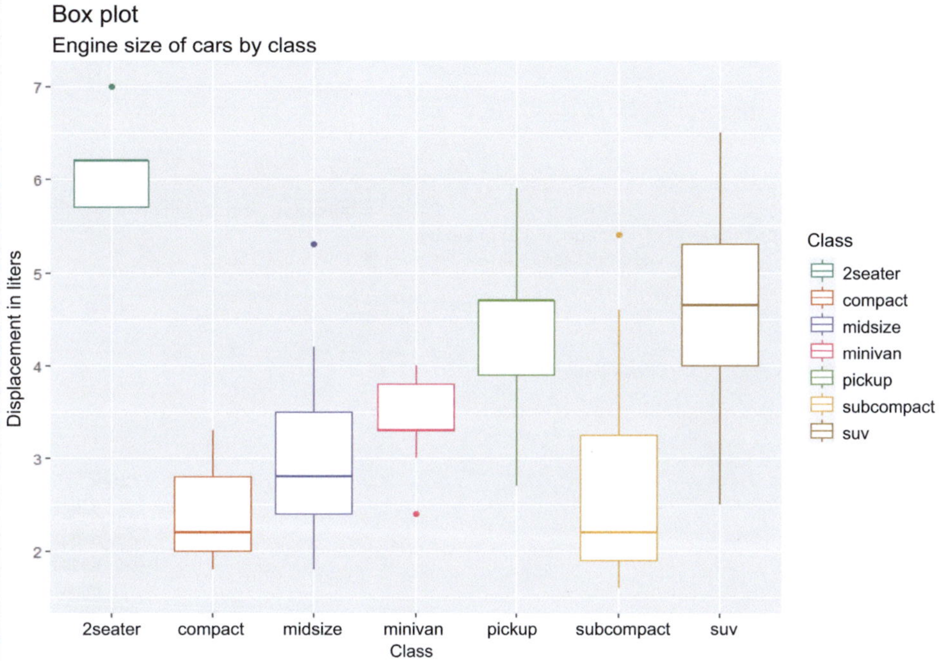

6.3.3.2.6 Scatter-Plot

A scatter plot is suitable for visualizing the relationship between numerical variables. The function `geom_point()` represents a point for each data set (row in the dataframe). Since the dataframe mpg describes 234 car models, there are numerous models with identical displacement and consumption. The representation of the points overlaps and therefore it is no longer recognizable how many models exist with a specific consumption. This is also referred to as **overplotting**. The function `geom_jitter()` solves the problem of overplotting by displaying the points next to each other so that all data points are recognizable. The parameter `width` can be used to determine how wide the scatter should be, the parameter `size` the size and the parameter `alpha` the transparency of the data points from 0=transparent to 1=non-transparent. The color can be assigned with the color parameter. If the color or another parameter such as shape (shape of the data point), alpha or size should be chosen depending on a variable, for example, the command `color=class` can be used if the variable class distinguishes different vehicle types.

With the function facet_wrap() you can also create your own illustration for each class with the following command: ggplot(data=mpg) + geom_point(mapping=aes(x=displacement,y=consumption,color= class))+facet_wrap(~class,nrow=3).

With ggplot() a regression line can also be displayed. This is done via the function geom_smooth(), e.g. geom_smooth(mapping=aes(x=displacement,y=consumption,group=class,color=class)).

Scatter-Plot

```
#- Scatter-Plot----------------------------------------------------------------
ggplot(data=mpg) +                                    # Ausgabe von Scatter-Plot
  geom_point(mapping=aes(x=Hubraum,y=Verbrauch,
                         color=Klasse))               # Farbzuweisung
```

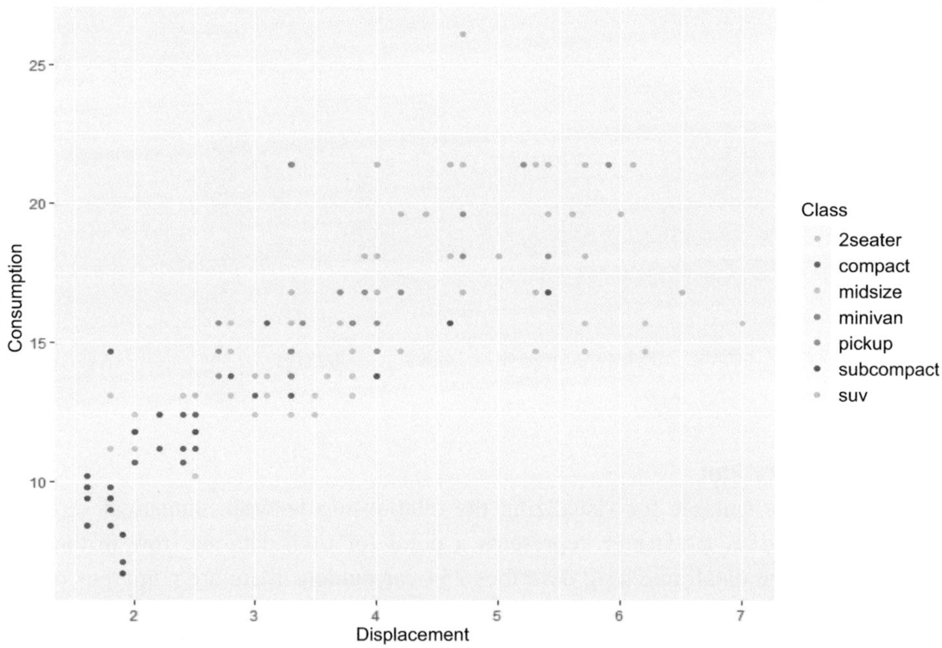


```
ggplot(data=mpg) +
  geom_jitter(mapping=aes(x=Hubraum,y=Verbrauch,     # Jitter
                          color=Klasse),
              width=0.3,alpha=0.3,size=3)            # Streuung,Transparenz,Größe
```

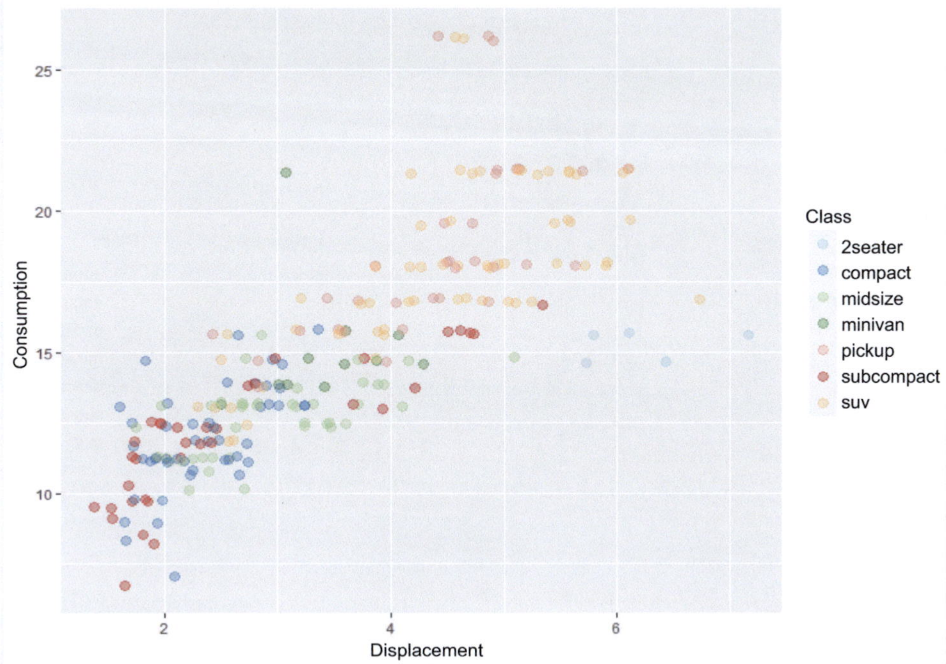


```
ggplot(data=mpg) +
  geom_point(mapping=aes(x=Hubraum,y=Verbrauch,
                         color=Klasse))+
  facet_wrap(~Klasse,nrow=3)                    # Abbildung je Klasse
```

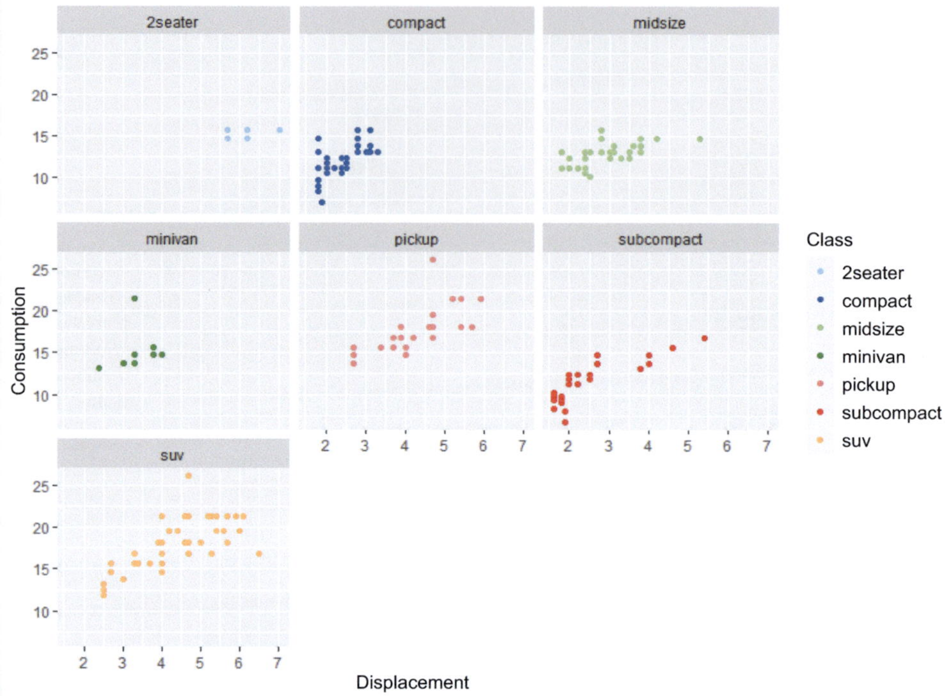


```
ggplot(data=mpg.teilmenge) +
  geom_point(mapping=aes(x=Hubraum,y=Verbrauch,
                         color=Klasse))+          # Mehrere geom_ durch +
  geom_smooth(mapping=aes(x=Hubraum,y=Verbrauch,  # Regressionslinie
                          group=Klasse,color=Klasse))
```

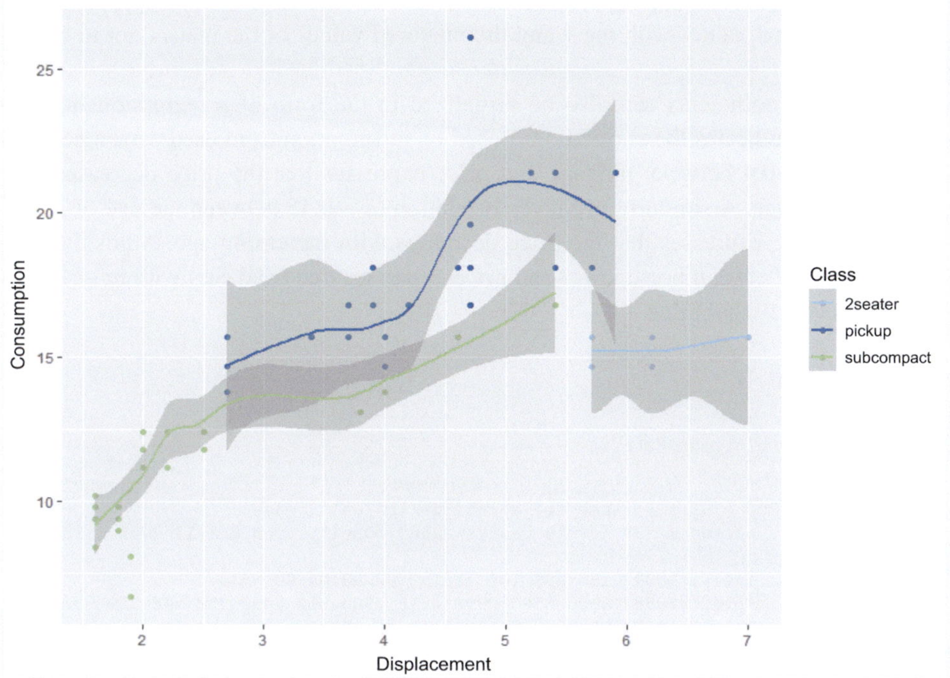

From the illustrations, it can be seen that there is a relationship between displacement and consumption and that consumption generally increases with increasing displacement. The differentiation of vehicle classes makes it clear that, for example, pickups generally have higher consumption than subcompact cars. All two-seaters of the car models have a large engine. Average values of engine size are therefore influenced by this vehicle class. Depending on the intended analyses, it may be important to exclude these vehicles from the analysis, for example, if only vehicles with 4 seats are to be examined.

6.3.3.2.7 Correlation Matrix

A correlation matrix (correlation chart) is suitable for representing the correlation of variables with each other in the form of a table with the Pearson correlation factors.

A correlation indicates the direction (positive or negative) and strength of a linear dependency between two variables. A positive correlation exists when high values of variable A coincide with high values of variable B. A negative correlation exists when high values of variable A coincide with low values of variable B. The strength of a correlation is expressed through the correlation coefficient, which lies between -1 (strong negative correlation) and +1 (strong positive correlation).

In R, the function `corPlot()` from the package `psych` is used for the representation of correlation matrices. The function is informed of the dataframe to be analyzed in the first parameter. Furthermore, the parameter scale = FALSE can be used to display the numerical values in a uniform font size. The parameter stars=TRUE causes the significance levels to be displayed and the parameters diag= FALSE and upper=FALSE cause the diagonal values with the 1 and the mirrored values of the matrix not to be displayed.

Correlation coefficients can also be visualized in the form of a regression line, the slope of which indicates the correlation. Based on information from used cars, a negative correlation of −0.4 between mileage and price expresses that the price decreases with increasing mileage. A similarly negative correlation of −0.45 between the age of a used car and the price expresses that the price decreases with increasing age. A positive correlation of 0.63 between horsepower and price expresses that used cars with more horsepower achieve a higher price.

Correlation Matrix

```
# Daten für Korrelations-Matrix -------------------------------------------------
str(kfz)

## spc_tbl_ [1,000 × 6] (S3: spec_tbl_df/tbl_df/tbl/data.frame)
##  $ Marke     : Factor w/ 39 levels "volkswagen","opel",..: 8 5 1 11 24 1 1 12 7 3
...
##  $ PS        : int [1:1000] 158 286 102 71 150 90 90 NA 90 NA ...
##  $ Preis     : int [1:1000] 5000 8500 8990 4350 7900 300 1990 250 590 350 ...
##  $ Alter     : int [1:1000] 12 19 7 9 10 21 18 16 19 16 ...
##  $ Kategorie: Ord.factor w/ 4 levels "Neuwertig"<"Mittelalt"<..: 3 3 2 2 3 4 3 3
3 3 ...
##  $ Kilometer: int [1:1000] 150 150 70 70 150 150 150 150 150 150 ...
##  - attr(*, "spec")=
##   .. cols(
##   ..   Marke = col_factor(levels = NULL, ordered = FALSE, include_na = FALSE),
##   ..   PS = col_integer(),
##   ..   Preis = col_integer(),
##   ..   Alter = col_integer(),
##   ..   Kategorie = col_factor(levels = NULL, ordered = FALSE, include_na = FALSE)
,
##   ..   Kilometer = col_integer()
##   .. )
##  - attr(*, "problems")=<externalptr>

# Ausgabe Korrelations-Matrix -------------------------------------------------
psych::corPlot(kfz[,c(2:4,6)],scale = FALSE,stars=TRUE,diag= FALSE,upper=FALSE)
```

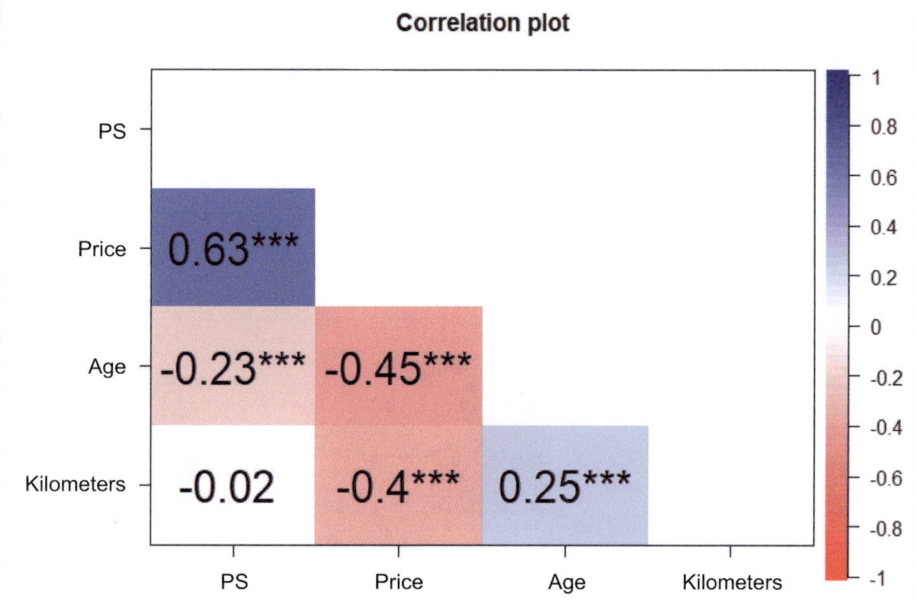


```
# Funktion für Korrelations-Plot basierend auf ggplot ---------------------------
ggplot.cor <- function(x,t) {
  colnames(x) <-c("x","y")
  ggplot(x,aes(x,y)) +
    geom_point(size=3,na.rm=TRUE) +
    geom_smooth(method="lm",se=FALSE,na.rm=TRUE,formula=y~x) +
    ggtitle("Korrelation:",round(cor(x,use="na.or.complete")[2],2)) +
    theme_minimal() +
    ylim(0,65000)
}
# Ausgabe Korrelations-Plot -----------------------------------------------------
p1 <-ggplot.cor(kfz[,c("Kilometer","Preis")]) + xlab("Kilometer in 1000") +
     ylab("Preis")
p2 <-ggplot.cor(kfz[,c("Alter","Preis")]) + xlab("Alter") +
     ylab("Preis")
p3 <-ggplot.cor(kfz[,c("PS","Preis")]) + xlab("PS") +
     ylab("Preis")
grid.arrange(p1,p2,p3,ncol=3)
```

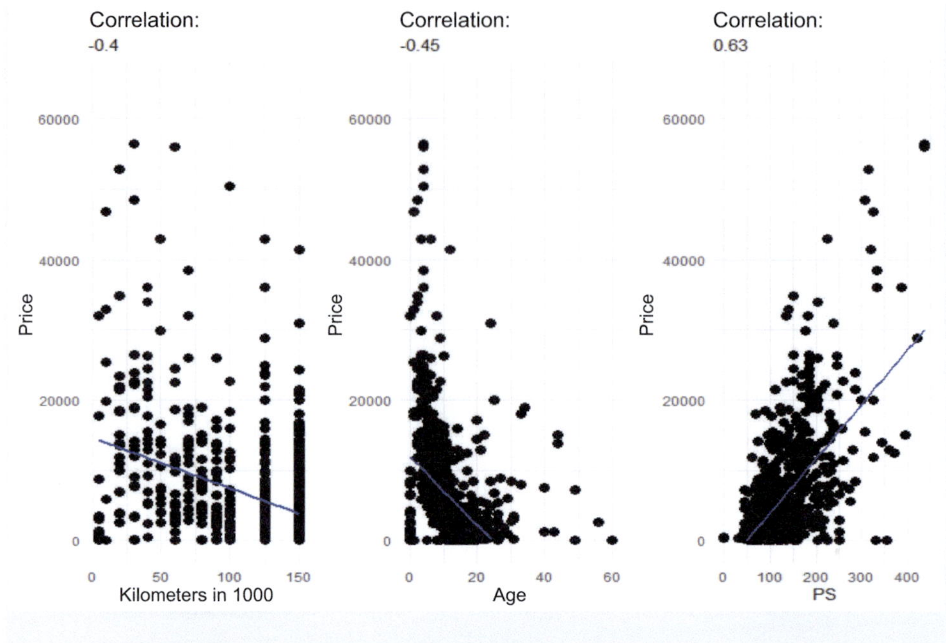

Another way to visualize the relationship between variables is a scatter plot matrix, which for metric variables shows not only the correlation coefficient but also a scatter plot and a distribution curve in the diagonal. Depending on the variable type, for ordinal variables instead of the scatter plot and the distribution curve, a bar chart is shown and instead of the correlation factors, a box plot is shown. In R, this is supported by the function `ggpairs()` from the package `GGally`.

Scatter Plot Matrix

```
#- Ausgabe Scatter-Plot-Matrix nur metrische Variablen ---------------------------
GGally::ggpairs(kfz,columns=c("PS","Preis","Alter","Kilometer"))
```

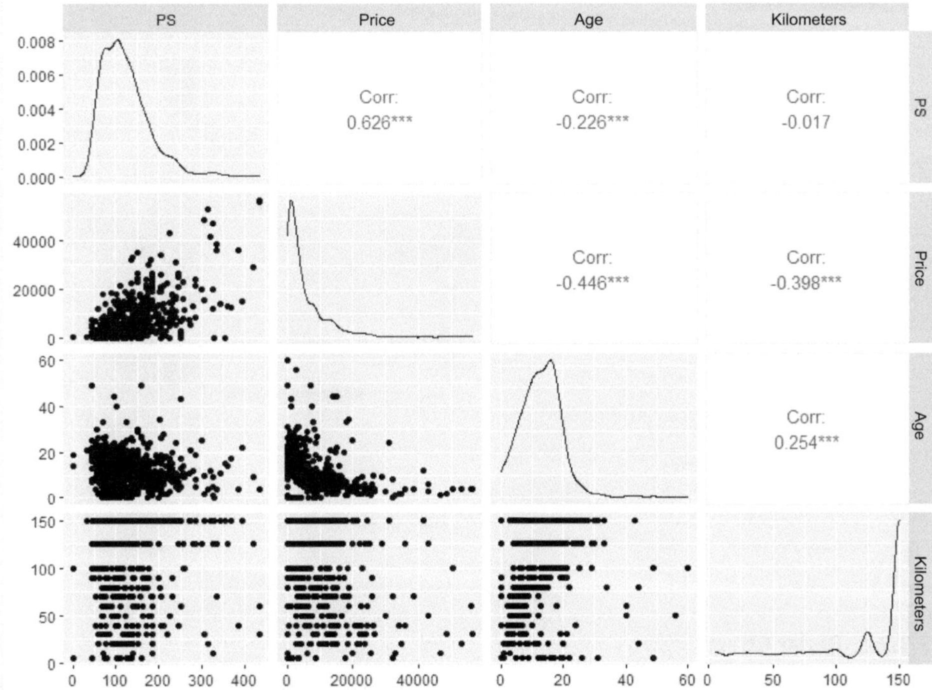


```
kfz.sub <- kfz[1:50,]                                    # Reduktion auf 50 Datensätze
#- Ausgabe Scatter-Plot-Matrix metrische und ordinale Variablen ------------------
GGally::ggpairs(kfz.sub,columns = c("PS","Preis","Alter","Kilometer","Kategorie"),
                aes(color=Kategorie),
                upper=list(continuous=GGally::wrap("cor",size=2)),
                diag =list(continuous=GGally::wrap("densityDiag", alpha=0.5)))
```

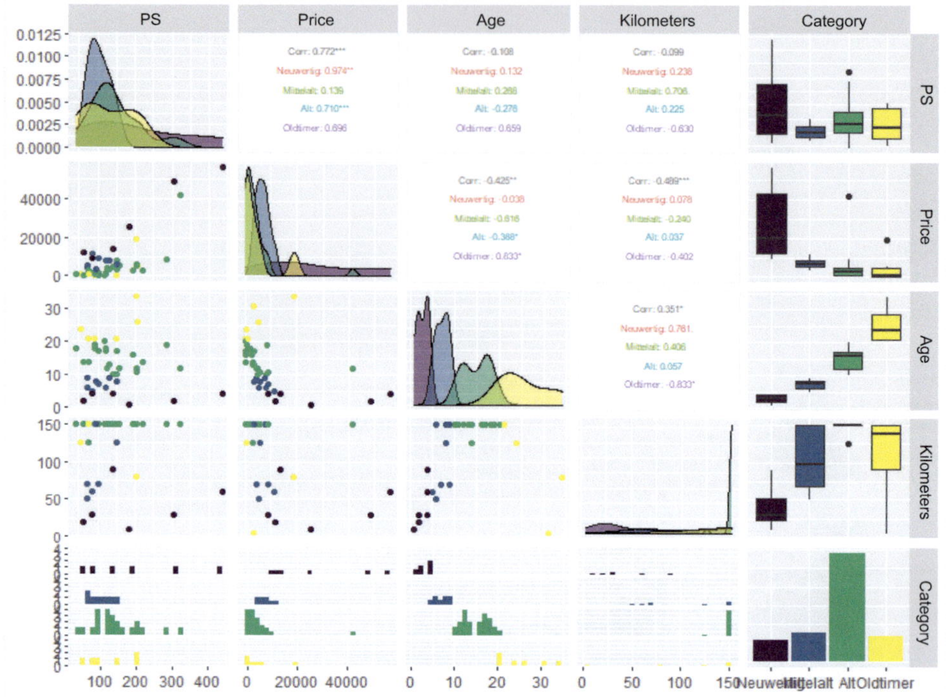

6.3.3.2.8 Radar Chart

A radar chart (radar diagram) is suitable for visualizing the relationship of independent variables to a dependent variable.

In R, a radar chart is supported by the function `radarchart()` from the package `fmsb`. The parameter `plty` allows a specific line type, `pcol` a line color and `pfcol` a fill color for the area to be selected. The function `legend()` enables the display of a legend.

Radar Chart

```
#- Radar-Chart ----------------------------------------------------------------------
set.seed(126)
df <- data.frame(rbind(rep(10,7), rep(0,7),
                 matrix(sample(0:10,21,replace=TRUE),
                        nrow = 3)))
colnames(df) <- paste0("Variable-", 1:7)
# Organisationsspezifische Farbpaletten laden-----------------------------------
library(fmsb)
farbe<-colour.own.nomin.2[1:3]
farbe.alpha<-paste0(colour.own.nomin.2[1:3],40)       # 0.25 durchsichtig
radarchart(df,plty=1,pcol=farbe,pfcol=farbe.alpha)
legend("topright", legend = paste0("Kunde-", 1:3),
       pch=20, col = farbe)
```

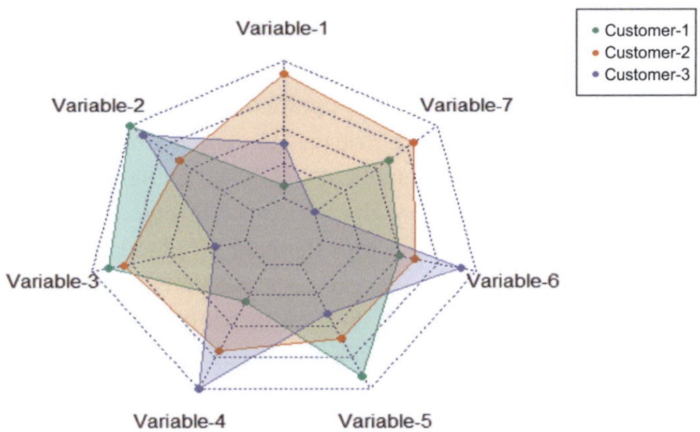

6.4 Modeling and Evaluation

We often use analytics applications that employ statistics in everyday life as a matter of course, whether it's the navigation system in the car, facial recognition on the phone, or the weather forecast. Artificial intelligence and machine learning enable ever new areas of application. These applications are based on models that should be as accurate as possible. After modeling and repeatedly thereafter, the evaluation of the model is necessary to determine its quality. How exactly does a model work and how can it be evaluated?

For example, what benefit does a model provide for retail that uses weather information among other things (Hertweck and Kinitzki 2015)? During the hurricane season in the USA, retailers have repeatedly found that the usually demanded range of goods changes as soon as a hurricane is expected. In the case of an approaching hurricane, significantly more canned goods and flashlight batteries are purchased, among other things. To verify this observation in the stores, the sales data were analyzed afterwards and the observation could be confirmed. To derive a benefit from this knowledge, a decision model for customer buying behavior in anticipation of a hurricane was created to enable a forecast of sales per product group and product. From this, in connection with the contribution margins of the products, it was determined which change in the product range would result in an optimal increase in sales and profit. Thus, perishable fresh goods are now reduced in the range and the proportion of canned goods is increased when hurricanes are approaching, which increases both sales and profit. Based on weather forecasts, orders are even automatically initiated so that the goods arrive in the stores in time from the suppliers. In addition to climate and weather data, the perceived threat level of consumers can also be taken into account via text analytics, based on comments in social networks. The perceived threat has a significant influence on buying behavior, in addition to the official threat level of a hurricane. In conjunction with data on current inventory levels, orders can then be automatically initiated using artificial intelligence to be prepared for the expected demand with an adequate inventory in the store.

In order to answer a question with the help of a model, data is needed that relates to the research subjects being investigated, for example, sales data and weather data in the above example. Unfortunately, it is often difficult or impossible to collect all relevant data for a complete population, the entire group of all objects considered, for example, all customers. On the one hand, it is very time-consuming and on the other hand, it is too expensive or simply impossible, for example, for a buyer to find the best price for a product, as there are very many providers of a product worldwide. Therefore, in reality as well as in research, one is usually forced to work with a sample or subset of the population. It is important that a sample is as representative as possible for the population, as one wants to be able to infer from the findings of the sample to the population, which is referred to as inference. Descriptive statistics and an understanding of data form the basis for inferential or inductive statistics, which deal with models, hypotheses, and predictions.

The goal of inductive statistics is to make statements that apply to more than just the individuals or objects for which data is available. For example, it may be about estimating the risk of a disease or making a prediction about what percentage of the population intends to vote for a party or how the demand for a product changes when the price is adjusted. Weather forecasts, estimates of arrival time by the navigation system, and many other functions are applications of inductive statistics that are based on models.

Of course, statements about the future are always fraught with uncertainty, and the same applies to statements about entire populations such as all students, all citizens, or all products, as these statements are based on incomplete information. The information is incomplete, among other things, when it relates to the future, because the data cannot be available at all. The information is also incomplete when it refers to a total population, because in reality it is often impossible to collect complete information about a total population. A current survey of all eligible voters in a constituency on a specific date would already fail because some members of this group of people are not reachable, for example, on vacation, in the hospital, do not own a phone, or do not have a telecommunications connection.

Therefore, inductive statistics are satisfied with incomplete information. However, this also means that the results of inductive statistics are always imprecise. The nice thing, however, is that the risk can be quantified and thus one is able to correctly interpret the results of inductive statistics and to determine the acceptable level of risk in self-applied procedures of inductive statistics. All of this also applies to the models of machine learning and the applications of artificial intelligence based on them. **Machine learning and applications of artificial intelligence are based on statistics and accordingly, predictions are always imprecise and risky**.

The level of risk depends on how a sample (see Sect. 6.4.2) is drawn, the so-called drawing procedure. The simple availability of data can be tempting, even if it may be unsuitable. Therefore, always pay attention to the quality of the data underlying the analyses when interpreting results and applying inductive statistics yourself.

6.4.1 Modeling

Modeling is always aimed at answering a question or solving a problem.

Suppose we want to find out which factors influence the number of flights at New York airports in order to be able to make as accurate a prediction as possible of the number of flights. If we first want to predict the number of flights per month, we can use the average of all months as a simple model for the prediction. This is 28,064 flights per month. If one looks at the prediction error (residual), it is 1245 per month. The mean model can also be applied to predict flights per day. The number of flights averages 923. If the mean is used for prediction, a prediction error (residual) of 91 flights per day results.

A closer examination of the number of flights per day shows that this strongly depends on the day of the week. There are significantly fewer flights on Saturdays. With the function `lm()`, a linear regression model for prediction can be created, which also takes into account the day of the week. With the function `add_predictions()` from the modelr package, a prediction value can then be generated for this model for each day of the week (the prediction values are represented in the following coding with illustrations as red dots). The prediction errors, the so-called residuals, can be output with the function `add_residuals()`. Residuals indicate the error for an observation, i.e., the difference between the predicted value and the real value. The prediction error is now only 48 flights per day instead of 91, after the weekday averages were used for the prediction. This represents an improvement over the mean value model by 43. What can be seen from this example is that modeling is a creative activity, where it may be possible to improve the results with increasing knowledge gain, and that this is significantly influenced by the consideration of suitable predictors. It has proven valuable to supplement the model with the day of the week. The basis for this addition to the model was a better understanding of the data through exploratory data analysis. **Modeling is a creative process that benefits from a good understanding of data and can thus be continuously improved**.

If the EDA is continued, further refinements of the model are possible. If you display the outliers, where the deviation from the real flights for a day is more than 200 flights, these include the 4.7. (July 4th), 28.+29.11. (Thanksgiving) and 25.12. (Christmas), i.e., holidays in the USA. The linear model does not yet take these into account in a prediction based on weekdays. One of many further refinements of the model could therefore be made by also taking holidays into account.

Modeling Flight Data

```
## Modellierung ---------------------------------------------------------------
library(lubridate)                              # Paket für Datumsoperationen
library(modelr)                                 # Paket zur Modellierung
options(na.action = na.warn)                    # Verhinderung von Warnung
Sys.setlocale("LC_TIME", "German")              # Windows Betriebssystem, Datum=D

## [1] "German_Germany.1252"

flug <- flug %>%
  mutate(Datum = make_date(year,month,day),
         Wochentag = wday(Datum,label=TRUE,locale = Sys.getlocale("LC_TIME"),
                          week_start=getOption("lubridate.week.start", 1)),
         Monat=month(Datum, label = TRUE, locale = Sys.getlocale("LC_TIME")))
# Einfaches Vorhersagemodell basierend auf Mittelwert je Monat--------------------
flug.monat <- flug[!is.na(flug$Monat),] %>%
  group_by(Monat) %>%
  summarise(Anzahl = n())
mod.mw<-mean(flug.monat$Anzahl)                  # Mittelwert-Modell
mod.mw                                           # Vorhersagewert für alle Monate

## [1] 28064.67

vorhersage.mw<-flug.monat                        # Vorhersage
vorhersage.mw$Anzahl<-mod.mw
ggplot(flug.monat,aes(Monat, Anzahl)) +
  geom_point(colour="blue") +
  geom_linerange(aes(ymin=mod.mw,ymax=Anzahl),colour="grey40") +
  labs(title="Flüge je Monat", x="Monat",y="Anzahl Flüge")+
  geom_point(data=vorhersage.mw,colour="red",size=2)
```

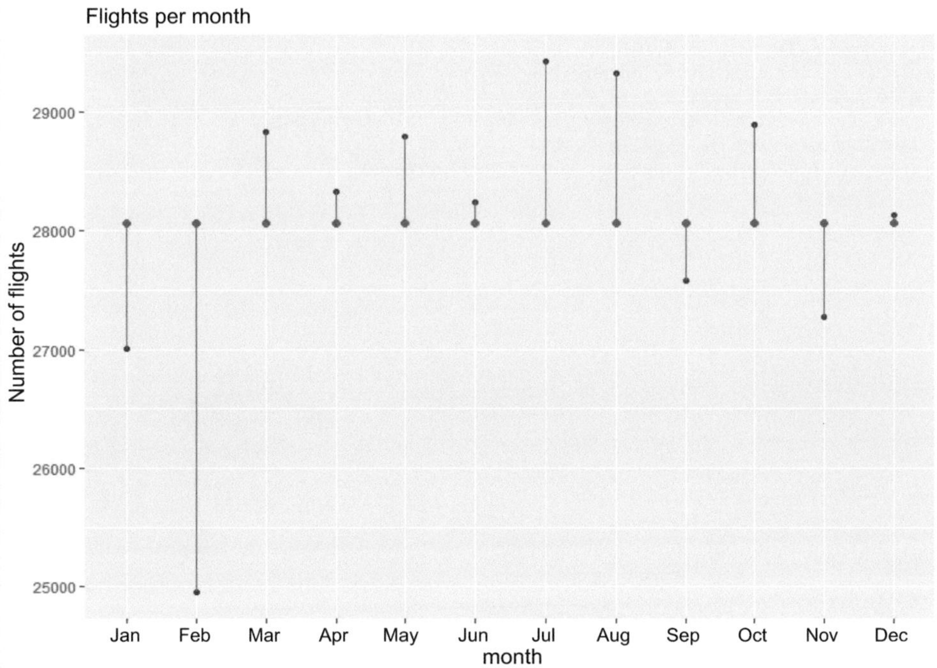

```
flug.monat$Residual<-vorhersage.mw$Anzahl-flug.monat$Anzahl
flug.monat                                              # Reale Flüge und Residual

## # A tibble: 12 × 3
##    Monat Anzahl Residual
##    <ord>  <int>    <dbl>
##  1 Jan    27004   1061.
##  2 Feb    24951   3114.
##  3 Mrz    28834   -769.
##  4 Apr    28330   -265.
##  5 Mai    28796   -731.
##  6 Jun    28243   -178.
##  7 Jul    29425  -1360.
##  8 Aug    29327  -1262.
##  9 Sep    27574    491.
## 10 Okt    28889   -824.
## 11 Nov    27268    797.
## 12 Dez    28135    -70.3

mod.mw.sd<-sd(flug.monat$Residual)                      # Standardabweichung Residual
mod.mw.sd                                               # Mittelwert-Modell

## [1] 1245.091

# Einfaches Vorhersagemodell basierend auf Mittelwert je Tag----------------------
flug.tag <- flug[!is.na(flug$Datum),] %>%
  group_by(Datum,Wochentag) %>%
  summarise(Anzahl = n())
mod.mw<-mean(flug.tag$Anzahl)                           # Mittelwert-Modell
mod.mw                                                  # Vorhersagewert für alle Tage

## [1] 922.674

vorhersage.mw<-flug.tag                                 # Vorhersage
vorhersage.mw$Anzahl<-mod.mw
ggplot(flug.tag,aes(Datum, Anzahl)) +
  geom_point(colour="blue") +
  geom_linerange(aes(ymin=mod.mw,ymax=Anzahl),colour="grey40") +
  labs(title="Flüge je Tag", x="Datum",y="Anzahl Flüge")+
  geom_point(data=vorhersage.mw,colour="red",size=2)
```

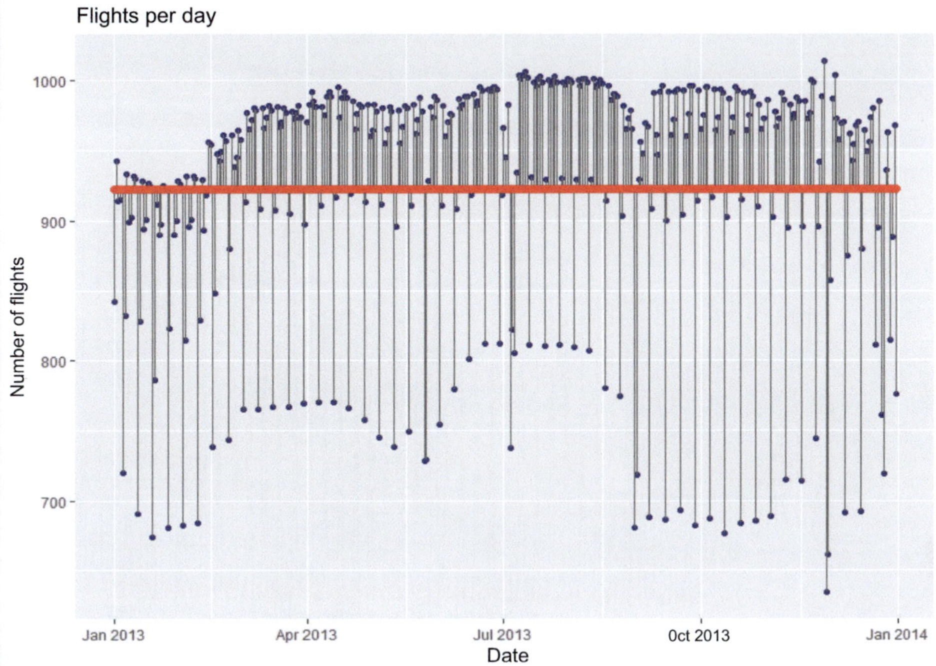


```
flug.tag$Residual<-vorhersage.mw$Anzahl-flug.tag$Anzahl
flug.tag %>%
  ggplot(aes(Datum,Residual)) +
```

```
  geom_ref_line(h = 0) +
  geom_line()
```

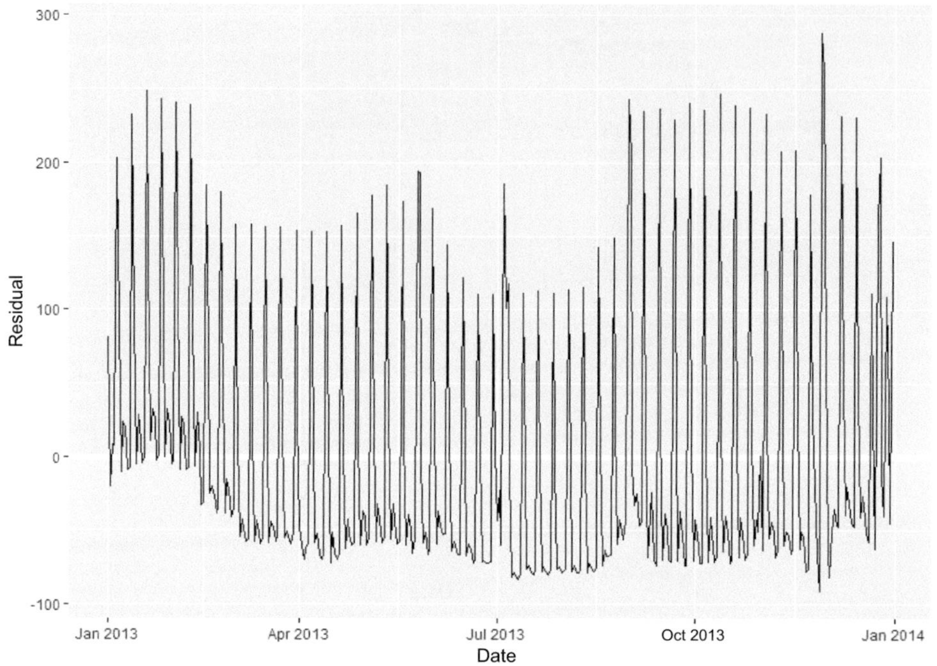


```
head(flug.tag)                                    # Reale Flüge und Residual

## # A tibble: 6 × 4
## # Groups:   Datum [6]
##   Datum      Wochentag Anzahl Residual
##   <date>     <ord>      <int>    <dbl>
## 1 2013-01-01 Di           842    80.7
## 2 2013-01-02 Mi           943   -20.3
## 3 2013-01-03 Do           914     8.67
## 4 2013-01-04 Fr           915     7.67
## 5 2013-01-05 Sa           720   203.
## 6 2013-01-06 So           832    90.7

mod.mw.sd<-sd(flug.tag$Residual)                  # Standardabweichung Residual
mod.mw.sd                                         # Mittelwert-Modell

## [1] 91.09677

# Einfaches Vorhersagemodell basierend auf Mittelwert je Wochentag--------------
ggplot(flug.tag, aes(Datum, Anzahl)) +
  geom_line()+
  labs(title="Flüge je Tag", x="Datum",y="Anzahl Flüge")
```

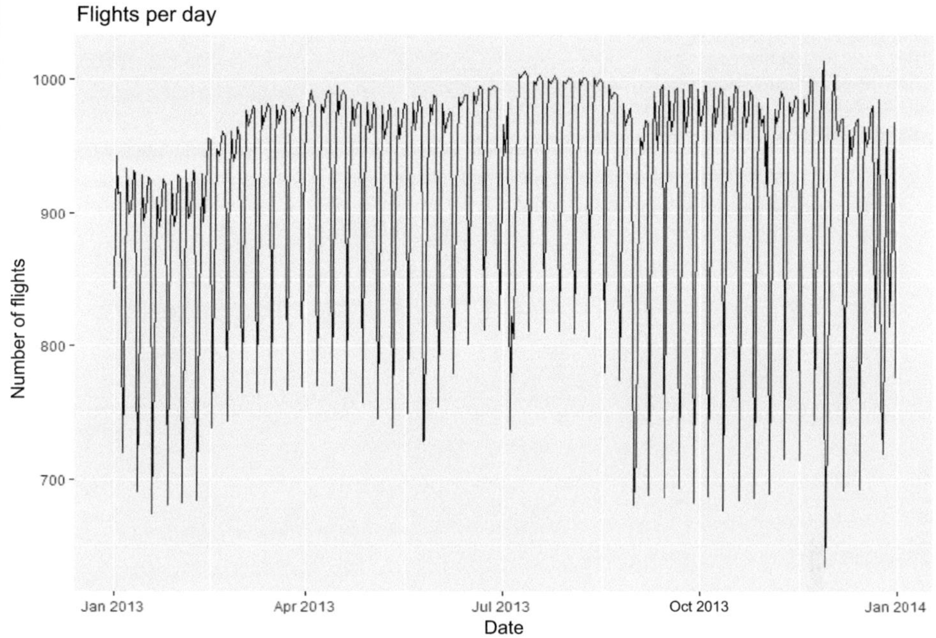


```
ggplot(flug.tag, aes(Wochentag, Anzahl)) +
  geom_boxplot()+
  labs(title="Flüge je Wochentag", x="Wochentag",y="Anzahl Flüge")
```

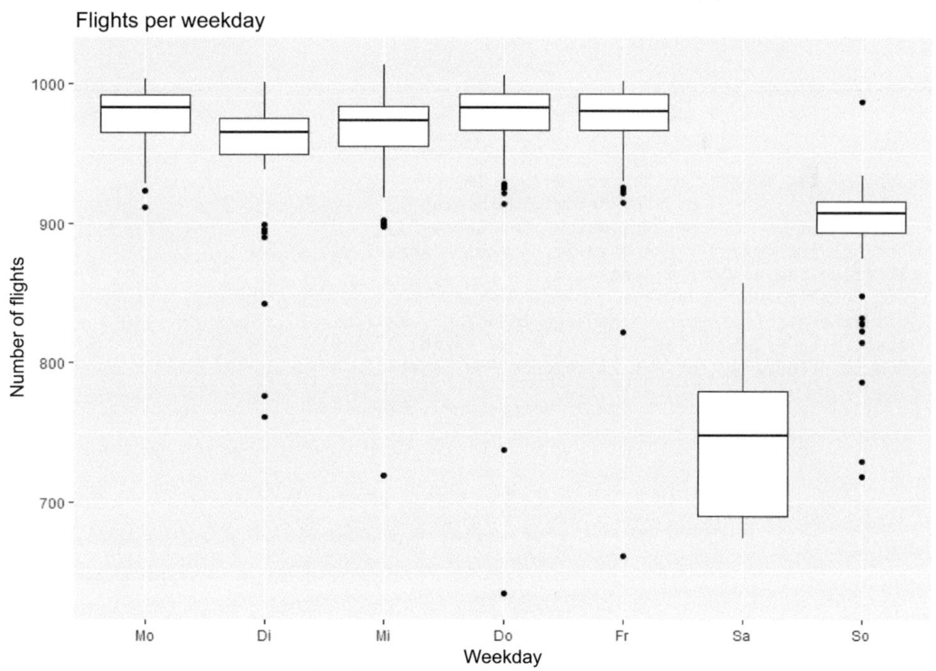

```
mod.rg.lm<-lm(data=flug.tag,Anzahl~Wochentag)  # Lineares Modell
vorhersage.lm<-flug.tag %>%                    # Vorhersage
  data_grid(Wochentag) %>%
  add_predictions(mod.rg.lm,"Anzahl")
ggplot(flug.tag,aes(Wochentag, Anzahl)) +
  geom_boxplot() +
  labs(title="Flüge je Wochentag mit Mittelwert", x="Wochentag",y="Anzahl Flüge")+
  geom_point(data=vorhersage.lm,colour="red",size=5)
```

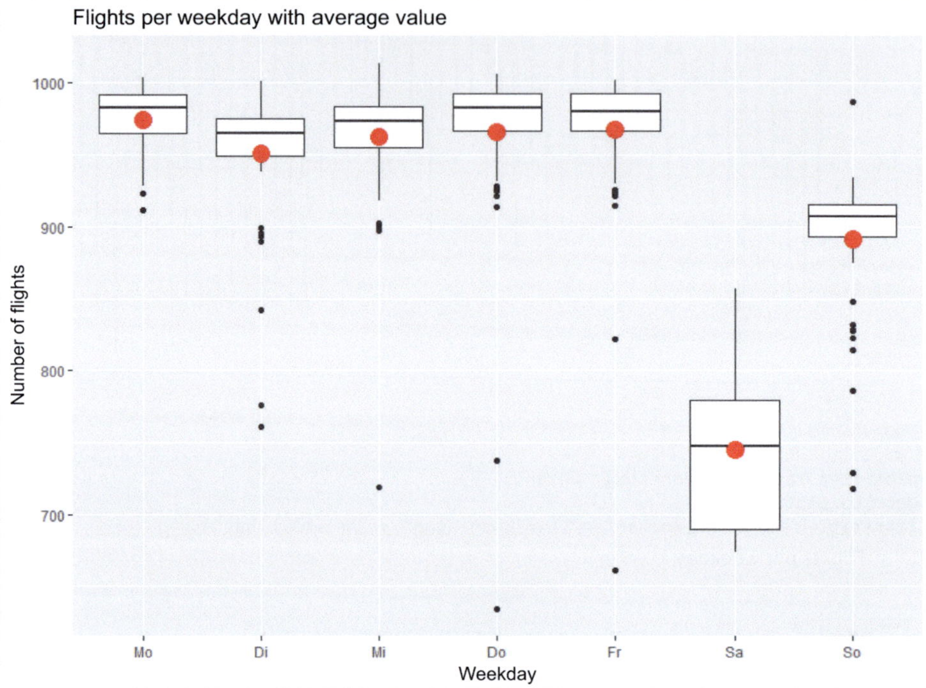


```
vorhersage.lm<-merge(flug.tag,vorhersage.lm,
                     by.x=c("Datum","Wochentag"),
                     by.y=c("Datum","Wochentag"))
vorhersage.lm<-rename(vorhersage.lm,"Anzahl"="Anzahl.y")
ggplot(flug.tag,aes(Datum, Anzahl)) +
  geom_point(colour="blue") +
  geom_linerange(aes(ymin=vorhersage.lm$Anzahl,ymax=Anzahl),colour="grey40") +
  labs(title="Flüge je Tag", x="Datum",y="Anzahl Flüge")+
  geom_point(data=vorhersage.lm,colour="red",size=2)
```

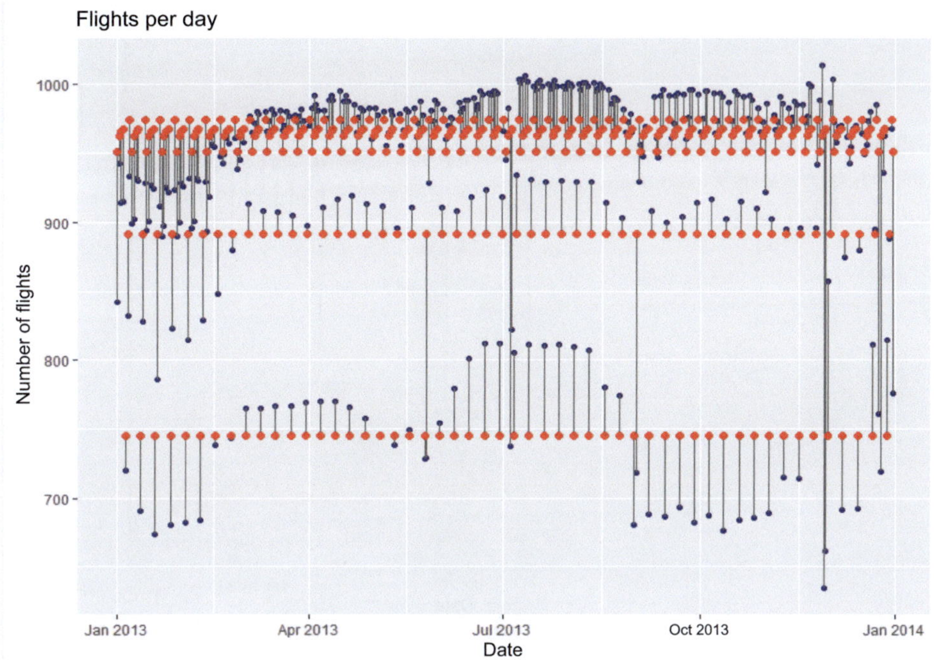


```
flug.tag <- flug.tag %>%                    # Neue Spalte für Residual
  add_residuals(mod.rg.lm,"Residual")
flug.tag %>%
  ggplot(aes(Datum,Residual)) +
  geom_ref_line(h = 0) +
  geom_line()
```

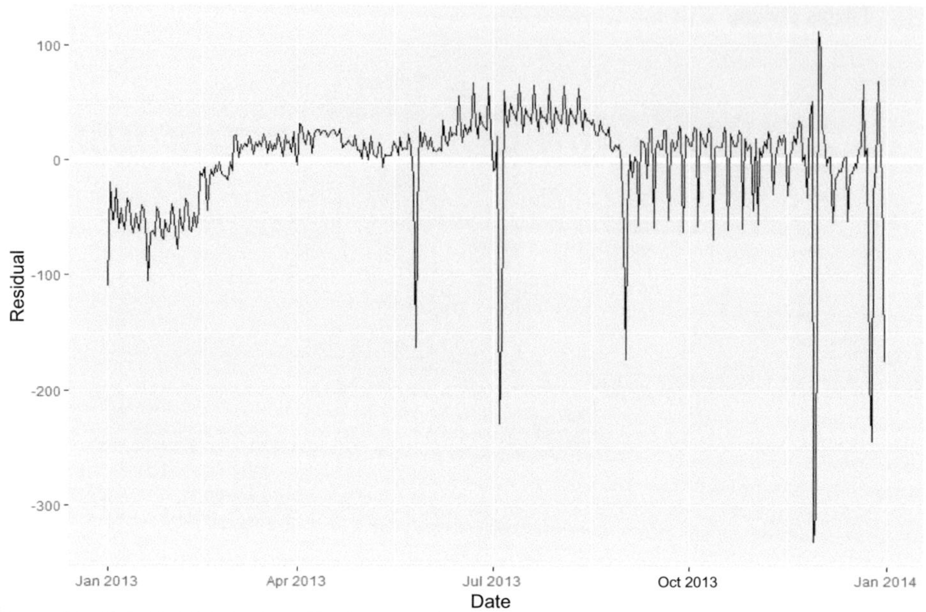


```
filter(flug.tag,Residual>200 | Residual< -200)# Schlechte Vorhersage

## # A tibble: 4 × 4
## # Groups:   Datum [4]
##   Datum      Wochentag Anzahl Residual
##   <date>     <ord>      <int>    <dbl>
## 1 2013-07-04 Do           737    -229.
## 2 2013-11-28 Do           634    -332.
## 3 2013-11-29 Fr           661    -306.
## 4 2013-12-25 Mi           719    -244.

modell.lm.sd<-sd(flug.tag$Residual)          # Standardabweichung Residual
modell.lm.sd                                 # Lineares Modell: 48 Flüge

## [1] 48.39272
```

Another example of modeling is the prediction of used car prices. The modeling will be deepened using this example in the following.

Before starting with the modeling, it is helpful to take a closer look at the data. The histogram and a box plot of used car prices can be displayed using the function `ml.summary()`. It shows that the majority of cars have a price between 1000 and 5000 US$ and there are many outliers, which are visible in the box plot as points above the box. A scatter plot of the variables age and horsepower shows that there is a correlation between these variables and the price. The function `cor.test()` allows calculating the correlation coefficients including confidence intervals and significance. The correlation coefficients of 0.6 for horsepower, −0.67 for age, and -0.48 for kilometers confirm that these variables could be suitable predictors for the price.

Since PS and age have the strongest correlations of the variables considered, a simple regression model can be created with the function `lm()`. The command `autos.modell.lr<-lm(data=autos,Preis ~ (Alter+PS))` calculates the formula for a linear regression function with the prediction variable price based on the predictor variables age and PS and the data in the dataframe autos. The resulting formula of the model autos.modell.lr (coefficients in the following coding) is Price = 9052.7 –859.64 * Age + 58.52 * PS. This means that the price decreases by 860 US$ each year and increases by 59 US$ for each additional PS.

If one displays a scatter plot of the model with the predicted prices on the x-axis and the prediction errors (residuals = predicted price – real price) on the y-axis for all 34,864 used cars using the command `plot(autos.modell.lr,which=1)`, it can be seen that the prediction error (residual) for many data sets is approximately zero (black area), but there are still quite a few data sets with a large deviation that are far from zero.

With the function `predict()`, prediction values for the price of used cars can also be generated, e.g. `predict(autos.modell.lr,new_data=autos)`. The **prediction errors**, the so-called **residuals**, can again be determined with the function `add_residuals()`. If one now looks at the real price, the prediction and the prediction error, this simple model results in an average prediction error of 4426 US$ per used car. Among the predictions, there were 224 predictions with a deviation of more than 20,000 US$.

Modeling Used Car Data (Part 1)

```
## Modellierung -------------------------------------------------------------------
# Funktion zum Plot der Korrelation---------------------------------------------------
corplot <- function(df) {
  colnames(df) <-c("x","y")
  ggplot(df,aes(x,y)) +
    geom_jitter(mapping=aes(x=x,y=y),   # Jitter
           width=0.3,alpha=0.3,size=3,na.rm=TRUE)+
    geom_smooth(method="lm",se=FALSE,na.rm=TRUE,formula=y~x) +
    labs(title="Scatterplot",
      subtitle=paste("Korrelation:",round(cor(df,use="na.or.complete")[2],2)))+
    theme_minimal() +
    ylim(0,75000)
}
# Visualisierung der Daten-----------------------------------------------------------
ml.summary(autos$Preis,                               # Daten verstehen
        "Gebrauchtwagenpreis","Preis in US$",box=T)
```

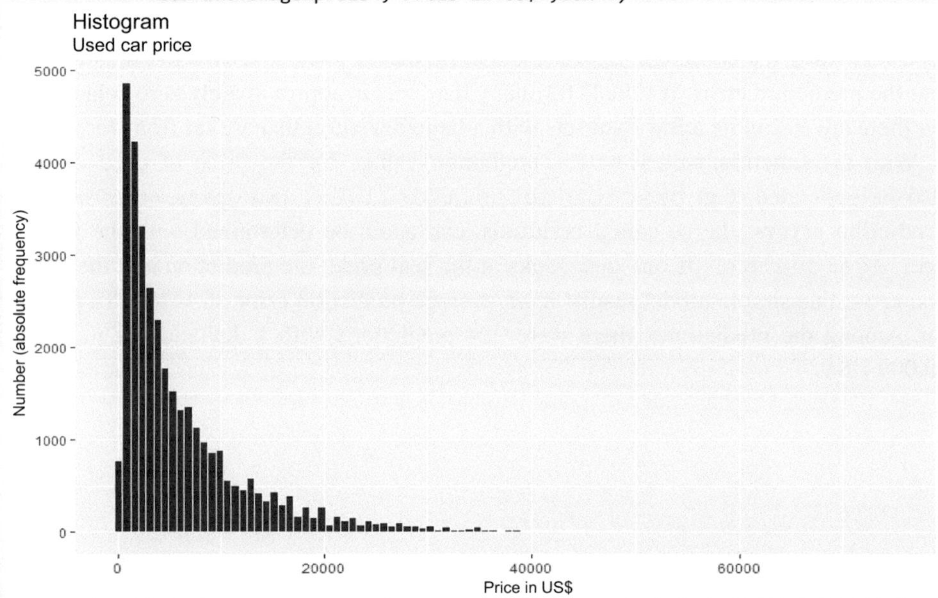

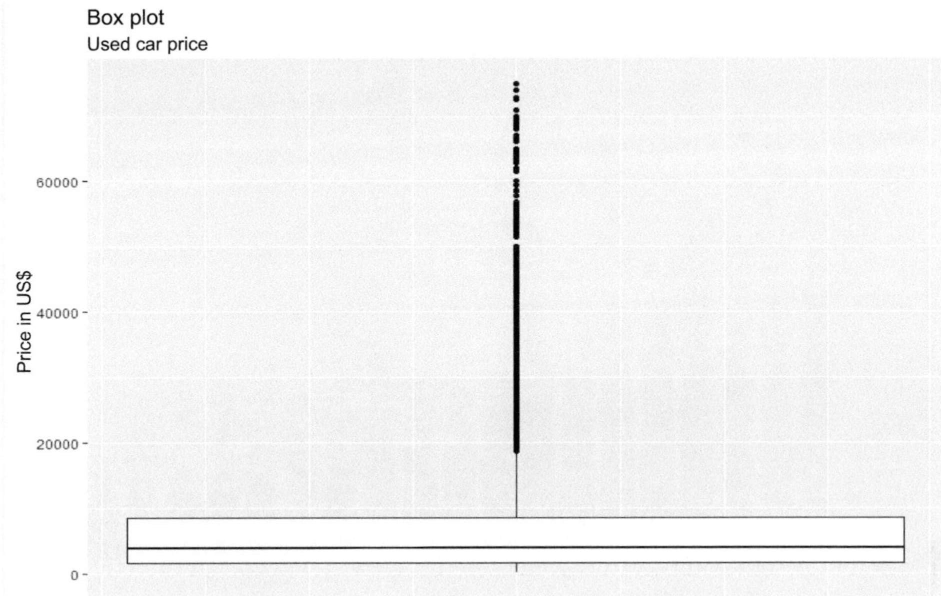


```
##                25%
## anzahl         "34864"
## anzahl.na      "0"
## modus          "1500"
## median         "3900"
## mean           "6540.34"
## min            "110"
## max            "74999"
## sd             "7476.86"
## q1             "1650"
## q3             "8500"
## iqr            "6850"
## whisker.min    "-8625"
## whisker.max    "18775"
## skewness       "2.66"
## skewness.txt "Rechte Schiefe: 2.66 > 0, positive Schiefe,linkssteil,rechtsschief"
## kurtosis       "10.76"
## kurtosis.txt "Steilgipflig mit Exzess Kurtosis 10.76 > 0"

corplot(autos[,c("Alter","Preis")]) + xlab("Alter") + ylab("Preis in US$")
```

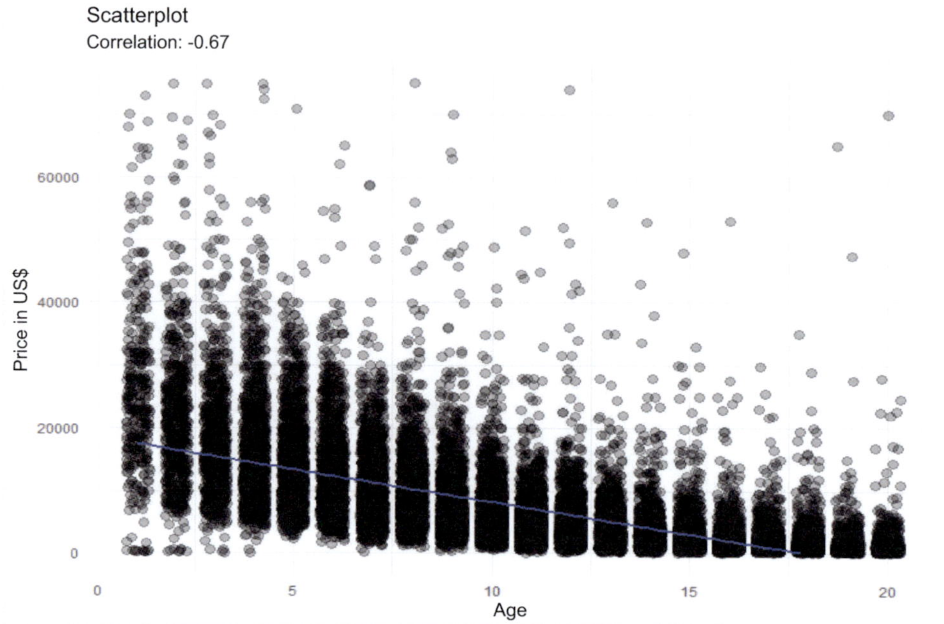


```
corplot(autos[,c("PS","Preis")]) + xlab("PS") + ylab("Preis in US$")
```

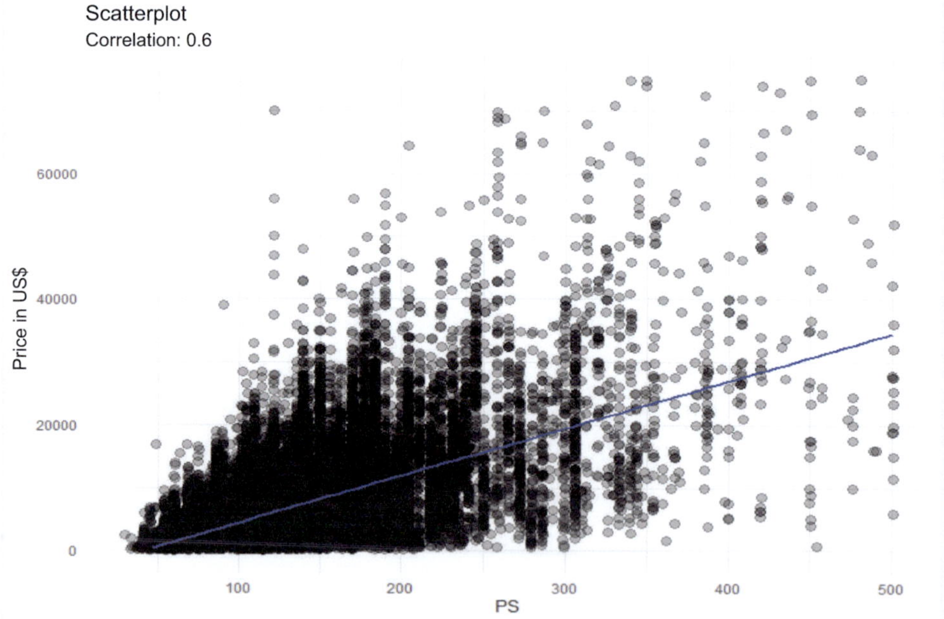

```
# Analyse möglicher Faktoren für die Vorhersage des Preises------------------------
autos.korr<-map(autos %>% select(PS,Alter,Kilometer),cor.test,y=autos$Preis)
autos.korr                                                  # Korrelationen

## $PS
##
##  Pearson's product-moment correlation
##
## data:  .x[[i]] and autos$Preis
## t = 139.2, df = 34862, p-value < 2.2e-16
## alternative hypothesis: true correlation is not equal to 0
## 95 percent confidence interval:
##  0.5909264 0.6044204
## sample estimates:
##       cor
## 0.5977157
##
##
## $Alter
##
##  Pearson's product-moment correlation
##
## data:  .x[[i]] and autos$Preis
## t = -167.34, df = 34862, p-value < 2.2e-16
## alternative hypothesis: true correlation is not equal to 0
## 95 percent confidence interval:
##  -0.6731983 -0.6615555
## sample estimates:
##        cor
## -0.6674176
##
##
## $Kilometer
##
##  Pearson's product-moment correlation
##
## data:  .x[[i]] and autos$Preis
## t = -102.93, df = 34862, p-value < 2.2e-16
## alternative hypothesis: true correlation is not equal to 0
## 95 percent confidence interval:
##  -0.4907729 -0.4746714
## sample estimates:
##        cor
## -0.4827629

tidy(autos.korr[[1]])                                       # Dataframe ein Element

## # A tibble: 1 × 8
##   estimate statistic p.value parameter conf.low conf.high method     alternative
##      <dbl>     <dbl>   <dbl>     <int>    <dbl>     <dbl> <chr>      <chr>
## 1    0.598      139.       0     34862    0.591     0.604 Pearson's… two.sided

autos.korr.df<-autos.korr %>%                               # Df alle Elemente
  map_dfr(tidy, .id = "Prädiktor")
autos.korr.df

## # A tibble: 3 × 9
##   Prädiktor estimate statistic p.value parameter conf.low conf.high method
##   <chr>        <dbl>     <dbl>   <dbl>     <int>    <dbl>     <dbl> <chr>
## 1 PS           0.598      139.       0     34862    0.591     0.604 Pearson's p…
## 2 Alter       -0.667     -167.       0     34862   -0.673    -0.662 Pearson's p…
```

```
## 3 Kilometer    -0.483      -103.      0      34862   -0.491     -0.475 Pearson's p…
## # ℹ 1 more variable: alternative <chr>

autos.korr.df %>%
  ggplot(aes(x=Prädiktor,y=estimate)) +
  geom_point(aes(y=estimate)) +
  geom_errorbar(aes(ymin = conf.low, ymax = conf.high)) +
  labs(title="Prädiktorenanalyse für Preis",
       subtitle="Variablen mit Konfidenzintervall",
       x="Variable",y="Korrelation")
```

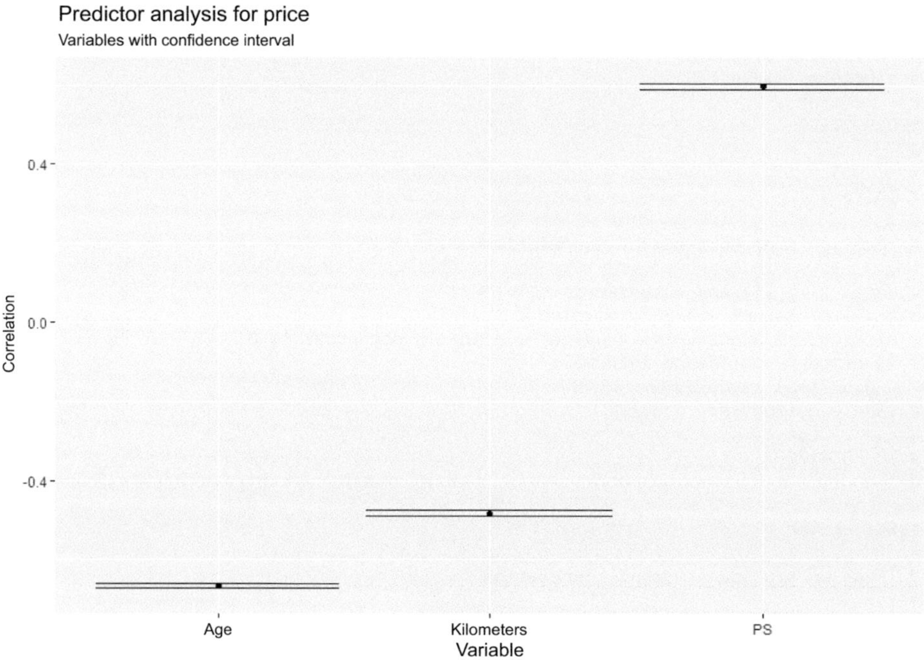


```
# Modell Lineare Regression------------------------------------------------------
mod.rg.lm<- lm(data=autos,Preis ~ (Alter+PS))                     # Modell
mod.rg.lm                                                          # +1 Jahr -> - US$851

##
## Call:
## lm(formula = Preis ~ (Alter + PS), data = autos)
##
## Coefficients:
## (Intercept)        Alter           PS
##     9052.70      -859.64        58.52

#                                                                  # +1 PS   -> + US$ 58
p1<-plot(mod.rg.lm,which=1) +                                      # Residuals ı Predicted
        title("Vorhersage und Vorhersagefehler")
```

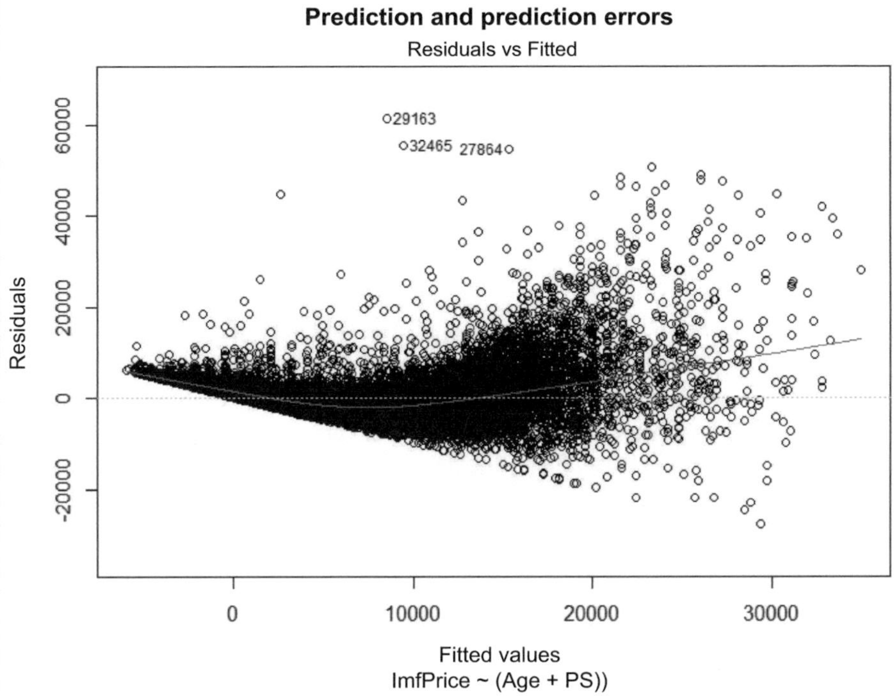


```
# Anzeige der Koeffizienten des Modells-----------------------------------------
tidy(mod.rg.lm)                                              # Koeffizienten

## # A tibble: 3 × 5
##   term        estimate std.error statistic p.value
##   <chr>          <dbl>     <dbl>     <dbl>   <dbl>
## 1 (Intercept)   9053.     91.0       99.5        0
## 2 Alter         -860.      5.04     -171.        0
## 3 PS              58.5     0.411     142.        0

# Vorhersagen machen-------------------------------------------------------------
vorhersage<-round(predict(mod.rg.lm,                         # Vorhersage Preis
                          new_data=autos),0)
autos.eval<-mod.rg.lm$model %>%                              # Daten
  add_residuals(mod.rg.lm,"Residual")                        # Ergänzen: Residual
autos.eval$Vorhersage<-vorhersage                            # Ergänzen: Vorhersage
head(autos.eval)

##   Preis Alter  PS   Residual Vorhersage
## 1  5000    12 158 -2983.3076       7983
## 2  8500    19 286  -956.5043       9457
## 3  8990     7 102   -14.3460       9004
## 4  4350     9  71 -1120.8998       5471
## 5  7900    10 150 -1334.4279       9234
## 6  1990    18  90  3144.0015      -1154

cat("Standardabweichung der Residuals:",                     # Stichprobe /(n-1)
    sd(autos.eval$Residual,na.rm=T))
```

```
## Standardabweichung der Residuals: 4426.222

cat("Standardabweichung der Residuals der Population:",    # Population /n
    sd(autos.eval$Residual,na.rm=T)/
      (nrow(autos.eval)-1)*nrow(autos.eval))

## Standardabweichung der Residuals der Population: 4426.349

fehler<-filter(autos.eval,Residual>  20000 |                # Schlechte Vorhersage
                          Residual< -20000)
cat("Anzahl Vorhersagen mit Abweichung > $20.000:",nrow(fehler))

## Anzahl Vorhersagen mit Abweichung > $20.000: 224
```

In the previous model, only the predictors age and PS were used for the price. If one now integrates further predictors in other models and then compares the prediction errors of these models, it becomes apparent which of the models makes the best predictions.

To output the scatter plot for any model, calculate the standard deviation of the residuals, and determine the number of used cars whose deviation is more than 20,000 US$, the function `modell.analyse()` is used. The following creates six different models. Model 1 is based on the predictors age and PS, model 2 is based on age+PS and brand, model 3 is based on age+PS and kilometers, model 4 is based on age+PS+kilometers and brand, model 5 is based on age+PS+kilometers+brand and transmission, and model 6 is based on age+PS+kilometers+brand and model.

The evaluation of these simple models with different predictors shows that model 6 has the smallest average prediction errors of 3482 US$ per used car, thus achieving an improvement of almost 1000 US$ compared to the predictions of the first model. Also, among the 34,864 predictions, the number of predictions with a deviation of more than 20,000 US$ improved from 224 to only 107. Evaluating different predictors proves to be valuable in order to be able to select the models with the best prediction quality in the process of modeling.

Modeling Used Car Data (Part 2)

```
# Funktion zur Modellanalyse-----------------------------------------------------
modell.analyse <- function(modell) {
  var.analyse<-NA
  p1<-plot(modell,which=1) +                               # Residuals + Predicted
      title("Vorhersagefehler (Residuals) und Vorhersage (Fitted values)")
  koeff<-tidy(modell)                                      # Koeffizienten
  vorhersage<-round(predict(modell,                        # Vorhersage Preis
                            new_data=autos),0)
  autos.eval<-modell$model %>%                             # Daten
    add_residuals(modell,"Residual")                       # Ergänzen: Residual
  autos.eval$Vorhersage<-vorhersage                        # Ergänzen: Vorhersage
  stdabw<-sd(autos.eval$Residual,na.rm=T)/                 # Fehler
        (nrow(autos.eval)-1)*nrow(autos.eval)
  txt.stdabw<-paste("\n","  Standardabweichung:",round(stdabw,2))
  var.stdabw<-round(stdabw,2)
  fehler<-filter(autos.eval,Residual> 20000 | Residual< -20000) # Schlechte Vorhersa
ge
  txt.fehler<-paste("\n","  Anzahl Vorhersagen mit Abw. > $20.000:",nrow(fehler))
  var.fehler<-nrow(fehler)
  txt.head<-paste0('Analyse von Modell: "',deparse(substitute(modell)),'"')
  var.head<-deparse(substitute(modell))
  txt.gesamt<-rbind(txt.head,txt.stdabw,txt.fehler)
  var.gesamt<-rbind(var.head,var.stdabw,var.fehler)
  cat(txt.gesamt)
  return(var.gesamt)
}
mod.rg.lm.1 <-mod.rg.lm
ergebnis1<-modell.analyse(mod.rg.lm.1)
```

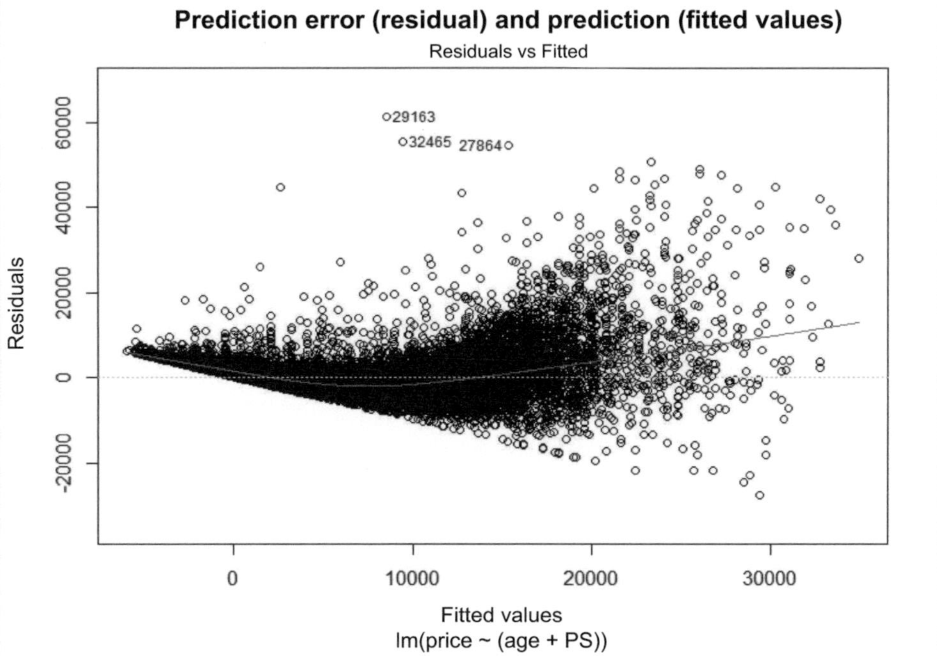

```
## Analyse von Modell: "mod.rg.lm.1"
##    Standardabweichung: 4426.35
##    Anzahl Vorhersagen mit Abw. > $20.000: 224

mod.rg.lm.2 <-lm(data=autos,Preis ~ (Alter+PS+Marke))
ergebnis2<-modell.analyse(mod.rg.lm.2)
```

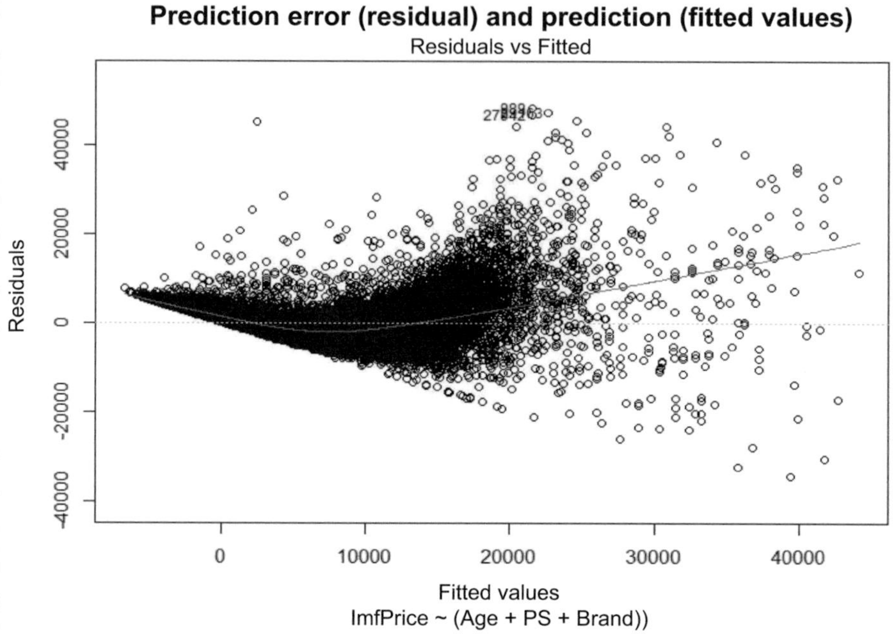


```
## Analyse von Modell: "mod.rg.lm.2"
##    Standardabweichung: 4168.39
##    Anzahl Vorhersagen mit Abw. > $20.000: 188

mod.rg.lm.3 <-lm(data=autos,Preis ~ (Alter+PS+Kilometer))
ergebnis3<-modell.analyse(mod.rg.lm.3)
```

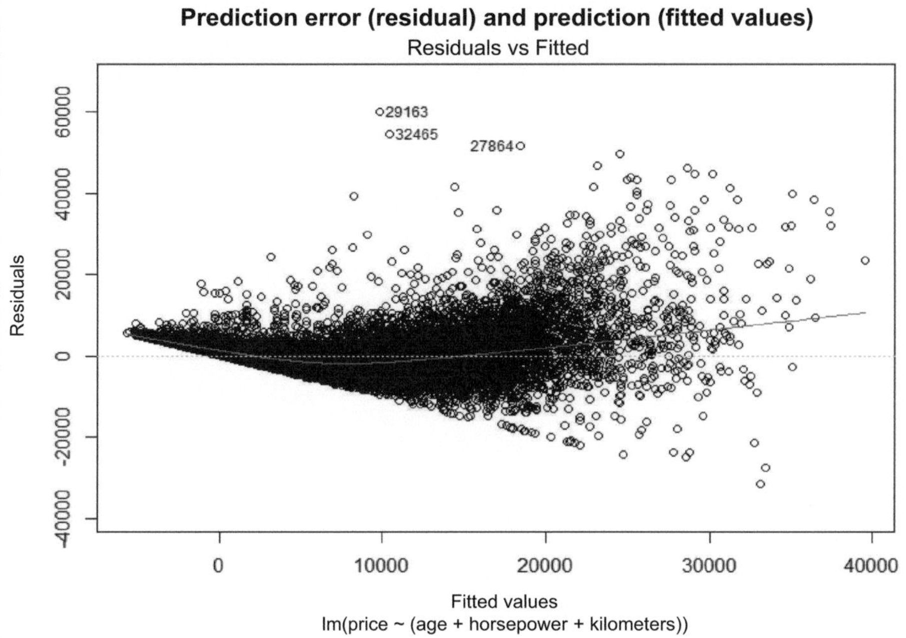


```
## Analyse von Modell: "mod.rg.lm.3"
##    Standardabweichung: 4177.44
##    Anzahl Vorhersagen mit Abw. > $20.000: 197
```

```
mod.rg.lm.4 <-lm(data=autos,Preis ~ (Alter+PS+Kilometer+Marke))
ergebnis4<-modell.analyse(mod.rg.lm.4)
```

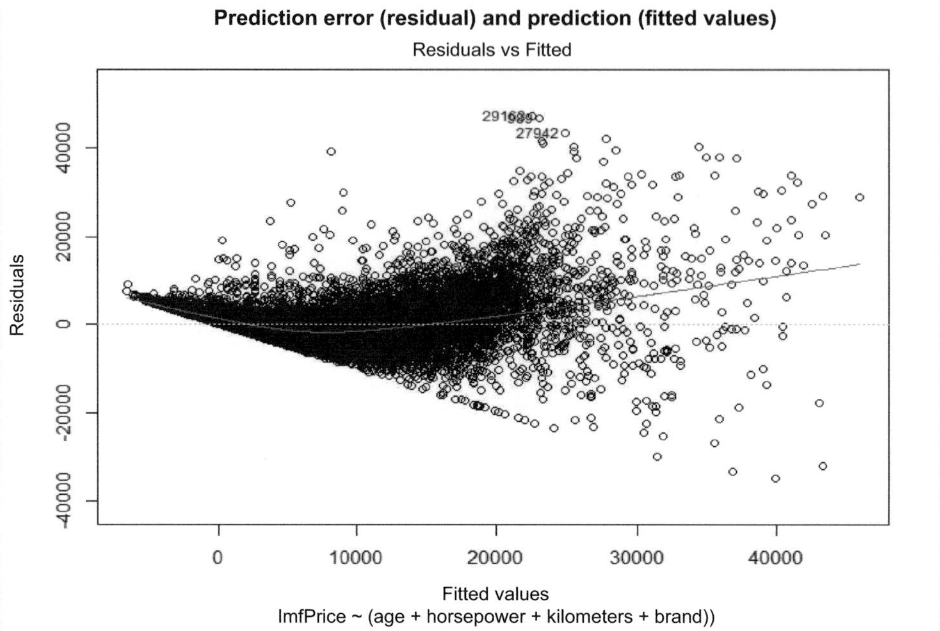


```
## Analyse von Modell: "mod.rg.lm.4"
##    Standardabweichung: 3903.77
##    Anzahl Vorhersagen mit Abw. > $20.000: 158
```

```
mod.rg.lm.5 <-lm(data=autos,Preis ~ (Alter+PS+Kilometer+Marke+Getriebe))
ergebnis5<-modell.analyse(mod.rg.lm.5)
```

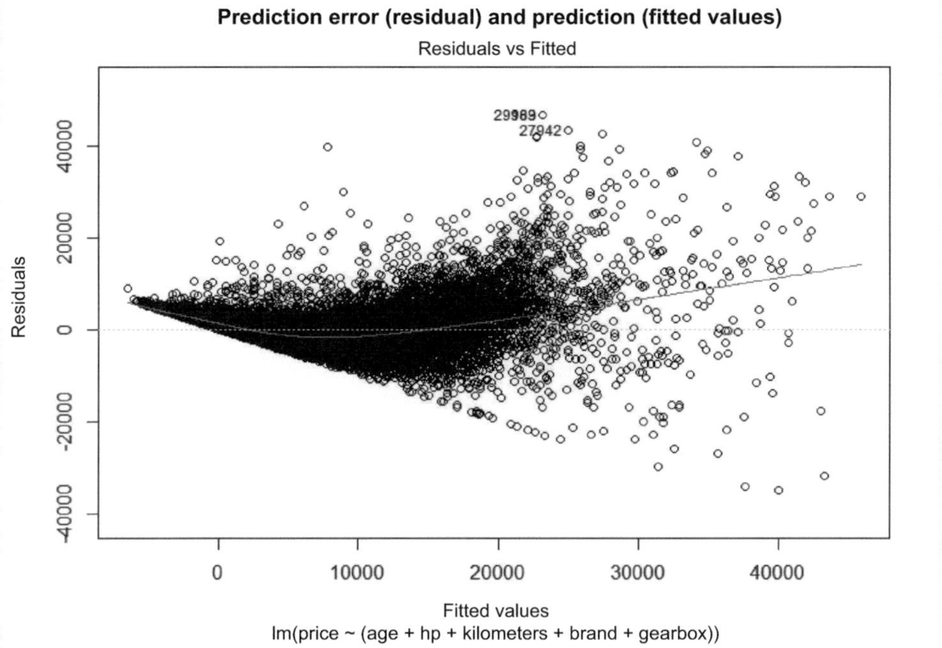


```
## Analyse von Modell: "mod.rg.lm.5"
##    Standardabweichung: 3885.07
##    Anzahl Vorhersagen mit Abw. > $20.000: 160
```

```
mod.rg.lm.6 <-lm(data=autos,Preis ~ (Alter+PS+Kilometer+Marke+Modell))
ergebnis6<-modell.analyse(mod.rg.lm.6)
```

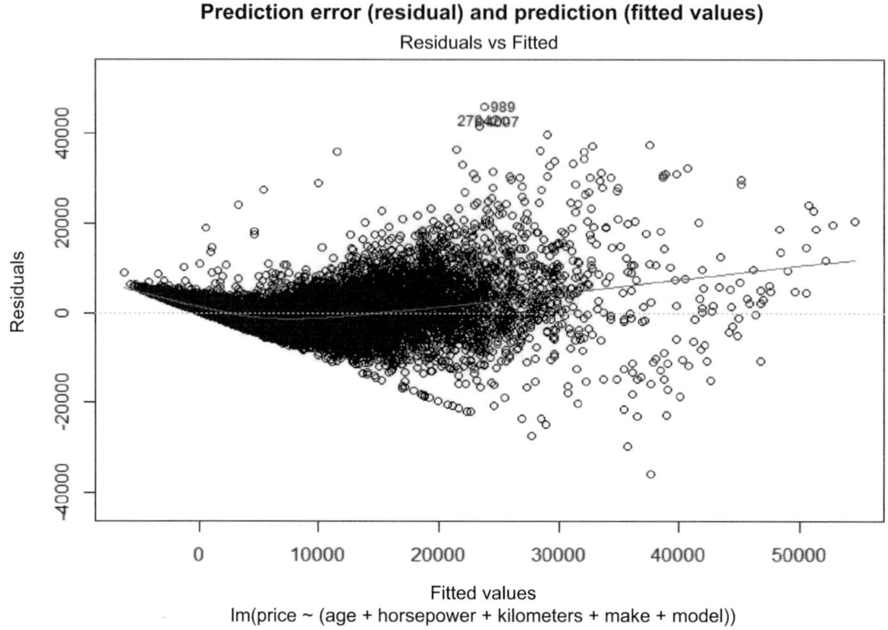


```
## Analyse von Modell: "mod.rg.lm.6"
##    Standardabweichung: 3482.39
##    Anzahl Vorhersagen mit Abw. > $20.000: 107

# Evaluation ------------------------------------------------------------------
summary.ergebnis<-as.data.frame(t(cbind(ergebnis1,ergebnis2,ergebnis3,
                                        ergebnis4,ergebnis5,ergebnis6)))
colnames(summary.ergebnis)<-(c("Modell","Standardabweichung","Autos mit Abweichung >
 $20.000"))
summary.ergebnis %>%
  arrange(Standardabweichung) %>%
  mutate(Platzierung=1:6) %>%
  select(Platzierung,Modell,Standardabweichung,"Autos mit Abweichung > $20.000")

##   Platzierung      Modell Standardabweichung Autos mit Abweichung > $20.000
## 1           1 mod.rg.lm.6            3482.39                            107
## 2           2 mod.rg.lm.5            3885.07                            160
## 3           3 mod.rg.lm.4            3903.77                            158
## 4           4 mod.rg.lm.2            4168.39                            188
## 5           5 mod.rg.lm.3            4177.44                            197
## 6           6 mod.rg.lm.1            4426.35                            224
```

The evaluation of various predictors is also referred to as **factor analysis**. It serves to infer from a multitude of variables to a few underlying variables (factors) which are of great relevance. The aim of factor analysis in modeling is to reduce the number of variables in order to make a decent prediction with as few variables as possible.

The results of the modeling always depend on the data. It has already been mentioned that in almost all cases of modeling, samples are used. Therefore, this will be discussed in more detail below.

6.4.2 Samples

Samples represent an incomplete selection of observations from the total population. It is often assumed that a large sample is better than a smaller one. However, in addition to size, the **representativeness** of a sample is also important. An election forecast based on a survey of a large number of people who were in a shopping center between 10 and 11 a.m. on a weekday is an example of a non-representative sample. At this time, there are likely to be fewer working people in the shopping center, and thus the sample is not representative of the population of eligible voters. This sample would be biased, which in English is referred to as biased, expressing that some individuals or entities had a higher chance of being included in the sample, while other individuals or entities had a lower chance. Therefore, examining these data would not be an adequate basis for an election forecast. A sample is considered representative when the distribution of all characteristics relevant to the study corresponds to those of the total population.

The following sample selection methods determine whether the methods of inductive statistics and thus also of machine learning may be applied if one intends to generalize the sample results, i.e., to make statements about the entire population based on the sample results (Cleff 2019, p. 9):

- **Conscious selection**: Selection is not made at random. This results in a subjective selection (e.g., what is easily identifiable, what is available, who the selecting person knows, arbitrary, etc.), which reduces or even eliminates the ability to generalize results. **The data is NOT suitable for inductive statistics and machine learning**!
- **Probability selection**: Elements are selected at random, with selection probabilities being specified in advance. **The data is suitable for inductive statistics and machine learning**!
  - Simple random selection: Each element of the population has the same probability and the selection is made using random numbers.
  - Stratified selection: The population is divided into disjoint sets (strata) based on certain characteristics, from which individual samples are drawn. If the individual samples of the disjoint sets are drawn according to their proportion of the population, it is a proportional stratified sample, otherwise it is a disproportionate stratified sample.
  - Cluster selection: The population is divided into clusters, e.g., by geography. Clusters are selected at random and their elements are fully examined. Cluster effects are possible if the clusters differ significantly.
  - Multi-stage selection: Random selection through a serial application of several probability selection methods.

Suppose you want to estimate the average height for the population of students. The real average can be determined with the function `mean()`, but this assumes that you would have conducted a total survey of the population. A total survey is, as already mentioned, rarely possible in reality. In the following example (see following code), the average is only used to see how well the chosen samples are suitable for generalizing the sample average to the entire population. You can use the function `sample()` in R to perform a simple random selection. If the function `set.seed()` is used beforehand, then the random selection of the function `sample()` will produce reproducible results. The average height of the population of students in the dataframe students is 172.04 cm. With a sample of 5 students, the sample average is 176 cm. If one were to infer the average size of the population based on the sample average, one would be off by 4.96 cm. The error is therefore 4.96 cm in this case. With a sample of 10 and 20 students, the error can already be more than halved. From a sample size of 30, the error is already less than 0.5 cm. This shows that the **sample size** has a significant influence on the results of a prediction.

The following example also shows that a subjective selection of the sample participants distorts the result. In the example (see following code), it is assumed that the selector himself has the hobby of football and for simplicity's sake forms the sample in such a way that he only asks students he knows from football, who have also stated football as a hobby. Out of the 357 students, this is 23 people. What is not transparent to the selector is that the average height of these 23 students is 179.35 cm, which is more than 7 cm away from the real average of the population. If a sample of 5 students is chosen, the sample average is 177.8 cm. If one were to infer the average size of the population based on the sample average, one would be off by 5.76 cm. The error is therefore 5.76 cm in this case. With a sample of 10 and 20 students, the error here does not decrease significantly due to the distortion. Increasing the sample size only solidifies the error. This is probably due to the fact that the group of students with the hobby of football (74% men + 26% women) contains more men than women in relation to the total population (30% men + 70% women) and men are on average taller than women. This example shows why a conscious selection, even in combination with a random selection of a sample from the group of selected individuals, is not ideal for inductive statistics and machine learning compared to a pure probability selection. The aspect of the representativeness of a sample is significant.

Sampling

```
#- Mittelwert der Population schätzen------------------------------------------
mw.alle<-round(mean(studierende$Größe),2)          # Mittelwert der Population
mw.alle

## [1] 172.04

fussball<-filter(studierende,studierende$Hobby       # Studierende Hobby Fußball
             %in% c("Fußball","fußball","fußball","fussball","Fussball"))
mw.fussball<-round(mean(fussball$Größe),2)          # Mittelwert Stud. Hobby Fußb.
mw.fussball

## [1] 179.35
# Funktion zur Berechnung des Mittelwerts einer Stichprobe
mw.stichprobe <- function (data,n,mw) { # data=Vektor mit Daten, n=Stichprobengröße
                                        # mw=korrekter Mittelwert
  set.seed(123)                         # Reproduzierbarkeit der Zufallsauswahl
  stichprobe<-sample(data,size=n,replace=FALSE)
  stichprobe
  mittelwert<-round(mean(stichprobe),2) # Mittelwert der Stichprobe
  mittelwert
  fehler<-round(abs(mw - mittelwert),2) # Fehler bei Schätzung des Mittelwerts
  fehler
  var.gesamt <- paste("Stichprobenumfang:",n,", Mittelwert:",mittelwert,", Fehler:",
fehler)
  return(var.gesamt)
}
#- Stichproben von allen Studierenden
mw.stichprobe(studierende$Größe,5,mw.alle)    # Fehler groß

## [1] "Stichprobenumfang: 5 , Mittelwert: 176 , Fehler: 3.96"

mw.stichprobe(studierende$Größe,20,mw.alle)  # Fehler kleiner als 1.Stichprobe

## [1] "Stichprobenumfang: 20 , Mittelwert: 173.5 , Fehler: 1.46"

mw.stichprobe(studierende$Größe,30,mw.alle)  # Fehler bereits optimal

## [1] "Stichprobenumfang: 30 , Mittelwert: 172.3 , Fehler: 0.26"

mw.stichprobe(studierende$Größe,100,mw.alle) # Fehler bleibt optimal

## [1] "Stichprobenumfang: 100 , Mittelwert: 171.49 , Fehler: 0.55"

mw.stichprobe(studierende$Größe,500,mw.alle) # Fehler bleibt optimal

## Error in sample.int(length(x), size, replace, prob): cannot take a sample larger
than the population when 'replace = FALSE'

#- Stichproben von Studierenden mit dem Hobby Fußball
mw.stichprobe(fussball$Größe,5,mw.alle)       # Fehler sehr groß

## [1] "Stichprobenumfang: 5 , Mittelwert: 177.8 , Fehler: 5.76"

mw.stichprobe(fussball$Größe,20,mw.alle)      # Fehler bleibt sehr groß

## [1] "Stichprobenumfang: 20 , Mittelwert: 177.85 , Fehler: 5.81"
```

```
#- Anteil Frauen/Männer an der Population und der Gruppe mit Hobby Fußball
table(studierende$Geschlecht)                    # 30% Männer in Population

##
## Frau Mann
##  247  110

table(fussball$Geschlecht)                       # 74% Männer in Gruppe Fußball

##
## Frau Mann
##    6   17

#- Durchschnittliche Körpergröße nach Geschlecht
mean(studierende$Größe[studierende$Geschlecht=="Mann"]) # MW Größe von Männern

## [1] 181.6273

mean(studierende$Größe[studierende$Geschlecht=="Frau"]) # MW Größe von Frauen

## [1] 167.7692
```

In inductive statistics and machine learning, samples are often used, and the moderndive package offers supportive functions to create and work with these samples. The function `rep_sample_n()` allows for a dataframe x to create samples of size n by specifying the parameter `reps=x` (number of samples) and `size=n`. The function `stichproben.virtuell<-rep_sample_n(studierende,size=100,reps=300)` creates 300 samples with 100 data sets each from the dataframe studierende and stores the result in the dataframe stichproben.virtuell. In the new dataframe, an additional column named replicate is inserted, which contains the sample number. Each sample can then be evaluated using the function `group_by()`. The command `stichproben.virtuell.mw<-stichproben.virtuell %>% group_by(replicate) %>% summarize(summe=sum(Größe)) %>% mutate(mittelwert=summe/n)` creates the new dataframe stichproben.virtuell.mw based on the contents of the dataframe stichproben.virtuell, which calculates the sum of the body sizes of the sample sets according to the sample number (replicate) and divides it by the sample size to calculate the sample mean. As can be seen in the following code example, the dispersion of the mean value (mean body size of male students of 181.63 cm) also decreases here with increasing sample size. The function `sd()` can be used to calculate the average deviation of the 300 samples from the mean value. This is still more than 3 cm for a sample size of 5, only 1 cm for a sample size of 30, and only 0.22 cm for a sample size of 100. This confirms, even with 300 repetitions, the fact that a larger sample contributes to less variation in the result and reduces the so-called standard error (see Fig. 6.29).

The calculation of statistical parameters, which are based on samples, are always an estimation of population values with a certain inaccuracy or a certain **standard error**. Nevertheless, the estimate will be centered around the real value on average. Determining and keeping the standard error as low as possible is important in this process and also in machine learning. The prerequisite for this is a large random sample.

Fig. 6.29 Standard error

Sample size	Standard error
5	3.30
30	1.08
100	0.22

Samples with the moderndive package

```
#- Stichproben mit dem Paket moderndive-----------------------------------------
library(moderndive)                          # Paket erforderlich
# Funktion zur Berechnung des Mittelwerts von n Stichproben
mw.stichproben <- function (data,n,w) {      # data=Dataframe mit Daten
                                             # n=Stichprobengröße, w=Wiederholungen
  stichproben.virtuell<-rep_sample_n(data,size=n,reps=w)
  stichproben.virtuell.mw <- stichproben.virtuell %>%
    group_by(replicate) %>%
    summarize(summe = sum(Größe)) %>%
    mutate(mittelwert = summe / n)
  sd<-sd(stichproben.virtuell.mw$mittelwert)
  plot<-ggplot(stichproben.virtuell.mw, aes(x = mittelwert)) +
     geom_histogram(binwidth = 0.25, color = "white") +
     scale_x_continuous(limits = c(175,190)) +
     scale_y_continuous(limits = c(0,150)) +
     labs(x="Mittlere Körpergröße in cm",y="Anzahl",
          title=paste("Stichprobenumfang:",n)) +
     geom_vline(aes(xintercept=181.62,colour="Red",alpha=0.5),show.legend=F) +
     geom_vline(aes(xintercept=181.62-sd,colour="Blue",linetype="dotted"),
                show.legend=F) +
     geom_vline(aes(xintercept=181.62+sd,colour="Blue",linetype="dotted"),
                show.legend=F)
  var.gesamt<-list(plot=plot,mw=stichproben.virtuell.mw)# Liste aus plot+mw
  return(var.gesamt)
}
studenten<-studierende[studierende$Geschlecht=="Mann",] # Dataframe nur männlich
mittelwert<-mean(studenten$Größe)                       # MW Größe von Männern
mittelwert

## [1] 181.6273

anzahl<-300                                             # Anzahl Wiederholungen
var1<-mw.stichproben(studenten,5,anzahl)                # 300 Stichproben je 5
var2<-mw.stichproben(studenten,30,anzahl)               # 300 Stichproben je 30
var3<-mw.stichproben(studenten,100,anzahl)              # 300 Stichproben je 100
# Ausgabe der Verteilungen der 300 Stichproben mit den vertikalen Linien in der
# Mitte der Mittelwert und links und rechts davon der Mittelwert +/- eine
# Standardabweichung
gridExtra::grid.arrange(var1$plot,var2$plot,var3$plot,ncol=3)
```

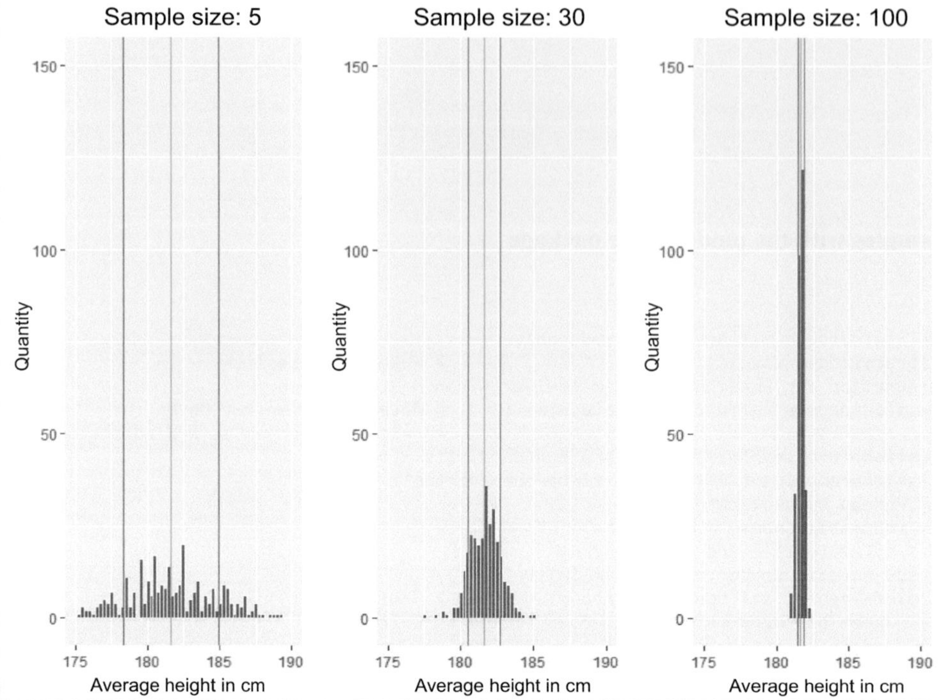


```
sd(var1$mw$mittelwert)           # Standardabweichung der 300 Stichproben je 5
## [1] 3.29895
sd(var2$mw$mittelwert)           # Standardabweichung der 300 Stichproben je 30
## [1] 1.078198
sd(var3$mw$mittelwert)           # Standardabweichung der 300 Stichproben je 100
## [1] 0.2242235
```

As can be seen from the figures in the code, the standard error decreases with increasing sample size. This is also stated by the so-called **Central Limit Theorem**according to Lindeberg-Lévy.

In the previous example, 300 samples each with a size of 5, 30, and 100 were taken to estimate the average body size of the population. In practice, it would be too cumbersome to take 300 samples each time, instead, one usually works with as large a sample as possible. In practice, the exact, correct mean value of a population would also be unknown, because if one had all the data of a population available, as in the above example of the students, then one could save oneself the trouble of working with samples to make a prediction. The example was only meant to illustrate that many and large samples reduce the standard error.

But how do you determine the standard error if you only have a single sample? The **Bootstrap method** or bootstrapping method is suitable for this, in which repeated analyses are calculated based on just a single sample. This involves drawing n bootstrap samples with replacement from the given sample. This corresponds to the repeated random drawing from a set of values, as was already introduced at the beginning of the section and can be realized with the function `sample()` in R.

To estimate the average body height of the student population, one initially selects a single sample at random, e.g., with a sample size of 50 (sample without replacement). The mean value in the subsequent code example is 181.6 cm. In principle, one could always collect a new sample of 50. However, in reality, this is often too cumbersome, e.g., to interview 50 people. This is where the bootstrap method comes in handy, in which 50 values are randomly selected from the original sample, with each value being replaced to be able to randomly select from all values in subsequent selections. If the value in a bootstrap sample of 50 from an original sample of 50 were not replaced, one would always get the same 50 values. By replacing, however, new sample constellations always arise. In our example, the first bootstrap sample has a mean value of 181.2 cm and is 0.4 cm away from the original sample.

However, a new constellation will arise with each new bootstrap sample. If, for example, 30 bootstrap samples with a sample size of 50 are formed, one can estimate how high the standard error is, i.e., how much the estimated values of the mean vary. For this, the function `rep_sample_n()` can be applied again. If one then looks at the 30 means of the bootstrap samples in a histogram, one does not yet see a normal distribution. An approximation to a normal distribution can also be achieved here by a higher number of samples. If the bootstrap procedure is repeated with 1000 samples, a normal distribution can be recognized in the histogram.

If one calculates the mean of the means from the 1000 bootstrap samples, the result is 181.6003 cm, which is almost exactly the real mean of the original sample of 181.6 cm. This shows how valuable the bootstrapping method is for prediction and also as a basis for machine learning.

Bootstrap Samples

```
#- Basis-Stichprobe -----------------------------------------------------------------
set.seed(123)                                       # Reproduzierbarkeit der Zufallsauswahl
stichprobe<-sample(studenten$Größe,size=50,replace=FALSE)    # 1 Stichprobe mit 50
stichprobe<-data.frame(Nr=1:50,"Körpergröße"=stichprobe)
stichprobe$Körpergröße

##  [1] 190 177 184 175 183 183 173 184 182 185 176 173 193 188 180 190 187 195 186
## [20] 193 180 178 176 180 181 187 177 170 179 175 185 191 175 185 185 203 170 185
## [39] 164 175 178 180 190 175 181 179 175 180 179 185

mw.orig<-round(mean(stichprobe$Körpergröße),2)  # Mittelwert der Stichprobe
mw.orig

## [1] 181.6

sd.orig<-sd(stichprobe$Körpergröße)              # Standardabweichung der Stichprobe
sd.orig

## [1] 7.219588

p1<-ggplot(stichprobe, aes(x=Körpergröße)) +
  geom_histogram(color = "white",boundary=180, binwidth=2)+
  scale_x_continuous(limits = c(160,210))+
  labs(y="Anzahl",title="Verteilung",subtitle="Original-Stichprobe:50") +
  geom_vline(aes(xintercept=mw.orig,colour="Red",alpha=0.5),show.legend=F) +
  geom_vline(aes(xintercept=mw.orig-sd.orig,colour="Blue",linetype="dotted"),
             show.legend=F)+
  geom_vline(aes(xintercept=mw.orig+sd.orig,colour="Blue",linetype="dotted"),
             show.legend=F)
#- Erste Bootstrap-Stichprobe ---------------------------------------------------
set.seed(123)                                       # Reproduzierbarkeit
bs<-sample(stichprobe$Körpergröße,size=50,replace=TRUE)   # 1 BS-Stichprobe mit 50
bs<-data.frame(Nr=1:50,"Körpergröße"=bs)
bs$Körpergröße

##  [1] 185 180 188 184 180 185 190 170 188 181 187 177 183 177 170 182 179 185 184
## [20] 187 173 180 182 186 203 188 187 190 164 173 180 191 180 181 173 182 178 185
## [39] 176 177 173 177 191 185 181 185 179 183 184 173

mw<-round(mean(bs$Körpergröße),2)                # Mittelwert der Bootstrap-Stichprobe
mw

## [1] 181.64

sd<-sd(bs$Körpergröße)                            # Standardabweichung der BS-Stichprobe
sd

## [1] 6.702908

p2<-ggplot(bs, aes(x=Körpergröße)) +
  geom_histogram(color = "white", boundary=180, binwidth=2)+
  scale_x_continuous(limits = c(160,210))+
  labs(y="Anzahl",title="Verteilung",subtitle="Bootstrap-Stichprobe:50")+
  geom_vline(aes(xintercept=mw,colour="Red",alpha=0.5),show.legend=F)+
  geom_vline(aes(xintercept=mw-sd,colour="Blue",linetype="dotted"),show.legend=F)+
  geom_vline(aes(xintercept=mw+sd,colour="Blue",linetype="dotted"),show.legend=F)
#- Vergleich von Original-Stichprobe und Bootstrap-Stichprobe --------------------
gridExtra::grid.arrange(p1,p2,ncol=2)
```

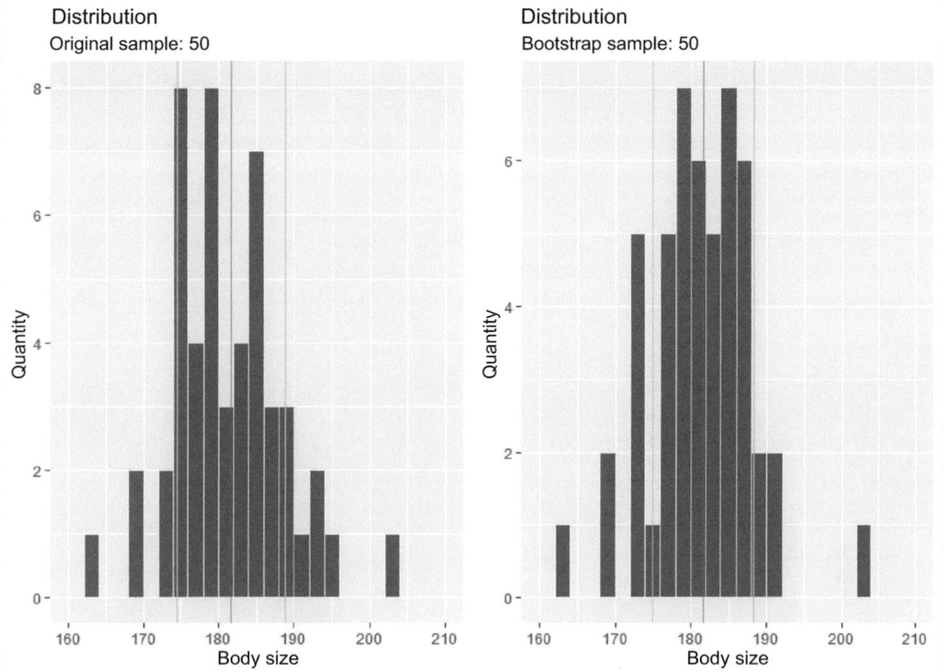


```
#- Mehrere Bootstrap-Stichproben -----------------------------------------------
set.seed(123)                                # Reproduzierbarkeit
bs.n<-rep_sample_n(stichprobe,size=50,reps=30, replace = TRUE)
bs.n.mw <- bs.n %>%
  group_by(replicate) %>%
  summarize(summe = sum(Körpergröße)) %>%
  mutate(mittelwert = summe/50)
sd<-sd(bs.n.mw$mittelwert)
p1<-ggplot(bs.n.mw, aes(x = mittelwert)) +
 geom_histogram(color = "white", boundary=181, binwidth=0.5) +
 scale_x_continuous(limits = c(178,186)) +
 scale_y_continuous(limits = c(0,9)) +
 labs(x="Mittlere Körpergröße in cm",y="Anzahl",
      title="Bootstrap-Verteilung",subtitle=("30 Bootstrap-Stichproben:50"))
set.seed(123)                                # Reproduzierbarkeit
bs.n<-rep_sample_n(stichprobe,size=50,reps=1000, replace = TRUE)
bs.n.mw <- bs.n %>%
  group_by(replicate) %>%
  summarize(summe = sum(Körpergröße)) %>%
  mutate(mittelwert = summe/50)
sd<-sd(bs.n.mw$mittelwert)
p2<-ggplot(bs.n.mw, aes(x = mittelwert)) +
 geom_histogram(color = "white", boundary=181, binwidth=0.5) +
 scale_x_continuous(limits = c(178,186)) +
 scale_y_continuous(limits = c(0,210)) +
 labs(x="Mittlere Körpergröße in cm",y="Anzahl",
      title="Bootstrap-Verteilung",subtitle=("1000 Bootstrap-Stichproben:50"))
#- Vergleich von 30 mit 1000 Bootstrap-Stichproben ------------------------------
gridExtra::grid.arrange(p1,p2,ncol=2)
```

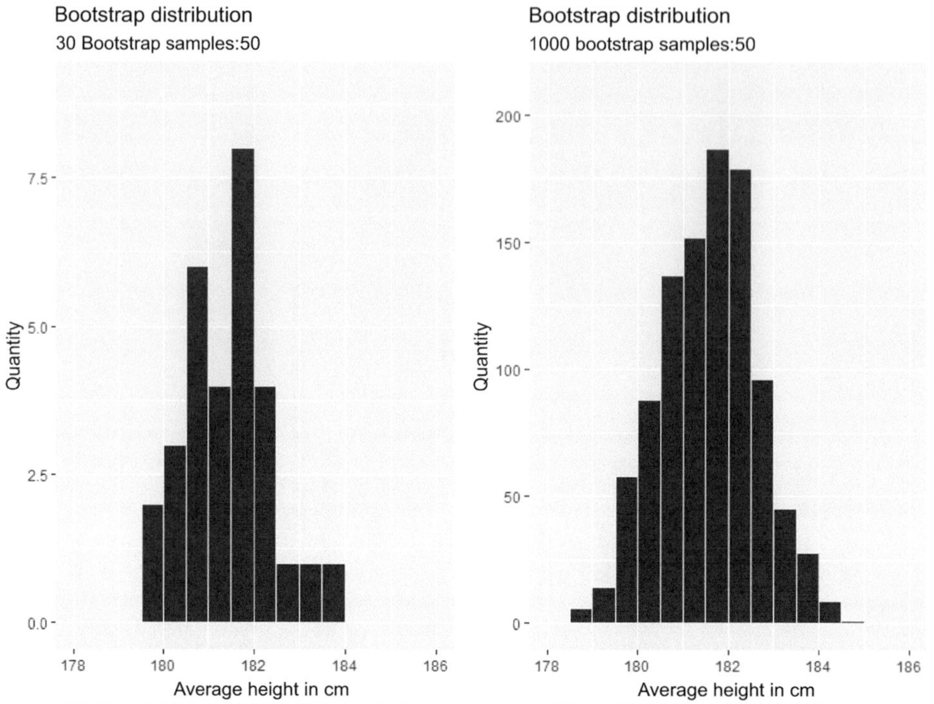


```
#- Mittelwert der Mittelwerte--------------------------------------------------------
mw.mw<-mean(bs.n.mw$mittelwert)
mw.mw                                  # Mittelwert der 1000 BS-Stichproben

## [1] 181.5824

mw.orig                                # Mittelwert der Originalstichprobe

## [1] 181.6
```

6.4.3 Training and Test Data

For machine learning, samples are formed in a special way. To ensure that the order of data collection or data acquisition does not make a difference for the analysis, all data should be randomized in machine learning. Furthermore, it is important to strictly separate the data on which a machine learning model is supposed to learn (**training data**) from the data that should serve the evaluation of the model (**test data**). If this were not done, it would contribute to the evaluation yielding more favorable results than justified. The separation of training and test data usually occurs in a ratio of 80:20 or 70:30.

The function `initial_split()` from the rsample package supports drawing a random sample of the training and test data in R. The prop parameter specifies what proportion of the data should be selected as training data. The command `flug.`

split<-initial_split(flug,prop=0.80) creates a new object flug.split from the dataframe flug, which marks 80% of the records as training records and 20% as test records. The functions flug.train<-training(flug.split) and flug.test<-testing(flug.split) are then used to create the dataframes flug.train and flug.test as a basis for machine learning.

With the strata parameter, the function inital_split() can draw a stratified random sample (**stratified random sample**) instead of a simple random sample. The strata parameter forms four strata, one for each of the quartiles of the specified variable. A stratified random sample is recommended when the population can be divided into meaningful groups, the so-called strata. Meaningful here means that the strata are relatively homogeneous in terms of one or more characteristics that also influence the expression of the ultimately interesting characteristic. The purely random selection of sample elements is now restricted in that the sample sizes per stratum are specified and a pure random sample is then drawn in each stratum. This prevents extreme samples that, for example, randomly contain almost only elements from one stratum, and as a consequence, better point estimators are obtained. With the command flug.split<-initial_split(flug,prop=0.80,strata= dep_delay) a stratified random sample is formed, in which the delay at departure is divided into four quartiles and a division into training and test data sets is then made from each of these quartiles in a ratio of 80 to 20.

There are also application areas where a random sample or stratified random sample is unsuitable, namely for time series data. For time series, the most recent data (period) should always be used as test data. The functions sliding_window(), sliding_index() and sliding_period() from the rsample package support the division of time series data into training and test data.

A disadvantage when models learn based on training data is the fact that it is only clear after the application of the test data whether the model also works well. To support an evaluation of the model in advance, so-called **validation data** can be used. If problems are detected when using the model on the validation data, this can be used to adjust the model and thus achieve better results. How this can be implemented with R will be described in Sect. 6.5.2.

Training and Test Data

```
## Trainings- und Testdaten für flug ----------------------------------------
set.seed(123)
flug.split<-initial_split(flug,prop=0.80)  # Split Trainings/Testdaten 80:20
flug.train<-training(flug.split)
flug.test<-testing(flug.split)
rm(flug.split)                             # Split löschen, nicht mehr benötigt
dim(flug)                                  # 100% Alle Datensätze

## [1] 336788     33
dim(flug.train)                            # 80%  Trainingsdaten

## [1] 269430     33

dim(flug.test)                             # 20%  Testdaten

## [1] 67358    33

set.seed(123)
flug.split<-initial_split(flug,prop=0.80,  # Split Trainings/Testdaten 80:20
                          strata=dep_delay)
flug.train<-training(flug.split)
flug.test<-testing(flug.split)
rm(flug.split)                             # Split löschen, nicht mehr benötigt
dim(flug)                                  # 100% Alle Datensätze

## [1] 336788     33

dim(flug.train)                            # 80%  Trainingsdaten

## [1] 269429     33

dim(flug.test)                             # 20%  Testdaten

## [1] 67359    33

# Trainings- und Testdaten für autos erzeugen-------------------------------------
set.seed(123)
autos.split<-initial_split(autos,prop=0.80,     # Split Trainings/Testdaten 80:20
                           strata=Preis)
autos.train<-training(autos.split)
autos.test<-testing(autos.split)
rm(autos.split)                                 # Split löschen, nicht mehr benötigt
dim(autos)                                      # 100% Alle Datensätze

## [1] 34864    23

dim(autos.train)                                # 80%  Trainingsdaten

## [1] 27890    23

dim(autos.test)                                 # 20%  Testdaten

## [1] 6974   23
```

6.4.4 Evaluation

To assess the quality of a machine learning model, the quality of the model is evaluated. For the evaluation of models, in the case of a numerical prediction, the **RMSE**(root mean squared error) and the coefficient of determination (also called determination coefficient or referred to as $\mathbf{R^2}$) are suitable as metrics, and in the case of a classification, the **ROC**curve (Receiver Operating Characteristic curve) including the Area Under the Curve (**AUC**).

For numerical prediction models, the **RMSE** is determined as the root of the mean square deviation of the predicted values from the real values. The RMSE indicates the accuracy of the model and corresponds to the **standard deviation of the residuals** (Residual=Real value – Predicted value) of the population. The RMSE can be calculated using the function `rmse()` from the yardstick package, e.g., `yardstick::rmse(autos.eval[,c("Preis","Vorhersage")], truth =Preis,estimate=Vorhersage)`.

The coefficient of determination $\mathbf{R^2}$ can be calculated with the function `rsq()` from the yardstick package, e.g. `yardstick::rsq(data= df,Price,.pred)`. The coefficient of determination is a measure on a scale from 0 to 1 for assessing the goodness of fit of a regression, i.e., to evaluate how well a regression model describes the data points. The smaller the deviation of the predicted values from the real values, the closer the coefficient of determination approaches 1. An R-square value of more than 0.7 indicates a high correlation between the data, a value of 0.4 to 0.699 shows a mediocre relationship, and a value below 0.4 is considered an insignificant correlation. If several measures are to be determined, this is possible with a function call of `metric_set()`.

To get the RMSE and R^2 displayed in conjunction with a scatter plot of the real values and predicted values, the function `ml.eval.lr()` from the machinelearning package can be used, e.g. `ml.eval.lr(cars.eval[,c("Price","Prediction","Used car price"," real price"," predicted price",scatter=T)`.

Since in Sect. 5.4.1 the model with the predictors age, horsepower, kilometers, brand, and model performed best, this will be used in the following. However, unlike the previous approach, the model is not trained with all data, but only with the training data that are in the dataframe cars.train. Since now 27,890 instead of 34,864 data sets are used for training, it can be expected that the predictions might be a bit worse.

As can be seen in the scatter plot of the evaluation for the used car price (see following coding), the prediction of the price for many data sets is negative and for several data sets significantly deviating from the displayed ideal line (reality=prediction). The value for the RMSE is now 3496, which expresses that the price is predicted wrong on average by 3496 US$. The R^2 of 0.78 indicates a high correlation and expresses that many data points are already predicted quite well by the model. However, the scatter plot also shows that several data sets are not predicted accurately (large distance from the bisector).

Evaluation Regression

```
## Evaluation ------------------------------------------------------------------
# Modell mit Vorhersage und Residual trainiert basierend auf Trainingsdaten ------
mod.rg.lm<-lm(data=autos.train,Preis ~ (Alter+PS+Kilometer+Marke+Modell))
autos.eval<-mod.rg.lm$model %>%                                      # Daten
  add_residuals(mod.rg.lm,"Residual") %>%                            # Residual
  mutate(Vorhersage=predict(mod.rg.lm,new_data=autos.train))# Vorhersage

# Aufruf Funktion rmse() und rsq() aus dem Paket yardstick-----------------------
yardstick::rmse(autos.eval[,c("Preis","Vorhersage")],
                truth=Preis,estimate=Vorhersage)

## # A tibble: 1 × 3
##   .metric .estimator .estimate
##   <chr>   <chr>          <dbl>
## 1 rmse    standard       3496.

yardstick::rsq(autos.eval[,c("Preis","Vorhersage")],
               truth=Preis,estimate=Vorhersage)

## # A tibble: 1 × 3
##   .metric .estimator .estimate
##   <chr>   <chr>          <dbl>
## 1 rsq     standard       0.783

# Aufruf Funktion ml.evaluation() -----------------------------------------------
ml.eval.lr(autos.eval[,c("Preis","Vorhersage")],                     # Scatter-Plot
           "Gebrauchtwagen","realer Preis",
           "vorhergesagter Preis",scatter=T)
```

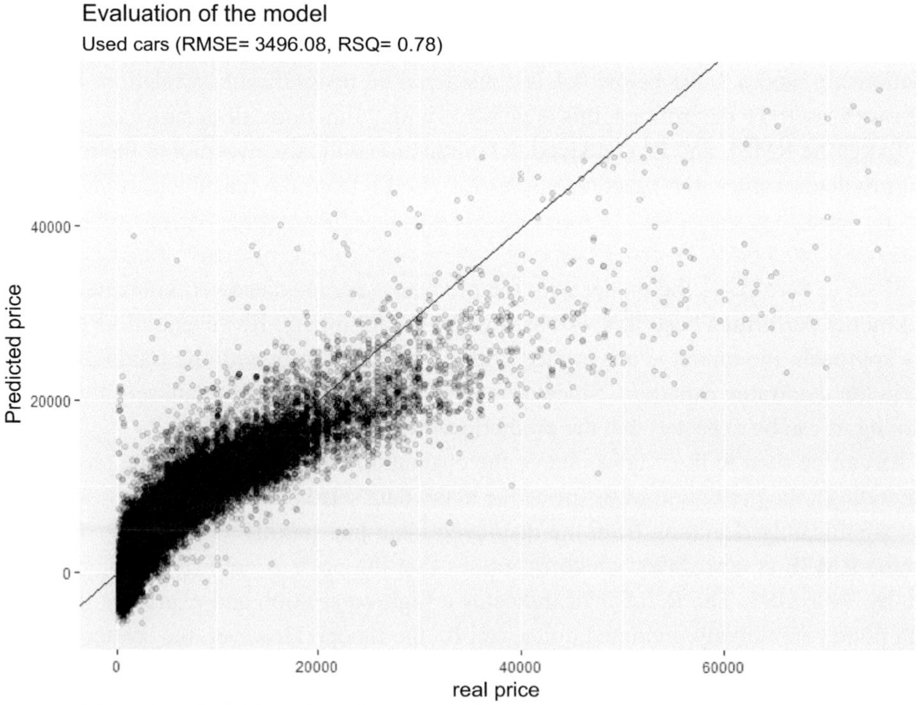


```
##   Kennzahl         Wert
## 1     rmse 3496.0846248
## 2      rsq    0.7826101
```

To evaluate a classifier, it must be applied repeatedly, with knowledge of the correct classification being obtained afterwards. An example would be a Corona test, which is used to determine whether a person has Corona. Later, it is actually determined whether the person is sick. The test classifies people into the categories sick and healthy. Depending on the test result, it is said that the test is positive (classification sick) or negative (classification healthy). To assess how well the test is suitable for classification, the test result is compared with the actual finding for each person. There are four possible cases, which can be represented in a confusion matrix:

True Positive: The patient is sick, and the test has correctly indicated this.
False Negative: The patient is sick, but the test has incorrectly classified him as healthy.
False Positive: The patient is healthy, but the test has incorrectly classified him as sick.
True Negative: The patient is healthy, and the test has correctly indicated this.

As shown, misclassifications can also occur in classification. This is referred to as a type I error or "false positive" and type II error or "false negative" (see Fig. 6.30).

The number of correct classifications (True Positive and True Negative) in relation to all cases describes the accuracy of the classification. The **True Positive Rate** (True Positive cases to all Positive cases) is referred to as **Sensitivity** and the **True Negative Rate** (True Negative cases to all Negative cases) as **Specificity**.

Assuming out of 1000 people, 10 have Corona. If the test's sensitivity is 80%, then 8 out of the 10 people with Corona will be detected (True-Positive) and 2 will not be detected (False-Negative). If the test has a specificity of 95%, then out of the 990 healthy people, 95%, or 940 people, will be correctly classified as such (True-Negative), while the remaining 5%, or 50 people, will be classified as Corona-infected (False-Positive). In total, this means that 942 people have a negative test result, which is incorrect for 2 of them, while 58 people have a positive test result, which is incorrect for 50 people. The accuracy or correct classification rate of the test as a classifier, based on these data, is then 94.8% (948 out of 1000 people were correctly classified).

Graphically, this accuracy of a classification can also be well represented for a specific parameter value, e.g., diagnosis sick or healthy after a Corona test, in a ROC curve or operating characteristic. **In a ROC curve, the sensitivity (True-Positive-Rate) is plotted on the y-axis and the specificity (True-Negative-Rate) on the x-axis**. A ROC curve close to the diagonal indicates a random process, as values close to the diagonal

	Sick	Healthy
Test result: Positive (sick)	Correct	**Type II error**
Test result: Negative (healthy)	**Type 1 error**	Correct

Fig. 6.30 Error types

mean an approximately half false positive rate. The ideal ROC curve rises (high sensitivity) vertically first, describing a hit rate close to 100% while the error rate is close to 0%, and then only moves to the right at the top (high specificity). A ROC curve significantly below the diagonal suggests that the values were probably misinterpreted.

The area under the curve can also be used as a numerical indicator **AUC (Area Under the Curve)** and is usually between 0.5 and 1, with an increasing approach to 1 indicating that the classification has been more successful. If the AUC were 1, this would mean that 100% of the classifications were correct. The yardstick package includes the functions `roc_curve()` for calculating the data points of the ROC curve and the function `roc_auc()` for calculating the area under the curve. In the following example, the AUC is 0.94. This means that about 94% of the classifications are correct.

In the following coding, we work with the data two_class_example from the package modeldata (Kuhn 2022), which have predicted class 1 or 2 for datasets. The confusion matrix, which can be displayed with the function `conf_mat()`, states that of the predictions which have resulted in class1, 227 cases were classified as True-Positive and 50 cases as False-Positive. In total, 419 out of 500 cases were correctly classified. The accuracy of the classification is therefore 0.838 and can be determined with the function `accuracy()`. The classifier therefore has an accuracy of approximately 84%.

Other suitable coefficients for evaluating binary classifiers are the Matthews correlation coefficient, which can be calculated with the function `mcc()`, and the **F-Score** as a combination measure of accuracy and hit rate, which can be calculated with the function `f_maes()`. An example of the use of key figures in machine learning models is described by Schulte (2019). Several key figures can be determined with a function call from `metric_set()`. The **Matthews correlation coefficient (MCC), also known as Phi coefficient or Yule-Phi coefficient** and identical to the Pearson-Phi coefficient, is suitable as a measure of the quality of binary (two-class) classifications. The MCC returns a value between −1 and +1. A coefficient of +1 describes a perfect prediction, 0 describes a random prediction, and −1 describes a complete disagreement between prediction and observation. Positive values from 0.3 are interpreted as weakly accurate prediction, from 0.5 as moderately accurate prediction, and from 0.8 as highly accurate prediction. In the case study in the following coding, a value of 0.677 is determined, which speaks for a moderately accurate prediction.

To calculate both the AUC and output the AUC curve, the function `ml.eval.kl()` from the machinelearning package can be used, e.g., `ml.eval.kl(two_class_example,"Classification Class1","truth","Class1",scatter=T)`. For classifications with more than two classes, the functions `accuracy()` and `mcc()` can also be applied. They indicate multiclass in the result in the .estimator column to indicate that the calculation is based on more than two classes. The function `sensitivity()` in connection with the estimator parameter allows the calculation of sensitivity. The functions `roc_auc()` and `roc_curve()` are also applicable, but it is necessary to include the characteristics of all classes as parameters (RStudio 2022).

Evaluation Classification

```
## Evaluation -----------------------------------------------------------------
# Laden der Daten--------------------------------------------------------------
data(two_class_example, package="modeldata")
head(two_class_example)

##    truth      Class1       Class2 predicted
## 1 Class2 0.003589243 0.9964107574    Class2
## 2 Class1 0.678621054 0.3213789460    Class1
## 3 Class2 0.110893522 0.8891064779    Class2
## 4 Class1 0.735161703 0.2648382969    Class1
## 5 Class2 0.016239960 0.9837600397    Class2
## 6 Class1 0.999275071 0.0007249286    Class1

# Aufruf Funktion conf_mat() aus dem Paket yardstick----------------------------
yardstick::conf_mat(two_class_example,          # Konfusions-Matrix
         truth = truth, estimate = predicted)

##           Truth
## Prediction Class1 Class2
##     Class1    227     50
##     Class2     31    192

# Aufruf Funktion accuracy(), mcc() und f_meas() aus dem Paket yardstick---------
accuracy(two_class_example, truth, predicted)   # Genauigkeit

## # A tibble: 1 × 3
##   .metric  .estimator .estimate
##   <chr>    <chr>          <dbl>
## 1 accuracy binary         0.838

mcc(two_class_example, truth, predicted)        # Matthews Korrelationskoeffizient

## # A tibble: 1 × 3
##   .metric .estimator .estimate
##   <chr>   <chr>          <dbl>
## 1 mcc     binary         0.677

f_meas(two_class_example, truth, predicted)     # F1 Metrik

## # A tibble: 1 × 3
##   .metric .estimator .estimate
##   <chr>   <chr>          <dbl>
## 1 f_meas  binary         0.849

# Aufruf der Kombinantionsfunktion für accuracy(), mcc() und f_meas() ------------
kz.kl<-metric_set(accuracy,mcc,f_meas)          # Kennzahlen für Klassifikation
kz.kl

## # A tibble: 3 × 3
##   metric   class        direction
##   <chr>    <chr>        <chr>
## 1 accuracy class_metric maximize
## 2 mcc      class_metric maximize
## 3 f_meas   class_metric maximize

kz.kl(two_class_example,truth=truth,estimate=predicted)

## # A tibble: 3 × 3
##   .metric  .estimator .estimate
##   <chr>    <chr>          <dbl>
## 1 accuracy binary         0.838
## 2 mcc      binary         0.677
## 3 f_meas   binary         0.849
```

```
# Aufruf der ROC-Kurve ------------------------------------------------------------
roc.werte<-roc_curve(two_class_example,truth,Class1)        # ROC-Werte für Class1
head(roc.werte)

## # A tibble: 6 × 3
##   .threshold specificity sensitivity
##        <dbl>       <dbl>       <dbl>
## 1 -Inf           0                 1
## 2    1.79e-7     0                 1
## 3    4.50e-6     0.00413           1
## 4    5.81e-6     0.00826           1
## 5    5.92e-6     0.0124            1
## 6    1.22e-5     0.0165            1

roc_auc(two_class_example, truth, Class1)                   # AUC für Class1

## # A tibble: 1 × 3
##   .metric .estimator .estimate
##   <chr>   <chr>          <dbl>
## 1 roc_auc binary         0.939

# Aufruf Funktion ml.eval.kl() ------------------------------------------------------
auc<-ml.eval.kl(two_class_example,                          # Scatter
                "Klassifikation Class1","truth","Class1",
                scatter=T)
```

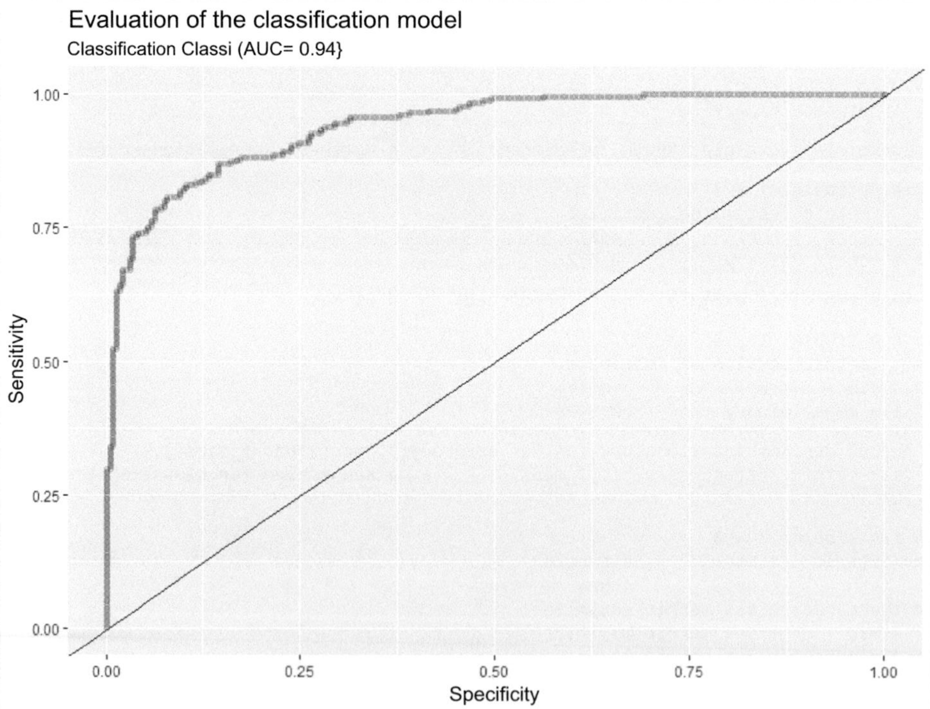


```
auc                                                          # AUC für Class1

## [1] 0.9393139
```

6.5 The Tidymodels Package

Now that the basics of machine learning with R have been introduced, it is also important to achieve a certain efficiency in the process of machine learning. For this purpose, it is advisable to follow a standardized process and to rely on established functions.

There are a multitude of packages in R that support machine learning. A disadvantage for users of machine learning is that each package uses its own procedures and parameters for its functions. This inconsistency has motivated the development of the caret package, which offers a uniform functionality for machine learning and uses established functions from other packages. The supported models can be found in the package's documentation (Kuhn 2019). A newer project involving one of the fathers of caret, Max Kuhn, which also aims to simplify and standardize machine learning with R, is the tidymodels package (RStudio 2022). Tidymodels, like tidyverse, is a combination of several packages. Numerous machine learning models are supported by the parsnip package included in tidymodels; a list of the models can be found on the package's website (RStudio 2022). The following will introduce how to easily apply machine learning with the help of the tidymodels package.

6.5.1 Model Selection, Model Training, and Model Application

Machine learning with the tidymodels package begins with the selection of the model types to be used, e.g., Linear Regression, Random Forest, KNN, etc. For Linear Regression, for example, the function `linear_reg()` and for Random Forest the function `rand_forest()` are suitable. The main arguments of these functions can be displayed using the help, e.g., `?linear_reg` or `?rand_forest`. After selecting the models, the underlying functions and the package from which this function should be used are specified. This is done with the function `set_engine()`. If a model can produce different results, it is determined which result is desired, e.g., Linear Regression, Classification, etc. If you were not working with tidymodels and wanted to know what the equivalent function call in the original package would be, this can be displayed using the function `translate()`.

In the following example, we will work with the ames data from the modeldata package (Kuhn 2022). The ames dataframe contains house prices of 2930 houses and 73 descriptive variables about the houses from the town of Ames in Iowa. These data are again split into training and test data in preparation for machine learning.

A model can then be trained with the function `fit()` or `fit_xy()`, e.g., to predict the house price based on the longitude and latitude, either as `mod.rg.lm.fit<-fit(data=ames.train,Sale_Price~Longitude+Latitude)` or `mod.rg.lm.fit<-fit_xy(x=select(ames.train,Longitude,Latitude),y=pull(ames.train,Sale_Price))`. If you save the generated model in mod.rg.lm.fit, it can then be displayed with the command `mod.rg.lm.fit`.

With the help of the function `vcov(extract_fit_engine(mod.rg.lm.fit))`, the variance-covariance matrix can be displayed for the model parameters. The diagonal elements of the matrix contain the variances of the variables, the non-diagonal elements contain the covariances between all possible pairs of variables. The function `extract_fit_engine()` in combination with the function `summary()` is also valuable for displaying the parameter values, the p-values, and significance, e.g., `extract_fit_engine(mod.rg.lm.fit) %>% summary()`. In addition, the coefficients of the model can be displayed with the function `tidy(mod.rg.lm.fit)`.

If we want to predict the price of a house and use the test data for this, this is possible with the function `predict()`. With the parameter type="pred_int", the prediction interval or with type="conf_int" the confidence interval can be displayed. To create a dataframe that includes both the real price of a house, the predicted price, and the 95% confidence interval, the following command can be used: `ames.pred<-ames.test %>% select(Sale_Price) %>% bind_cols(predict(mod.rg.lm.fit,ames.test)) %>% bind_cols(predict(mod.rg.lm.fit,ames.test,type="conf_int"))`.

If instead of linear regression a tree-based method is to be used for machine learning, the same functions can be applied in the same order, starting with the definition of the method, e.g. `mod.rg.rpart <-decision_tree(min_n=2) %>% set_engine("rpart") %>% set_mode ("regression")`, then the function `fit()` and finally the function `predict()`.

In the following coding for model selection and model training, the data from the package is first loaded and then split into training and test data in an 80/20 ratio. The function `ml.summary()` is again suitable for better understanding the data and getting a histogram and a box plot displayed. Most houses are obviously in the price range between $80,000 and $250,000, with quite a few outliers at the top.

In the following coding, three model types are applied. Model1 serves for linear regression based on the engine lm, Model2 serves for linear regression based on the engine stan, and Model3 supports the Random Forest algorithm based on the engine ranger. Using the lm engine as an example, it is shown how a model can be trained with the predictors longitude and latitude using the functions `fit()` or `fit_xy()` to predict the house price. In the USA, the location of the neighborhood and the school district is often a crucial factor for the house price, and the longitude and latitude identify the neighborhood and the school district quite well. Both methods lead to the regression equation: Price = -124970600 -774125 * Longitude + 1252758 * Latitude. The function `extract_fit_engine()` also shows the significance (p-values), which is given at a level of 0.001. A significance level less than 5% is usually sufficient to consider the results as relevant.

With the function `predict()`, the house prices, the prediction interval, and the confidence interval are predicted for 5 data sets of the test data. Subsequently, the prediction data for all test data is stored in the dataframe lr.pred and the first data sets are displayed.

To also train a tree method, the engine rpart is used and trained with the training data. Here again, the function extract_fit_engine() is used to display the trained model, which however looks quite different this time, as it describes the formulas that are to be used under certain conditions in the decision tree to predict the price. Here again, predictions for the first five data sets are generated using the function predict().

Model Selection and Model Training

```
## Das Paket Tidymodels -----------------------------------------------------------
tidymodels_prefer()                                # Konflikt zwischen Paketen anzeigen
# Laden der Daten-------------------------------------------------------------------
data(ames, package = "modeldata")
# Trainings- und Testdaten erzeugen-------------------------------------------------
set.seed(123)
ames.split<-initial_split(ames,prop=0.80,          # Split Trainings/Testdaten 80:20
                          strata=Sale_Price)
ames.train<-training(ames.split)
ames.test<-testing(ames.split)
rm(ames.split)                                     # Split löschen, nicht mehr benötigt
dim(ames)                                          # 100% Alle Datensätze

## [1] 2930   74

dim(ames.train)                                    # 80%  Trainingsdaten

## [1] 2342   74

dim(ames.test)                                     # 20%  Testdaten

## [1] 588  74

# Visualisierung der Trainingsdaten-------------------------------------------------
ml.summary(ames.train$Sale_Price/1000,             # Daten verstehen
             "Verkaufspreis des Hauses","Preis in 1000 US Dollar",box=T)
```

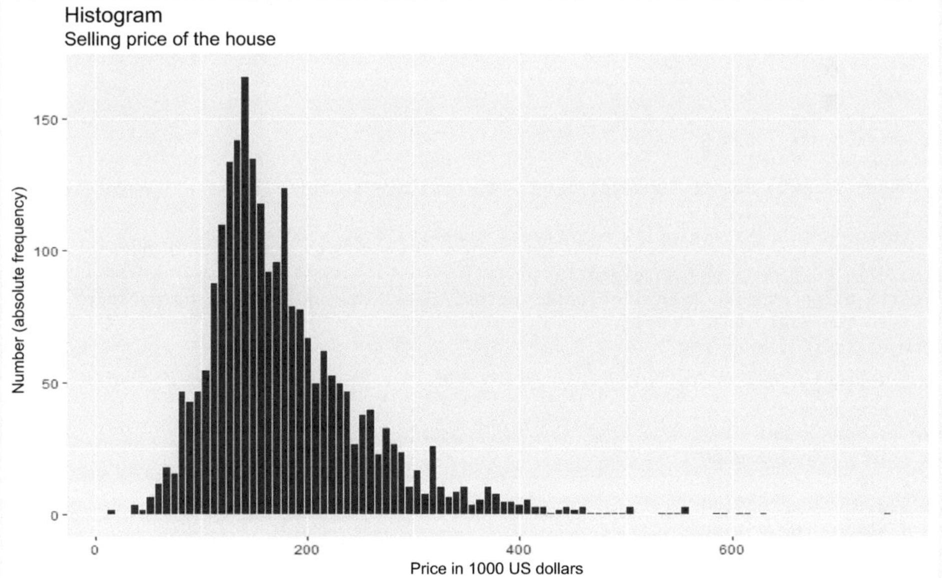

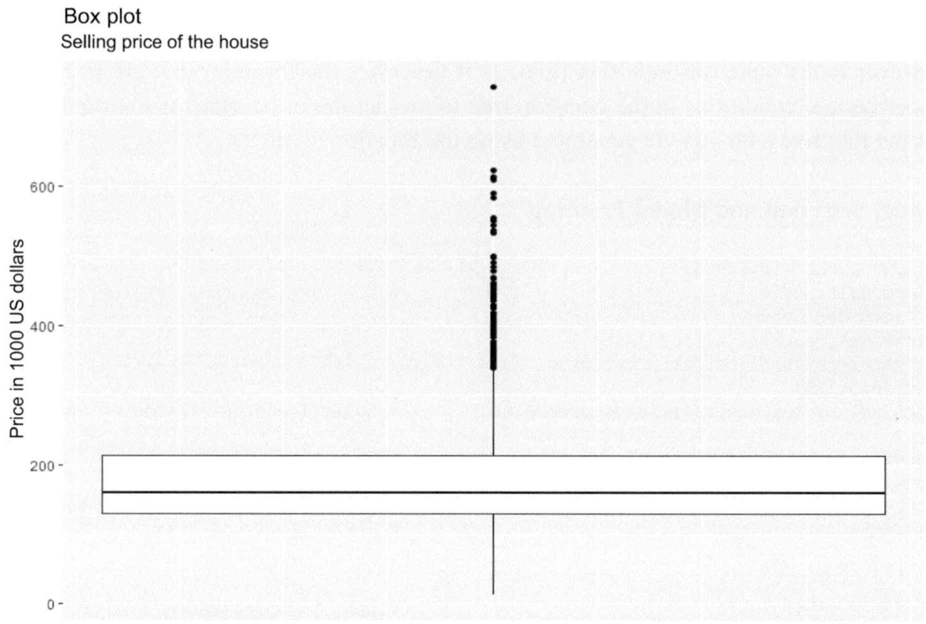


```
##                25%
## anzahl         "2342"
## anzahl.na      "0"
## modus          "140"
## median         "160"
## mean           "179.8"
## min            "12.789"
## max            "745"
## sd             "78.23"
## q1             "129.5"
## q3             "213.5"
## iqr            "84"
## whisker.min    "3.5"
## whisker.max    "339.5"
## skewness       "1.69"
## skewness.txt "Rechte Schiefe: 1.69 > 0, positive Schiefe,linkssteil,rechtsschief"
## kurtosis       "4.89"
## kurtosis.txt "Steilgipflig mit Exzess Kurtosis 4.89 > 0"

# Modelle bestimmen und anzeigen lassen------------------------------------------
# Liste aller Modelle & Engines unter https://www.tidymodels.org/find/parsnip/
# Liste von Engines je Modell
show_engines("linear_reg")

## # A tibble: 7 × 2
##   engine mode
##   <chr>  <chr>
## 1 lm     regression
## 2 glm    regression
## 3 glmnet regression
## 4 stan   regression
## 5 spark  regression
## 6 keras  regression
```

```
## 7 brulee regression

mod.rg.lm<-linear_reg() %>% set_engine("lm")    # Modellauswahl
translate(mod.rg.lm)                             # Verwendung der Parameter

## Linear Regression Model Specification (regression)
##
## Computational engine: lm
##
## Model fit template:
## stats::lm(formula = missing_arg(), data = missing_arg(), weights = missing_arg())

mod.rg.stan<-linear_reg() %>% set_engine("stan")# Modellauswahl
translate(mod.rg.stan)                           # Verwendung der Parameter

## Linear Regression Model Specification (regression)
##
## Computational engine: stan
##
## Model fit template:
## rstanarm::stan_glm(formula = missing_arg(), data = missing_arg(),
##     weights = missing_arg(), family = stats::gaussian, refresh = 0)

mod.rg.rf<-rand_forest(trees=200,min_n=5) %>%   # Modellauswahl
  set_engine("ranger") %>%
  set_mode("regression")
translate(mod.rg.rf)

## Random Forest Model Specification (regression)
##
## Main Arguments:
##   trees = 200
##   min_n = 5
##
## Computational engine: ranger
##
## Model fit template:
## ranger::ranger(x = missing_arg(), y = missing_arg(), weights = missing_arg(),
##     num.trees = 200, min.node.size = min_rows(~5, x), num.threads = 1,
##     verbose = FALSE, seed = sample.int(10^5, 1))

## Lineares Regressionsmodell anwenden############################################
mod.rg.lm<-linear_reg() %>% set_engine("lm")    # Modellauswahl
# Trainieren des Modells---------------------------------------------------------
mod.rg.lm.fit <- mod.rg.lm %>%
  fit(data=ames.train,Sale_Price~Longitude+Latitude)
mod.rg.lm.fit

## parsnip model object
##
##
## Call:
## stats::lm(formula = Sale_Price ~ Longitude + Latitude, data = data)
##
## Coefficients:
## (Intercept)    Longitude     Latitude
##  -124970600      -774125      1252758

mod.rg.lm.fit2 <- mod.rg.lm %>%
  fit_xy(x=select(ames.train,Longitude,Latitude),
         y=pull(ames.train,Sale_Price))
mod.rg.lm.fit2
```

```
## parsnip model object
##
##
## Call:
## stats::lm(formula = ..y ~ ., data = data)
##
## Coefficients:
## (Intercept)    Longitude     Latitude
##  -124970600      -774125      1252758

# Anzeige des trainierten Modells----------------------------------------------
extract_fit_engine(mod.rg.lm.fit)

##
## Call:
## stats::lm(formula = Sale_Price ~ Longitude + Latitude, data = data)
##
## Coefficients:
## (Intercept)    Longitude     Latitude
##  -124970600      -774125      1252758

# Anzeige der Varianz/Kovarianzmatrix des trainierten Modells--------------------
vcov(extract_fit_engine(mod.rg.lm.fit))

##               (Intercept)    Longitude      Latitude
## (Intercept)  4.271370e+13 323697368287 -295037063744
## Longitude    3.236974e+11   3378313946    -174669832
## Latitude    -2.950371e+11   -174669832    6629806177

# Anzeige der Summary für Modellparameter und p-Werte/Signifikanz----------------
mod.rg.lm.sum<-extract_fit_engine(mod.rg.lm.fit) %>% summary()
mod.rg.lm.sum
##
## Call:
## stats::lm(formula = Sale_Price ~ Longitude + Latitude, data = data)
##
## Residuals:
##     Min      1Q  Median      3Q     Max
## -142269  -46064  -18811   32771  530253
##
## Coefficients:
##               Estimate Std. Error t value Pr(>|t|)
## (Intercept) -124970601    6535572  -19.12   <2e-16 ***
## Longitude      -774125      58123  -13.32   <2e-16 ***
## Latitude       1252758      81424   15.39   <2e-16 ***
## ---
## Signif. codes:  0 '***' 0.001 '**' 0.01 '*' 0.05 '.' 0.1 ' ' 1
##
## Residual standard error: 72330 on 2339 degrees of freedom
## Multiple R-squared:  0.1459, Adjusted R-squared:  0.1452
## F-statistic: 199.8 on 2 and 2339 DF,  p-value: < 2.2e-16

# Anzeige der Koeffizienten des Modells----------------------------------------
tidy(mod.rg.lm.fit)

## # A tibble: 3 × 5
##   term           estimate std.error statistic  p.value
##   <chr>             <dbl>     <dbl>     <dbl>    <dbl>
## 1 (Intercept) -124970600.  6535572.     -19.1 7.59e-76
## 2 Longitude      -774125.    58123.     -13.3 4.60e-39
## 3 Latitude       1252758.    81424.      15.4 5.87e-51
```

```
# Vorhersagen machen---------------------------------------------------------
ames.test.n <- ames.test[1:5,]
predict(mod.rg.lm.fit,new_data=ames.test.n)                  # Vorhersage des Preises

## # A tibble: 5 × 1
##     .pred
##     <dbl>
## 1 185073.
## 2 184342.
## 3 209796.
## 4 191615.
## 5 221546.

predict(mod.rg.lm.fit,new_data=ames.test.n,type="pred_int")# 95% Vorhersageintervall

## # A tibble: 5 × 2
##   .pred_lower .pred_upper
##         <dbl>       <dbl>
## 1      43143.     327002.
## 2      42414.     326271.
## 3      67857.     351735.
## 4      49680.     333551.
## 5      79600.     363492.

predict(mod.rg.lm.fit,new_data=ames.test.n,type="conf_int")# 95% Konfidenzintervall

## # A tibble: 5 × 2
##   .pred_lower .pred_upper
##         <dbl>       <dbl>
## 1     180200.     189945.
## 2     179480.     189204.
## 3     204653.     214939.
## 4     186566.     196665.
## 5     216198.     226894.

mod.rg.lm.pred<-ames.test %>%
  select(Sale_Price) %>%
  bind_cols(predict(mod.rg.lm.fit,ames.test)) %>%
  bind_cols(predict(mod.rg.lm.fit,ames.test, type="conf_int"))
dim(mod.rg.lm.pred)

## [1] 588   4

head(mod.rg.lm.pred)

## # A tibble: 6 × 4
##   Sale_Price   .pred .pred_lower .pred_upper
##        <int>   <dbl>       <dbl>       <dbl>
## 1     105000 185073.     180200.     189945.
## 2     172000 184342.     179480.     189204.
## 3     189900 209796.     204653.     214939.
## 4     115000 191615.     186566.     196665.
## 5     395192 221546.     216198.     226894.
## 6     214000 219396.     214195.     224597.

## Baumbasiertes Modell anwenden##################################################
mod.rg.rpart<-decision_tree(min_n=2) %>%                   # Modellauswahl
  set_engine("rpart") %>%
  set_mode("regression")
# Trainieren des Modells-------------------------------------------------------
mod.rg.rpart.fit <- mod.rg.rpart %>%
  fit(data=ames.train,Sale_Price~Longitude+Latitude)
# Anzeige des trainierten Modells----------------------------------------------
```

```
extract_fit_engine(mod.rg.rpart.fit)

## n= 2342
##
## node), split, n, deviance, yval
##       * denotes terminal node
##
##   1) root 2342 1.432853e+13 179795.2
##     2) Latitude< 42.04931 1731 5.811501e+12 159652.6
##       4) Longitude>=-93.62997 788 1.393106e+12 134702.5 *
##       5) Longitude< -93.62997 943 3.517947e+12 180501.7
##        10) Latitude>=41.99424 928 3.169731e+12 178300.0
##          20) Latitude>=42.01873 669 2.276173e+12 169507.3
##            40) Latitude< 42.02408 197 3.551576e+11 137584.3 *
##            41) Latitude>=42.02408 472 1.636465e+12 182831.1
##              82) Latitude< 42.03673 310 1.002950e+12 171319.5
##               164) Longitude>=-93.68073 206 5.161266e+11 151185.6 *
##               165) Longitude< -93.68073 104 2.379093e+11 211200.0 *
##              83) Latitude>=42.03673 162 5.138247e+11 204859.5 *
##          21) Latitude< 42.01873 259 7.082396e+11 201011.7 *
##        11) Latitude< 41.99424 15 6.540196e+10 316716.1 *
##     3) Latitude>=42.04931 611 5.825049e+12 236860.3
##       6) Longitude>=-93.65119 477 2.685636e+12 205957.8
##        12) Longitude>=-93.6277 87 6.540643e+11 165991.3
##          24) Latitude< 42.06072 84 2.464888e+11 153248.6 *
##          25) Latitude>=42.06072 3 1.202205e+10 522789.0 *
##        13) Longitude< -93.6277 390 1.861605e+12 214873.4
##          26) Longitude< -93.62781 388 1.675826e+12 213308.7 *
##          27) Longitude>=-93.62781 2 5.498923e+08 518418.5 *
##       7) Longitude< -93.65119 134 1.062382e+12 346864.1
##        14) Longitude>=-93.65544 83 2.519449e+11 312103.6 *
##        15) Longitude< -93.65544 51 5.469346e+11 403435.1 *

# Anzeige der Summary für Modellparameter und p-Werte/Signifikanz-----------------
mod.rg.rpart.sum<-extract_fit_engine(mod.rg.rpart.fit) %>% summary()

## Call:
## rpart::rpart(formula = Sale_Price ~ Longitude + Latitude, data = data,
##     minsplit = min_rows(2, data))
##   n= 2342
##
##            CP nsplit rel error    xerror       xstd
## 1  0.18787580      0 1.0000000 1.0012059 0.05426279
## 2  0.14495771      1 0.8121242 0.8160012 0.04130225
## 3  0.06284292      2 0.6671665 0.6743967 0.03397796
## 4  0.01973782      3 0.6043236 0.6136748 0.03292286
## 5  0.01973406      4 0.5845857 0.5757240 0.03276036
## 6  0.01839003      6 0.5451176 0.5692803 0.03248566
## 7  0.01639626      7 0.5267276 0.5551877 0.03171346
## 8  0.01292730      9 0.4939351 0.5278144 0.03006900
## 9  0.01286262     10 0.4810078 0.5213344 0.03010606
## 10 0.01000000     12 0.4552826 0.4868092 0.02756613
##
## Variable importance
##  Latitude Longitude
##        52        48
##
## Node number 1: 2342 observations,    complexity param=0.1878758
##   mean=179795.2, MSE=6.118076e+09
##   left son=2 (1731 obs) right son=3 (611 obs)
```

```
##    Primary splits:
##        Latitude  < 42.04931  to the left,  improve=0.1878758, (0 missing)
##        Longitude < -93.62859 to the right, improve=0.1543356, (0 missing)
##
## Node number 2: 1731 observations,    complexity param=0.06284292
##    mean=159652.6, MSE=3.357308e+09
##    left son=4 (788 obs) right son=5 (943 obs)
##    Primary splits:
##        Longitude < -93.62997 to the right, improve=0.15494220, (0 missing)
##        Latitude  < 42.0188   to the right, improve=0.08857629, (0 missing)
##    Surrogate splits:
##        Latitude < 42.0267   to the right, agree=0.642, adj=0.213, (0 split)
##
## Node number 3: 611 observations,    complexity param=0.1449577
##    mean=236860.3, MSE=9.533632e+09
##    left son=6 (477 obs) right son=7 (134 obs)
##    Primary splits:
##        Longitude < -93.65119 to the right, improve=0.35656890, (0 missing)
##        Latitude  < 42.05848  to the left,  improve=0.06864547, (0 missing)
##
##
## … …
##
## Node number 164: 206 observations
##    mean=151185.6, MSE=2.505469e+09
##
## Node number 165: 104 observations
##    mean=211200, MSE=2.287589e+09
# Vorhersagen machen------------------------------------------------------------
predict(mod.rg.rpart.fit,new_data=ames.test.n)                 # Vorhersage des Prei
ses

## # A tibble: 5 × 1
##     .pred
##     <dbl>
## 1 153249.
## 2 153249.
## 3 213309.
## 4 153249.
## 5 312104.

mod.rg.rpart.pred<-ames.test %>%
  select(Sale_Price) %>%
  bind_cols(predict(mod.rg.rpart.fit, ames.test))
head(mod.rg.rpart.pred)

## # A tibble: 6 × 2
##   Sale_Price   .pred
##        <int>   <dbl>
## 1     105000 153249.
## 2     172000 153249.
## 3     189900 213309.
## 4     115000 153249.
## 5     395192 312104.
## 6     214000 312104.
```

So far, the modeling has only used two of the independent variables and has not taken into account, among other things, the number of rooms, the size of the living area, the age of the house, etc. Therefore, it is hardly surprising that the estimates of house prices are currently still far from the real prices. However, the basic procedure of creating a model, training it, and making predictions has already been presented.

To further simplify the process of modeling, the function `workflow()` from the workflow package can be used, which can include a function for preprocessing the data. Details on the possible preprocessing functions will follow in Sect. 6.5.3. In the following coding, only one preprocessing function is used, namely the definition of the dependent and independent variables with the function `add_formula(Sale_Price~Longitude+Latitude)`. The function `add_model()` from the parsnip package adds a previously defined model to the workflow, e.g. `wfl.rg.lm<-workflow() %>% add_model(mod.rg.lm)`. The function `fit()` can then be used again to train the model, e.g. `fit(wfl.rg.lm, ames.train)` and the function `predict()` for prediction, e.g. `predict(wfl.rg.lm.fit, ames.test)`. An update of the model or the formula is possible with the help of the functions `update_model()` and `update_formula()`. Alternatively to using a formula as a preprocessor, it is also possible to define the predictor and outcome variables with the function `add_Variables()`. The following trains four models using a single workflow. Predictions for the respective models can again be made using the function `predict()`. At the end of the following coding, the predicted prices for the first five data sets of the test data for the four models are displayed.

Model selection and model training with workflows

```
# Trainings- und Testdaten erzeugen----------------------------------------------
data(ames, package = "modeldata")
set.seed(123)
ames.split<-initial_split(ames,prop=0.80,             # Split Train/Test 80:20
                          strata=Sale_Price)
ames.train<-training(ames.split)
ames.test<-testing(ames.split)
rm(ames.split)                                        # Split löschen
# Modell mit Workflow------------------------------------------------------------
mod.rg.lm<-linear_reg() %>% set_engine("lm")          # Modellauswahl
wfl.rg.lm<-workflow() %>% add_model(mod.rg.lm)        # Workflow
wfl.rg.lm

## ══ Workflow ═══════════════════════════════════════════════════════════════════
## Preprocessor: None
## Model: linear_reg()
##
## ── Model ──────────────────────────────────────────────────────────────────────
────────────────────────────────────────────────────────────

## Linear Regression Model Specification (regression)
##
## Computational engine: lm
wfl.rg.lm<-wfl.rg.lm %>%                              # Preprocessor
  add_formula(Sale_Price~Longitude+Latitude)          # Formel
wfl.rg.lm

## ══ Workflow ═══════════════════════════════════════════════════════════════════
## Preprocessor: Formula
## Model: linear_reg()
##
## ── Preprocessor ───────────────────────────────────────────────────────────────
───────────────────────────────────────────────────────

## Sale_Price ~ Longitude + Latitude
##
## ── Model ──────────────────────────────────────────────────────────────────────
────────────────────────────────────────────────────────────

## Linear Regression Model Specification (regression)
##
## Computational engine: lm

wfl.rg.lm.fit.1<-fit(wfl.rg.lm,ames.train)            # Modell trainieren
wfl.rg.lm.fit.1

## ══ Workflow [trained] ═════════════════════════════════════════════════════════
## Preprocessor: Formula
## Model: linear_reg()
##
## ── Preprocessor ───────────────────────────────────────────────────────────────
───────────────────────────────────────────────────────

## Sale_Price ~ Longitude + Latitude
##
## ── Model ──────────────────────────────────────────────────────────────────────
────────────────────────────────────────────────────────────

##
## Call:
## stats::lm(formula = ..y ~ ., data = data)
```

```
##
## Coefficients:
## (Intercept)    Longitude     Latitude
##  -124970600      -774125      1252758

p1<-predict(wfl.rg.lm.fit.1,ames.test.n)              # Vorhersage
# Update von Workflow (Modell)-----------------------------------------------
mod.rg.stan<-linear_reg() %>% set_engine("stan")      # Modellauswahl
wfl.rg.stan<-wfl.rg.lm %>%                            # Update Modell
  update_model(mod.rg.stan)
wfl.rg.stan

## ══ Workflow ════════════════════════════════════════════════════════════════
## Preprocessor: Formula
## Model: linear_reg()
##
## ── Preprocessor ────────────────────────────────────────────────────────────
────────────────────────────────────────────

## Sale_Price ~ Longitude + Latitude
##
## ── Model ───────────────────────────────────────────────────────────────────
────────────────────────────────────────────────

## Linear Regression Model Specification (regression)
##
## Computational engine: stan

wfl.rg.stan.fit<-fit(wfl.rg.stan,ames.train)          # Modell trainieren
wfl.rg.stan.fit

## ══ Workflow [trained] ══════════════════════════════════════════════════════
## Preprocessor: Formula
## Model: linear_reg()
##
## ── Preprocessor ────────────────────────────────────────────────────────────
────────────────────────────────────────────

## Sale_Price ~ Longitude + Latitude
##
## ── Model ───────────────────────────────────────────────────────────────────
────────────────────────────────────────────────

## stan_glm
##  family:       gaussian [identity]
##  formula:      ..y ~ .
##  observations: 2342
##  predictors:   3
## ------
##             Median       MAD_SD
## (Intercept) -124910123.0    6455936.2
## Longitude      -773903.4      58766.2
## Latitude       1253944.6      82527.3
##
## Auxiliary parameter(s):
##       Median  MAD_SD
## sigma 72347.8  1092.7
##
## ------
## * For help interpreting the printed output see ?print.stanreg
## * For info on the priors used see ?prior_summary.stanreg
```

```
p2<-predict(wfl.rg.stan.fit,ames.test.n)              # Vorhersage
# Update von Workflow (Formel/Preprocessor)------------------------------------
wfl.rg.lm.2<-wfl.rg.lm %>%
  update_formula(Sale_Price~Longitude)                # Update Formel
wfl.rg.lm.2

## ══ Workflow ══════════════════════════════════════════════════════════════════
## Preprocessor: Formula
## Model: linear_reg()
##
## ── Preprocessor ──────────────────────────────────────────────────────────────
────────────────────────────────────────────────────────
## Sale_Price ~ Longitude
##
## ── Model ─────────────────────────────────────────────────────────────────────
──────────────────────────────────────────────────────────────
## Linear Regression Model Specification (regression)
##
## Computational engine: lm
wfl.rg.lm.fit.2<-fit(wfl.rg.lm.2,ames.train)         # Modell trainieren
wfl.rg.lm.fit.2

## ══ Workflow [trained] ════════════════════════════════════════════════════════
## Preprocessor: Formula
## Model: linear_reg()
##
## ── Preprocessor ──────────────────────────────────────────────────────────────
────────────────────────────────────────────────────────
## Sale_Price ~ Longitude
##
## ── Model ─────────────────────────────────────────────────────────────────────
──────────────────────────────────────────────────────────────
##
## Call:
## stats::lm(formula = ..y ~ ., data = data)
##
## Coefficients:
## (Intercept)    Longitude
##   -69220842      -741120

p3<-predict(wfl.rg.lm.fit.2,ames.test.n)              # Vorhersage
# Mit Variablen anstatt Formel arbeiten (Alternative für Preprocessor)------------
wfl.rg.lm.3<-wfl.rg.lm.2 %>%
  remove_formula() %>%
  add_variables(outcome=Sale_Price, predictors=c(Longitude, Latitude))
wfl.rg.lm.3

## ══ Workflow ══════════════════════════════════════════════════════════════════
## Preprocessor: Variables
## Model: linear_reg()
##
## ── Preprocessor ──────────────────────────────────────────────────────────────
────────────────────────────────────────────────────────
## Outcomes: Sale_Price
## Predictors: c(Longitude, Latitude)
##
```

```
## ── Model ─────────────────────────────────────────────────────────
──────────────────────────────────────────────────────────

## Linear Regression Model Specification (regression)
##
## Computational engine: lm

wfl.rg.lm.fit.3<-fit(wfl.rg.lm.3,ames.train)         # Modell trainieren
wfl.rg.lm.fit.3

## ══ Workflow [trained] ═══════════════════════════════════════════
## Preprocessor: Variables
## Model: linear_reg()
##
## ── Preprocessor ──────────────────────────────────────────────────
─────────────────────────────────────────────────────

## Outcomes: Sale_Price
## Predictors: c(Longitude, Latitude)
##
```

```
## ── Model ─────────────────────────────────────────────────────────────────
──────────────────────────────────────────────────────────────

## Linear Regression Model Specification (regression)
##
## Computational engine: lm

wfl.rg.lm.fit.3<-fit(wfl.rg.lm.3,ames.train)        # Modell trainieren
wfl.rg.lm.fit.3

## ══ Workflow [trained] ════════════════════════════════════════════════════
## Preprocessor: Variables
## Model: linear_reg()
##
## ── Preprocessor ──────────────────────────────────────────────────────────
──────────────────────────────────────────────────────────

## Outcomes: Sale_Price
## Predictors: c(Longitude, Latitude)
##
## ── Model ─────────────────────────────────────────────────────────────────
──────────────────────────────────────────────────────────────

##
## Call:
## stats::lm(formula = ..y ~ ., data = data)
##
## Coefficients:
## (Intercept)    Longitude     Latitude
##  -124970600      -774125      1252758

p4<-predict(wfl.rg.lm.fit.3,ames.test.n)               # Vorhersage
# Vergleich der Vorhersagen---------------------------------------------------
vorhersagen<-cbind(p1,p2,p3,p4)
colnames(vorhersagen)<-c("wfl.rg.lm.fit.1","wfl.rg.stan.fit",
                         "wfl.rg.lm.fit.2","wfl.rg.lm.fit.3")
vorhersagen

##   wfl.rg.lm.fit.1 wfl.rg.stan.fit wfl.rg.lm.fit.2 wfl.rg.lm.fit.3
## 1        185072.5        185087.3        162621.4        185072.5
## 2        184342.4        184357.1        162348.1        184342.4
## 3        209795.9        209805.5        176833.8        209795.9
## 4        191615.5        191630.7        164854.3        191615.5
## 5        221546.2        221548.3        187802.3        221546.2
```

The above example has demonstrated that the training of models can also be facilitated with the help of workflows. Several models can be applied to the same data and the results can then be used to compare the predictive quality of the models.

If several models with many combinations of variables and formulas are to be trained and tested for modeling, there is another suitable function for this, the function `workflow_set()` from the workflowsets package. The instruction `wflow.set<-workflow_set(preproc=formeln, models=list(lr= mod.rg.lm))` creates a workflow set with the four formulas created in the list formeln, based on the model mod.rg.lm, which was previously defined as a linear regression model with the engine lm. The formulas use different variables in the example and in this way the group of most relevant variables can be determined. With the function `extract_workflow()` *the individual workflows of the workflow set can be displayed. A single model from the workflow set can be trained with the instruction* `fit(wflow.set$workflow[[1]], autos.train)`.

To determine the metrics RMSE and R^2 for a model, the function `ml.eval()` from the machinelearning package can be used, e.g. `ml.eval( wflow.set$fit[[i]],autos.train)`, to determine the metrics for the first model in a workflow set based on the training data for used cars. The comparison of the four models results in a ranking in which the model alter.ps.kilometer has the best value for RMSE and RSQ. A model comparison based on workflow sets can be effectively realized with a few lines of coding.

Model selection, model training and model evaluation with workflow sets

```
# Trainings- und Testdaten erzeugen--------------------------------------------------
set.seed(123)
autos.split<-initial_split(autos,prop=0.80,          # Split Train/Test 80:20
                           strata=Preis)
autos.train<-training(autos.split)
autos.test<-testing(autos.split)
rm(autos.split)                                      # Split löschen
# Workflow-Sets: Modellauswahl und Training mit verschiedenen Formeln------------
mod.rg.lm<-linear_reg() %>% set_engine("lm")       # Modellauswahl
formeln <- list(
  alter              = Preis ~ Alter,
  ps                 = Preis ~ PS,
  alter.ps           = Preis ~ Alter + PS,
  alter.ps.kilometer = Preis ~ Alter + PS + Kilometer
)
wfl.set<-workflow_set(preproc=formeln, models=list(lr=mod.rg.lm))
wfl.set                                            # Anzeige Wflow-Sets

## # A workflow set/tibble: 4 × 4
##   wflow_id             info             option     result
##   <chr>                <list>           <list>     <list>
## 1 alter_lr             <tibble [1 × 4]> <opts[0]> <list [0]>
## 2 ps_lr                <tibble [1 × 4]> <opts[0]> <list [0]>
## 3 alter.ps_lr          <tibble [1 × 4]> <opts[0]> <list [0]>
## 4 alter.ps.kilometer_lr <tibble [1 × 4]> <opts[0]> <list [0]>

wfl.set$info[[1]]                                  # Anzeige Modell

## # A tibble: 1 × 4
##   workflow   preproc model      comment
##   <list>     <chr>   <chr>      <chr>
## 1 <workflow> formula linear_reg ""

extract_workflow(wfl.set, id="alter.ps_lr")        # Anzeige Wflow id
## ══ Workflow ════════════════════════════════════════════════════════════════
## Preprocessor: Formula
## Model: linear_reg()
##
## ── Preprocessor ────────────────────────────────────────────────────────────
────────────────────────────────────────────────────
## Preis ~ Alter + PS
##
## ── Model ───────────────────────────────────────────────────────────────────
─────────────────────────────────────────────────────────
## Linear Regression Model Specification (regression)
##
## Computational engine: lm

wfl.set<-wfl.set %>%                               # Trainieren der wflows
   mutate(fit=map(info, ~ fit(.x$workflow[[1]], autos.train)))
wfl.set

## # A workflow set/tibble: 4 × 5
##   wflow_id             info             option     result     fit
##   <chr>                <list>           <list>     <list>     <list>
## 1 alter_lr             <tibble [1 × 4]> <opts[0]> <list [0]> <workflow>
## 2 ps_lr                <tibble [1 × 4]> <opts[0]> <list [0]> <workflow>
## 3 alter.ps_lr          <tibble [1 × 4]> <opts[0]> <list [0]> <workflow>
## 4 alter.ps.kilometer_lr <tibble [1 × 4]> <opts[0]> <list [0]> <workflow>
```

```
wfl.set$fit[[1]]                                    # Anzeige trainiertes Mod.1

## ══ Workflow [trained] ══════════════════════════════════════════════════════
## Preprocessor: Formula
## Model: linear_reg()
##
## ── Preprocessor ──────────────────────────────────────────────────────────────
## Preis ~ Alter
##
## ── Model ─────────────────────────────────────────────────────────────────────
##
## Call:
## stats::lm(formula = ..y ~ ., data = data)
##
## Coefficients:
## (Intercept)        Alter
##       18603        -1033

# Vorhersagen machen-------------------------------------------------------------
autos.test.n <- autos.test[1:5,]
predict(wfl.set$fit[[1]],autos.test.n)              # Vorhersage Mod.1

## # A tibble: 5 × 1
##    .pred
##    <dbl>
## 1 -1021.
## 2  6209.
## 3  6209.
## 4  7241.
## 5  6209.

# Workflow-Sets: Evaluation------------------------------------------------------
ergebnis<-data.frame()
formel.n<-length(wfl.set$wflow_id)
for (i in 1:formel.n){                              # Evaluation Modelle
  ergebnis<-rbind(ergebnis,
                  ml.eval(wfl.set$fit[[i]],autos.train,22) %>%
                    mutate(Formel=wfl.set$wflow_id[i]))

}
# Ranking der Modelle nach RMSE (Trainingsdaten)
rmse.ergebnis<-filter(ergebnis,Kennzahl == "rmse") %>% arrange(Wert) %>%
  mutate(Platzierung=1:formel.n) %>%
  select(Platzierung,Formel,Wert,Kennzahl,Daten)
rmse.ergebnis

##   Platzierung                Formel     Wert Kennzahl       Daten
## 1           1 alter.ps.kilometer_lr 4199.061     rmse autos.train
## 2           2           alter.ps_lr 4445.956     rmse autos.train
## 3           3              alter_lr 5586.854     rmse autos.train
## 4           4                 ps_lr 6004.990     rmse autos.train

# Ranking der Modelle nach Bestimmtheitsmaß (Trainingsdaten)
rsq.ergebnis<-filter(ergebnis,Kennzahl == "rsq") %>% arrange(desc(Wert)) %>%
  mutate(Platzierung=1:formel.n) %>%
  select(Platzierung,Formel,Wert,Kennzahl,Daten)
rsq.ergebnis

##   Platzierung                Formel      Wert Kennzahl       Daten
## 1           1 alter.ps.kilometer_lr 0.6863973      rsq autos.train
## 2           2           alter.ps_lr 0.6484349      rsq autos.train
## 3           3              alter_lr 0.4448504      rsq autos.train
## 4           4                 ps_lr 0.3586428      rsq autos.train
```

If, after the evaluation of various models, the most suitable model is to be validated again with all the training and test data, the function `last_fit()` is suitable. The function `extract_workflow()` allows the trained model to be displayed. The functions `collect_metrics()` and `collect_predictions()` from the tune package enable the display of metrics and predictions.

At this point in model development, the price is predicted to deviate from the actual price by an average of $4199, and the R^2 of 0.69 indicates a mediocre correlation of the predictions with the actual prices. Further optimization is recommended in this case. In the currently tested model, many of the possible predictors have not yet been used, including brand and car model type, which makes the deviation understandable.

Model Test

```
## Tidymodels Komplетttest -------------------------------------------------------
set.seed(123)
autos.split <-initial_split(autos,prop=0.80,            # Split Train/Test 80:20
                            strata=Preis)
mod.rg.lm<-linear_reg() %>% set_engine("lm")            # Modellauswahl
wfl.rg.lm <-workflow() %>% add_model(mod.rg.lm)         # Workflow
wfl.rg.lm <-wfl.rg.lm %>%                               # Preprocessor
  add_formula(Preis ~ Alter+PS+Kilometer)               # Formel
wfl.rg.lm.fit.final<-last_fit(wfl.rg.lm, autos.split)  # Training und Test
                                                        # Daher autos.split
                                                        # anstatt nur autos.train
wfl.rg.lm.fit.final

## # Resampling results
## # Manual resampling
## # A tibble: 1 × 6
##   splits              id               .metrics .notes   .predicti…¹ .workflow
##   <list>              <chr>            <list>   <list>   <list>      <list>
## 1 <split [27890/6974]> train/test split <tibble> <tibble> <tibble>    <workflow>
## # … with abbreviated variable name ¹.predictions

extract_workflow(wfl.rg.lm.fit.final )                  # Anzeige trainiertes Mod.

## ══ Workflow [trained] ══════════════════════════════════════════════════════
## Preprocessor: Formula
## Model: linear_reg()
##
## ── Preprocessor ────────────────────────────────────────────────────────────
```

```
## Preis ~ Alter + PS + Kilometer
##
## —— Model ————————————————————————————————————————————————
————————————————————————————————————————————
##
## Call:
## stats::lm(formula = ..y ~ ., data = data)
##
## Coefficients:
## (Intercept)        Alter           PS    Kilometer
##   1.166e+04   -6.428e+02    6.237e+01   -4.509e-02

collect_metrics(wfl.rg.lm.fit.final)                 # Metriken

## # A tibble: 2 × 4
##   .metric .estimator .estimate .config
##   <chr>   <chr>          <dbl> <chr>
## 1 rmse    standard    4089.    Preprocessor1_Model1
## 2 rsq     standard       0.694 Preprocessor1_Model1

collect_predictions(wfl.rg.lm.fit.final) %>% slice(1:5)# Vorhersage
## # A tibble: 5 × 5
##   id                .pred  .row Preis .config
##   <chr>             <dbl> <int> <dbl> <chr>
## 1 train/test split 10518.     2  8500 Preprocessor1_Model1
## 2 train/test split  4914.    11  4150 Preprocessor1_Model1
## 3 train/test split  9903.    17  4999 Preprocessor1_Model1
## 4 train/test split 10546.    33  6000 Preprocessor1_Model1
## 5 train/test split  5542.    35  2500 Preprocessor1_Model1
```

6.5.2 Resampling

In the following example, two models are selected, trained, and evaluated to predict the prices of used cars. The Model-1 lr.fit of Linear Regression and a Model-2 rf.fit based on Random Forest decision trees are estimated, initially for the training data, the RMSE for the linear model is 3928 and for the Random Forest model 1871. The average deviation of the prediction from the actual price is more than $2000 lower for the Random Forest model for the training data. Also, the coefficient of determination $R^2$is significantly better at 94% for the Random Forest model than the Linear Regression model at 73%. Therefore, it initially seems obvious that only the Random Forest model will be used for further proceedings.

Model Evaluation Based on Training Data

```
## Resampling Modellvergleich ---------------------------------------------------
# Modelltyp: Lineare Regression--------------------------------------------------
mod.rg.lm<-linear_reg() %>%                                # Modell lm
  set_engine("lm")
wfl.rg.lm <-workflow() %>%                                 # Workflow
  add_model(mod.rg.lm) %>%
  add_formula(Preis~Kilometer+Alter+PS+Marke)
wfl.rg.lm

## == Workflow ==================================================================
## Preprocessor: Formula
## Model: linear_reg()
##
## —— Preprocessor ——————————————————————————————————————————————————————————————
——————————————————————————————————————————————————————
## Preis ~ Kilometer + Alter + PS + Marke
##
## —— Model ——————————————————————————————————————————————————————————————————————
———————————————————————————————————————————————————————————

## Linear Regression Model Specification (regression)
##
## Computational engine: lm

wfl.rg.lm.fit<-fit(wfl.rg.lm,autos.train)                  # Modell trainieren
# Modelltyp: Random Forest-------------------------------------------------------
mod.rg.rf<-rand_forest(trees = 1000) %>%                   # Modell rf
  set_engine("ranger") %>%
  set_mode("regression")
wfl.rg.rf <-workflow() %>%
  add_formula(Preis~Kilometer+Alter+PS+Marke) %>%
  add_model(mod.rg.rf)
wfl.rg.rf

## == Workflow ==================================================================
## Preprocessor: Formula
## Model: rand_forest()
##
## — Preprocessor ———————————————————————————————————————————————————————————————
## Preis ~ Kilometer + Alter + PS + Marke
##
## — Model ——————————————————————————————————————————————————————————————————————
## Random Forest Model Specification (regression)
##
## Main Arguments:
##   trees = 1000
##
## Computational engine: ranger

wfl.rg.rf.fit<-fit(wfl.rg.rf,autos.train)                  # Modell trainieren
# Evaluation---------------------------------------------------------------------
#autos.train$x<-autos.train$Preis                          # x für ml.eval erforderlich, we
nn #nicht bereits vorhanden!
ml.eval(wfl.rg.lm.fit,autos.train,22)

##          Modell       Daten Kennzahl         Wert
## 1 wfl.rg.lm.fit autos.train     rmse 3927.6179003
## 2 wfl.rg.lm.fit autos.train      rsq    0.7256317

ml.eval(wfl.rg.rf.fit,autos.train,22)
```

```
##           Modell       Daten Kennzahl         Wert
## 1 wfl.rg.rf.fit autos.train     rmse 1871.0866202
## 2 wfl.rg.rf.fit autos.train      rsq    0.9386506
```

Models that were trained based on the training data are often very good at predicting, based on the training data, if they have learned these data well and have a so-called bias that positively influences the prediction. However, the predictions for the test data are not always equally good, as these were unknown to the model at the time of training. This often applies to Random Forest models as well. If the prediction is now also evaluated for both models based on the test data, it turns out that the RMSE for the linear model has improved to 3809 and for the Random Forest model has worsened to 2538. The same applies to the RSQ. This confirms the need to evaluate the prediction quality not only based on the training data but also based on the test data.

Model Evaluation Based on Test Data

```
# Evaluation ------------------------------------------------------------------
ergebnis<-rbind(ml.eval(wfl.rg.lm.fit,autos.train,22),
                ml.eval(wfl.rg.rf.fit,autos.train,22),
                ml.eval(wfl.rg.lm.fit,autos.test,22),
                ml.eval(wfl.rg.rf.fit,autos.test,22))
filter(ergebnis,Kennzahl == "rmse")    # RMSE

##           Modell       Daten Kennzahl     Wert
## 1 wfl.rg.lm.fit autos.train     rmse 3927.618
## 2 wfl.rg.rf.fit autos.train     rmse 1871.087
## 3 wfl.rg.lm.fit  autos.test     rmse 3808.690
## 4 wfl.rg.rf.fit  autos.test     rmse 2538.387

filter(ergebnis,Kennzahl == "rsq")     # Bestimmtheitsmaß

##           Modell       Daten Kennzahl      Wert
## 1 wfl.rg.lm.fit autos.train      rsq 0.7256317
## 2 wfl.rg.rf.fit autos.train      rsq 0.9386506
## 3 wfl.rg.lm.fit  autos.test      rsq 0.7344065
## 4 wfl.rg.rf.fit  autos.test      rsq 0.8821269
```

In this example, the Random Forest model remains the superior model. The insight from this example is that one cannot necessarily infer the test data from the metrics that relate to the training data. However, to improve learning, it is recommended that the training data be further divided into groups of sub-samples or sample repetitions (Resample Folds) during learning, each composed of analysis data and validation data (see Fig. 6.31). These sub-samples are drawn exclusively from the original sample of the training data, so the clear separation between the training and test data is maintained. The process of forming the sub-samples is referred to as **Resampling**. Although the learning process becomes more complex and comprehensive due to the number of sub-samples formed, it contributes to better results.

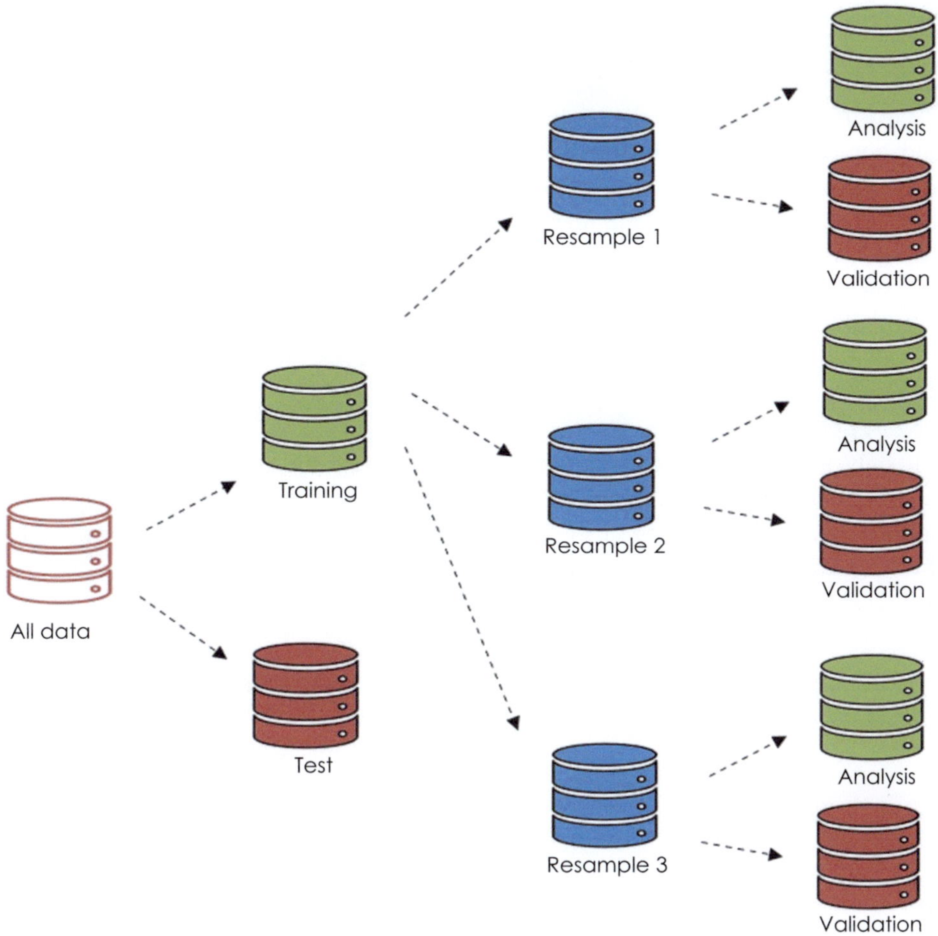

Fig. 6.31 Resampling Samples

The most widespread resampling methods are **Cross-validation** and **Bootstrapping**. In V-Fold Cross-validation, v sub-samples are formed, in which the training data is divided into analysis data of size v-1/v and validation data of size 1/v for each of the v folds by random sampling from the total population. All validation data of the folds together are identical to the total population of the training data, which means that no validation data set is used more than once for validation. To significantly improve the test results, a v of 10 or greater is typically used. The function `vfold_cv()` in R supports the formation of the sets. The command `autos.vfolds<-vfold_cv(autos.train,v=10)` creates 10 folds with 25,101 analysis data sets and 2,789 validation data sets each. With 10 folds, 10 models are created from which the "best" can be selected. To further enhance the quality of model selection, the method of repeated cross-validation can be applied, in which n times v folds are formed. Based on the **Central Limit Theorem**, the standard error decreases with increasing repetition. The number of repetitions is achieved in the function `vfold_cv()` by the parameter repeats, e.g., `autos.nvfolds<-vfold_cv(autos.train,v=10,repeats=10)`.

A special variant of cross-validation is Monte Carlo cross-validation, in which the selection of validation data sets is not disjoint, as the data sets are randomly selected from all data sets. The function to generate the folds is `mc_cv()`, where the proportion of the split can be specified as a parameter, e.g., `autos.mcfolds<-mc_cv(autos.train,prop=7/10, times=10)`.

If only a single fold is to be generated to separate the training data into analysis data and validation data apart from the test data, this is also possible. The function `validation_split()` is used for this, e.g., `autos.fold<-validation_split(autos.train,prop=4/5)`, to use 80% of the training data for analysis and 20% for validation.

Cross-validation

```
set.seed(123)
autos.vfolds<-vfold_cv(autos.train,v=10)                    # 10 Folds erzeugen
autos.vfolds

## #  10-fold cross-validation
## # A tibble: 10 × 2
##    splits               id
##    <list>               <chr>
##  1 <split [25101/2789]> Fold01
##  2 <split [25101/2789]> Fold02
##  3 <split [25101/2789]> Fold03
##  4 <split [25101/2789]> Fold04
##  5 <split [25101/2789]> Fold05
##  6 <split [25101/2789]> Fold06
##  7 <split [25101/2789]> Fold07
##  8 <split [25101/2789]> Fold08
##  9 <split [25101/2789]> Fold09
## 10 <split [25101/2789]> Fold10

autos.vfold01<-autos.vfolds$splits[[1]]                     # Fold01 erzeugen
dim(autos.vfold01 %>% analysis())                           # Fold01-Analyse

## [1] 25101    23

dim(autos.vfold01 %>% assessment())                         # Fold01-Validierung

## [1] 2789   23

# Wiederholte Kreuzvalidierung ----------------------------------------------------
set.seed(123)
autos.nvfolds<-vfold_cv(autos.train,v=10,repeats=10) # 10x10 Folds erzeugen
autos.nvfolds

## #  10-fold cross-validation repeated 10 times
## # A tibble: 100 × 3
##    splits               id       id2
##    <list>               <chr>    <chr>
##  1 <split [25101/2789]> Repeat01 Fold01
##  2 <split [25101/2789]> Repeat01 Fold02
##  3 <split [25101/2789]> Repeat01 Fold03
##  4 <split [25101/2789]> Repeat01 Fold04
##  5 <split [25101/2789]> Repeat01 Fold05
##  6 <split [25101/2789]> Repeat01 Fold06
##  7 <split [25101/2789]> Repeat01 Fold07
##  8 <split [25101/2789]> Repeat01 Fold08
##  9 <split [25101/2789]> Repeat01 Fold09
## 10 <split [25101/2789]> Repeat01 Fold10
## # … with 90 more rows

autos.repeat01.fold01<-autos.nvfolds$splits[[1]]     # Repeat01/Fold01 erzeugen
dim(autos.repeat01.fold01 %>% analysis())            # Repeat01/Fold01-Analyse

## [1] 25101    23
dim(autos.repeat01.fold01 %>% assessment())          # Repeat01/Fold01-Validierung

## [1] 2789   23

# Monte-Carlo Kreuzvalidierung ----------------------------------------------------
set.seed(123)
autos.mcfolds<-mc_cv(autos.train,prop=7/10,times=10) # 10 MC-Folds erzeugen
autos.mcfolds

## # Monte Carlo cross-validation (0.7/0.3) with 10 resamples
## # A tibble: 10 × 2
```

```
##     splits                  id
##     <list>                  <chr>
##   1 <split [19523/8367]> Resample01
##   2 <split [19523/8367]> Resample02
##   3 <split [19523/8367]> Resample03
##   4 <split [19523/8367]> Resample04
##   5 <split [19523/8367]> Resample05
##   6 <split [19523/8367]> Resample06
##   7 <split [19523/8367]> Resample07
##   8 <split [19523/8367]> Resample08
##   9 <split [19523/8367]> Resample09
##  10 <split [19523/8367]> Resample10

autos.mcfold01<-autos.mcfolds$splits[[1]]              # Fold01 erzeugen
dim(autos.mcfold01 %>% analysis())                     # Fold01-Analyse

## [1] 19523    23

dim(autos.mcfold01 %>% assessment())                   # Fold01-Validierung

## [1] 8367   23

# Einzelner Fold mit Trennung in Analyse- und Validierung -----------------------
set.seed(123)
autos.fold<-validation_split(autos.train, prop=4/5)  # Split in 80% Anal.+20% Val.
autos.fold

## # Validation Set Split (0.8/0.2)
## # A tibble: 1 × 2
##   splits                 id
##   <list>                 <chr>
## 1 <split [22312/5578]> validation
```

The **Bootstrap method**, or Bootstrapping method, draws the analysis data at random with replacement from a single sample, here the total population of the training data. In this way, n sets of analysis data are generated from a pool of training data. The corresponding validation data is not explicitly generated. The validation data are all the data sets that are not contained in the analysis data (so-called Out-Of-Bag sample). In Bootstrap, n Bootstrap samples are drawn with replacement from the given sample. It is possible that individual data sets flow into the samples multiple times or not at all. The function `bootstraps()` from the rsample package supports this method. The command `autos.boots<-bootstraps(autos.train,times=10)` generates, for example, 10 Bootstrap samples.

Bootstrap Method

```
# Bootstrap-Verfahren -------------------------------------------------------------
autos.boots<-bootstraps(autos.train,times=10)        # 10 Bootstraps erzeugen
autos.boots

## # Bootstrap sampling
## # A tibble: 10 × 2
##    splits                id
##    <list>                <chr>
##  1 <split [27890/10256]> Bootstrap01
##  2 <split [27890/10276]> Bootstrap02
##  3 <split [27890/10183]> Bootstrap03
##  4 <split [27890/10335]> Bootstrap04
##  5 <split [27890/10257]> Bootstrap05
##  6 <split [27890/10273]> Bootstrap06
##  7 <split [27890/10272]> Bootstrap07
##  8 <split [27890/10268]> Bootstrap08
##  9 <split [27890/10193]> Bootstrap09
## 10 <split [27890/10254]> Bootstrap10

autos.boot01<-autos.boots$splits[[1]]                # Fold01 erzeugen
```

There are also application areas where a random sample or stratified random sample is unsuitable, namely for time series data. With time series, the historical data should always be used as training data and the most recent data (periods) as test data. The functions `sliding_window()`, `sliding_index()` and `sliding_period()` from the rsample package support the division of time series data into training and test data.

After the resampling samples have been generated, they can be used for training the models. This is supported by the function `fit_resamples()`, which works analogously to the function `fit()`, except that it expects an object with resampling samples as data. Previously, the options for execution can be defined with the function `control_resamples()`. The parameter save_pred = TRUE is helpful, among other things, to save the predicted values. The function call could look like this: `fit_resamples(resamples=autos.vfolds,control=control_resamples(save_pred=TRUE))`. In addition, the standard for regressions stores the metrics RMSE and R^2 and for classifications AUC and accuracy as a result of the function. Afterwards, these metrics can be displayed with the function `collect_metrics()`, e.g. `collect_metrics(wfl.rg.lm.fitres)` for the averages of all resample runs. If the parameter summarize=FALSE is used, the metrics of the resample samples are displayed individually.

While the linear regression model trained with the classic training data results in an RMSE of 3928 and with 10 folds an RMSE of 3933 and no significant difference is noticeable, the RMSE for the Random Forest model has changed from 1836 to a more realistic 2751. The values do not differ significantly in this example between cross-validation and bootstrap resampling. The scatterplot output by the function ml.eval.lr() from the machinelearning package also visually shows that the predicted values for the Random Forest model are much closer to the line of equality than for the linear regression with lm.

Use of the Resample Samples

```
# Verwendung der Resample-Stichproben für Lineare Regression --------------------
set.seed(123)
wfl.rg.lm

## ══ Workflow ══════════════════════════════════════════════════════════════════
## Preprocessor: Formula
## Model: linear_reg()
##
## ── Preprocessor ────────────────────────────────────────────────────────────
────────────────────────────────────────────────────────
## Preis ~ Kilometer + Alter + PS + Marke
##
## ── Model ───────────────────────────────────────────────────────────────────
──────────────────────────────────────────────────────────────
## Linear Regression Model Specification (regression)
##
## Computational engine: lm

wfl.rg.lm.fit<-fit(wfl.rg.lm,autos.train)      # Modell mit 1 Traindaten trainieren
ml.eval(wfl.rg.lm.fit,autos.train,22)             # -> RMSE=3928, RSQ=0.73

##            Modell        Daten Kennzahl           Wert
## 1 wfl.rg.lm.fit autos.train      rmse 3927.6179003
## 2 wfl.rg.lm.fit autos.train       rsq    0.7256317

wfl.rg.lm.fitres <- wfl.rg.lm %>%                # Modell mit 10 Resamples trainieren
  fit_resamples(resamples=autos.vfolds,control=control_resamples(save_pred=TRUE))
wfl.rg.lm.fitres

## # Resampling results
## # 10-fold cross-validation
## # A tibble: 10 × 5
##    splits               id     .metrics         .notes           .predictions
##    <list>               <chr>  <list>           <list>           <list>
##  1 <split [25101/2789]> Fold01 <tibble [2 × 4]> <tibble [0 × 3]> <tibble>
##  2 <split [25101/2789]> Fold02 <tibble [2 × 4]> <tibble [0 × 3]> <tibble>
##  3 <split [25101/2789]> Fold03 <tibble [2 × 4]> <tibble [0 × 3]> <tibble>
##  4 <split [25101/2789]> Fold04 <tibble [2 × 4]> <tibble [0 × 3]> <tibble>
##  5 <split [25101/2789]> Fold05 <tibble [2 × 4]> <tibble [0 × 3]> <tibble>
##  6 <split [25101/2789]> Fold06 <tibble [2 × 4]> <tibble [0 × 3]> <tibble>
##  7 <split [25101/2789]> Fold07 <tibble [2 × 4]> <tibble [0 × 3]> <tibble>
##  8 <split [25101/2789]> Fold08 <tibble [2 × 4]> <tibble [0 × 3]> <tibble>
##  9 <split [25101/2789]> Fold09 <tibble [2 × 4]> <tibble [0 × 3]> <tibble>
## 10 <split [25101/2789]> Fold10 <tibble [2 × 4]> <tibble [0 × 3]> <tibble>

collect_metrics(wfl.rg.lm.fitres)               # -> RMSE=3933, RSQ=0.73

## # A tibble: 2 × 6
##   .metric .estimator     mean     n  std_err .config
##   <chr>   <chr>         <dbl> <int>    <dbl> <chr>
## 1 rmse    standard   3933.       10 40.6     Preprocessor1_Model1
## 2 rsq     standard      0.725    10  0.00203 Preprocessor1_Model1

collect_metrics(wfl.rg.lm.fitres,summarize=FALSE)

## # A tibble: 20 × 5
##    id     .metric .estimator .estimate .config
##    <chr>  <chr>   <chr>          <dbl> <chr>
##  1 Fold01 rmse    standard    4074.    Preprocessor1_Model1
```

```
##  2 Fold01 rsq     standard       0.724 Preprocessor1_Model1
##  3 Fold02 rmse    standard    3896.    Preprocessor1_Model1
##  4 Fold02 rsq     standard       0.729 Preprocessor1_Model1
##  5 Fold03 rmse    standard    3692.    Preprocessor1_Model1
##  6 Fold03 rsq     standard       0.727 Preprocessor1_Model1
##  7 Fold04 rmse    standard    3823.    Preprocessor1_Model1
##  8 Fold04 rsq     standard       0.722 Preprocessor1_Model1
##  9 Fold05 rmse    standard    3865.    Preprocessor1_Model1
## 10 Fold05 rsq     standard       0.736 Preprocessor1_Model1
## 11 Fold06 rmse    standard    4069.    Preprocessor1_Model1
## 12 Fold06 rsq     standard       0.734 Preprocessor1_Model1
## 13 Fold07 rmse    standard    4014.    Preprocessor1_Model1
## 14 Fold07 rsq     standard       0.720 Preprocessor1_Model1
## 15 Fold08 rmse    standard    4034.    Preprocessor1_Model1
## 16 Fold08 rsq     standard       0.717 Preprocessor1_Model1
## 17 Fold09 rmse    standard    4024.    Preprocessor1_Model1
## 18 Fold09 rsq     standard       0.722 Preprocessor1_Model1
## 19 Fold10 rmse    standard    3841.    Preprocessor1_Model1
## 20 Fold10 rsq     standard       0.719 Preprocessor1_Model1
```

```
wfl.rg.lm.fitres.val<-collect_predictions(wfl.rg.lm.fitres)
wfl.rg.lm.fitres.val                          # Vorhersagen Validierungsdaten
```

```
## # A tibble: 27,890 × 5
##    id      .pred  .row Preis .config
##    <chr>   <dbl> <int> <dbl> <chr>
##  1 Fold01 -3080.     8   449 Preprocessor1_Model1
##  2 Fold01  -858.    10   800 Preprocessor1_Model1
##  3 Fold01  9110.    15  1100 Preprocessor1_Model1
##  4 Fold01 -1380.    23   450 Preprocessor1_Model1
##  5 Fold01 -2187.    37   400 Preprocessor1_Model1
##  6 Fold01   149.    40  1199 Preprocessor1_Model1
##  7 Fold01   602.    41   730 Preprocessor1_Model1
##  8 Fold01  2201.    43  1250 Preprocessor1_Model1
##  9 Fold01  3122.    48  1500 Preprocessor1_Model1
## 10 Fold01   180.    50  1100 Preprocessor1_Model1
## # … with 27,880 more rows
```

```
# Verwendung der Resample-Stichproben für Random-Forest -------------------------
set.seed(123)
wfl.rg.rf
```

```
## ══ Workflow ═══════════════════════════════════════════════════════════════════
## Preprocessor: Formula
## Model: rand_forest()
##
## ── Preprocessor ────────────────────────────────────────────────────────────────
## Preis ~ Kilometer + Alter + PS + Marke
##
## ── Model ───────────────────────────────────────────────────────────────────────
## Random Forest Model Specification (regression)
##
## Main Arguments:
##   trees = 1000
##
## Computational engine: ranger
```

```
wfl.rg.rf.fit<-fit(wfl.rg.rf,autos.train)     # Modell mit 1 Traindaten trainieren
ml.eval(wfl.rg.rf.fit,autos.train,22)            # -> RMSE=1868, RSQ=0.94
```

```
##           Modell       Daten Kennzahl         Wert
```

```
## 1 wfl.rg.rf.fit autos.train     rmse 1868.9446961
## 2 wfl.rg.rf.fit autos.train      rsq    0.9387952

wfl.rg.rf.fitres <- wfl.rg.rf %>%                # Modell mit 10 Resamples trainieren
  fit_resamples(resamples=autos.vfolds,control=control_resamples(save_pred=TRUE))
wfl.rg.rf.fitres

## # Resampling results
## # 10-fold cross-validation
## # A tibble: 10 × 5
##    splits                id     .metrics         .notes           .predictions
##    <list>                <chr>  <list>           <list>           <list>
##  1 <split [25101/2789]> Fold01 <tibble [2 × 4]> <tibble [0 × 3]> <tibble>
##  2 <split [25101/2789]> Fold02 <tibble [2 × 4]> <tibble [0 × 3]> <tibble>
##  3 <split [25101/2789]> Fold03 <tibble [2 × 4]> <tibble [0 × 3]> <tibble>
##  4 <split [25101/2789]> Fold04 <tibble [2 × 4]> <tibble [0 × 3]> <tibble>
##  5 <split [25101/2789]> Fold05 <tibble [2 × 4]> <tibble [0 × 3]> <tibble>
##  6 <split [25101/2789]> Fold06 <tibble [2 × 4]> <tibble [0 × 3]> <tibble>
##  7 <split [25101/2789]> Fold07 <tibble [2 × 4]> <tibble [0 × 3]> <tibble>
##  8 <split [25101/2789]> Fold08 <tibble [2 × 4]> <tibble [0 × 3]> <tibble>
##  9 <split [25101/2789]> Fold09 <tibble [2 × 4]> <tibble [0 × 3]> <tibble>
## 10 <split [25101/2789]> Fold10 <tibble [2 × 4]> <tibble [0 × 3]> <tibble>

collect_metrics(wfl.rg.rf.fitres)                # -> RMSE=2751, RSQ=0.87

## # A tibble: 2 × 6
##   .metric .estimator     mean     n  std_err .config
##   <chr>   <chr>         <dbl> <int>    <dbl> <chr>
## 1 rmse    standard   2751.       10 42.1     Preprocessor1_Model1
## 2 rsq     standard      0.865    10  0.00439 Preprocessor1_Model1

wfl.rg.rf.fitres.val<-collect_predictions(wfl.rg.rf.fitres)
wfl.rg.rf.fitres.val                             # Vorhersagen Validierungsdaten

## # A tibble: 27,890 × 5
##    id     .pred  .row Preis .config
##    <chr>  <dbl> <int> <dbl> <chr>
##  1 Fold01  755.     8   449 Preprocessor1_Model1
##  2 Fold01 1414.    10   800 Preprocessor1_Model1
##  3 Fold01 1840.    15  1100 Preprocessor1_Model1
##  4 Fold01 1055.    23   450 Preprocessor1_Model1
##  5 Fold01 1008.    37   400 Preprocessor1_Model1
##  6 Fold01 1191.    40  1199 Preprocessor1_Model1
##  7 Fold01 2163.    41   730 Preprocessor1_Model1
##  8 Fold01 1602.    43  1250 Preprocessor1_Model1
##  9 Fold01 2865.    48  1500 Preprocessor1_Model1
## 10 Fold01 1081.    50  1100 Preprocessor1_Model1
## # … with 27,880 more rows

# Scatterplot der Resample-Stichproben ------------------------------------------
ml.eval.lr(wfl.rg.lm.fitres.val[,c("Preis",".pred")],   # Scatter für lm
                 "Gebrauchtwagen","realer Preis",
                 "vorhergesagter Preis",scatter=T)
```

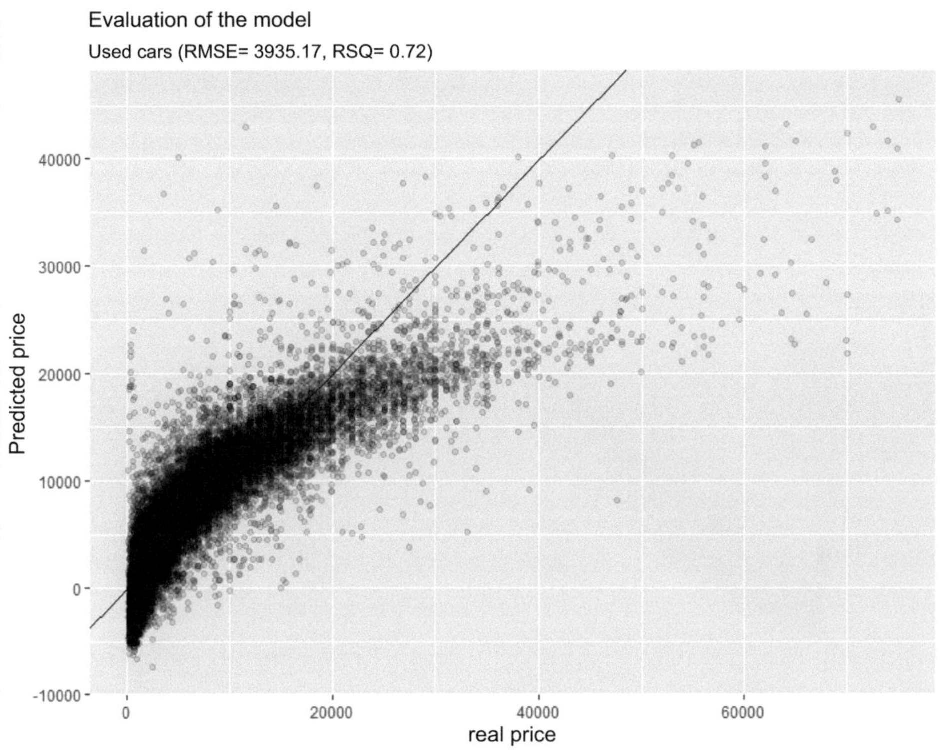


```
##    Kennzahl         Wert
## 1      rmse 3935.1688831
## 2       rsq    0.7245762
```

```
ml.eval.lr(wfl.rg.rf.fitres.val[,c("Preis",".pred")],    # Scatter für rf
                "Gebrauchtwagen","realer Preis",
                "vorhergesagter Preis",scatter=T)
```

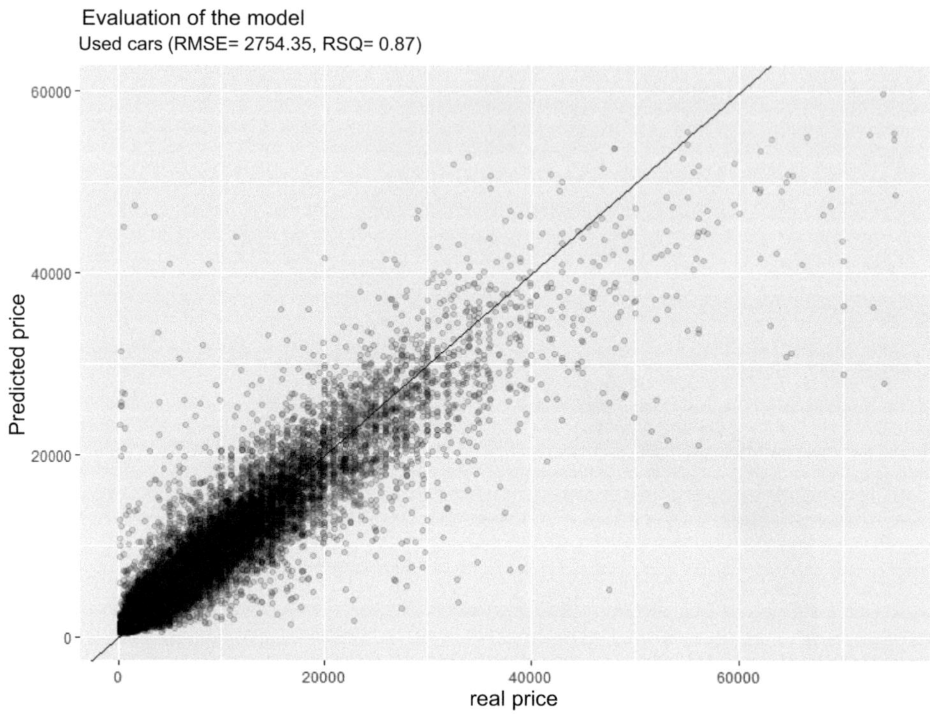


```
##   Kennzahl         Wert
## 1     rmse 2754.3470406
## 2      rsq    0.8651238

# Ausreißer der Resample-Stichproben -------------------------------------------
wfl.rg.lm.ausr.pred <- wfl.rg.lm.fitres.val %>%         # TOP3 Ausreißer lm
  mutate(Residual=Preis - .pred,Vorhersage=.pred) %>%
  arrange(desc(abs(Residual))) %>%
  slice(1:3)
wfl.rg.lm.ausr.pred[,c("Residual","Preis","Vorhersage")]

## # A tibble: 3 × 3
##   Residual Preis Vorhersage
##      <dbl> <dbl>      <dbl>
## 1   48141. 69993     21852.
## 2   46784. 69900     23116.
## 3   42690. 70000     27310.

wfl.rg.lm.ausr.train<-autos.train[wfl.rg.lm.ausr.pred$.row,]
wfl.rg.lm.ausr<-cbind(wfl.rg.lm.ausr.pred,wfl.rg.lm.ausr.train)
wfl.rg.lm.ausr[,c("Residual","Preis","Vorhersage","Marke","Modell",
                  "Typ","Alter","Kilometer","PS","Unfall")]

##   Residual Preis Vorhersage         Marke   Modell       Typ Alter Kilometer
## 1 48141.49 69993   21851.51       porsche      911     coupe    20    150000
## 2 46783.59 69900   23116.41 mercedes_benz s_klasse limousine     3     60000
## 3 42690.23 70000   27309.77    land_rover defender       suv     1      5000
##    PS Unfall
## 1 286   nein
```

```
## 2 258   nein
## 3 122   nein

wfl.rg.rf.ausr.pred <- wfl.rg.rf.fitres.val %>%        # TOP3 Ausreißer rf
  mutate(Residual=Preis - .pred,Vorhersage=.pred) %>%
  arrange(desc(abs(Residual))) %>%
  slice(1:3)
wfl.rg.rf.ausr.pred[,c("Residual","Preis","Vorhersage")]

## # A tibble: 3 × 3
##   Residual Preis Vorhersage
##      <dbl> <dbl>      <dbl>
## 1   46078. 73996     27918.
## 2  -45714.  1642     47356.
## 3  -44545.   580     45125.

wfl.rg.rf.ausr.train<-autos.train[wfl.rg.rf.ausr.pred$.row,]
wfl.rg.rf.ausr<-cbind(wfl.rg.rf.ausr.pred,wfl.rg.rf.ausr.train)
wfl.rg.rf.ausr[,c("Residual","Preis","Vorhersage","Marke","Modell",
                  "Typ","Alter","Kilometer","PS","Unfall")]

##    Residual Preis Vorhersage   Marke Modell       Typ Alter Kilometer  PS
## 1  46077.52 73996   27918.48 porsche    911    cabrio    12    125000 420
## 2 -45713.85  1642   47355.85     bmw andere     coupe     1     10000 362
## 3 -44544.57   580   45124.57     bmw    5er limousine     2     40000 258
##   Unfall
## 1   nein
## 2   nein
## 3   nein

# Einzelner Fold mit Trennung in Analyse- und Validierung ------------------------
wfl.rg.rf.fitres.single<-wfl.rg.rf %>%                 # Modell mit 1 Fold
  fit_resamples(resamples=autos.fold,control=control_resamples(save_pred=TRUE))
wfl.rg.rf.fitres.single

## # Resampling results
## # Validation Set Split (0.8/0.2)
## # A tibble: 1 × 5
##   splits              id         .metrics         .notes           .predictions
##   <list>              <chr>      <list>           <list>           <list>
## 1 <split [22312/5578]> validation <tibble [2 × 4]> <tibble [0 × 3]> <tibble>

collect_metrics(wfl.rg.rf.fitres.single)               # -> RMSE=2756, RSQ=0.87

## # A tibble: 2 × 6
##   .metric .estimator     mean     n std_err .config
##   <chr>   <chr>         <dbl> <int>   <dbl> <chr>
## 1 rmse    standard   2756.        1      NA Preprocessor1_Model1
## 2 rsq     standard      0.866     1      NA Preprocessor1_Model1

# Bootstrap ----------------------------------------------------------------------
wfl.rg.rf.fitres.boots<-wfl.rg.rf %>%                  # Modell mit 10 Bootstraps
  fit_resamples(resamples=autos.boots,control=control_resamples(save_pred=TRUE))
wfl.rg.rf.fitres.boots

## # Resampling results
## # Bootstrap sampling
## # A tibble: 10 × 5
##    splits                id          .metrics         .notes           .predic…¹
##    <list>                <chr>       <list>           <list>           <list>
##  1 <split [27890/10285]> Bootstrap01 <tibble [2 × 4]> <tibble [0 × 3]> <tibble>
##  2 <split [27890/10205]> Bootstrap02 <tibble [2 × 4]> <tibble [0 × 3]> <tibble>
```

```
##  3 <split [27890/10344]> Bootstrap03 <tibble [2 × 4]> <tibble [0 × 3]> <tibble>
##  4 <split [27890/10250]> Bootstrap04 <tibble [2 × 4]> <tibble [0 × 3]> <tibble>
##  5 <split [27890/10264]> Bootstrap05 <tibble [2 × 4]> <tibble [0 × 3]> <tibble>
##  6 <split [27890/10270]> Bootstrap06 <tibble [2 × 4]> <tibble [0 × 3]> <tibble>
##  7 <split [27890/10205]> Bootstrap07 <tibble [2 × 4]> <tibble [0 × 3]> <tibble>
##  8 <split [27890/10260]> Bootstrap08 <tibble [2 × 4]> <tibble [0 × 3]> <tibble>
##  9 <split [27890/10251]> Bootstrap09 <tibble [2 × 4]> <tibble [0 × 3]> <tibble>
## 10 <split [27890/10276]> Bootstrap10 <tibble [2 × 4]> <tibble [0 × 3]> <tibble>
## # … with abbreviated variable name ¹.predictions

collect_metrics(wfl.rg.rf.fitres.boots)              # -> RMSE=2853, RSQ=0.86

## # A tibble: 2 × 6
##   .metric .estimator     mean     n  std_err .config
##   <chr>   <chr>         <dbl> <int>    <dbl> <chr>
## 1 rmse    standard   2853.       10 29.1     Preprocessor1_Model1
## 2 rsq     standard      0.855    10  0.00251 Preprocessor1_Model1
```

The prediction of the prices is still better with the Random Forest model. However, the scatterplot shows data points in both model types where the predicted price deviates significantly from the real price. A detailed view of the outliers suggests that age, horsepower, and brand as the sole predictors of the machine learning model are not sufficient and should perhaps be supplemented by the car model. A Mercedes Benz S-Class that is accident-free and 3 years old should certainly be valued at more than $22,000. From the outliers, one can not only learn to adjust the considered variables, but also to find errors in the data. It is possible that the price of the 20-year-old Porsche 911 at $69,000 was recorded incorrectly and should be corrected.

Training the models during resampling can put a heavy load on the CPU. To reduce response time, it can be valuable to distribute tasks across multiple CPUs and processes and run them simultaneously, i.e., in parallel. The parallel package offers functions for this. First, the function `detectCores()` with the parameter logical=TRUE can be used to display the number of CPUs of the computer and with logical=FALSE the number of maximum parallel executable processes. The function `makePSOCKcluster()` from the doParallel package defines a number of parallel clusters, which then activates the parallel execution of processes with the function `registerDoParallel()`. The deactivation of parallel execution can later be done with the function `stopCluster()`. It should be considered that parallel execution also changes the memory requirement and roughly increases the required memory requirement per parallel execution by the data to be processed. In the case of 1 GB of data and 4 parallel processes, for example, you would then need about 5 GB (1x as a base and 4x for each process) of memory. Too many processes can also cause a slowdown if memory becomes scarce. In the following coding, the performance is determined using the function `system.time()`.

Parallel Processing

```
# Verwendung der Resample-Stichproben für Lineare Regression --------------------
parallel::detectCores(logical = FALSE)   # Anzahl CPUs

## [1] 6

parallel::detectCores(logical = TRUE)    # Anzahl max. gleichzeitiger Prozesse

## [1] 12

library(doParallel)
# Performanz ohne Parallelisierung ----------------------------------------------
ml.normal.cpu.sec<-system.time({# CPU Zeit
  rf.res.boots<-rf.wflow %>%                          # Modell mit 10 Bootstraps train.
    fit_resamples(resamples=autos.boots,control=control_resamples(save_pred=TRUE))
})
# Performanz mit Parallelisierung -----------------------------------------------
ml.parallel<-makePSOCKcluster(3)                      # 3 Cluster
registerDoParallel(ml.parallel)                       # Starte Paralleles Prozessieren
ml.parallel3.cpu.sec<-system.time({ # CPU Zeit
  rf.res.boots<-rf.wflow %>%                          # Modell mit 10 Bootstraps train.
    fit_resamples(resamples=autos.boots,control=control_resamples(save_pred=TRUE))
})
ml.parallel<-makePSOCKcluster(6)                      # 6 Cluster
registerDoParallel(ml.parallel)                       # Starte Paralleles Prozessieren
ml.parallel6.cpu.sec<-system.time({ # CPU Zeit
  rf.res.boots<-rf.wflow %>%                          # Modell mit 10 Bootstraps train.
    fit_resamples(resamples=autos.boots,control=control_resamples(save_pred=TRUE))
})
ml.parallel<-makePSOCKcluster(12)                     # 12 Cluster
registerDoParallel(ml.parallel)                       # Starte Paralleles Prozessieren
ml.parallel12.cpu.sec<-system.time({ # CPU Zeit
  rf.res.boots<-rf.wflow %>%                          # Modell mit 10 Bootstraps train.
    fit_resamples(resamples=autos.boots,control=control_resamples(save_pred=TRUE))
})
stopCluster(ml.parallel)                              # Ende paralleles Prozessieren
# Performanzvergleich -----------------------------------------------------------
performanz<-data.frame(Cluster =c("ohne","3","6","12"),
                       Sekunden=c(ml.normal.cpu.sec[1],
                                  ml.parallel3.cpu.sec[1],
                                  ml.parallel6.cpu.sec[1],
                                  ml.parallel12.cpu.sec[1]))
colnames(performanz)<-c("Cluster","CPU-Sekunden")

performanz

## # A tibble: 4 × 2
##   Cluster CPU-Sekunden
## 1    ohne       154.86
## 2       3         0.31
## 3       6         0.19
## 4      12         0.28
```

The performance of the CPU time can be accelerated by a factor of more than 100 times with the help of the doParallel package. The exact performance gains that can be realized, of course, depend on the specific task and the computer configuration (number of CPUs, memory …).

6.5.3 Preprocessing

Now that we know how to select, train, evaluate a model, support modeling with workflows and workflow sets, and supplement modeling through resampling, the next step is to prepare all data for use in modeling.

Many models can only work with numerical data, for example, to apply the least squares method. For this, categorical variables (character and factor variables) must be converted into numerical ones. Also, taking the logarithm of numbers can be helpful. Logarithms can be used to clearly represent very rapidly growing number series, as the logarithm rises much slower for large numbers than the numbers themselves. As a general rule for predictors, they should use a uniform scale. Using different scales such as meters and kilometers would disproportionately weight the higher numerical values of the meters in the model. The recipes package contains many helpful functions for preprocessing data.

The function `recipe()` is used to specify the formula with the variables used in a model. In addition, the functions `step_log()` logarithmizes a parameter and the function `step_dummy()` converts categorical parameters into numerical ones. To avoid having to specify all variables individually, functions like `all_nominal_predictors()` can be used. If the variables of the model have already been defined with the function `add_formula()`, this preprocessing must first be deleted before the function recipe() can be applied, because there can only be one specification of the variables of the model. The details of the model can be displayed if necessary with the function `extract_recipe()` and `extract_fit_parsnip()`.

Data Preprocessing with the recipe() Function

```
# Trainings- und Testdaten erzeugen----------------------------------------------
set.seed(123)
autos.split<-initial_split(autos,prop=0.80,     # Split Trainings/Testdaten 80:20
                           strata=Preis)
autos.train<-training(autos.split)
autos.test<-testing(autos.split)
# wfl.rg.lm.fit1 mit recipe()----------------------------------------------------
rec.1<-recipe(data=autos.train,
             Preis~Typ+Kilometer+Alter+Kategorie+PS+Marke+Modell) %>%
  step_log(PS,base=10) %>%                            # PS Logarithmieren
  step_dummy(all_nominal_predictors())                # Kategorial->Numerisch
mod.rg.lm<-linear_reg() %>% set_engine("lm")          # Modellauswahl
wfl.rg.lm <-workflow() %>% add_model(mod.rg.lm)       # Workflow
wfl.rg.lm

## == Workflow ====================================================================
## Preprocessor: None
## Model: linear_reg()
##
## ── Model ─────────────────────────────────────────────────────────────────────
────────────────────────────────────────────────────────────

## Linear Regression Model Specification (regression)
##
## Computational engine: lm

wfl.rg.lm <-wfl.rg.lm %>%                             # Preprocessor
  add_formula(Sale_Price~Longitude+Latitude)          # Formel
wfl.rg.lm <- wfl.rg.lm %>%                            # Fehler, da nur ein Preproc.
  add_recipe(rec.1)

## Error in `add_recipe()`:
## ! A recipe cannot be added when a formula already exists.

wfl.rg.lm <- wfl.rg.lm %>%                            # Erfolgreich Recipe ergänzt
  remove_formula() %>%
  add_recipe(rec.1)
wfl.rg.lm

## == Workflow ====================================================================
## Preprocessor: Recipe
## Model: linear_reg()
##
## ── Preprocessor ────────────────────────────────────────────────────────────────
## 2 Recipe Steps
##
## • step_log()
## • step_dummy()
##
## ── Model ───────────────────────────────────────────────────────────────────────
## Linear Regression Model Specification (regression)
##
## Computational engine: lm

wfl.rg.lm.fit1  <- fit(wfl.rg.lm, autos.train)        # Modell trainieren
wfl.rg.lm.fit1                                        # >230 Modell-Koeffizienten

## == Workflow [trained] ==========================================================
## Preprocessor: Recipe
## Model: linear_reg()
##
```

```
## — Preprocessor ————————————————————————————————————————
## 2 Recipe Steps
##
## • step_log()
## • step_dummy()
##
## — Model ———————————————————————————————————————————————
##
## Call:
## stats::lm(formula = ..y ~ ., data = data)
##
## Coefficients:
##            (Intercept)             Kilometer
##             -5.630e+03            -2.833e-02
##                  Alter                    PS
##             -4.393e+02             1.135e+04
##        Typ_kleinwagen             Typ_kombi
##              5.631e+02            -1.835e+02
##                Typ_bus           Typ_cabrio
##              2.007e+01             1.859e+03
##              Typ_coupe               Typ_suv
##              1.122e+03             1.325e+03
##            Kategorie_1           Kategorie_2
##             -4.954e+03             1.132e+03
##              Marke_bmw            Marke_opel
##              7.656e+03            -3.442e+03
##    Marke_mercedes_benz            Marke_audi
##              3.532e+03             7.937e+03
##             Marke_ford         Marke_renault
##             -2.747e+03            -2.303e+03
##          Marke_peugeot            Marke_fiat
##             -2.885e+03            -3.098e+03
##             Marke_seat           Marke_skoda
##             -1.413e+02            -1.829e+03
##            Marke_mazda          Marke_nissan
##             -2.553e+03            -7.967e+02
##          Marke_citroen           Marke_smart
##             -1.706e+03             7.217e+02
##           Marke_toyota         Marke_hyundai
##             -1.019e+03            -3.375e+03
##            Marke_volvo            Marke_mini
##             -1.820e+03             1.606e+03
##       Marke_mitsubishi           Marke_honda
##             -3.044e+03            -1.453e+03
##              Marke_kia      Marke_alfa_romeo
##             -3.905e+03            -2.952e+03
##          Marke_porsche          Marke_suzuki
##              1.642e+04            -2.782e+03
##        Marke_chevrolet        Marke_chrysler
##             -3.684e+03            -4.176e+03
##            Marke_dacia        Marke_daihatsu
##             -9.141e+03            -2.286e+03
##             Marke_jeep          Marke_subaru
##             -7.541e+02             3.662e+03
##       Marke_land_rover            Marke_saab
##              9.978e+03            -2.942e+03
##           Marke_jaguar          Marke_daewoo
##
## ...
## and 242 more lines.
```

```
wfl.rg.lm.fit1 %>% extract_fit_parsnip() %>% tidy()  # = 286 Koeffizienten

## # A tibble: 286 × 5
##    term               estimate  std.error statistic   p.value
##    <chr>                 <dbl>      <dbl>     <dbl>     <dbl>
##  1 (Intercept)     -5630.      889.           -6.33 2.49e- 10
##  2 Kilometer          -0.0283    0.000728   -38.9   8   e-323
##  3 Alter            -439.        9.04       -48.6   0
##  4 PS              11353.      225.          50.5   0
##  5 Typ_kleinwagen    563.       99.2          5.68  1.39e-  8
##  6 Typ_kombi        -183.       66.6         -2.75  5.90e-  3
##  7 Typ_bus            20.1     146.           0.138 8.90e-  1
##  8 Typ_cabrio       1859.      114.          16.4   8.04e- 60
##  9 Typ_coupe        1122.      119.           9.41  5.49e- 21
## 10 Typ_suv          1325.      205.           6.45  1.11e- 10
## # … with 276 more rows
```

In the case of qualitative data, which are present in character or factor variables, there are often a multitude of factor levels. In the dataframe autos, this applies to the variable Model, which has 238 car models as a manifestation and thus leads to a multitude of coefficients in the model wfl.rg.lm.fit1, a total of 286. If individual car models only affect a few data sets, then their use for modeling can make the model more complex without bringing significant advantages. Therefore, there is the function `step_other()`, which combines all the values to the value "other" which affect less than x percent of the data sets, e.g., `step_other(Model,threshold=0.01)`, to only consider the car models in the modeling that affect at least 1% of the data sets.

In the variable model in the dataframe cars, there are 238 car model types, some of which affect less than 6 of the more than 40,000 records (see Fig. 6.32), while a few car model types affect many records (see Fig. 6.33) and therefore have a high relevance for the modeling.

The model rg.lm.fit2 eliminates all models and brands that do not affect at least 1% of the records and combines them into the predictors marke_other and modell_other, so that the number of predictors of the model can be reduced from 286 to 56. This means that the prediction of the price is no longer based on a formula with 286 parameters, but only 56 parameters.

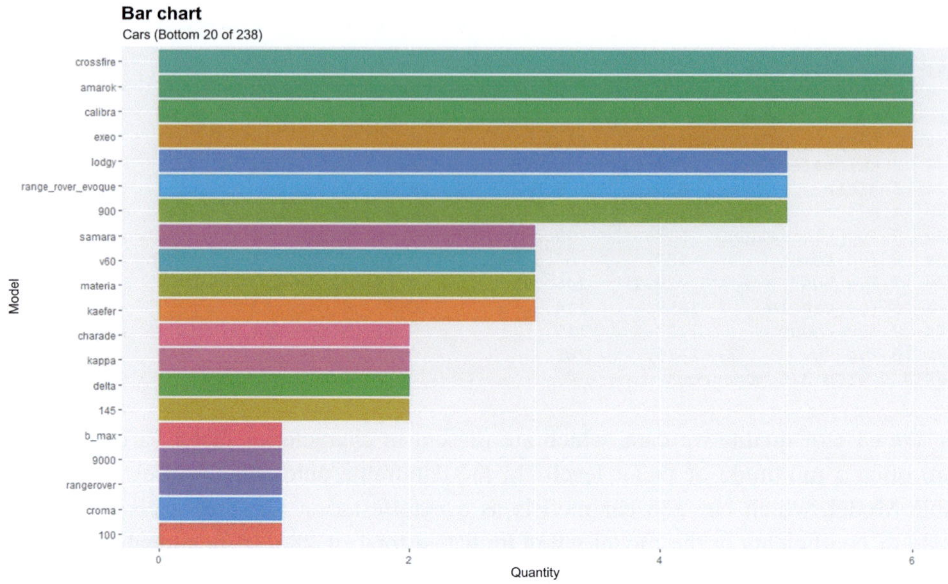

Fig. 6.32 Car model types with low occurrence

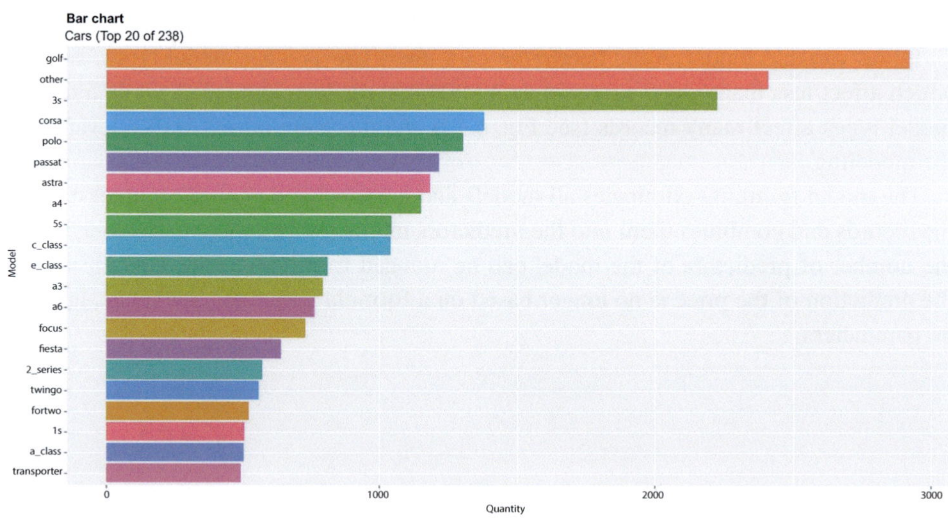

Fig. 6.33 Car model types with frequent occurrence

Preprocessing of data with the function step_other()

```
# wfl.rg.lm.fit2 mit step_other()-----------------------------------------------
rec.2<-recipe(data=autos.train,
              Preis~Typ+Kilometer+Alter+Kategorie+PS+Marke+Modell) %>%
  step_other(c(Marke,Modell),threshold=0.01) %>%      # Marke+Modell je > 1%
  step_log(PS,base=10) %>%                            # PS Logarithmieren
  step_dummy(all_nominal_predictors())                # Kategorial->Numerisch
wfl.rg.lm <- wfl.rg.lm %>%                            # Erfolgreich Recipe ergänzt
  update_recipe(rec.2)
wfl.rg.lm

## == Workflow ==================================================================
## Preprocessor: Recipe
## Model: linear_reg()
##
## ── Preprocessor ──────────────────────────────────────────────────────────────
──────────────────────────────────────────────────────
## 3 Recipe Steps
##
## • step_other()
## • step_log()
## • step_dummy()
##
## ── Model ─────────────────────────────────────────────────────────────────────
────────────────────────────────────────────────────────────
## Linear Regression Model Specification (regression)
##
## Computational engine: lm

wfl.rg.lm.fit2  <- fit(wfl.rg.lm, autos.train)        # Modell trainieren
wfl.rg.lm.fit2                                        # nur 56 Modell-Koeffizienten

## == Workflow [trained] ========================================================
## Preprocessor: Recipe
## Model: linear_reg()
##
## ── Preprocessor ──────────────────────────────────────────────────────────────
──────────────────────────────────────────────────────
## 3 Recipe Steps
##
## • step_other()
## • step_log()
## • step_dummy()
##
## ── Model ─────────────────────────────────────────────────────────────────────
────────────────────────────────────────────────────────────
##
## Call:
## stats::lm(formula = ..y ~ ., data = data)
##
## Coefficients:
##       (Intercept)           Kilometer                Alter
##        -1.536e+04          -2.922e-02           -3.939e+02
##                PS     Typ_kleinwagen            Typ_kombi
##         1.554e+04           7.640e+02           -1.702e+02
##           Typ_bus          Typ_cabrio            Typ_coupe
##         1.864e+02           2.401e+03            1.409e+03
##           Typ_suv         Kategorie_1          Kategorie_2
```

```
##           2.882e+03            -4.834e+03            1.060e+03
##           Marke_bmw            Marke_opel  Marke_mercedes_benz
##           3.735e+03            -2.214e+03            1.658e+03
##          Marke_audi            Marke_ford        Marke_renault
##           3.889e+03            -1.358e+03           -2.206e+03
##       Marke_peugeot            Marke_fiat           Marke_seat
##          -1.848e+03            -1.538e+03           -1.160e+03
##         Marke_skoda           Marke_mazda         Marke_nissan
##          -1.303e+03            -2.005e+03           -1.322e+03
##       Marke_citroen           Marke_smart         Marke_toyota
##          -1.884e+03            -5.047e+02           -1.793e+03
##       Marke_hyundai            Marke_mini          Marke_other
##          -3.987e+03            -9.747e+02           -1.057e+03
##         Modell_golf         Modell_fortwo         Modell_focus
##          -2.125e+02            -7.327e+02           -6.899e+02
##           Modell_a3         Modell_vectra          Modell_X3er
##          -3.811e+03             1.806e+02           -4.548e+03
##           Modell_a4          Modell_X5er          Modell_polo
##          -4.285e+03            -4.040e+03            7.293e+02
##     Modell_e_klasse      Modell_X2_reihe      Modell_c_klasse
##          -1.244e+03            -8.412e+02           -2.041e+03
##        Modell_corsa         Modell_mondeo        Modell_twingo
##           2.440e+03            -1.197e+03            2.042e+03
##     Modell_a_klasse      Modell_X3_reihe          Modell_clio
##          -2.024e+03            -2.347e+02            1.174e+03
##       Modell_passat         Modell_fiesta            Modell_a6
##          -1.105e+03             3.078e+02           -4.159e+03
##  Modell_transporter          Modell_astra          Modell_X1er
##           5.726e+03             6.620e+02           -3.955e+03
##       Modell_touran          Modell_other
##          -4.733e+02            -1.822e+02
```

```
wfl.rg.lm.fit2 %>% extract_recipe(estimated=TRUE)     # Anzeige Modell
```

```
## Recipe
##
## Inputs:
##
##       role #variables
##    outcome          1
##  predictor          7
##
## Training data contained 27890 data points and no missing data.
##
## Operations:
##
## Collapsing factor levels for Marke, Modell [trained]
## Log transformation on PS [trained]
## Dummy variables from Typ, Kategorie, Marke, Modell [trained]
```

```
wfl.rg.lm.fit2 %>% extract_fit_parsnip() %>% tidy()  # = 56 Koeffizienten
```

```
## # A tibble: 56 × 5
##    term              estimate  std.error statistic   p.value
##    <chr>                <dbl>      <dbl>     <dbl>     <dbl>
##  1 (Intercept)     -15360.    503.           -30.5 1.31e-201
##  2 Kilometer           -0.0292  0.000826     -35.4 2.52e-268
##  3 Alter             -394.      9.99         -39.4 0
##  4 PS               15544.    213.            73.1 0
##  5 Typ_kleinwagen     764.     96.6            7.91 2.68e- 15
```

```
## 6 Typ_kombi      -170.     73.6      -2.31 2.08e-  2
## 7 Typ_bus         186.    113.        1.65 9.93e-  2
## 8 Typ_cabrio     2401.    112.       21.4  1.33e-100
## 9 Typ_coupe      1409.    122.       11.5  1.30e- 30
## 10 Typ_suv       2882.    133.       21.6  5.38e-103
## # … with 46 more rows
```

A model produces a better result when dependencies among the predictors are reduced or eliminated. These interactions between predictors can be handled with the function `step_interact()`. In the following figure, it can be seen that the predictors PS and brand do not have a uniform dependency on the price, but depend on each other, which explains the different course of the regression curves in Fig. 6.34. The dependency can be taken into account with the following function: `step_interact(~ PS:starts_with ("Marke_"))`.

The model wfl.rg.lm.fit3 therefore combines the variables PS and brand and generates new predictors with the prefix `PS_x_Marke`.

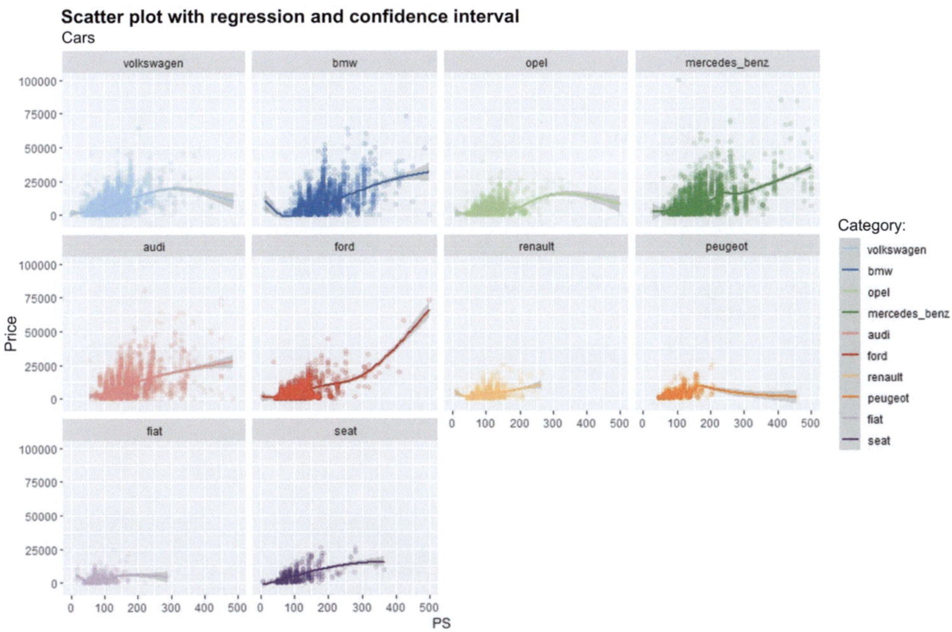

Fig. 6.34 Dependency of the price on the combination of PS and car brand

Preprocessing of data with the function step_interact()

```
# wfl.rg.lm.fit3 mit step_interact() PS mit Marke--------------------------------
rec.3<-recipe(data=autos.train,
              Preis~Typ+Kilometer+Alter+Kategorie+PS+Marke+Modell) %>%
  step_other(c(Marke,Modell),threshold=0.01) %>%     # Marke+Modell je > 1%
  step_log(PS,base=10) %>%                          # PS Logarithmieren
  step_dummy(all_nominal_predictors()) %>%          # Kategorial->Numerisch
  step_interact(~ PS:starts_with("Marke_"))         # Abhängigkeit Prädiktoren
wfl.rg.lm <- wfl.rg.lm %>%                          # Erfolgreich Recipe ergänzt
  update_recipe(rec.3)
wfl.rg.lm
## ══ Workflow ═════════════════════════════════════════════════════════════════
## Preprocessor: Recipe
## Model: linear_reg()
##
## ── Preprocessor ──────────────────────────────────────────────────────────────
──────────────────────────────────────────────────────
## 4 Recipe Steps
##
## • step_other()
## • step_log()
## • step_dummy()
## • step_interact()
##
## ── Model ─────────────────────────────────────────────────────────────────────
────────────────────────────────────────────────────────────
## Linear Regression Model Specification (regression)
##
## Computational engine: lm
wfl.rg.lm.fit3  <- fit(wfl.rg.lm, autos.train)      # Modell trainieren
wfl.rg.lm.fit3
## ══ Workflow [trained] ═══════════════════════════════════════════════════════
## Preprocessor: Recipe
## Model: linear_reg()
##
## ── Preprocessor ──────────────────────────────────────────────────────────────
## 4 Recipe Steps
##
## • step_other()
## • step_log()
## • step_dummy()
## • step_interact()
##
## ── Model ─────────────────────────────────────────────────────────────────────
##
## Call:
## stats::lm(formula = ..y ~ ., data = data)
##
## Coefficients:
##          (Intercept)            Kilometer                Alter
##           -6.883e+03           -2.985e-02           -4.300e+02
##                   PS       Typ_kleinwagen            Typ_kombi
##            1.165e+04            3.362e+02           -2.209e+02
##              Typ_bus           Typ_cabrio           Typ_coupe
##            3.705e+02            2.219e+03            1.189e+03
##              Typ_suv          Kategorie_1          Kategorie_2
##            2.564e+03           -4.743e+03            1.043e+03
```

```
##              Marke_bmw                    Marke_opel       Marke_mercedes_benz
##             -2.191e+04                     1.861e+04                -2.542e+04
##             Marke_audi                    Marke_ford             Marke_renault
##             -2.226e+04                     5.971e+03                 2.319e+04
##          Marke_peugeot                    Marke_fiat                Marke_seat
##              1.881e+04                     1.355e+04                 6.875e+03
##            Marke_skoda                   Marke_mazda              Marke_nissan
##             -8.806e+03                     6.010e+03                -4.232e+01
##          Marke_citroen                   Marke_smart              Marke_toyota
##              3.647e+03                     5.007e+04                -3.416e+03
##          Marke_hyundai                    Marke_mini               Marke_other
##              8.405e+01                    -4.665e+03                -3.685e+04
##            Modell_golf                 Modell_fortwo              Modell_focus
##             -2.663e+02                    -4.703e+03                -4.632e+02
##              Modell_a3                 Modell_vectra               Modell_X3er
##             -2.382e+03                     4.804e+02                -3.326e+03
##              Modell_a4                   Modell_X5er               Modell_polo
##             -3.348e+03                    -3.615e+03                 2.590e+02
##        Modell_e_klasse               Modell_X2_reihe           Modell_c_klasse
##             -1.235e+03                    -1.137e+03                -1.421e+03
##           Modell_corsa                 Modell_mondeo             Modell_twingo
##             -6.815e+02                    -4.231e+02                -1.598e+03
##        Modell_a_klasse               Modell_X3_reihe               Modell_clio
##              3.140e+02                     1.408e+02                -1.418e+03
##          Modell_passat                 Modell_fiesta                 Modell_a6
##             -8.108e+02                    -5.771e+02                -3.983e+03
##     Modell_transporter                  Modell_astra               Modell_X1er
##              5.724e+03                    -3.223e+01                -2.750e+03
##          Modell_touran                  Modell_other             PS_x_Marke_bmw
##             -5.496e+02                    -1.810e+02                 1.144e+04
##        PS_x_Marke_opel  PS_x_Marke_mercedes_benz            PS_x_Marke_audi
##             -1.000e+04                     1.236e+04                 1.182e+04
##        PS_x_Marke_ford          PS_x_Marke_renault         PS_x_Marke_peugeot
##             -3.695e+03                    -1.243e+04                -1.040e+04
##        PS_x_Marke_fiat             PS_x_Marke_seat           PS_x_Marke_skoda
##             -8.212e+03                    -4.179e+03                 3.654e+03
##       PS_x_Marke_mazda           PS_x_Marke_nissan         PS_x_Marke_citroen
##
## ...
## and 6 more lines.
```

```
wfl.rg.lm.fit3 %>% extract_recipe(estimated=TRUE)    # Anzeige Modell
```

```
## Recipe
##
## Inputs:
##
##       role #variables
##    outcome          1
##  predictor          7
##
## Training data contained 27890 data points and no missing data.
##
## Operations:
##
## Collapsing factor levels for Marke, Modell [trained]
## Log transformation on PS [trained]
## Dummy variables from Typ, Kategorie, Marke, Modell [trained]
## Interactions with PS:(Marke_bmw + Marke_opel + Marke_mercedes_benz +... [trained]
```

```
wfl.rg.lm.fit3 %>% extract_fit_parsnip() %>% tidy()  # = 74 Koeffizienten

## # A tibble: 74 × 5
##    term              estimate  std.error statistic   p.value
##    <chr>                <dbl>      <dbl>     <dbl>     <dbl>
##  1 (Intercept)      -6883.     808.          -8.52 1.69e- 17
##  2 Kilometer           -0.0298   0.000783   -38.1  8.98e-310
##  3 Alter             -430.        9.53      -45.1  0
##  4 PS               11652.     380.          30.7  2.54e-203
##  5 Typ_kleinwagen     336.       94.8         3.55 3.91e-  4
##  6 Typ_kombi         -221.       70.0        -3.16 1.60e-  3
##  7 Typ_bus            370.     109.           3.41 6.52e-  4
##  8 Typ_cabrio        2219.     107.          20.7  3.08e- 94
##  9 Typ_coupe         1189.     116.          10.2  1.87e- 24
## 10 Typ_suv           2564.     128.          20.0  3.99e- 88
## # … with 64 more rows
```

Another modification of the model is possible by setting the degree of non-linearity. The function `step_ns()` serves this purpose, which specifies a number of degrees of freedom that is suitable for describing the data as accurately as possible. In the following coding, different degrees of freedom are depicted in the trend line between PS and price. With a degree of freedom of 1, a straight line results, which represents the data inaccurately. With a degree of freedom of 20, the trend line is excessively non-linear.

The model wfl.rg.lm.fit4 will therefore prefer the function `step_ns()` with the parameter of 5 degrees of freedom for modeling, which leads to the five coefficients PS_ns_1 to PS_ns_5.

Preprocessing of Data with the Function step_ns()

```
# wfl.rg.lm.fit4 mit step_ns() PS------------------------------------------------
library(splines)
plot.ns <- function(degreefree) {
  ggplot(autos.train, aes(x=PS,y=Preis)) +
    geom_point(alpha=0.1) +
    scale_y_log10() +
    geom_smooth(method=lm,
      formula=y ~ ns(x,df=degreefree),
      color="red",se = FALSE
    ) +
    labs(title=paste(degreefree,"Splines Freiheitsgrade"),
         y="Preis")
}
p1<-plot.ns(1)
p2<-plot.ns(5)
p3<-plot.ns(10)
p4<-plot.ns(20)
grid.arrange(p1,p2,p3,p4,nrow=2,ncol=2)
```

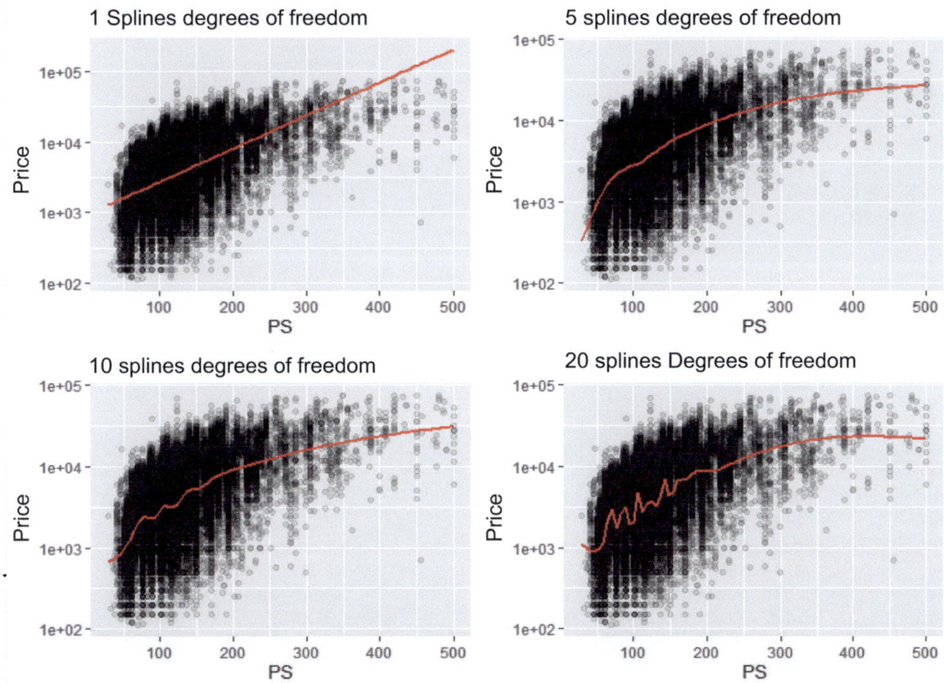


```
rec.4<-recipe(data=autos.train,
              Preis~Typ+Kilometer+Alter+Kategorie+PS+Marke+Modell) %>%
  step_other(c(Marke,Modell),threshold=0.01) %>%     # Marke+Modell je > 1%
  step_log(PS,base=10) %>%                           # PS Logarithmieren
  step_dummy(all_nominal_predictors()) %>%           # Kategorial->Numerisch
  step_ns(starts_with("PS"),deg_free=5)              # Nichtlinear Splines=5
wfl.rg.lm <- wfl.rg.lm %>%                           # Erfolgreich Recipe ergänzt
  update_recipe(rec.4)
wfl.rg.lm
```

```
## ══ Workflow ═══════════════════════════════════════════════════════════
## Preprocessor: Recipe
## Model: linear_reg()
##
## ── Preprocessor ───────────────────────────────────────────────────────
## 4 Recipe Steps
##
## • step_other()
## • step_log()
## • step_dummy()
## • step_ns()
##
## ── Model ──────────────────────────────────────────────────────────────
## Linear Regression Model Specification (regression)
##
## Computational engine: lm
```

```
wfl.rg.lm.fit4  <- fit(wfl.rg.lm, autos.train)       # Modell trainieren
wfl.rg.lm.fit4
## ══ Workflow [trained] ═════════════════════════════════════════════════
## Preprocessor: Recipe
## Model: linear_reg()
##
## ── Preprocessor ───────────────────────────────────────────────────────
## 4 Recipe Steps
##
## • step_other()
## • step_log()
## • step_dummy()
## • step_ns()
##
## ── Model ──────────────────────────────────────────────────────────────
##
## Call:
## stats::lm(formula = ..y ~ ., data = data)
##
## Coefficients:
##         (Intercept)            Kilometer                Alter
##          20944.3129              -0.0274            -452.4670
##      Typ_kleinwagen            Typ_kombi              Typ_bus
##           -757.1794              12.6557             851.5477
##          Typ_cabrio           Typ_coupe               Typ_suv
##           2438.2020            1010.7644            3123.9900
##         Kategorie_1          Kategorie_2            Marke_bmw
##          -4774.5989            1106.7729            2537.7520
##          Marke_opel  Marke_mercedes_benz           Marke_audi
##          -1427.3912            1468.4707            3184.7755
##          Marke_ford        Marke_renault        Marke_peugeot
##          -1142.1901           -1441.2464           -1515.7149
##          Marke_fiat           Marke_seat          Marke_skoda
##          -1603.3867            -862.2447            -900.6248
##         Marke_mazda         Marke_nissan        Marke_citroen
##          -1058.3421           -1000.4051           -1519.5758
##         Marke_smart         Marke_toyota        Marke_hyundai
##           -121.6541            -806.0763           -3196.7759
##          Marke_mini          Marke_other          Modell_golf
##           870.4825            -559.0038             651.2159
##      Modell_fortwo         Modell_focus            Modell_a3
##          -3035.7876             182.0642           -1520.1095
```

```
##        Modell_vectra             Modell_X3er               Modell_a4
##             944.0723              -1977.5566              -2184.3708
##          Modell_X5er             Modell_polo         Modell_e_klasse
##           -2145.3068                352.4940               -302.0025
##      Modell_X2_reihe         Modell_c_klasse            Modell_corsa
##              -1.5740               -414.3847               1220.2753
##        Modell_mondeo           Modell_twingo         Modell_a_klasse
##             137.5342                625.8869               -597.4251
##      Modell_X3_reihe             Modell_clio           Modell_passat
##             510.3906                791.2340                388.7182
##        Modell_fiesta               Modell_a6      Modell_transporter
##             279.1361              -2688.7317               6470.9477
##         Modell_astra             Modell_X1er           Modell_touran
##             896.4464              -1387.5105                424.1840
##         Modell_other                 PS_ns_1                 PS_ns_2
##               3.7025              -5935.7671              -4792.8443
##              PS_ns_3                 PS_ns_4                 PS_ns_5
##            1524.0060               3231.5817              22176.9683
wfl.rg.lm.fit4 %>% extract_recipe(estimated=TRUE)    # Anzeige Modell
```

```
## Recipe
##
## Inputs:
##
##       role #variables
##    outcome          1
##  predictor          7
##
## Training data contained 27890 data points and no missing data.
##
## Operations:
##
## Collapsing factor levels for Marke, Modell [trained]
## Log transformation on PS [trained]
## Dummy variables from Typ, Kategorie, Marke, Modell [trained]
## Natural splines on PS [trained]
```

```
wfl.rg.lm.fit4 %>% extract_fit_parsnip() %>% tidy()  # = 60 Koeffizienten
```

```
## # A tibble: 60 × 5
##    term              estimate  std.error statistic   p.value
##    <chr>                <dbl>      <dbl>     <dbl>     <dbl>
##  1 (Intercept)    20944.      508.          41.3   0
##  2 Kilometer         -0.0274   0.000759    -36.1   8.18e-279
##  3 Alter           -452.        9.22       -49.1   0
##  4 Typ_kleinwagen  -757.       92.8         -8.16  3.42e- 16
##  5 Typ_kombi         12.7      67.7          0.187 8.52e-  1
##  6 Typ_bus          852.      104.           8.15  3.77e- 16
##  7 Typ_cabrio      2438.      103.          23.6   2.62e-122
##  8 Typ_coupe       1011.      113.           8.98  2.81e- 19
##  9 Typ_suv         3124.      123.          25.5   1.95e-141
## 10 Kategorie_1    -4775.       97.4        -49.0   0
## # … with 50 more rows
```

Distributions of variables are often skewed. The function `step_normalize()` can correct this and transform the values of variables into normally distributed values, so that the `predict()` function can deliver more accurate results. The variables Kilometer and Age in the dataframe autos are left-skewed and the variable PS is right-skewed, which can be seen from the histograms displayed with the function `ml.summary()`. The function `step_normalize()` transforms the values of numerical variables so that they subsequently have a mean of zero and a standard deviation of 1, which is also referred to as Z-standardization. The function `step_normalize(Kilometer,Age,starts_with("PS"))` achieves the desired normalization for the variables Kilometer, Age, and PS. The following coding shows that the numerical values of the predictors for the corresponding variables change in the model as a result.

The model wfl.rg.lm.fit5 changes the coefficients for the normalized variables, e.g., `−2191` instead of previously −452 for Age. The coefficients of the non-normalized variables remain unchanged.

Preprocessing of Data with the Function step_normalize()

```
# wfl.rg.lm.fit5 mit step_normalize() -------------------------------------------
-----
ml.summary(autos$Kilometer,"Autos","Kilometer")    # Kilometer: Linke Schiefe
```

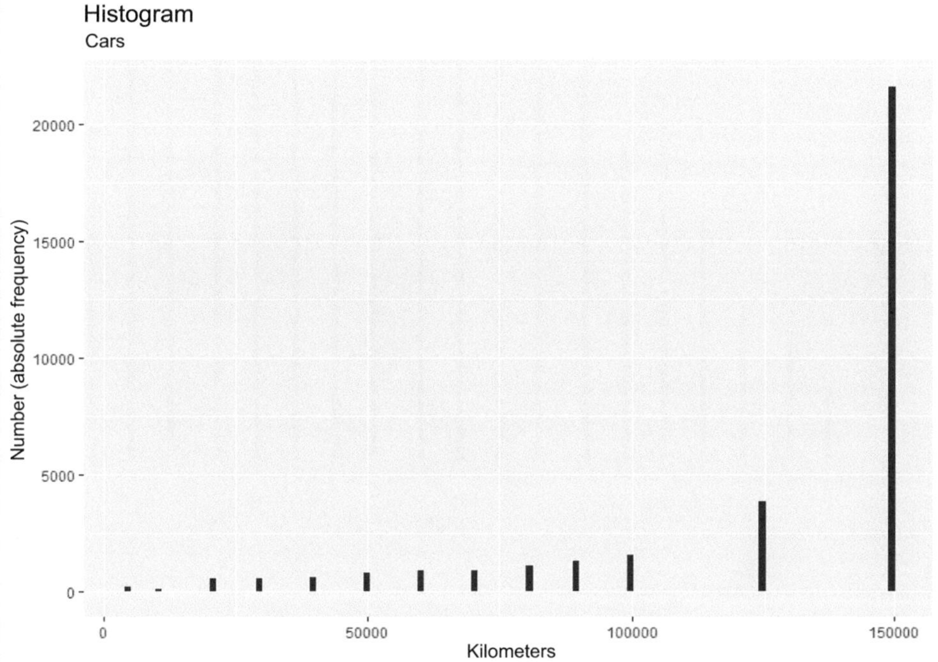


```
##             25%
## anzahl      "34864"
## anzahl.na   "0"
## modus       "150000"
## median      "150000"
## mean        "124614.79"
## min         "5000"
## max         "150000"
## sd          "39355.69"
## q1          "1e+05"
## q3          "150000"
## iqr         "50000"
## whisker.min "25000"
## whisker.max "225000"
## skewness    "-1.4"
## skewness.txt "Linke Schiefe: -1.4 < 0, negative Schiefe,linksschief,rechtssteil"
## kurtosis    "0.75"
## kurtosis.txt "Steilgipflig mit Exzess Kurtosis 0.75 > 0"

ml.summary(autos$Alter,"Autos","Alter")           # Alter: Linke Schiefe
```

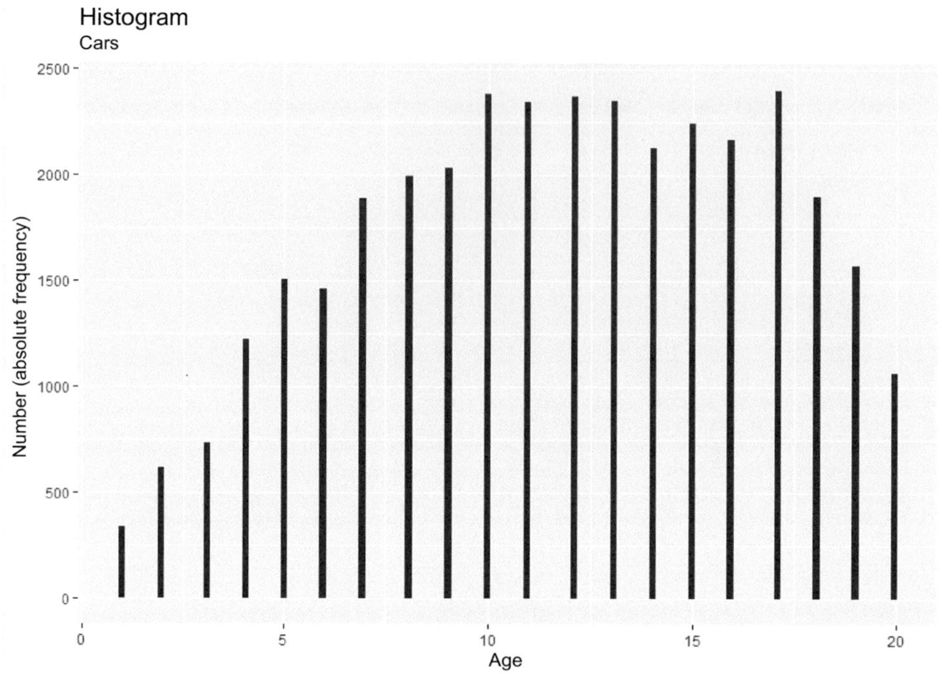


```
##               25%
## anzahl        "34864"
## anzahl.na     "0"
## modus         "17"
## median        "12"
## mean          "11.67"
## min           "1"
## max           "20"
## sd            "4.84"
## q1            "8"
## q3            "16"
## iqr           "8"
## whisker.min   "-4"
## whisker.max   "28"
## skewness      "-0.18"
## skewness.txt "Linke Schiefe: -0.18 < 0, negative Schiefe,linksschief,rechtssteil"
## kurtosis      "-0.9"
## kurtosis.txt "Flachgipflig mit Exzess Kurtosis -0.9 < 0"
```

```
ml.summary(autos$PS,"Autos","PS")                          # PS: Rechte Schiefe
```

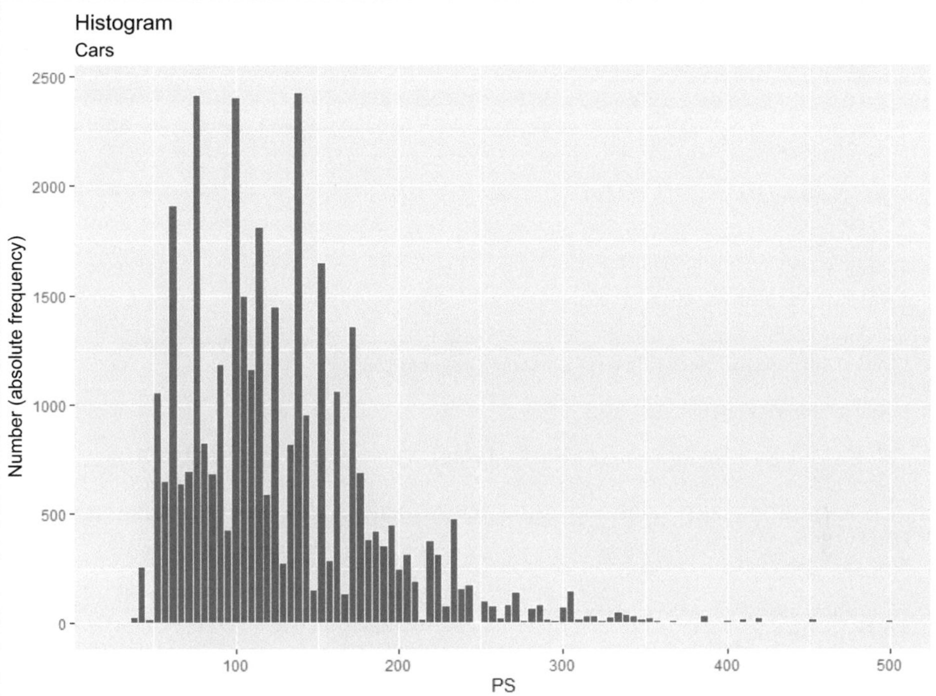


```
##               25%
## anzahl        "34864"
## anzahl.na     "0"
## modus         "75"
## median        "116"
## mean          "128.48"
## min           "29"
## max           "500"
## sd            "59.43"
## q1            "86"
## q3            "156"
## iqr           "70"
## whisker.min   "-19"
## whisker.max   "261"
## skewness      "1.47"
## skewness.txt "Rechte Schiefe: 1.47 > 0, positive Schiefe,linkssteil,rechtsschief"
## kurtosis      "3.7"
## kurtosis.txt "Steilgipflig mit Exzess Kurtosis 3.7 > 0"

rec.5<-recipe(data=autos.train,
              Preis~Typ+Kilometer+Alter+Kategorie+PS+Marke+Modell) %>%
  step_other(c(Marke,Modell),threshold=0.01) %>%      # Marke+Modell je > 1%
  step_log(PS,base=10) %>%                             # PS Logarithmieren
  step_dummy(all_nominal_predictors()) %>%             # Kategorial->Numerisch
  step_ns(starts_with("PS"),deg_free=5) %>%            # Nichtlinear Splines=5
  step_normalize(Kilometer,Alter,starts_with("PS"))   # Skalen normalisieren
wfl.rg.lm <- wfl.rg.lm %>%                             # Erfolgreich Recipe ergänzt
```

```
  update_recipe(rec.5)
wfl.rg.lm

## ══ Workflow ══════════════════════════════════════════════════════════
## Preprocessor: Recipe
## Model: linear_reg()
##
## ── Preprocessor ──────────────────────────────────────────────────────
## 5 Recipe Steps
##
## • step_other()
## • step_log()
## • step_dummy()
## • step_ns()
## • step_normalize()
##
## ── Model ─────────────────────────────────────────────────────────────
## Linear Regression Model Specification (regression)
##
## Computational engine: lm

wfl.rg.lm.fit5  <- fit(wfl.rg.lm, autos.train)       # Modell trainieren
wfl.rg.lm.fit5

## ══ Workflow [trained] ════════════════════════════════════════════════
## Preprocessor: Recipe
## Model: linear_reg()
##
## ── Preprocessor ──────────────────────────────────────────────────────
## 5 Recipe Steps
##
## • step_other()
## • step_log()
## • step_dummy()
## • step_ns()
## • step_normalize()
##
## ── Model ─────────────────────────────────────────────────────────────
##
## Call:
## stats::lm(formula = ..y ~ ., data = data)
##
## Coefficients:
##         (Intercept)            Kilometer                 Alter
##            8088.397            -1081.699             -2190.913
##      Typ_kleinwagen            Typ_kombi               Typ_bus
##            -757.179               12.656               851.548
##          Typ_cabrio            Typ_coupe               Typ_suv
##            2438.202             1010.764              3123.990
##         Kategorie_1          Kategorie_2             Marke_bmw
##           -4774.599             1106.773              2537.752
##          Marke_opel  Marke_mercedes_benz            Marke_audi
##           -1427.391             1468.471              3184.775
##          Marke_ford        Marke_renault         Marke_peugeot
##           -1142.190            -1441.246             -1515.715
##          Marke_fiat           Marke_seat           Marke_skoda
##           -1603.387             -862.245              -900.625
##         Marke_mazda         Marke_nissan         Marke_citroen
##           -1058.342            -1000.405             -1519.576
##         Marke_smart         Marke_toyota         Marke_hyundai
```

```
##          -121.654           -806.076          -3196.776
##        Marke_mini        Marke_other        Modell_golf
##           870.483           -559.004            651.216
##     Modell_fortwo       Modell_focus          Modell_a3
##         -3035.788            182.064          -1520.110
##     Modell_vectra        Modell_X3er          Modell_a4
##           944.072          -1977.557          -2184.371
##       Modell_X5er        Modell_polo   Modell_e_klasse
##         -2145.307            352.494           -302.003
##   Modell_X2_reihe    Modell_c_klasse       Modell_corsa
##            -1.574           -414.385           1220.275
##     Modell_mondeo      Modell_twingo   Modell_a_klasse
##           137.534            625.887           -597.425
##   Modell_X3_reihe        Modell_clio      Modell_passat
##           510.391            791.234            388.718
##     Modell_fiesta          Modell_a6 Modell_transporter
##           279.136          -2688.732           6470.948
##      Modell_astra        Modell_X1er      Modell_touran
##           896.446          -1387.510            424.184
##      Modell_other            PS_ns_1            PS_ns_2
##             3.703          -1605.491          -1415.834
##           PS_ns_3            PS_ns_4            PS_ns_5
##           345.792            507.360           2560.389
```

```
wfl.rg.lm.fit5 %>% extract_recipe(estimated=TRUE)    # Anzeige Modell
```

```
## Recipe
##
## Inputs:
##
##       role #variables
##    outcome          1
##  predictor          7
##
## Training data contained 27890 data points and no missing data.
##
## Operations:
##
## Collapsing factor levels for Marke, Modell [trained]
## Log transformation on PS [trained]
## Dummy variables from Typ, Kategorie, Marke, Modell [trained]
## Natural splines on PS [trained]
## Centering and scaling for Kilometer, Alter, PS_ns_1, PS_ns_2, PS_ns_3, PS... [tra
ined]
```

```
wfl.rg.lm.fit5 %>% extract_fit_parsnip() %>% tidy()  # = 60 Koeffizienten
```

```
## # A tibble: 60 × 5
##    term            estimate std.error statistic   p.value
##    <chr>              <dbl>     <dbl>     <dbl>     <dbl>
##  1 (Intercept)      8088.      162.      49.9   0
##  2 Kilometer       -1082.       30.0    -36.1   8.18e-279
##  3 Alter           -2191.       44.6    -49.1   0
##  4 Typ_kleinwagen   -757.       92.8     -8.16  3.42e- 16
##  5 Typ_kombi          12.7      67.7      0.187 8.52e-  1
##  6 Typ_bus           852.      104.       8.15  3.77e- 16
##  7 Typ_cabrio       2438.      103.      23.6   2.62e-122
##  8 Typ_coupe        1011.      113.       8.98  2.81e- 19
##  9 Typ_suv          3124.      123.      25.5   1.95e-141
```

```
## 10 Kategorie_1      -4775.        97.4   -49.0   0
## # … with 50 more rows
```

It makes sense to use as few predictors as possible in a model to keep the complexity of the model low. Models should be kept as lean as possible, i.e., only use as many predictors as are crucial for a good result. Therefore, the method of Principal Component Analysis (PCA) is used to reduce the number of predictors by calculating new predictors that are as little correlated with each other as possible. In the example of the dataframe autos, the variable Category as an indication of "New", "Middle-aged" and "Old" is in a strong dependency with the variable Age or directly derived from it (see Fig. 6.35).

The model wfl.rg.lm.fit6 is supplemented by the function `step_pca(Age,starts_with("Category_"))`, which results in the three coefficients PC1 to PC3 being added to the model.

To display the effects of all preprocessing steps of the function recipe, the function `tidy()` is useful, e.g., the function `process.recipe<- wfl.rg.lm.fit6 %>% extract_recipe(estimated=TRUE)` followed by the function `tidy(process.recipe)` to display the process steps. Individual process steps can also be viewed in detail when the function `tidy(process.recipe,number=n)` is called to take a closer look at process step n. This is demonstrated in the following coding for process step 6 step_pca.

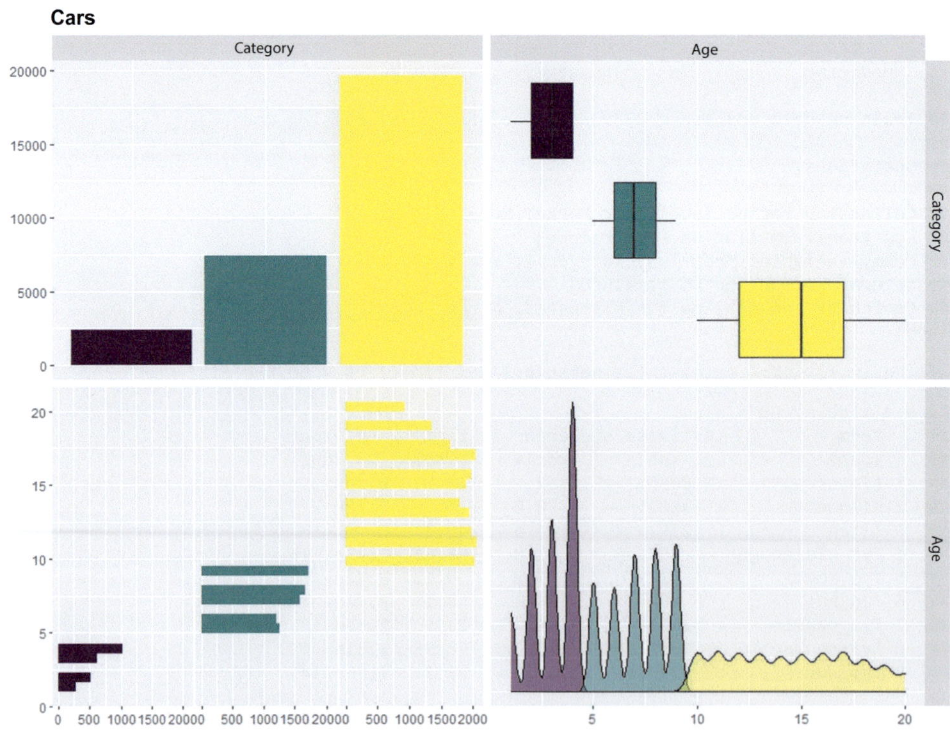

Fig. 6.35 Relationship Category and Age

Preprocessing of Data with the Function step_pca()

```
# wfl.rg.lm.fit6 mit step_pca() ---------------------------------------------------
-----
levels(autos$Kategorie)                                    # Variable Kategorie

## [1] "Neuwertig" "Mittelalt" "Alt"

rec.6<-recipe(data=autos.train,
              Preis~Typ+Kilometer+Alter+Kategorie+PS+Marke+Modell) %>%
  step_other(c(Marke,Modell),threshold=0.01) %>%     # Marke+Modell je > 1%
  step_log(PS,base=10) %>%                           # PS Logarithmieren
  step_dummy(all_nominal_predictors()) %>%           # Kategorial->Numerisch
  step_ns(starts_with("PS"),deg_free=5) %>%          # Nichtlinear Splines=5
  step_normalize(Kilometer,Alter,starts_with("PS")) %>% # Skalen normalisieren
  step_pca(Alter,starts_with("Kategorie_"))          # Principal Component Analysis
wfl.rg.lm <- wfl.rg.lm %>%                           # Erfolgreich Recipe ergänzt
  update_recipe(rec.6)
wfl.rg.lm

## ══ Workflow ════════════════════════════════════════════════════════════════════
## Preprocessor: Recipe
## Model: linear_reg()
##
## ── Preprocessor ────────────────────────────────────────────────────────────────
## 6 Recipe Steps
##
## • step_other()
## • step_log()
## • step_dummy()
## • step_ns()
## • step_normalize()
## • step_pca()
##
## ── Model ───────────────────────────────────────────────────────────────────────
## Linear Regression Model Specification (regression)
##
## Computational engine: lm

wfl.rg.lm.fit6  <- fit(wfl.rg.lm, autos.train)       # Modell trainieren
wfl.rg.lm.fit6

## ══ Workflow [trained] ══════════════════════════════════════════════════════════
## Preprocessor: Recipe
## Model: linear_reg()
##
## ── Preprocessor ────────────────────────────────────────────────────────────────
## 6 Recipe Steps
##
## • step_other()
## • step_log()
## • step_dummy()
## • step_ns()
## • step_normalize()
## • step_pca()
##
## ── Model ───────────────────────────────────────────────────────────────────────
##
## Call:
## stats::lm(formula = ..y ~ ., data = data)
##
```

```
## Coefficients:
##         (Intercept)            Kilometer      Typ_kleinwagen
##            8088.397            -1081.699            -757.179
##           Typ_kombi              Typ_bus          Typ_cabrio
##              12.656              851.548            2438.202
##          Typ_coupe               Typ_suv           Marke_bmw
##            1010.764             3123.990            2537.752
##          Marke_opel  Marke_mercedes_benz          Marke_audi
##           -1427.391             1468.471            3184.775
##          Marke_ford        Marke_renault       Marke_peugeot
##           -1142.190            -1441.246           -1515.715
##          Marke_fiat           Marke_seat         Marke_skoda
##           -1603.387             -862.245            -900.625
##         Marke_mazda         Marke_nissan       Marke_citroen
##           -1058.342            -1000.405           -1519.576
##         Marke_smart         Marke_toyota       Marke_hyundai
##            -121.654             -806.076           -3196.776
##          Marke_mini          Marke_other         Modell_golf
##             870.483             -559.004             651.216
##       Modell_fortwo         Modell_focus           Modell_a3
##           -3035.788              182.064           -1520.110
##       Modell_vectra         Modell_X3er            Modell_a4
##             944.072            -1977.557           -2184.371
##         Modell_X5er          Modell_polo      Modell_e_klasse
##           -2145.307              352.494            -302.003
##     Modell_X2_reihe      Modell_c_klasse         Modell_corsa
##              -1.574             -414.385            1220.275
##       Modell_mondeo        Modell_twingo     Modell_a_klasse
##             137.534              625.887            -597.425
##     Modell_X3_reihe          Modell_clio        Modell_passat
##             510.391              791.234             388.718
##       Modell_fiesta            Modell_a6  Modell_transporter
##             279.136            -2688.732            6470.948
##        Modell_astra          Modell_X1er        Modell_touran
##             896.446            -1387.510             424.184
##        Modell_other              PS_ns_1             PS_ns_2
##               3.703            -1605.491           -1415.834
##             PS_ns_3              PS_ns_4             PS_ns_5
##             345.792              507.360            2560.389
##                 PC1                  PC2                 PC3
##            3488.831            -1028.185           -3948.763
```

```
wfl.rg.lm.fit6 %>% extract_recipe(estimated=TRUE)     # Anzeige Modell
```

```
## Recipe
##
## Inputs:
##
##       role #variables
##    outcome          1
##  predictor          7
##
## Training data contained 27890 data points and no missing data.
##
## Operations:
##
## Collapsing factor levels for Marke, Modell [trained]
## Log transformation on PS [trained]
## Dummy variables from Typ, Kategorie, Marke, Modell [trained]
## Natural splines on PS [trained]
```

```
## Centering and scaling for Kilometer, Alter, PS_ns_1, PS_ns_2, PS_ns_3, PS... [tra
ined]
## PCA extraction with Alter, Kategorie_1, Kategorie_2 [trained]
wfl.rg.lm.fit6 %>% extract_fit_parsnip() %>% tidy()  # = 60 Koeffizienten
## # A tibble: 60 × 5
##    term          estimate std.error statistic   p.value
##    <chr>            <dbl>     <dbl>     <dbl>     <dbl>
##  1 (Intercept)     8088.      162.      49.9  0
##  2 Kilometer      -1082.       30.0    -36.1  8.18e-279
##  3 Typ_kleinwagen  -757.       92.8     -8.16 3.42e- 16
##  4 Typ_kombi         12.7      67.7      0.187 8.52e-  1
##  5 Typ_bus          852.      104.       8.15 3.77e- 16
##  6 Typ_cabrio      2438.      103.      23.6  2.62e-122
##  7 Typ_coupe       1011.      113.       8.98 2.81e- 19
##  8 Typ_suv         3124.      123.      25.5  1.95e-141
##  9 Marke_bmw       2538.      217.      11.7  1.25e- 31
## 10 Marke_opel     -1427.      173.      -8.26 1.53e- 16
## # … with 50 more rows
# Details zu den Prozessschritten der Funktion recipe()--------------------------
prozess.recipe <- wfl.rg.lm.fit6 %>%                         # Auswahl des Modells
  extract_recipe(estimated = TRUE)
tidy(prozess.recipe)                                  # Liste der Prozessschritte
## # A tibble: 6 × 6
##   number operation type      trained skip  id
##    <int> <chr>     <chr>     <lgl>   <lgl> <chr>
## 1      1 step      other     TRUE    FALSE other_svPzs
## 2      2 step      log       TRUE    FALSE log_2gESe
## 3      3 step      dummy     TRUE    FALSE dummy_sWC34
## 4      4 step      ns        TRUE    FALSE ns_47hZI
## 5      5 step      normalize TRUE    FALSE normalize_rVf1P
## 6      6 step      pca       TRUE    FALSE pca_0heKM
tidy(prozess.recipe,number=6)                         # Details zu Prozess Nr. 6
## # A tibble: 9 × 4
##   terms          value component id
##   <chr>          <dbl> <chr>     <chr>
## 1 Alter       -0.854   PC1       pca_0heKM
## 2 Kategorie_1 -0.412   PC1       pca_0heKM
## 3 Kategorie_2 -0.317   PC1       pca_0heKM
## 4 Alter       -0.516   PC2       pca_0heKM
## 5 Kategorie_1  0.595   PC2       pca_0heKM
## 6 Kategorie_2  0.616   PC2       pca_0heKM
## 7 Alter       -0.0654  PC3       pca_0heKM
## 8 Kategorie_1  0.690   PC3       pca_0heKM
## 9 Kategorie_2 -0.721   PC3       pca_0heKM
```

In the recipes package, there are additional functions for transformation such as `step_mutate()` for calculating new variables, `step_filter()` and `step_slice()` for selection, or `step_arrange()` for sorting data records. The textrecipes package also offers special functions for Natural Language Processing, which are not further discussed in this book.

The prediction of the used car price based on the trained models wfl.rg.lm.fit1 to wfl.rg.lm.fit6 can again be achieved with the function `predict()`. It is worth mentioning the error message in the following coding when the model wfl.rg.lm.fit1 is used. Because 238 different car models are used for the prediction, the model runs the risk of not being able to provide a prediction if only a single rarely occurring car model was not part of the training data. Since no predictor exists for this model, a prediction of the price cannot be made for a used car of this car model; the corresponding coefficient for this car model is then missing. Instead of a value, the function then returns the prediction value NA. For this reason alone, it is advisable to use the function `step_other()`.

When evaluating the metrics RMSE and R^2 in comparison of the models, it turns out that the model wfl.rg.lm.fit1 with an RMSE of 3435 and an RSQ of 0.79 provides the best prediction of the models considered here.

Preprocessing of Data: Prediction and Evaluation

```
# Vorhersage-----------------------------------------------------------------
n<-5
p0<-autos.test[1:n,"Preis"]                                # Realer Preis
p1<-predict(wfl.rg.lm.fit1, autos.test %>% slice(1:n)) %>%  # Vorhersagen...
      rename(wfl.rg.lm.fit1 = .pred)

## Warning in predict.lm(object = object$fit, newdata = new_data, type =
## "response"): prediction from a rank-deficient fit may be misleading

p2<-predict(wfl.rg.lm.fit2, autos.test %>% slice(1:n)) %>%
      rename(wfl.rg.lm.fit2 = .pred)
p3<-predict(wfl.rg.lm.fit3, autos.test %>% slice(1:n)) %>%
      rename(wfl.rg.lm.fit3 = .pred)
p4<-predict(wfl.rg.lm.fit4, autos.test %>% slice(1:n)) %>%
      rename(wfl.rg.lm.fit4 = .pred)
p5<-predict(wfl.rg.lm.fit5, autos.test %>% slice(1:n)) %>%
      rename(wfl.rg.lm.fit5 = .pred)
p6<-predict(wfl.rg.lm.fit6, autos.test %>% slice(1:n)) %>%
      rename(wfl.rg.lm.fit6 = .pred)
vorhersage<-cbind(p0,p1,p2,p3,p4,p5,p6)
vorhersage

##   Preis wfl.rg.lm.fit1 wfl.rg.lm.fit2 wfl.rg.lm.fit3 wfl.rg.lm.fit4
## 1  8500       7332.595      11522.243      12172.252      13201.600
## 2  4150       4346.644       5961.952       5470.092       5447.947
## 3  4999       8488.647       8689.496       9131.196       8889.139
## 4  6000      12921.848      10315.710      10836.441       9634.655
## 5  2500       3250.110       3317.650       3231.052       2816.842
##   wfl.rg.lm.fit5 wfl.rg.lm.fit6
## 1      13201.600      13201.600
## 2       5447.947       5447.947
## 3       8889.139       8889.139
## 4       9634.655       9634.655
## 5       2816.842       2816.842

# Evaluation -----------------------------------------------------------------
# Trainingsdaten
ergebnis<-rbind(ml.eval(wfl.rg.lm.fit1,autos.train,22),
                ml.eval(wfl.rg.lm.fit2,autos.train,22),
                ml.eval(wfl.rg.lm.fit3,autos.train,22),
                ml.eval(wfl.rg.lm.fit4,autos.train,22),
                ml.eval(wfl.rg.lm.fit5,autos.train,22),
                ml.eval(wfl.rg.lm.fit6,autos.train,22))
## Warning in predict.lm(object = object$fit, newdata = new_data, type =
## "response"): prediction from a rank-deficient fit may be misleading

# Ranking der Modelle nach RMSE (Trainingsdaten)
rmse.ergebnis<-filter(ergebnis,Kennzahl == "rmse") %>% arrange(Wert) %>%
  mutate(Platzierung=1:6) %>%
  select(Platzierung,Modell,Wert,Kennzahl,Daten)
rmse.ergebnis

##   Platzierung         Modell     Wert Kennzahl       Daten
## 1           1 wfl.rg.lm.fit1 3435.204     rmse autos.train
## 2           2 wfl.rg.lm.fit4 3649.877     rmse autos.train
## 3           3 wfl.rg.lm.fit5 3649.877     rmse autos.train
## 4           4 wfl.rg.lm.fit6 3649.877     rmse autos.train
## 5           5 wfl.rg.lm.fit3 3767.595     rmse autos.train
## 6           6 wfl.rg.lm.fit2 3977.315     rmse autos.train
```

```
# Ranking der Modelle nach Bestimmtheitsmaß (Trainingsdaten)
rsq.ergebnis<-filter(ergebnis,Kennzahl == "rsq") %>% arrange(desc(Wert)) %>%
  mutate(Platzierung=1:6) %>%
  select(Platzierung,Modell,Wert,Kennzahl,Daten)
rsq.ergebnis

##   Platzierung          Modell      Wert Kennzahl       Daten
## 1           1 wfl.rg.lm.fit1 0.7901154      rsq autos.train
## 2           2 wfl.rg.lm.fit4 0.7630635      rsq autos.train
## 3           3 wfl.rg.lm.fit5 0.7630635      rsq autos.train
## 4           4 wfl.rg.lm.fit6 0.7630635      rsq autos.train
## 5           5 wfl.rg.lm.fit3 0.7475335      rsq autos.train
## 6           6 wfl.rg.lm.fit2 0.7186445      rsq autos.train

# Testdaten
ergebnis<-rbind(ml.eval(wfl.rg.lm.fit1,autos.test,22),
                ml.eval(wfl.rg.lm.fit2,autos.test,22),
                ml.eval(wfl.rg.lm.fit3,autos.test,22),
                ml.eval(wfl.rg.lm.fit4,autos.test,22),
                ml.eval(wfl.rg.lm.fit5,autos.test,22),
                ml.eval(wfl.rg.lm.fit6,autos.test,22))

## Warning in predict.lm(object = object$fit, newdata = new_data, type =
## "response"): prediction from a rank-deficient fit may be misleading

rmse.ergebnis<-filter(ergebnis,Kennzahl == "rmse") %>% arrange(Wert) %>%
  mutate(Platzierung=1:6) %>%
  select(Platzierung,Modell,Wert,Kennzahl,Daten)
rmse.ergebnis

##   Platzierung          Modell     Wert Kennzahl      Daten
## 1           1 wfl.rg.lm.fit1 3440.628     rmse autos.test
## 2           2 wfl.rg.lm.fit5 3534.251     rmse autos.test
## 3           3 wfl.rg.lm.fit6 3534.251     rmse autos.test
## 4           4 wfl.rg.lm.fit4 3534.251     rmse autos.test
## 5           5 wfl.rg.lm.fit3 3651.170     rmse autos.test
## 6           6 wfl.rg.lm.fit2 3858.186     rmse autos.test

# Ranking der Modelle nach Bestimmtheitsmaß (Trainingsdaten)
rsq.ergebnis<-filter(ergebnis,Kennzahl == "rsq") %>% arrange(desc(Wert)) %>%
  mutate(Platzierung=1:6) %>%
  select(Platzierung,Modell,Wert,Kennzahl,Daten)
rsq.ergebnis
##   Platzierung          Modell      Wert Kennzahl      Daten
## 1           1 wfl.rg.lm.fit1 0.7833991      rsq autos.test
## 2           2 wfl.rg.lm.fit5 0.7713365      rsq autos.test
## 3           3 wfl.rg.lm.fit4 0.7713365      rsq autos.test
## 4           4 wfl.rg.lm.fit6 0.7713365      rsq autos.test
## 5           5 wfl.rg.lm.fit3 0.7559299      rsq autos.test
## 6           6 wfl.rg.lm.fit2 0.7274744      rsq autos.test
```

6.5.4 Comparison of Models

Workflow sets were previously created based on formulas, but this is also possible based on recipe() functions. The command `wflset.1<-workflow_set(preproc=rec.komb, models=list( lr=mod.rg.lm), cross=FALSE))` creates a workflow set with the four recipe() functions defined in the list rec.komb for preprocessing based on the model mod.rg.lm. The parameter cross=FALSE determines that the list of preprocessing arguments and the list of model arguments should contain an identical number. The function `workflow_map()` then allows the workflow sets to be trained based on a function, e.g., "fit_resamples". The command `workflow_map("fit_resamples",seed=123,verbose=TRUE, resamples=autos.fold,control=control_resamples (save_pred=TRUE))` applies the function fit_resamples() to the resamples autos.fold. The seed parameter ensures that the same random numbers are used for all models and the verbose parameter determines that the progress and CPU time for function execution are displayed.

To compare the models based on the resample data, the function `perf_mod()` from the tidyposterior package can be used, which calculates the ANOVA (Analysis of Variance) between the models. In the following coding, another function `student_t()` from the rstanarm package is used. The function `contrast_models()` compares the models and estimates the ROPE (Region Of Practical Equivalence) with a deviation of the RSQ of 3%, when the parameter size=0.03 is used. If the RSQ deviates less than 3%, then the practical equivalence (pract_eqiv) is rated with 1. The columns pract_neg and pract_pos are marked with 1 if there is a deviation downwards or upwards, respectively.

A comparison of the models of a workflow set in terms of the metrics RMSE and RSQ is done via the function `ml.eval.wflset()`, e.g., `ml.eval.wflset(mod.alle,"autos.fold")`, to compare all models contained in the workflow set mod.alle based on the data autos.fold. The model normalized_rf of the type Random Forest performs best, both in the test with the dataset autos.fold and autos.boots.

Comparison of Models

```
library(ggrepel)
# Trainings- und Testdaten erzeugen-------------------------------------------
set.seed(123)
autos.split<-initial_split(autos,prop=0.80,          # Split Train/Test 80:20
                           strata=Preis)
autos.train<-training(autos.split)
autos.test<-testing(autos.split)
# Bootstrap-Samples ---------------------------------------------------------
set.seed(123)
autos.boots<-bootstraps(autos.train,times=10)       # 10 Bootstraps erzeugen
# Einzelner Fold mit Trennung in Analyse- und Validierung -----------------------
set.seed(123)
autos.fold<-validation_split(autos.train, prop=4/5)  # Split in 80% Anal.+20% Val.
## Workflow-Set Lineare Regression ##############################################
# Modelltyp und Engine für Lineare Regression definieren ------------------------
mod.rg.lm<-linear_reg() %>% set_engine("lm")
# Recipes-Kombinationen für Lineare Regression definieren -----------------------
rec.1<-recipe(data=autos.train,
              Preis~Kilometer+Alter+PS+Marke) %>%
  step_log(PS,base=10) %>%                          # PS Logarithmieren
  step_dummy(all_nominal_predictors())              # Kategorial->Numerisch
rec.2<-recipe(data=autos.train,
              Preis~Kilometer+Alter+PS+Marke+Modell) %>%
  step_other(c(Marke,Modell),threshold=0.001) %>%   # Marke+Modell je > 0,1%
  step_log(PS,base=10) %>%                          # PS Logarithmieren
  step_dummy(all_nominal_predictors())              # Kategorial->Numerisch
rec.3<-recipe(data=autos.train,
              Preis~Kilometer+Alter+PS+Marke+Modell+Typ) %>%
  step_other(c(Marke,Modell),threshold=0.001) %>%   # Marke+Modell je > 0,1%
  step_log(PS,base=10) %>%                          # PS Logarithmieren
  step_dummy(all_nominal_predictors())              # Kategorial->Numerisch
rec.4<-recipe(data=autos.train,
              Preis~Kilometer+Alter+PS+Marke+Modell+Typ) %>%
  step_other(c(Marke,Modell),threshold=0.001) %>%   # Marke+Modell je > 0,1%
  step_log(PS,base=10) %>%                          # PS Logarithmieren
  step_dummy(all_nominal_predictors()) %>%          # Kategorial->Numerisch
  step_normalize(Kilometer,Alter,starts_with("PS")) # Skalen normalisieren
# Recipes-Kombinationen definieren -------------------------------------------
rec.komb <- list(einfach      = rec.1,
                 mitModell    = rec.2,
                 mitTyp       = rec.3,
                 normalisiert = rec.4)
# Workflow-Set basierend auf Recipe-Kombinationen definieren --------------------
wflset.1<-workflow_set(preproc=rec.komb,
                       models=list(lr=mod.rg.lm),cross=FALSE)
wflset.1

## # A workflow set/tibble: 4 × 4
##   wflow_id          info              option     result
##   <chr>             <list>            <list>     <list>
## 1 einfach_lr        <tibble [1 × 4]> <opts[0]> <list [0]>
## 2 mitModell_lr      <tibble [1 × 4]> <opts[0]> <list [0]>
## 3 mitTyp_lr         <tibble [1 × 4]> <opts[0]> <list [0]>
## 4 normalisiert_lr <tibble [1 × 4]> <opts[0]> <list [0]>

# Modelle aus Workflow-Set basierend Bootstrap-Resamples trainieren -------------
mod.wflset.1<-wflset.1 %>%
  workflow_map("fit_resamples",seed=123,verbose = TRUE,
               resamples=autos.fold,control=control_resamples(save_pred=TRUE))
```

```
## i 1 of 4 resampling: einfach_lr

## ✓ 1 of 4 resampling: einfach_lr (13.6s)

## i 2 of 4 resampling: mitModell_lr

## ✓ 2 of 4 resampling: mitModell_lr (1.5s)

## i 3 of 4 resampling: mitTyp_lr

## ✓ 3 of 4 resampling: mitTyp_lr (1.3s)

## i 4 of 4 resampling: normalisiert_lr

## ✓ 4 of 4 resampling: normalisiert_lr (1.3s)

mod.wflset.1

## # A workflow set/tibble: 4 × 4
##   wflow_id        info             option    result
##   <chr>           <list>           <list>    <list>
## 1 einfach_lr      <tibble [1 × 4]> <opts[2]> <rsmp[+]>
## 2 mitModell_lr    <tibble [1 × 4]> <opts[2]> <rsmp[+]>
## 3 mitTyp_lr       <tibble [1 × 4]> <opts[2]> <rsmp[+]>
## 4 normalisiert_lr <tibble [1 × 4]> <opts[2]> <rsmp[+]>
# Modellevaluation -------------------------------------------------------------
evaluation<-ml.eval.wflset(mod.wflset.1,"autos.fold")
evaluation$Kennzahlen                                      # Tabelle

##        WorkflowID Platzierung.RMSE     RMSE Platzierung.RSQ       RSQ
## 1 normalisiert_lr                1 3751.711               1 0.7519323
## 2       mitTyp_lr                2 3751.711               2 0.7519323
## 3    mitModell_lr                3 3783.677               3 0.7476598
## 4      einfach_lr                4 4167.318               4 0.6937394
##                Model      Daten
## 1 linear_reg\nrecipe autos.fold
## 2 linear_reg\nrecipe autos.fold
## 3 linear_reg\nrecipe autos.fold
## 4 linear_reg\nrecipe autos.fold

evaluation$rmse                                            # Scatter-Plot RMSE
```

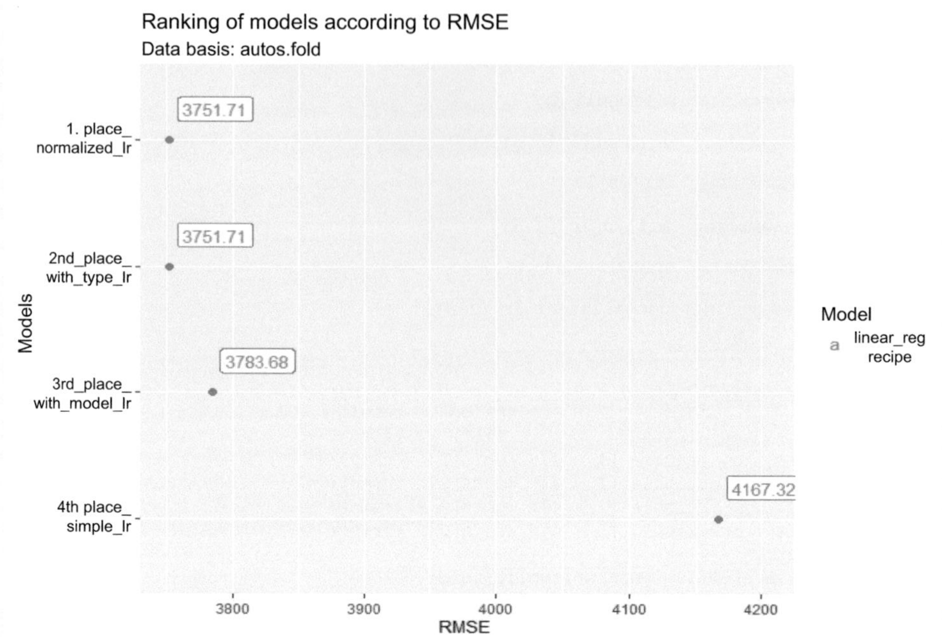


```
evaluation$rsq                                          # Scatter-Plot RSQ
```

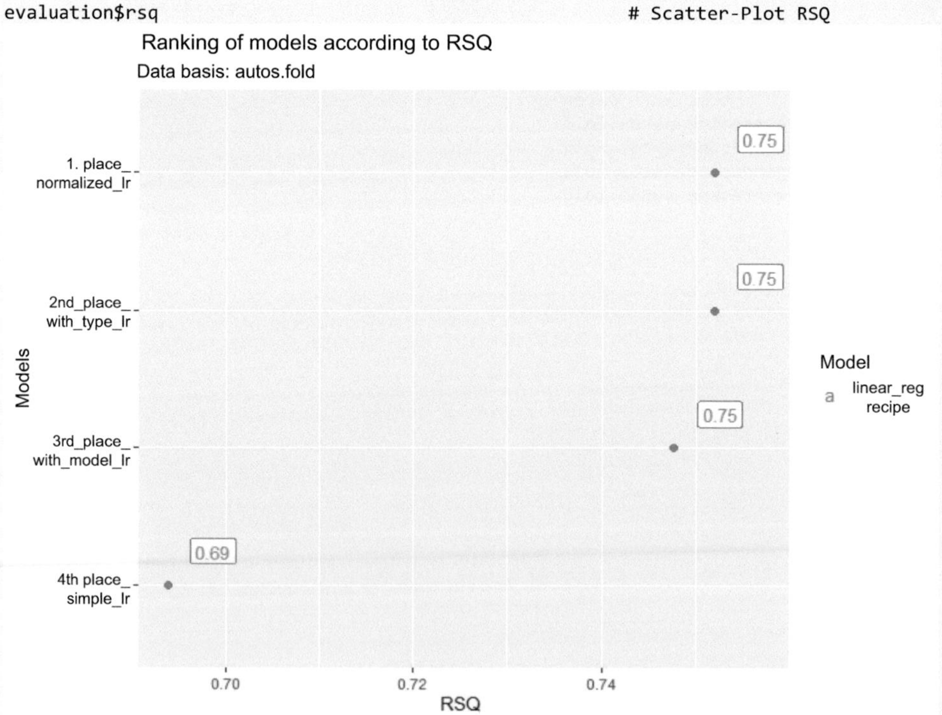

```
## Workflow-Set Random Forest ##############################################################
# Modelltyp und Engine für Random Forest definieren -----------------------------
mod.rg.rf<-rand_forest(trees = 1000) %>%
  set_engine("ranger") %>%
  set_mode("regression")
# Recipes-Varianten für Random Forest definieren --------------------------------
rec.5<-recipe(data=autos.train,
              Preis~Kilometer+Alter+PS+Marke+Modell+Typ) %>%
  step_log(PS,base=10) %>%                               # PS Logarithmieren
  step_dummy(all_nominal_predictors())                   # Kategorial->Numerisch
rec.6<-recipe(data=autos.train,
              Preis~Kilometer+Alter+PS+Marke+Modell+Typ) %>%
  step_other(c(Marke,Modell),threshold=0.001) %>%        # Marke+Modell je > 0,1%
  step_log(PS,base=10) %>%                               # PS Logarithmieren
  step_dummy(all_nominal_predictors())                   # Kategorial->Numerisch
rec.varianten<- list(einfach      = rec.5,
                    normalisiert = rec.6)
# Workflow-Set basierend auf Formula und Recipe-Varianten definieren ------------
wflset.2<-workflow_set(preproc=rec.varianten,
                       models=list(rf=mod.rg.rf),cross=FALSE)
wflset.2

## # A workflow set/tibble: 2 × 4
##   wflow_id        info             option    result
##   <chr>           <list>           <list>    <list>
## 1 einfach_rf      <tibble [1 × 4]> <opts[0]> <list [0]>
## 2 normalisiert_rf <tibble [1 × 4]> <opts[0]> <list [0]>
## Workflow-Set.2 ergänzen ##############################################################
# Modell mit Formel anstatt recipe zum Workflow-Set hinzufügen -------------------
wfl.rg.lm <-workflow() %>% add_model(mod.rg.lm)          # Workflow
wfl.rg.lm

## ══ Workflow ════════════════════════════════════════════════════════════════════
## Preprocessor: None
## Model: linear_reg()
##
## ── Model ───────────────────────────────────────────────────────────────────────
## Linear Regression Model Specification (regression)
##
## Computational engine: lm

wfl.rg.lm <-wfl.rg.lm %>%                                # Preprocessor
  add_formula(Preis~Kilometer+Alter+PS+Marke)            # Formel
wfl.rg.lm

## ══ Workflow ════════════════════════════════════════════════════════════════════
## Preprocessor: Formula
## Model: linear_reg()
##
## ── Preprocessor ────────────────────────────────────────────────────────────────
## Preis ~ Kilometer + Alter + PS + Marke
##
## ── Model ───────────────────────────────────────────────────────────────────────
## Linear Regression Model Specification (regression)
##
## Computational engine: lm

wflset.2 <-
  as_workflow_set(linearreg_lm=wfl.rg.lm) %>%
  bind_rows(wflset.2)
wflset.2
```

```
## # A workflow set/tibble: 3 × 4
##   wflow_id          info              option     result
##   <chr>             <list>            <list>     <list>
## 1 linearreg_lm      <tibble [1 × 4]> <opts[0]> <NULL>
## 2 einfach_rf        <tibble [1 × 4]> <opts[0]> <list [0]>
## 3 normalisiert_rf <tibble [1 × 4]> <opts[0]> <list [0]>

# Modelle aus Workflow-Set basierend Bootstrap-Resamples trainieren -------------
mod.wflset.2<-wflset.2 %>%
  workflow_map("fit_resamples",seed=123,verbose = TRUE,
             resamples=autos.fold,control=control_resamples(save_pred=TRUE))

## i 1 of 3 resampling: linearreg_lm

## ✓ 1 of 3 resampling: linearreg_lm (450ms)

## i 2 of 3 resampling: einfach_rf

## ✓ 2 of 3 resampling: einfach_rf (2m 2.6s)

## i 3 of 3 resampling: normalisiert_rf

## ✓ 3 of 3 resampling: normalisiert_rf (1m 33.4s)

mod.wflset.2

## # A workflow set/tibble: 3 × 4
##   wflow_id          info              option     result
##   <chr>             <list>            <list>     <list>
## 1 linearreg_lm      <tibble [1 × 4]> <opts[2]> <rsmp[+]>
## 2 einfach_rf        <tibble [1 × 4]> <opts[2]> <rsmp[+]>
## 3 normalisiert_rf <tibble [1 × 4]> <opts[2]> <rsmp[+]>

# Modellevaluation --------------------------------------------------------------
evaluation<-ml.eval.wflset(mod.wflset.2,"autos.fold")
evaluation$Kennzahlen                                    # Tabelle

##          WorkflowID Platzierung.RMSE     RMSE Platzierung.RSQ       RSQ
## 1 normalisiert_rf                 1 3246.495               1 0.8470527
## 2      einfach_rf                 2 3405.260               2 0.8369482
## 3    linearreg_lm                 3 3910.259               3 0.7304388
##                 Model      Daten
## 1 rand_forest\nrecipe autos.fold
## 2 rand_forest\nrecipe autos.fold
## 3 linear_reg\nformula autos.fold

evaluation$rmse                                          # Scatter-Plot RMSE
```

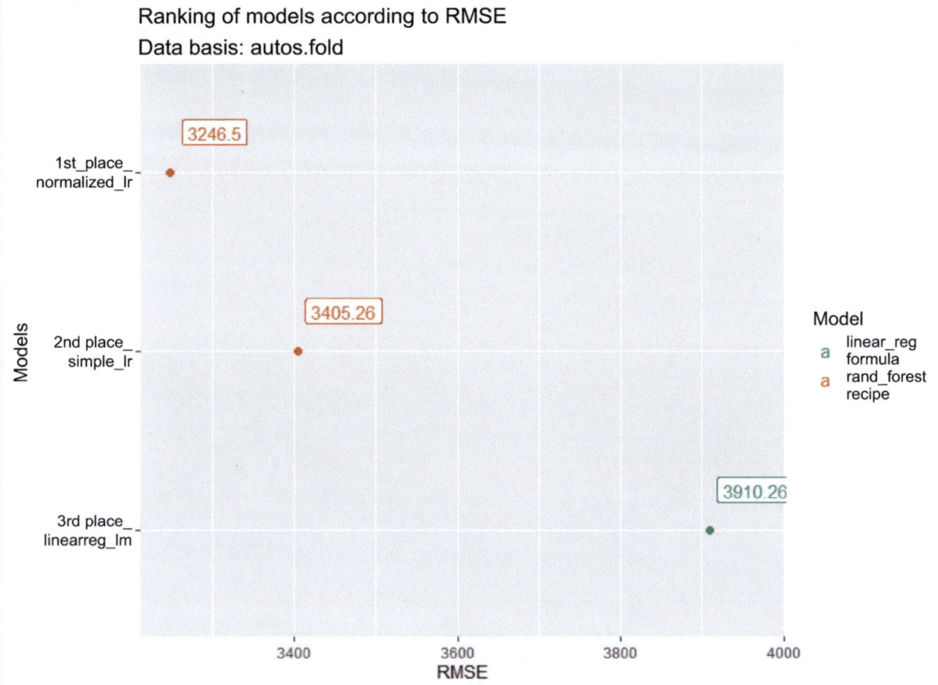


```
evaluation$rsq                                          # Scatter-Plot RSQ
```

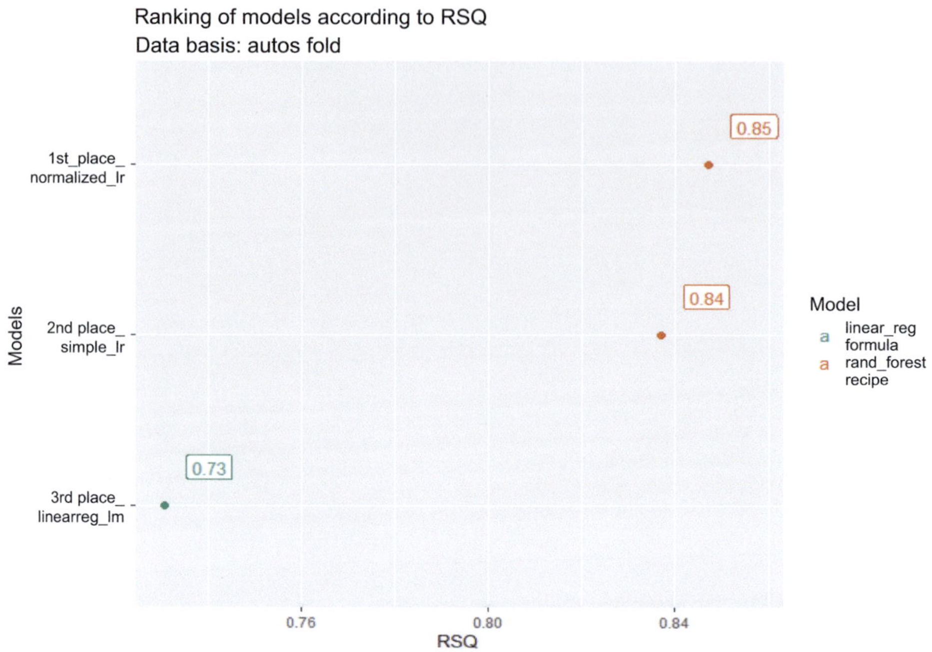


```
# Zusammenfassung von Workflow-Sets und Modellen ##################################
wflset.alle<-rbind(wflset.1,wflset.2)
wflset.alle

## # A workflow set/tibble: 7 × 4
##   wflow_id          info              option     result
##   <chr>             <list>            <list>     <list>
## 1 einfach_lr        <tibble [1 × 4]> <opts[0]> <list [0]>
## 2 mitModell_lr      <tibble [1 × 4]> <opts[0]> <list [0]>
## 3 mitTyp_lr         <tibble [1 × 4]> <opts[0]> <list [0]>
## 4 normalisiert_lr <tibble [1 × 4]> <opts[0]> <list [0]>
## 5 linearreg_lm      <tibble [1 × 4]> <opts[0]> <NULL>
## 6 einfach_rf        <tibble [1 × 4]> <opts[0]> <list [0]>
## 7 normalisiert_rf <tibble [1 × 4]> <opts[0]> <list [0]>

mod.alle<-rbind(mod.wflset.1,mod.wflset.2)
mod.alle

## # A workflow set/tibble: 7 × 4
##   wflow_id          info              option     result
##   <chr>             <list>            <list>     <list>
## 1 einfach_lr        <tibble [1 × 4]> <opts[2]> <rsmp[+]>
## 2 mitModell_lr      <tibble [1 × 4]> <opts[2]> <rsmp[+]>
## 3 mitTyp_lr         <tibble [1 × 4]> <opts[2]> <rsmp[+]>
## 4 normalisiert_lr <tibble [1 × 4]> <opts[2]> <rsmp[+]>
## 5 linearreg_lm      <tibble [1 × 4]> <opts[2]> <rsmp[+]>
## 6 einfach_rf        <tibble [1 × 4]> <opts[2]> <rsmp[+]>
## 7 normalisiert_rf <tibble [1 × 4]> <opts[2]> <rsmp[+]>

# Evaluation aller Modelle ##########################################################
evaluation<-ml.eval.wflset(mod.alle,"autos.fold")
evaluation$Kennzahlen                                    # Tabelle
```

```
##          WorkflowID Platzierung.RMSE     RMSE Platzierung.RSQ       RSQ
## 1 normalisiert_rf                1 3246.495               1 0.8470527
## 2       einfach_rf               2 3405.260               2 0.8369482
## 3 normalisiert_lr                3 3751.711               3 0.7519323
## 4        mitTyp_lr               4 3751.711               4 0.7519323
## 5     mitModell_lr               5 3783.677               5 0.7476598
## 6     linearreg_lm               6 3910.259               6 0.7304388
## 7       einfach_lr               7 4167.318               7 0.6937394
##                   Model      Daten
## 1 rand_forest\nrecipe autos.fold
## 2 rand_forest\nrecipe autos.fold
## 3  linear_reg\nrecipe autos.fold
## 4  linear_reg\nrecipe autos.fold
## 5  linear_reg\nrecipe autos.fold
## 6 linear_reg\nformula autos.fold
## 7  linear_reg\nrecipe autos.fold
evaluation$rmse                                          # Scatter-Plot RMSE
```

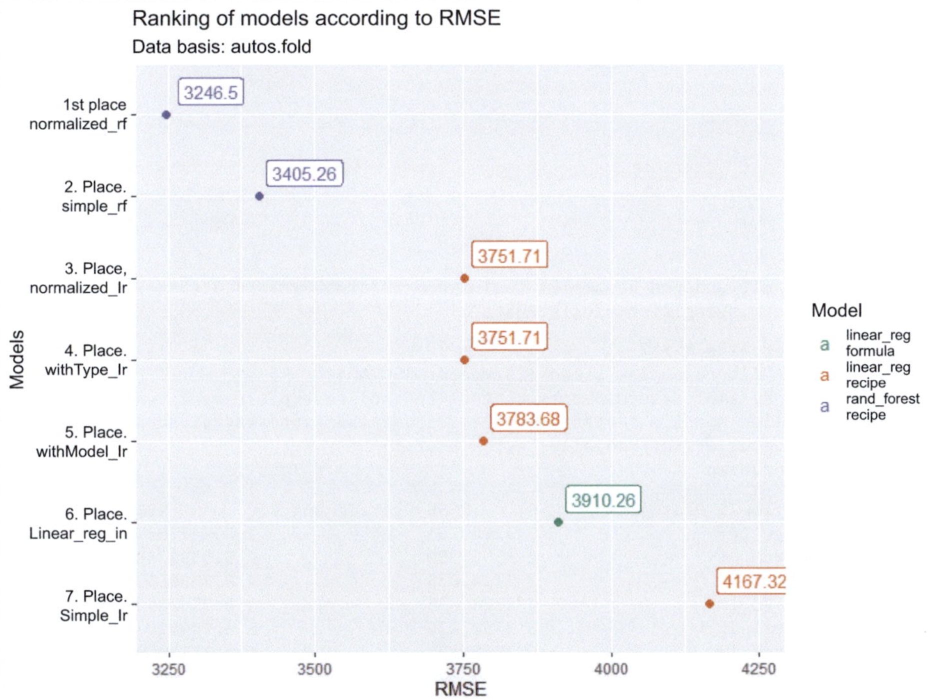


```
evaluation$rsq                                           # Scatter-Plot RSQ
```

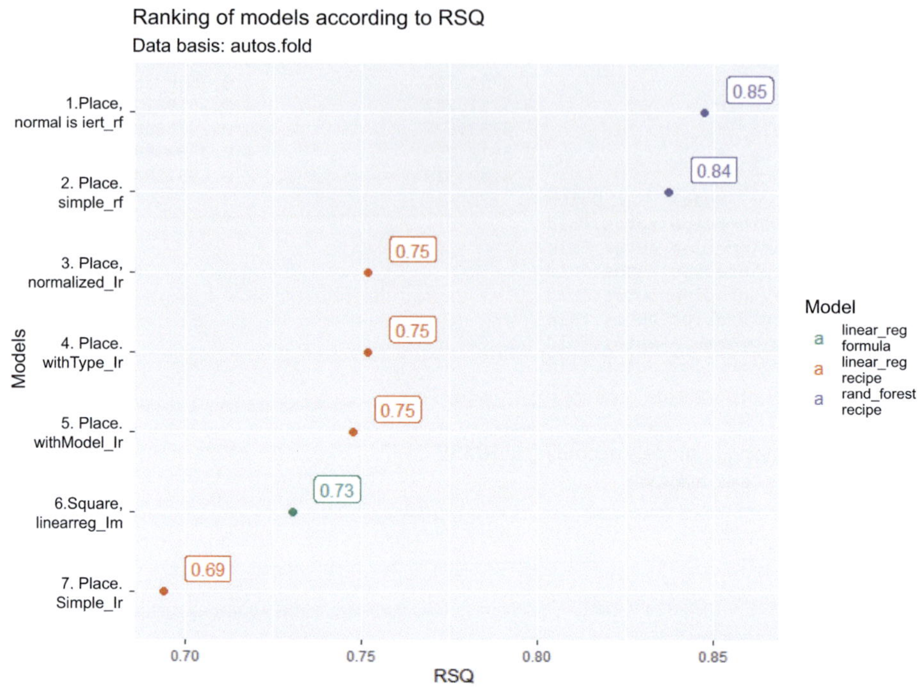


```
# Alle Modelle mit anderen Resamples trainieren ###################################
load(file="ErgebnisVergleichModelle.RData")

mod.alle<-wflset.alle %>%
  workflow_map("fit_resamples",seed=123,verbose = TRUE,
               resamples=autos.boots,control=control_resamples(save#pred=TRUE))
# Evaluation aller Modelle #######################################################
evaluation<-ml.eval.wflset(mod.alle,"autos.boots")
evaluation$Kennzahlen                                   # Tabelle

##        WorkflowID Platzierung.RMSE     RMSE Platzierung.RSQ       RSQ
## 1 normalisiert_rf                1 3208.791               1 0.8477019
## 2      einfach_rf                2 3355.386               2 0.8381348
## 3 normalisiert_lr                3 3724.835               4 0.7530653
## 4       mitTyp_lr                4 3724.835               3 0.7530653
## 5    mitModell_lr                5 3752.232               5 0.7494193
## 6    linearreg_lm                6 3946.335               6 0.7227969
## 7      einfach_lr                7 4175.446               7 0.6896643
##                  Model       Daten
## 1 rand_forest\nrecipe autos.boots
## 2 rand_forest\nrecipe autos.boots
## 3  linear_reg\nrecipe autos.boots
## 4  linear_reg\nrecipe autos.boots
## 5  linear_reg\nrecipe autos.boots
## 6 linear_reg\nformula autos.boots
## 7  linear_reg\nrecipe autos.boots

evaluation$rmse                                         # Scatter-Plot RMSE
```

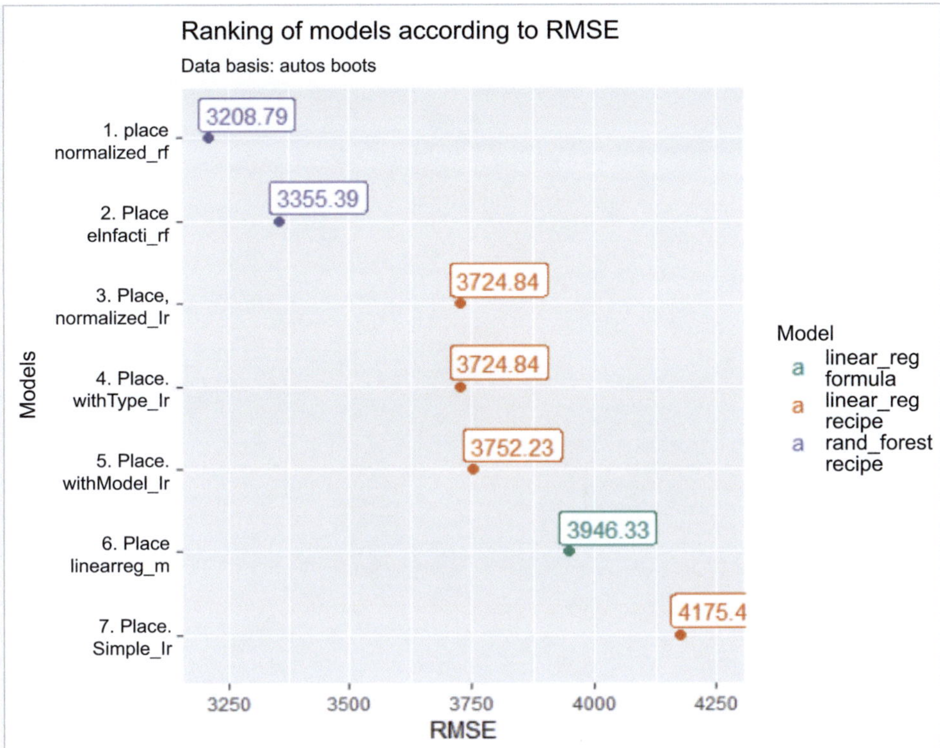


```
evaluation$rsq                                    # Scatter-Plot RSQ
```

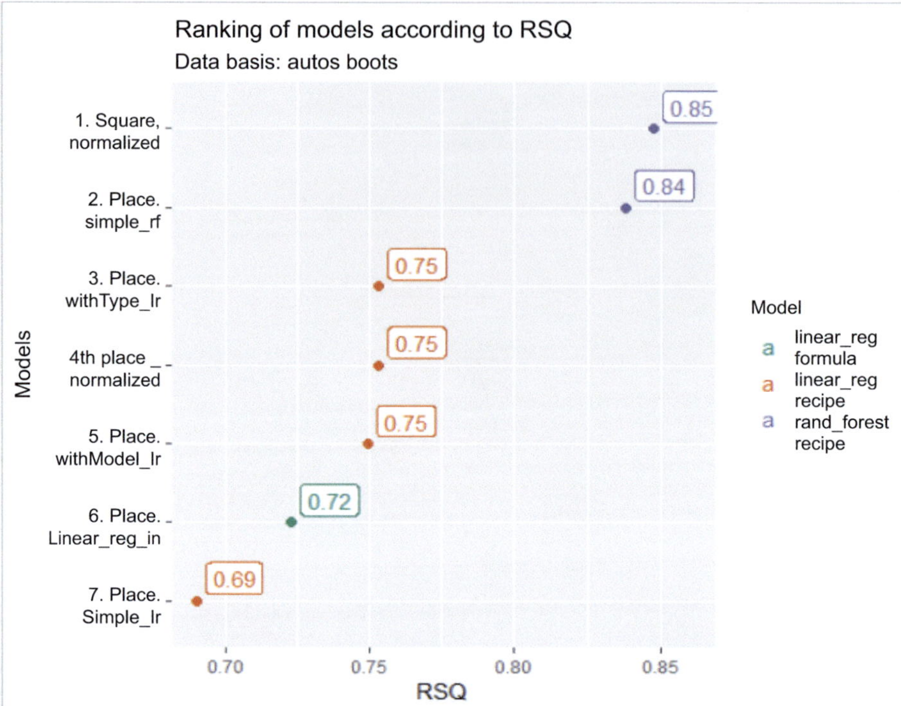


```
# Vergleich der Resample-Modelle ################################################
library(tidyposterior)
library(rstanarm)

anova.rsq<-perf_mod(mod.alle, metric="rsq",
                    prior_intercept=rstanarm::student_t(df=1),
                    chains=3,iter=3000,seed=123)

equiv.rsq<-summary(contrast_models(anova.rsq,          # Äquivalenztest alle M
                   seed=123),
                   size=0.03) %>%                       # 3 Prozent
  select(contrast,pract_neg,pract_equiv,pract_pos)
equiv.rsq

## # A tibble: 21 × 4
##    contrast                       pract_neg pract_equiv pract_pos
##    <chr>                              <dbl>       <dbl>     <dbl>
##  1 einfach_lr vs einfach_rf           1       0                 0
##  2 einfach_lr vs linear_reg           0.999   0.000667          0
##  3 einfach_lr vs mitModell_lr         1       0                 0
##  4 einfach_lr vs mitTyp_lr            1       0                 0
##  5 einfach_lr vs normalisiert_lr      1       0                 0
##  6 einfach_lr vs normalisiert_rf      1       0                 0
##  7 einfach_rf vs normalisiert_rf      0       1                 0
##  8 linear_reg vs einfach_rf           1       0                 0
##  9 linear_reg vs normalisiert_rf      1       0                 0
## 10 mitModell_lr vs einfach_rf         1       0                 0
## # … with 11 more rows
```

```
equiv.rsq<-summary(contrast_models(anova.rsq,          # Äquivalenz ausgewählte M.
                   list_1=c("normalisiert_rf","normalisiert_rf"),
                   list_2=c("einfach_rf","mitTyp_lr"),
                   seed=123),
                   size=0.03) %>%                       # 3 Prozent
  select(contrast,pract_neg,pract_equiv,pract_pos)
equiv.rsq

## # A tibble: 2 × 4
##   contrast                      pract_neg pract_equiv pract_pos
##   <chr>                             <dbl>       <dbl>     <dbl>
## 1 normalisiert_rf vs einfach_rf         0           1         0
## 2 normalisiert_rf vs mitTyp_lr          0           0         1
```

6.5.5 Parameter Tuning

Depending on the model type, models calculate parameters, (**model parameters**, which are determined by machine learning), e.g., the coefficients of predictors. However, other parameters are predefined for the models (**hyperparameters**, which are not determined by machine learning), e.g., which variables are considered as predictors in the equations, the number of branches in a decision tree, or the number of layers and hidden layers in a neural network. The aim of parameter tuning is to determine these parameters as optimally as possible for each model type. Examples of tuning parameters that were already used in Sect. 6.5 include the degrees of freedom in the function `step_ns()`, the number of trees for `rand_forest()`, and the minimum percentage value for the function `step_other()`.

In general, hyperparameters are differentiated into elementary tuning parameters and engine tuning parameters. While elementary tuning parameters are applicable to all models of a type, e.g., trees and min_n for all `rand_forest()` models, engine tuning parameters are only usable with specific engines, e.g., regularization.factor for the ranger engine. All of these parameters can be set for a model. However, it makes sense to test different parameter values to see which value produces the best result. Therefore, parameters are tuned in advance to determine the most favorable parameter values and then use these for modeling.

To enrich a model with parameters that are to be tuned, the function `tune()` is used. The function `extract_parameter_set_dials()` can display a model with the possible tuning parameters and simultaneously set the standard intervals for the parameters. The command `nn.k.para<-extract_parameter_set_dials(nn.k.modell)` stores, for example, the list of tuning parameters of the model nn.k.modell in the object nn.k.para. If you display these tuning parameters, a tuning parameter specification nparam[+] stands for a numerical tuning parameter, an nparam[?] for missing information on the tuning interval. Missing information on the tuning interval can be supplemented with the function `update()`. The function `extract_parameter_dials()` shows the standard intervals for the parameters, e.g., `nn.k.para %>% extract_parameter_dials("hidden_units")`, to display the interval of the number of hidden units.

For each tuning parameter, there is a corresponding function in the dials package with the corresponding name, e.g., the function `threshold()` and `spline_degree()` for the parameters threshold and spline_degree. The function `finalize()` allows estimating the limits of the not yet determined tuning parameters.

If the best possible parameter values are to be found, combinations of parameter values to be tested can be defined via so-called grids. Grids can be predefined with the function `crossing()` with specific values or generated with the function `grid_regular()` within the parameter limits into a number of values determined by the parameter levels. The function `grid_regular()` by default selects the values at equal intervals between the minimum and maximum of the parameter limits. If the values should rather be randomly selected from the spectrum of possibilities within the interval limits, this can be achieved with the function `grid_random()`. The parameter size determines the number of combinations generated. Since the values are generated randomly, the result of the function `grid_random()` can also cause similar value combinations and on the other hand large distances between the value combinations that are generated. An optimized design with good coverage of the space of all parameter-value combinations is generated by the function `grid_latin_hypercube()`. The function uses the maximum entropy design by default, which increases the chances of finding good results.

When working with parameter grids, a function equivalent to `fit_resamples()` with additional parameters for the grids is the function `tune_grid()`. The function `autoplot()` displays the results in figures. The function `show_best()` allows the best parameter configurations to be displayed. The best parameter values for Threshold are 0.0000388 and for Degree of Freedom are 5. The second-best parameter values for Threshold are 0.00071 and for Degree of Freedom are 6. The parameters with which the final model should then be executed are manually set to 0.0005 for the Threshold and 5 for Degree of Freedom.

To use the model with the best possible parameters and the test data, the parameters can be manually selected based on the results of the `tune_grid()` function or the combination with the best metric result can be selected with the function select_best(). The parameter metric="rsq" or similar can be used to specify which metric should be used for selection.

The final execution of the workflow can be done with the function `finalize_workflow()`. If no workflow was used, the functions `finalize_model()` or `finalize_recipe()` can be used similarly. The following coding presents the tuning parameters for the Random Forest, Neural Networks, and Linear Regression (lm) models. Although the Random Forest model performed best in the last step and would therefore offer itself for further optimization, the parameter tuning is demonstrated on the Linear Regression model. In the previous Sect. 6.5.4, the Linear Regression model resulted in an RSME of 3724 and an RSQ of 0.75. The tuning carried out here then contributed to both values being significantly improved to an RMSE of 3211 and an RSQ of 0.81.

Parameter Tuning (Part 1)

```
# Trainings- und Testdaten erzeugen-----------------------------------------------
set.seed(123)
autos.split<-initial_split(autos[,                          # alle Datensätze
                           c("Typ","Getriebe","PS",         # Relevante Variablen
                             "Kilometer","Motor","Marke",
                             "Unfall","Alter","Preis",
                             "Kategorie","Modell")],
                           prop=0.80,                       # Split Train/Test 80:20
                           strata=Preis)
autos.train<-training(autos.split)
autos.test<-testing(autos.split)
# Bootstrap-Samples ---------------------------------------------------------------
set.seed(123)
autos.boots<-bootstraps(autos.train,times=10)               # 10 Bootstraps erzeugen
# Einzelner Fold mit Trennung in Analyse- und Validierung -----------------------
set.seed(123)
autos.fold<-validation_split(autos.train, prop=4/5)  # Split in 80% Anal.+20% Val.
# Modelltyp und Engine für Neural Network nnet/Klassifikation definieren ---------
mod.kl.nnet<- mlp(hidden_units=tune(),penalty=tune(),       # Modell nnet
                  epochs=tune()) %>%
  set_mode("classification") %>%
  set_engine("nnet",trace=0)                                # trace=0->kein Logging
mod.kl.nnet.par<-extract_parameter_set_dials(mod.kl.nnet)  # 3 Tuningparameter
mod.kl.nnet.par

## Collection of 3 parameters for tuning
##
##    identifier         type    object
##  hidden_units hidden_units nparam[+]
##       penalty      penalty nparam[+]
##        epochs       epochs nparam[+]

mod.kl.nnet.par %>% extract_parameter_dials("hidden_units")# Details zu hidden_units


## # Hidden Units (quantitative)
## Range: [1, 10]

mod.kl.nnet.par %>% extract_parameter_dials("penalty")     # Details zu penalty

## Amount of Regularization (quantitative)
## Transformer: log-10 [1e-100, Inf]
## Range (transformed scale): [-10, 0]

mod.kl.nnet.par %>% extract_parameter_dials("epochs")      # Details zu epochs

## # Epochs (quantitative)
## Range: [10, 1000]

# Modelltyp und Engine für Neural Network keras/Regression definieren ------------
mod.rg.keras<- mlp(hidden_units=tune()) %>%                 # Modell keras
  set_mode("regression") %>%
  set_engine("keras")
extract_parameter_set_dials(mod.rg.keras)                   # 1 Tuningparameter

## Collection of 1 parameters for tuning
##
##    identifier         type    object
##  hidden_units hidden_units nparam[+]

# Modelltyp und Engine für Lineare Regression/Regression definieren --------------
mod.rg.lm<-linear_reg() %>%                                 # Modell lm
  set_engine("lm")
```

```
extract_parameter_set_dials(mod.rg.lm)                    # Kein Tuningparameter

## Collection of 0 parameters for tuning
##
## [1] identifier type       object
## <0 rows> (or 0-length row.names)

# Modelltyp und Engine für Random Forest/Regression definieren ------------------
mod.rg.rf<-rand_forest(mtry = tune()) %>%                 # Modell Random Forest
  set_engine("ranger",regularization.factor=tune("regularization")) %>%
  set_mode("regression")
extract_parameter_set_dials(mod.rg.rf)                    # 2 Tuningparameter

## Collection of 2 parameters for tuning
##
##      identifier                   type    object
##            mtry                   mtry nparam[?]
##  regularization regularization.factor nparam[+]
##
## Model parameters needing finalization:
##    # Randomly Selected Predictors ('mtry')
##
## See `?dials::finalize` or `?dials::update.parameters` for more information.

# Recipe definieren -------------------------------------------------------------
rec.1<-recipe(data=autos.train,
             Preis~Typ+Kilometer+Alter+Kategorie+PS+Marke+Modell) %>%
  step_other(c(Marke,Modell),threshold=tune()) %>%   # Marke+Modell je > ?%
  step_log(PS,base=10) %>%                            # PS Logarithmieren
  step_dummy(all_nominal_predictors()) %>%            # Kategorial->Numerisch
  step_ns(starts_with("PS"),deg_free=tune()) %>%      # Nichtlinear Splines=?
  step_normalize(Kilometer,Alter,starts_with("PS")) %>% # Skalen normalisieren
  step_pca(Alter,starts_with("Kategorie_"))           # Principal Component Analysis
rec.2<-recipe(data=autos.train,
             Preis~Kilometer+Alter+PS+Marke+Modell+Typ) %>%
  step_other(c(Marke,Modell),threshold=tune()) %>%   # Marke+Modell je > ?%
  step_log(PS,base=10) %>%                            # PS Logarithmieren
  step_dummy(all_nominal_predictors())                # Kategorial->Numerisch
# Tuningparameter von Recipe ----------------------------------------------------
rec.1.par<-extract_parameter_set_dials(rec.1)        # Tuningparameter von rec.1
rec.1.par                                             # 2 Tuningpara

## Collection of 2 parameters for tuning
##
##  identifier      type    object
##   threshold threshold nparam[+]
##    deg_free  deg_free nparam[+]

rec.1.par %>% extract_parameter_dials("threshold")   # Details zu Tuningpara

## Threshold (quantitative)
## Range: [0, 0.1]

rec.1.par %>% extract_parameter_dials("deg_free")    # Details zu Tuningpara

## Spline Degrees of Freedom (quantitative)
## Range: [1, 15]

rec.2.par<-extract_parameter_set_dials(rec.2)        # Tuningpara von rec.2
rec.2.par                                             # 1 Tuningpara
```

```
## Collection of 1 parameters for tuning
##
##  identifier      type    object
##   threshold threshold nparam[+]

# Workflow für Neuronales Netzwerk keras definieren ------------------------------
wfl.rg.keras <-workflow() %>%                          # Workflow keras
  add_model(mod.rg.keras) %>%
  add_recipe(rec.1)                                    # mit rec.1
# Workflow für Lineare Regression definieren -------------------------------------
wfl.rg.lm <-workflow() %>%                             # Workflow lm
  add_model(mod.rg.lm) %>%
  add_recipe(rec.1)                                    # mit rec.1
# Workflow für Random Forest definieren ------------------------------------------
wfl.rg.rf <-workflow() %>%                             # Workflow rf
  add_model(mod.rg.rf) %>%
  add_recipe(rec.2)                                    # mit rec.2
# Tuningparameter von Workflows --------------------------------------------------
extract_parameter_set_dials(wfl.rg.keras)              # 3 Tuningpara wfl.rg.keras

## Collection of 3 parameters for tuning
##
##    identifier         type    object
##  hidden_units hidden_units nparam[+]
##     threshold    threshold nparam[+]
##      deg_free     deg_free nparam[+]

extract_parameter_set_dials(wfl.rg.lm)                 # 2 Tuningpara wfl.rg.lm

## Collection of 2 parameters for tuning
##
##  identifier      type    object
##   threshold threshold nparam[+]
##    deg_free  deg_free nparam[+]

extract_parameter_set_dials(wfl.rg.rf)                 # 3 Tuningpara wfl.rg.rf

## Collection of 3 parameters for tuning
##
##      identifier                  type    object
##            mtry                  mtry nparam[?]
##  regularization regularization.factor nparam[+]
##       threshold             threshold nparam[+]
##
## Model parameters needing finalization:
##    # Randomly Selected Predictors ('mtry')
##
## See `?dials::finalize` or `?dials::update.parameters` for more information.

# Parametergrenzen für Workflow Random Forest anpassen ---------------------------
mod.rg.rf.par<-mod.rg.rf %>%                           # 2 Tuningpara mod.rg.rf
  extract_parameter_set_dials()
mod.rg.rf.par

## Collection of 2 parameters for tuning
##
##      identifier                  type    object
##            mtry                  mtry nparam[?]
##  regularization regularization.factor nparam[+]
##
```

```
## Model parameters needing finalization:
##    # Randomly Selected Predictors ('mtry')
##
## See `?dials::finalize` or `?dials::update.parameters` for more information.

wfl.rg.rf.par<-wfl.rg.rf %>%                          # 3 Tuningpara wfl.rg.rf
  extract_parameter_set_dials()
wfl.rg.rf.par

## Collection of 3 parameters for tuning
##
##      identifier                  type    object
##            mtry                  mtry nparam[?]
##  regularization regularization.factor nparam[+]
##       threshold             threshold nparam[+]
##
## Model parameters needing finalization:
##    # Randomly Selected Predictors ('mtry')
##
## See `?dials::finalize` or `?dials::update.parameters` for more information.

wfl.rg.rf.par %>% extract_parameter_dials("mtry")    # Details zu mtry

## # Randomly Selected Predictors (quantitative)
## Range: [1, ?]

wfl.rg.rf.par <-                                      # Update der Grenzen
  wfl.rg.rf %>%
  extract_parameter_set_dials() %>%
  update(mtry=mtry(c(1,50)))
wfl.rg.rf.par %>% extract_parameter_dials("mtry")    # Details zu mtry

## # Randomly Selected Predictors (quantitative)
## Range: [1, 50]

wfl.rg.rf.par.2 <- workflow() %>%                     # Update Wflow Tuningpara.
  add_model(mod.rg.rf) %>%                            # mit finalize()
  add_recipe(rec.2) %>%
  extract_parameter_set_dials() %>%
  finalize(autos.train)
wfl.rg.rf.par.2

## Collection of 3 parameters for tuning
##
##      identifier                  type    object
##            mtry                  mtry nparam[+]
##  regularization regularization.factor nparam[+]
##       threshold             threshold nparam[+]

wfl.rg.rf.par.2 %>% extract_parameter_dials("mtry")  # Details zu mtry

## # Randomly Selected Predictors (quantitative)
## Range: [1, 11]

# Parametergrenzen für Workflow Lineare Regression anpassen ----------------------
wfl.rg.lm.par<-extract_parameter_set_dials(wfl.rg.lm)# 2 Tuningpara wfl.rg.lm
wfl.rg.lm.par

## Collection of 2 parameters for tuning
##
##  identifier      type    object
##   threshold threshold nparam[+]
##    deg_free  deg_free nparam[+]
```

```
wfl.rg.lm.par %>% extract_parameter_dials("threshold")# Details zu threshold

## Threshold (quantitative)
## Range: [0, 0.1]

wfl.rg.lm.par %>% extract_parameter_dials("deg_free") # Details zu deg_free

## Spline Degrees of Freedom (quantitative)
## Range: [1, 15]

wfl.rg.lm.par <-                                     # Update der Grenzen
  wfl.rg.lm %>%
  extract_parameter_set_dials() %>%
  update(threshold=threshold(c(0,0.05)),
         deg_free =deg_free(c(1,10)))
wfl.rg.lm.par %>% extract_parameter_dials("threshold")# Details zu threshold
## Threshold (quantitative)
## Range: [0, 0.05]

wfl.rg.lm.par %>% extract_parameter_dials("deg_free") # Details zu deg_free

## Degrees of Freedom (quantitative)
## Range: [1, 10]

# Parameter-Grids für Workflow Lineare Regression erstellen ----------------------
grid.cr<-crossing(                                   # Grid mit 8=4x2 Kombis
  threshold=c(0.0005,0.001,0.002,0.003),             # Werte vorgegeben
  deg_free =c(1,5))                                  # Werte vorgegeben
grid.cr

## # A tibble: 8 × 2
##   threshold deg_free
##       <dbl>    <dbl>
## 1    0.0005        1
## 2    0.0005        5
## 3    0.001         1
## 4    0.001         5
## 5    0.002         1
## 6    0.002         5
## 7    0.003         1
## 8    0.003         5

grid.reg<-grid_regular(wfl.rg.lm.par,                # Grid mit 9=3x3 Kombis
  levels=c(threshold=3,deg_free=3))                  # Werte in Grenzen
grid.reg

## # A tibble: 9 × 2
##   threshold deg_free
##       <dbl>    <int>
## 1     0            1
## 2     0.025        1
## 3     0.05         1
## 4     0            5
## 5     0.025        5
## 6     0.05         5
## 7     0           10
## 8     0.025       10
## 9     0.05        10

grid.rand<-grid_random(wfl.rg.lm.par,size=10)        # Grid mit 10 Kombis
grid.rand                                            # Werte per Zufall in Grenzen
```

```
## # A tibble: 10 × 2
##    threshold deg_free
##        <dbl>    <int>
##  1   0.0280         9
##  2   0.0403         6
##  3   0.0188         6
##  4   0.00484        2
##  5   0.0431         4
##  6   0.0271         6
##  7   0.00572        5
##  8   0.0307         7
##  9   0.00347       10
## 10   0.0273         5

grid.lh<-grid_latin_hypercube(wfl.rg.lm.par,size=10) # Grid mit 10 Kombis
grid.lh                                              # Werte per Zufall nach LH
## # A tibble: 10 × 2
##    threshold deg_free
##        <dbl>    <int>
##  1   0.0293         4
##  2   0.0191         6
##  3   0.0438         2
##  4   0.0245         9
##  5   0.0318        10
##  6   0.0133         5
##  7   0.0489         8
##  8   0.00245        3
##  9   0.00955        7
## 10   0.0396         2

# Modellevaluation mit Resamples und Parameter-Grids ----------------------------
tune<- wfl.rg.lm %>%
  tune_grid(autos.boots,
            grid=grid.lh,
            metrics=metric_set(rsq))
tune

## # Tuning results
## # Bootstrap sampling
## # A tibble: 10 × 4
##    splits                 id          .metrics          .notes
##    <list>                 <chr>       <list>            <list>
##  1 <split [27890/10309]> Bootstrap01 <tibble [10 × 6]> <tibble [0 × 3]>
##  2 <split [27890/10270]> Bootstrap02 <tibble [10 × 6]> <tibble [0 × 3]>
##  3 <split [27890/10229]> Bootstrap03 <tibble [10 × 6]> <tibble [0 × 3]>
##  4 <split [27890/10343]> Bootstrap04 <tibble [10 × 6]> <tibble [0 × 3]>
##  5 <split [27890/10267]> Bootstrap05 <tibble [10 × 6]> <tibble [0 × 3]>
##  6 <split [27890/10277]> Bootstrap06 <tibble [10 × 6]> <tibble [0 × 3]>
##  7 <split [27890/10290]> Bootstrap07 <tibble [10 × 6]> <tibble [0 × 3]>
##  8 <split [27890/10167]> Bootstrap08 <tibble [10 × 6]> <tibble [0 × 3]>
##  9 <split [27890/10270]> Bootstrap09 <tibble [10 × 6]> <tibble [0 × 3]>
## 10 <split [27890/10281]> Bootstrap10 <tibble [10 × 6]> <tibble [0 × 3]>

autoplot(tune)
```

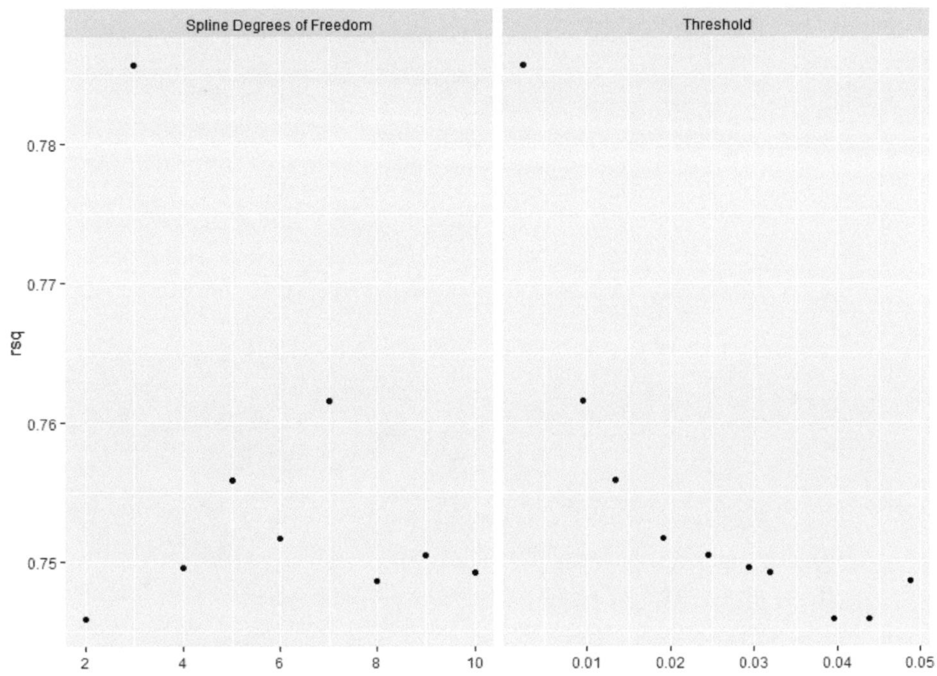


```
wfl.rg.lm.par<-wfl.rg.lm %>%                          # Update der Grenzen
  extract_parameter_set_dials() %>%
  update(threshold=threshold(c(0,0.01)),
         deg_free =deg_free(c(3,10)))
grid.lh<-grid_latin_hypercube(wfl.rg.lm.par,size=10) # Grid mit 10 Kombis
tune<-wfl.rg.lm %>%
  tune_grid(autos.boots,
            grid=grid.lh,
            metrics=metric_set(rsq))
autoplot(tune)
```

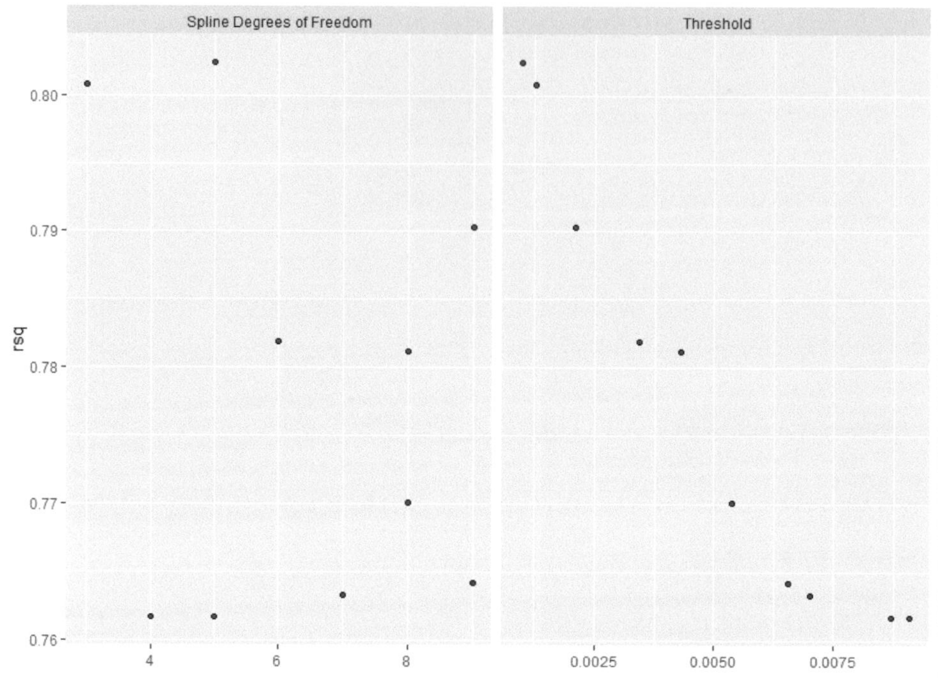


```
wfl.rg.lm.par<-wfl.rg.lm %>%                          # Update der Grenzen
  extract_parameter_set_dials() %>%
  update(threshold=threshold(c(0,0.005)),
         deg_free =deg_free(c(4,9)))
grid.lh<-grid_latin_hypercube(wfl.rg.lm.par,size=10) # Grid mit 10 Kombis
tune<- wfl.rg.lm %>%
  tune_grid(autos.boots,
            grid=grid.lh,
            metrics=metric_set(rsq))
autoplot(tune)
```

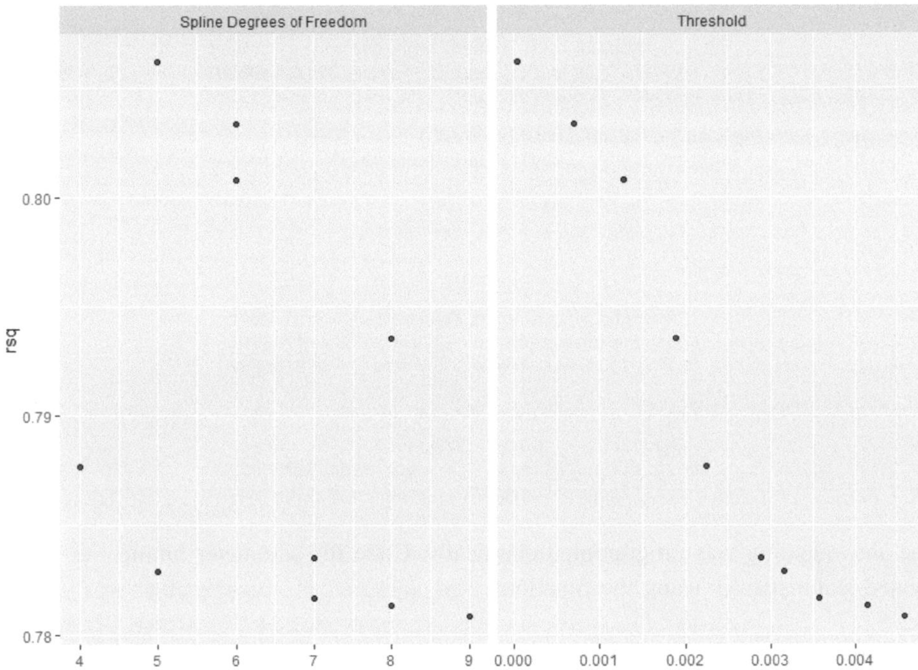


```
show_best(tune) %>% select(-.estimator)

## # A tibble: 5 × 7
##   threshold deg_free .metric  mean     n std_err .config
##       <dbl>    <int> <chr>   <dbl> <int>   <dbl> <chr>
## 1 0.0000388        5 rsq     0.806    10 0.00122 Preprocessor09_Model1
## 2 0.000710         6 rsq     0.803    10 0.00142 Preprocessor08_Model1
## 3 0.00129          6 rsq     0.801    10 0.00162 Preprocessor02_Model1
## 4 0.00189          8 rsq     0.794    10 0.00191 Preprocessor07_Model1
## 5 0.00225          4 rsq     0.788    10 0.00185 Preprocessor01_Model1

final.manuell.par<-tibble(                             # Manuelle Auswahl
    threshold=0.0005,
    deg_free =5)
final.manuell.par

## # A tibble: 1 × 2
##   threshold deg_free
##       <dbl>    <dbl>
## 1    0.0005        5

wfl.rg.lm.final.manuell<- wfl.rg.lm %>%
  finalize_workflow(final.manuell.par)
wfl.rg.lm.final.manuell.fit<-wfl.rg.lm.final.manuell %>%
  fit(autos)
ml.eval(wfl.rg.lm.final.manuell.fit,autos.train,9)

##                         Modell       Daten Kennzahl         Wert
## 1 wfl.rg.lm.final.manuell.fit autos.train     rmse 3247.5425248
## 2 wfl.rg.lm.final.manuell.fit autos.train      rsq    0.8124229
```

```
ml.eval(wfl.rg.lm.final.manuell.fit,autos.test,9)

##                           Modell       Daten Kennzahl         Wert
## 1 wfl.rg.lm.final.manuell.fit autos.test     rmse 3211.5401208
## 2 wfl.rg.lm.final.manuell.fit autos.test      rsq    0.8111773

final.best.par<-select_best(tune,metric="rsq")
final.best.par<-final.best.par[,-ncol(final.best.par)]
wfl.rg.lm.final.best<- wfl.rg.lm %>%
  finalize_workflow(final.best.par)
wfl.rg.lm.final.best.fit<-wfl.rg.lm.final.best %>%
  fit(autos)
ml.eval(wfl.rg.lm.final.best.fit,autos.train,9)

##                        Modell       Daten Kennzahl         Wert
## 1 wfl.rg.lm.final.best.fit autos.train     rmse 3226.1208895
## 2 wfl.rg.lm.final.best.fit autos.train      rsq    0.8148897

ml.eval(wfl.rg.lm.final.best.fit,autos.test,9)

##                        Modell      Daten Kennzahl         Wert
## 1 wfl.rg.lm.final.best.fit autos.test     rmse 3164.3023006
## 2 wfl.rg.lm.final.best.fit autos.test      rsq    0.8166978
```

The above coding was programmed classically. Code for parameter tuning can also be created automatically using the functions `use_glmnet()`, `use_xgboost()`, `use_kknn()`, `use_ranger()`, `use_earth()`, `use_cubist()`, `use_kernlab_svm_rbf()`, `use_kernlab_svm_poly()` or `use_C5.0()` from the usemodels package. Then in the function `tune_grid()`, only the parameters resamples and grid regarding the details of the data for resampling and the grid need to be added. In fact, the automatically created model with tuning based on the Random Forest type with the same seven predictors can achieve an RMSE of 1811 and an RSQ of 0.94.

Parameter Tuning (Part 2)

```
# Automatisierte Codegeneration für Parametertuning ------------------------------
library(usemodels)
use_ranger(Preis~Typ+Kilometer+Alter+Kategorie+PS+Marke+Modell,
           data=autos.train,
           verbose = TRUE)

## ranger_recipe <-
##   recipe(formula = Preis ~ Typ + Kilometer + Alter + Kategorie + PS + Marke +
##     Modell, data = autos.train)
##
## ranger_spec <-
##   rand_forest(mtry = tune(), min_n = tune(), trees = 1000) %>%
##   set_mode("classification") %>%
##   set_engine("ranger")
##
## ranger_workflow <-
##   workflow() %>%
##   add_recipe(ranger_recipe) %>%
##   add_model(ranger_spec)
##
## set.seed(65848)
## ranger_tune <-
##   tune_grid(ranger_workflow, resamples = stop("add your rsample object"), grid =
stop("add number of candidate points"))

## ANFANG von Funktion use_ranger() erzeugter Code ------------------------------
ranger_recipe <-
  recipe(formula = Preis ~ Typ + Kilometer + Alter + Kategorie + PS + Marke +
    Modell, data = autos.train)
ranger_spec <-
  rand_forest(mtry = tune(), min_n = tune(), trees = 1000) %>%
# Manuell Modifiziert set_mode zu "regression"
  set_mode("regression") %>%
  set_engine("ranger")
ranger_workflow <-
  workflow() %>%
  add_recipe(ranger_recipe) %>%
  add_model(ranger_spec)
# Manuell eingefügt ab hier zur Generierung von Grid
wfl.rg.lm.par <-
  ranger_workflow %>%
  extract_parameter_set_dials() %>%
  finalize(autos.train)
wfl.rg.lm.par

## Collection of 2 parameters for tuning
##
##  identifier  type    object
##        mtry  mtry nparam[+]
##       min_n min_n nparam[+]

grid.lh<-grid_latin_hypercube(wfl.rg.lm.par,size=10)# Grid mit 10 Kombis
set.seed(15546)
ranger_tune <-
# Manuell Modifiziert resamples und grid
  tune_grid(ranger_workflow, resamples = autos.boots, grid = grid.lh)
## ENDE von Funktion use_ranger() erzeugter Code --------------------------------
# Evaluation ---------------------------------------------------------------------
autoplot(ranger_tune)
```

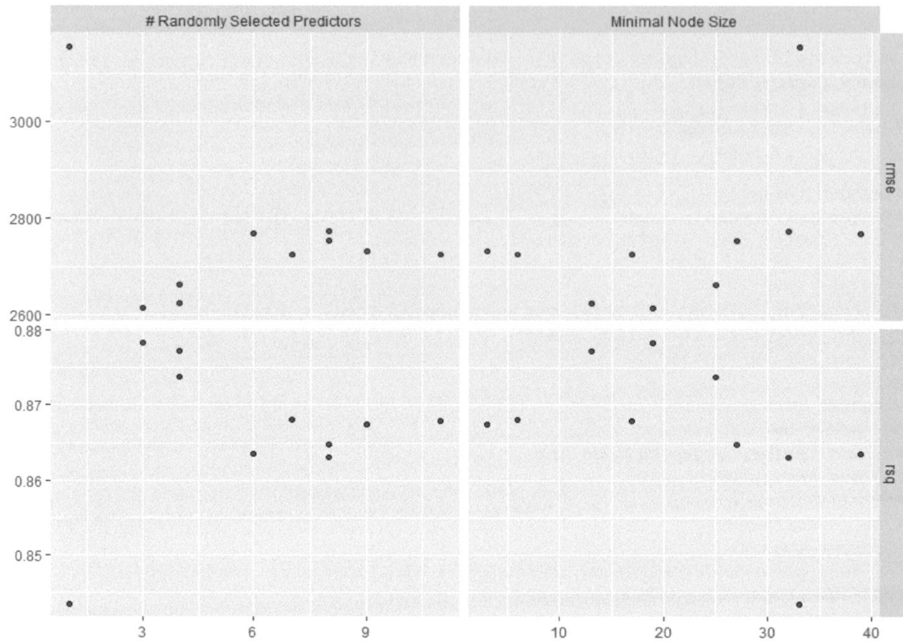


```
show_best(ranger_tune,metric="rmse") %>% select(-.estimator)

## # A tibble: 5 × 7
##    mtry min_n .metric  mean     n std_err .config
##   <int> <int> <chr>   <dbl> <int>   <dbl> <chr>
## 1     3    19 rmse    2616.    10    35.6 Preprocessor1_Model07
## 2     4    13 rmse    2625.    10    36.9 Preprocessor1_Model10
## 3     4    25 rmse    2663.    10    35.8 Preprocessor1_Model01
## 4    11    17 rmse    2726.    10    34.6 Preprocessor1_Model03
## 5     7     6 rmse    2726.    10    35.7 Preprocessor1_Model06

show_best(ranger_tune,metric="rsq") %>% select(-.estimator)

## # A tibble: 5 × 7
##    mtry min_n .metric  mean     n std_err .config
##   <int> <int> <chr>   <dbl> <int>   <dbl> <chr>
## 1     3    19 rsq     0.878    10 0.00257 Preprocessor1_Model07
## 2     4    13 rsq     0.877    10 0.00274 Preprocessor1_Model10
## 3     4    25 rsq     0.874    10 0.00268 Preprocessor1_Model01
## 4     7     6 rsq     0.868    10 0.00274 Preprocessor1_Model06
## 5    11    17 rsq     0.868    10 0.00267 Preprocessor1_Model03

# Finale Evaluation des Modells ---------------------------------------------------
final.best.par<-select_best(ranger_tune,metric="rmse")
final.best.par<-final.best.par[,-ncol(final.best.par)]
ranger_workflow.final.best<- ranger_workflow %>%
  finalize_workflow(final.best.par)
ranger_workflow.final.best.fit<-ranger_workflow.final.best %>%
  fit(autos)
ml.eval(ranger_workflow.final.best.fit,autos.train,9)
```

```
##                               Modell        Daten Kennzahl          Wert
## 1 ranger_workflow.final.best.fit autos.train     rmse 1913.5241076
## 2 ranger_workflow.final.best.fit autos.train      rsq    0.9361633
```

```
ml.eval(ranger_workflow.final.best.fit,autos.test,9)
```

```
##                               Modell       Daten Kennzahl          Wert
## 1 ranger_workflow.final.best.fit autos.test     rmse 1811.9385086
## 2 ranger_workflow.final.best.fit autos.test      rsq    0.9410589
```

In the following coding, parameter tuning will also be presented using an example of classification. The model is of the Support Vector Machines type. For this, data on the survival statistics of the Titanic from the titanic package is used. The variables used are the class (1–3), the gender, the ticket price, and the information whether the passengers had siblings or partners on board and whether they had parents or children on board. In addition, the port of embarkation is available as a variable. In the recipe rec.1, all variables are used. The function `step_YeoJohnson()` in the recipe is used in the following coding for preprocessing when a more symmetrical distribution of all numerical variables is desired.

In classification, the evaluation of a model can be carried out using the ROC (Area under the Curve) metric. The best ROC result of 0.829 is achieved with the parameter cost of 2 and rbf_sigma of 0.00001. With the help of the function tune_bayes(), further parameter simulations can be carried out, which ultimately lead to an improved ROC of 0.84 when cost 0.097 and rbf_sigma are set at 0.046.

Parameter Tuning (Part 3)

```
# Datenbasis ------------------------------------------------------------------
require(titanic)                                        # Paket erforderlich

## Loading required package: titanic

titanic<-select(titanic_train,                          # Variablenanzahl reduzieren
                -c("PassengerId","Name","Cabin","Ticket","Age"))
titanic <- titanic %>%
  mutate(Survived=factor(Survived,
                         levels=c(0,1),
                         labels=c("Gestorben","Überlebt")),
         Pclass  =factor(Pclass,
                         levels=c(1,2,3),
                         ordered=TRUE),
         Embarked=factor(Embarked,
                         levels=c("C","Q","S"),
                         labels=c("Cherbourg","Queenstown","Southampton")),
         Sex     =factor(Sex,
                         levels=c("female","male"),
                         labels=c("Frau","Mann")))
colnames(titanic)<-c("Gestorben","Klasse","Geschlecht","Geschwister-Partner",
                     "Eltern-Kinder","Preis","Einschiffung")
head(titanic)                                           # Anzeige

##   Gestorben Klasse Geschlecht Geschwister-Partner Eltern-Kinder   Preis
## 1 Gestorben      3       Mann                   1             0  7.2500
## 2  Überlebt      1       Frau                   1             0 71.2833
## 3  Überlebt      3       Frau                   0             0  7.9250
## 4  Überlebt      1       Frau                   1             0 53.1000
## 5 Gestorben      3       Mann                   0             0  8.0500
## 6 Gestorben      3       Mann                   0             0  8.4583
##   Einschiffung
## 1  Southampton
## 2    Cherbourg
## 3  Southampton
## 4  Southampton
## 5  Southampton
## 6   Queenstown

titanic %>%
  ggplot(aes(x=Geschlecht,fill=Gestorben)) +
  geom_bar(width = 0.4) +
  facet_wrap(~ Klasse) +
  theme_test() +
  labs(title="Titanic",
       subtitle="Überlebende je Passagierklasse und Geschlecht",
       x="Klasse",y="Anzahl")
```

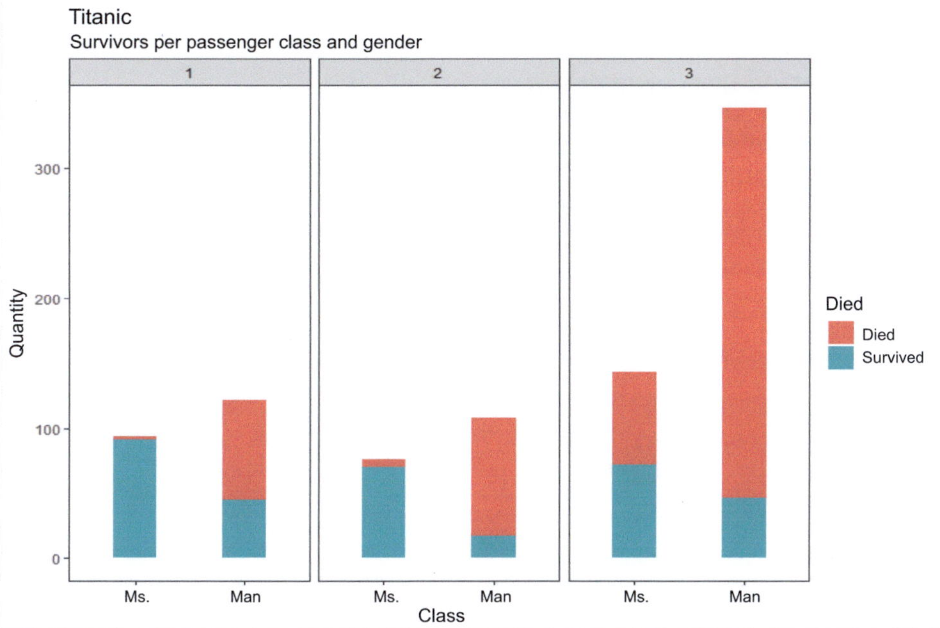


```
# Trainings- und Testdaten erzeugen---------------------------------------------------
set.seed(123)
tit.split<-initial_split(titanic,
                         prop=0.80,                        # Split Train/Test 80:20
                         strata=Gestorben)
tit.train<-training(tit.split)
tit.test<-testing(tit.split)
# Bootstrap-Samples -------------------------------------------------------------
set.seed(123)
tit.boots<-bootstraps(tit.train,times=10)              # 10 Bootstraps erzeugen
# Einzelner Fold mit Trennung in Analyse- und Validierung -----------------------
set.seed(123)
tit.fold<-validation_split(tit.train, prop=4/5)        # Split in 80% Anal.+20% Val.
# Modelltyp und Engine für Support Vector Machines/Klassifikation definieren -----
mod.kl.svm<-svm_rbf(cost = tune(), rbf_sigma = tune()) %>%
  set_engine("kernlab") %>%
  set_mode("classification")
# Recipe definieren -------------------------------------------------------------
rec.1<-recipe(data=titanic,Gestorben~.) %>%
  step_dummy(all_nominal_predictors()) %>%             # Kategorial->Numerisch
  step_normalize(all_numeric_predictors())
rec.2<-recipe(data=titanic,Gestorben~Geschlecht+Preis+Klasse+Einschiffung) %>%
  step_dummy(all_nominal_predictors()) %>%             # Kategorial->Numerisch
  step_YeoJohnson(all_numeric_predictors()) %>%
  step_normalize(all_numeric_predictors())
# Workflow definieren -----------------------------------------------------------
wfl.kl.svm<- workflow() %>%
  add_model(mod.kl.svm) %>%
  add_recipe(rec.1)
# Tuningparameter von Workflow --------------------------------------------------
wfl.kl.svm.par<-extract_parameter_set_dials(wfl.kl.svm) # Tuningparameter von wfl
wfl.kl.svm.par                                            # 2 Tuningpara
```

```
## Collection of 2 parameters for tuning
##
##  identifier      type    object
##        cost      cost nparam[+]
##   rbf_sigma rbf_sigma nparam[+]

wfl.kl.svm.par %>% extract_parameter_dials("cost")         # Details zu cost

## Cost (quantitative)
## Transformer: log-2 [1e-100, Inf]
## Range (transformed scale): [-10, 5]

wfl.kl.svm.par %>% extract_parameter_dials("rbf_sigma")    # Details zu rbf_sigma

## Radial Basis Function sigma (quantitative)
## Transformer: log-10 [1e-100, Inf]
## Range (transformed scale): [-10, 0]

#
set.seed(123)
grid.reg<-wfl.kl.svm.par %>%
  update(cost=cost(c(-10,1))) %>%
  grid_regular(levels=3)
grid.reg

## # A tibble: 9 × 2
##        cost    rbf_sigma
##       <dbl>        <dbl>
## 1 0.000977 0.0000000001
## 2 0.0442   0.0000000001
## 3 2        0.0000000001
## 4 0.000977 0.00001
## 5 0.0442   0.00001
## 6 2        0.00001
## 7 0.000977 1
## 8 0.0442   1
## 9 2        1

set.seed(123)
tune<- wfl.kl.svm %>%
  tune_grid(resamples=tit.boots,
            grid=grid.reg,
            metrics=metric_set(roc_auc))
autoplot(tune)
```

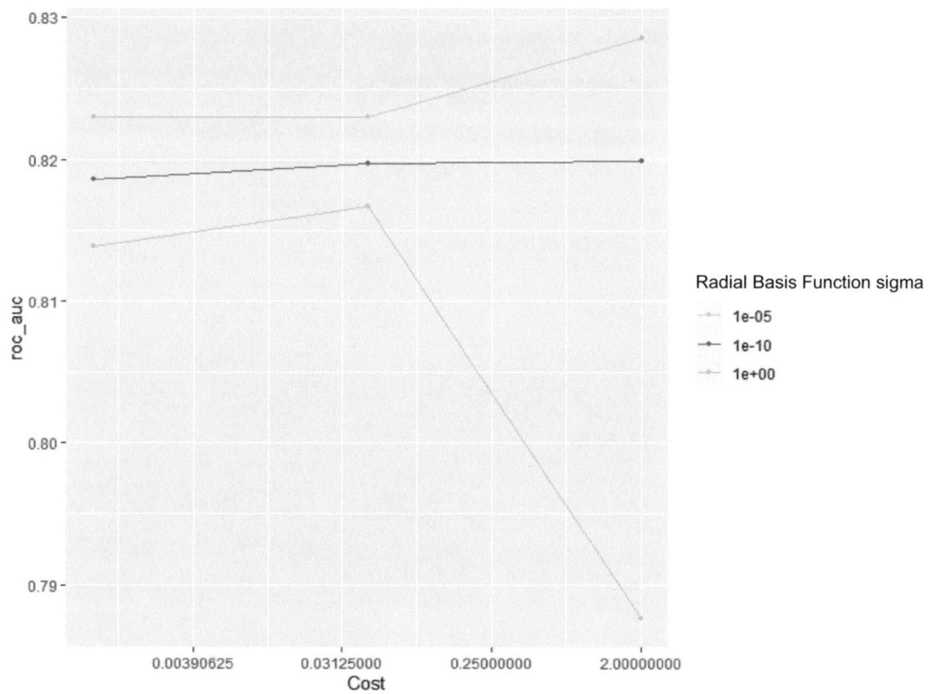


```
show_best(tune) %>% select(-.estimator)

## # A tibble: 5 × 7
##       cost    rbf_sigma .metric  mean     n std_err .config
##      <dbl>        <dbl> <chr>   <dbl> <int>   <dbl> <chr>
## 1 2        0.00001      roc_auc 0.829    10 0.00754 Preprocessor1_Model6
## 2 0.0442   0.00001      roc_auc 0.823    10 0.00550 Preprocessor1_Model5
## 3 0.000977 0.00001      roc_auc 0.823    10 0.00550 Preprocessor1_Model4
## 4 2        0.0000000001 roc_auc 0.820    10 0.00655 Preprocessor1_Model3
## 5 0.0442   0.0000000001 roc_auc 0.820    10 0.00654 Preprocessor1_Model2

collect_metrics(tune)

## # A tibble: 9 × 8
##       cost    rbf_sigma .metric .estimator  mean     n std_err .config
##      <dbl>        <dbl> <chr>   <chr>      <dbl> <int>   <dbl> <chr>
## 1 0.000977 0.0000000001 roc_auc binary     0.819    10 0.00660 Preprocessor1_Mo…
## 2 0.0442   0.0000000001 roc_auc binary     0.820    10 0.00654 Preprocessor1_Mo…
## 3 2        0.0000000001 roc_auc binary     0.820    10 0.00655 Preprocessor1_Mo…
## 4 0.000977 0.00001      roc_auc binary     0.823    10 0.00550 Preprocessor1_Mo…
## 5 0.0442   0.00001      roc_auc binary     0.823    10 0.00550 Preprocessor1_Mo…
## 6 2        0.00001      roc_auc binary     0.829    10 0.00754 Preprocessor1_Mo…
## 7 0.000977 1            roc_auc binary     0.814    10 0.00893 Preprocessor1_Mo…
## 8 0.0442   1            roc_auc binary     0.817    10 0.00778 Preprocessor1_Mo…
## 9 2        1            roc_auc binary     0.788    10 0.00844 Preprocessor1_Mo…

set.seed(123)
tune.bayesoptimized<-wfl.kl.svm %>%
  tune_bayes(resamples=tit.boots,
           iter=40,
```

```
            metrics=metric_set(roc_auc),
            param_info=wfl.kl.svm.par,
            initial= tune,
            control=control_bayes(verbose = TRUE))
##
## — Iteration 1 ─────────────────────────────────────────────────
##
## i Current best:      roc_auc=0.8286 (@iter 0)
## i Gaussian process model
## √ Gaussian process model
## i Generating 5000 candidates
## i Predicted candidates
## i cost=0.934, rbf_sigma=9.7e-06
## i Estimating performance
## √ Estimating performance
## ⓧ Newest results:    roc_auc=0.8284 (+/-0.00719)
## … …
## — Iteration 40 ────────────────────────────────────────────────
##
## i Current best:      roc_auc=0.8396 (@iter 39)
## i Gaussian process model
## √ Gaussian process model
## i Generating 5000 candidates
## i Predicted candidates
## i cost=11, rbf_sigma=1.98e-07
## i Estimating performance
## √ Estimating performance
## ⓧ Newest results:    roc_auc=0.826 (+/-0.00707)
show_best(tune.bayesoptimized) %>% select(-.estimator)
## # A tibble: 5 × 8
##     cost rbf_sigma .metric  mean     n std_err .config .iter
##    <dbl>     <dbl> <chr>   <dbl> <int>   <dbl> <chr>   <int>
## 1 0.0970    0.0460 roc_auc 0.840    10 0.00774 Iter39     39
## 2 0.0864    0.0442 roc_auc 0.839    10 0.00773 Iter30     30
## 3 0.0971    0.0374 roc_auc 0.839    10 0.00805 Iter28     28
## 4 0.0908    0.0361 roc_auc 0.838    10 0.00799 Iter33     33
## 5 0.105     0.0385 roc_auc 0.838    10 0.00816 Iter25     25
collect_metrics(tune.bayesoptimized)
```

```
## # A tibble: 49 × 9
##        cost    rbf_sigma .metric .estimator  mean     n std_err .config    .iter
##       <dbl>        <dbl> <chr>   <chr>      <dbl> <int>   <dbl> <chr>      <int>
##  1 0.000977 0.0000000001 roc_auc binary     0.819    10 0.00660 Preproces…     0
##  2 0.0442   0.0000000001 roc_auc binary     0.820    10 0.00654 Preproces…     0
##  3 2        0.0000000001 roc_auc binary     0.820    10 0.00655 Preproces…     0
##  4 0.000977 0.00001      roc_auc binary     0.823    10 0.00550 Preproces…     0
##  5 0.0442   0.00001      roc_auc binary     0.823    10 0.00550 Preproces…     0
##  6 2        0.00001      roc_auc binary     0.829    10 0.00754 Preproces…     0
##  7 0.000977 1            roc_auc binary     0.814    10 0.00893 Preproces…     0
##  8 0.0442   1            roc_auc binary     0.817    10 0.00778 Preproces…     0
##  9 2        1            roc_auc binary     0.788    10 0.00844 Preproces…     0
## 10 0.934    0.00000970   roc_auc binary     0.828    10 0.00719 Iter1          1
## # … with 39 more rows
```

```
autoplot(tune.bayesoptimized,type="performance")
```

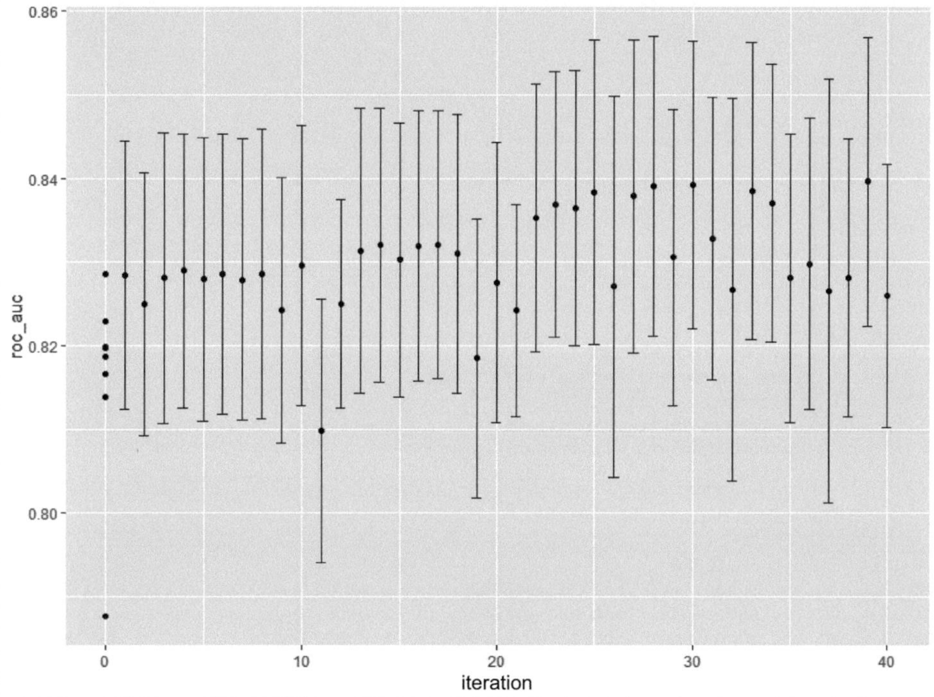


```
autoplot(tune.bayesoptimized,type="parameters")
```

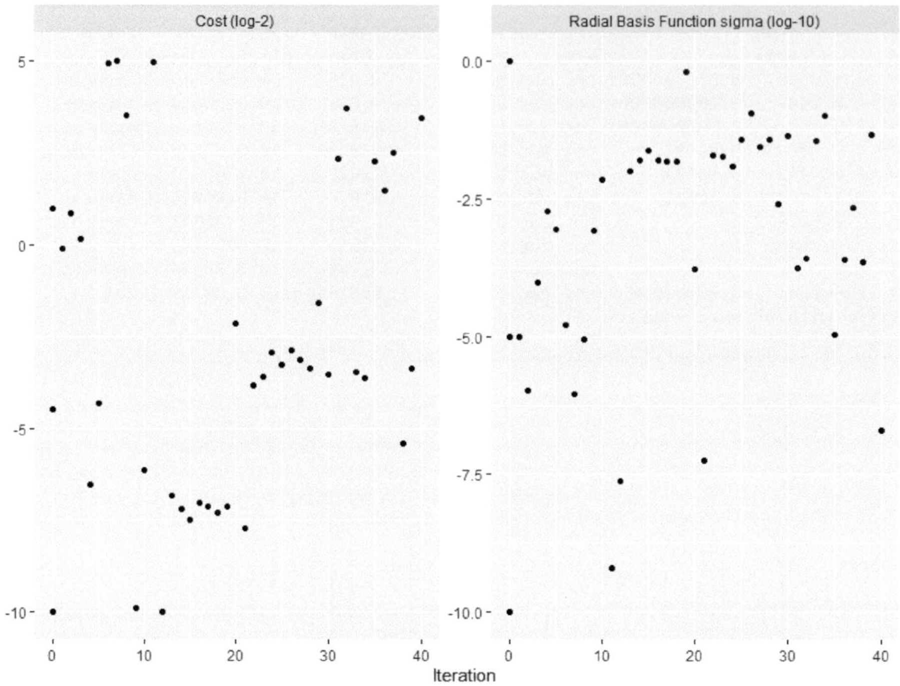


```
set.seed(123)
tune.simanneal<-wfl.kl.svm %>%
  finetune::tune_sim_anneal(resamples=tit.boots,
           iter=40,
           metrics=metric_set(roc_auc),
           param_info=wfl.kl.svm.par,
           initial= tune,
           control=finetune::control_sim_anneal(verbose=TRUE,no_improve=10L))
```

```
## Optimizing roc_auc
```

```
## Initial best: 0.82860
##  1 ◯ accept suboptimal  roc_auc=0.82839  (+/-0.007417)
##  2 ◯ accept suboptimal  roc_auc=0.82433  (+/-0.006817)
##  3 ◯ accept suboptimal  roc_auc=0.82407  (+/-0.006003)
##  4 + better suboptimal  roc_auc=0.82461  (+/-0.006732)
##  5 + better suboptimal  roc_auc=0.82841  (+/-0.007389)
##  6 ◯ accept suboptimal  roc_auc=0.82313  (+/-0.005645)
##  7 ◯ accept suboptimal  roc_auc=0.82295  (+/-0.005513)
##  8 ✕ restart from best  roc_auc=0.82458  (+/-0.006686)
##  9 ◯ accept suboptimal  roc_auc=0.82421  (+/-0.007044)
## 10 + better suboptimal  roc_auc=0.8286   (+/-0.007233)
## 11 ◯ accept suboptimal  roc_auc=0.82599  (+/-0.007173)
## 12 ♥ new best           roc_auc=0.82881  (+/-0.007369)
## 13 ◯ accept suboptimal  roc_auc=0.82423  (+/-0.005898)
## 14 + better suboptimal  roc_auc=0.82547  (+/-0.005811)
## 15 ◯ accept suboptimal  roc_auc=0.82476  (+/-0.006067)
## 16 ◯ accept suboptimal  roc_auc=0.82376  (+/-0.005932)
## 17 ◯ accept suboptimal  roc_auc=0.82359  (+/-0.006004)
```

```
## 18 + better suboptimal  roc_auc=0.82447  (+/-0.006007)
## 19 ◯ accept suboptimal  roc_auc=0.82371  (+/-0.006071)
## 20 ✕ restart from best  roc_auc=0.82337  (+/-0.005981)
## 21 ◯ accept suboptimal  roc_auc=0.82772  (+/-0.007675)
## 22 + better suboptimal  roc_auc=0.82826  (+/-0.007677)
## 23 ♥ new best           roc_auc=0.82902  (+/-0.00737)
## 24 ◯ accept suboptimal  roc_auc=0.82597  (+/-0.007123)
## 25 + better suboptimal  roc_auc=0.82837  (+/-0.007251)
## 26 ◯ accept suboptimal  roc_auc=0.82455  (+/-0.006888)
## 27 + better suboptimal  roc_auc=0.82553  (+/-0.006643)
## 28 – discard suboptimal roc_auc=0.80306  (+/-0.006816)
## 29 – discard suboptimal roc_auc=0.82123  (+/-0.007202)
## 30 – discard suboptimal roc_auc=0.81603  (+/-0.00674)
## 31 ✕ restart from best  roc_auc=0.82406  (+/-0.006009)
## 32 ◯ accept suboptimal  roc_auc=0.82896  (+/-0.007489)
## 33 ◯ accept suboptimal  roc_auc=0.82513  (+/-0.006844)
## 34 – discard suboptimal roc_auc=0.82389  (+/-0.006238)
## 35 ◯ accept suboptimal  roc_auc=0.82475  (+/-0.00697)
## 36 ◯ accept suboptimal  roc_auc=0.81478  (+/-0.008174)
## 37 – discard suboptimal roc_auc=0.80434  (+/-0.006868)
## Stopping; no best in 10 iterations.
## ✕ Optimization stopped prematurely; returning current results.
```

```
show_best(tune.bayesoptimized) %>% select(-.estimator)
```

```
## # A tibble: 5 × 8
##     cost rbf_sigma .metric  mean     n std_err .config .iter
##    <dbl>     <dbl> <chr>   <dbl> <int>   <dbl> <chr>   <int>
## 1 0.0970    0.0460 roc_auc 0.840    10 0.00774 Iter39     39
## 2 0.0864    0.0442 roc_auc 0.839    10 0.00773 Iter30     30
## 3 0.0971    0.0374 roc_auc 0.839    10 0.00805 Iter28     28
## 4 0.0908    0.0361 roc_auc 0.838    10 0.00799 Iter33     33
## 5 0.105     0.0385 roc_auc 0.838    10 0.00816 Iter25     25
```

```
collect_metrics(tune.bayesoptimized)
```

```
## # A tibble: 49 × 9
##        cost    rbf_sigma .metric .estimator  mean     n std_err .config    .iter
##       <dbl>        <dbl> <chr>   <chr>      <dbl> <int>   <dbl> <chr>      <int>
##  1 0.000977 0.0000000001 roc_auc binary     0.819    10 0.00660 Preproces…     0
##  2 0.0442   0.0000000001 roc_auc binary     0.820    10 0.00654 Preproces…     0
##  3 2        0.0000000001 roc_auc binary     0.820    10 0.00655 Preproces…     0
##  4 0.000977 0.00001      roc_auc binary     0.823    10 0.00550 Preproces…     0
##  5 0.0442   0.00001      roc_auc binary     0.823    10 0.00550 Preproces…     0
##  6 2        0.00001      roc_auc binary     0.829    10 0.00754 Preproces…     0
##  7 0.000977 1            roc_auc binary     0.814    10 0.00893 Preproces…     0
##  8 0.0442   1            roc_auc binary     0.817    10 0.00778 Preproces…     0
##  9 2        1            roc_auc binary     0.788    10 0.00844 Preproces…     0
## 10 0.934    0.00000970   roc_auc binary     0.828    10 0.00719 Iter1          1
## # … with 39 more rows
```

```
autoplot(tune.simanneal,type="performance")
```

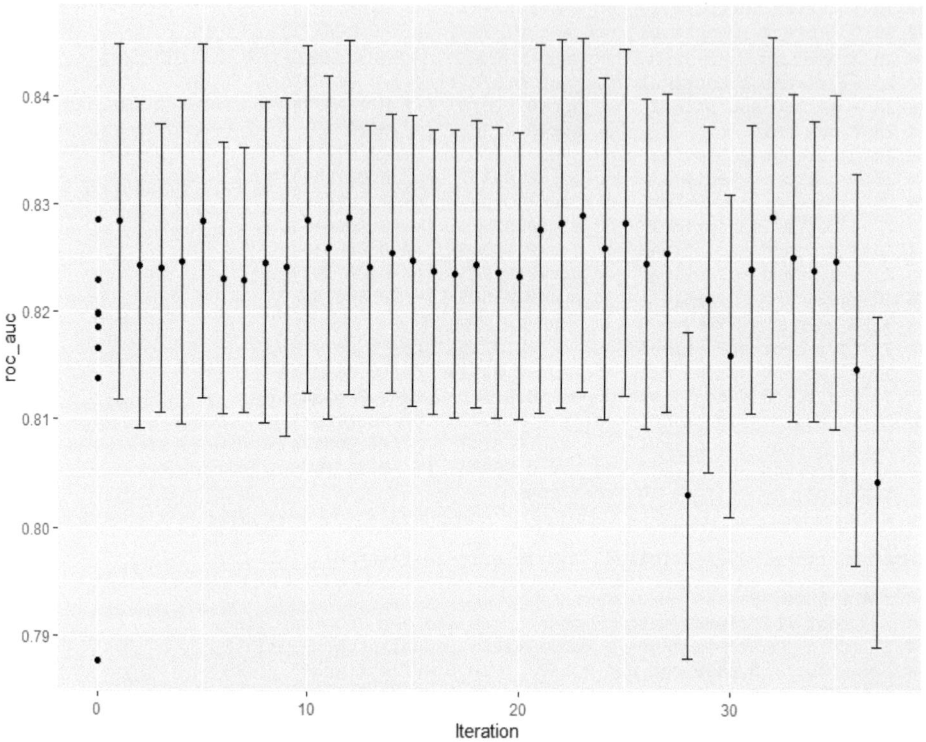


```
autoplot(tune.simanneal,type="parameters")
```

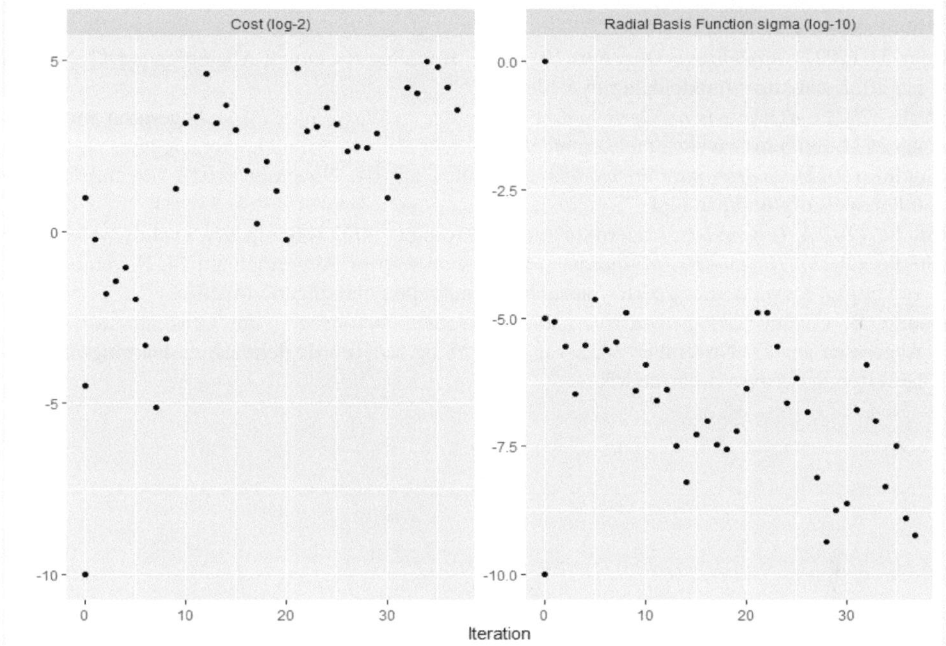

6.5.6 Conclusion

The tidymodels package allows a selection of many model types, a generation of suitable data for training, validation, and testing, the creation of rules for preprocessing to prepare the data for machine learning. In addition, workflows and workflow sets support the selection of the most relevant predictors and the selection of the best models. The metrics for evaluation can be compared and displayed for each model, and the optimization of parameters, whether generic or engine-specific, is possible with the help of the package's functions. Even though getting used to the methods of machine learning may take a moment, the time has come to consider machine learning in application.

References

Cleff, T. (2019). *Angewandte Induktive Statistik und Statistische Testverfahren: Eine computergestützte Einführung mit Excel, SPSS und Stata.* Wiesbaden: Springer Gabler.

Hertweck, D., & Kinitzki, M. (2015). Datenorientierung statt Bauchentscheidung: Führungs- und Organisationskultur in der datenorientierten Unternehmung. In J. Dorschel, *Praxishandbuch Big Data: Wirtschaft, Recht, Technik* (pp. 19–20). Wiesbaden: Springer Gabler.

Kuhn, M. (27. März 2019). *The caret Package*. Von https://topepo.github.io/caret/available-models.html abgerufen

Kuhn, M. (2022). *modeldata: Data Sets Useful for Modeling Examples*. Abgerufen am 17. November 2022 von https://modeldata.tidymodels.org/

Oracle. (2021). *ROracle is an Open Source R Package available on CRAN*. Abgerufen am 8. Januar 2021 von https://www.oracle.com/de/database/technologies/appdev/roracle.html

RStudio. (2022). *Search parsnip models*. Abgerufen am 04. November 2022 von https://www.tidymodels.org/find/parsnip/

RStudio. (2022). *Tidymodels*. Abgerufen am 04. November 2022 von https://www.tidymodels.org/

RStudio. (2022). *Tidymodels: Multiclass classification metrics*. Abgerufen am 14. November 2022 von https://www.tmwr.org/performance.html#multiclass-classification-metrics

Schulte, B. (1. Juli 2019). *Machine-Learning-Modelle bewerten – die Crux mit der Metrik*. Abgerufen am 17. November 2022 von https://blog.codecentric.de/machine-learning-modelle-bewerten-die-crux-mit-der-metrik

Application of Machine Learning with R

7

Machine learning can be divided into the three types shown in Fig. 7.1.

The following presents implementations of supervised and unsupervised learning with R. Covering reinforcement learning would have exceeded the scope of this book.

7.1 Supervised Learning

In supervised learning, the two most essential applications are regression and classification. Therefore, these application areas are presented below.

7.1.1 Regression

Regression analysis is a statistical analysis method that aims to model relationships between a dependent variable (often also called the explained variable or regressand) and one or more independent variables (often also called explanatory variables or regressors). Performing a regression is used to quantitatively describe relationships or predict values of the dependent variable. The most common form of regression analysis is linear regression, in which the user finds a linear function that best fits the data. For example, the ordinary least squares method calculates a unique straight line that minimizes the sum of the squared deviations between the true data and this line, i.e., the sum of squared residuals. The user can then estimate the expected value of the dependent variable when all independent variables are known. Regression analysis is therefore about determining a function that best describes the data and can be used for prediction.

B. Heesen, *Artificial Intelligence and Machine Learning with R*,
https://doi.org/10.1007/978-3-658-45392-3_7

Fig. 7.1 Machine learning methods

Predictions based on regression functions can be applied in many contexts, for example, to predict how a price increase affects sales or how the dose of a medication affects the success of healing.

The **Linear Regression** describes a linear function $y = f(x) = b_0 + b_1 * x$with an independent variable X as a so-called predictor or regressor and the dependent variable Y as a response or regressand. If you have more than one influencing factor, you use **nonlinear regression** or **multiple regression,** which is also capable of mapping curves and, for example, with three independent variables can be described by a function like $y = f(x_1, x_2, x_3) = b_0 + b_1 * x_1 + b_2 * x_2 + b_3 * x_3$ with several independent variables.

To determine the parameters for the regression, you can use the function `lm()` from the stats package in R. This function can be defined as a model named mod.rg.lm with the following command: `mod.rg.lm<-linear_reg() %>% set_engine("lm")`. The fact that this model uses the function from the stats package can be displayed with the command `translate(mod.rg.lm)`.

The function determines the regression line so that the squared deviation of the points described by the function from the real values is minimal, i.e., the function describes the data as accurately as possible. The function `coef()` from the same package shows the calculated coefficients b_0 for the intersection with the Y-axis (intercept) and b_1 the slope of the line or the coefficient for the independent variable, e.g., `coef(lm(formula=Price~Age + PS,data=data))`. In the moderndive package, there is the function `get_regression_table()`, which in addition to the intercept and the coefficients also gives the standard error, the p-value, and the confidence interval in the variables lower_ci and upper_ci. The standard error depends on the sample size, as already described in Sect. 6.4.2, and decreases with increasing size of the sample. To display the regression line, the function `geom_smooth()` can be used, e.g., `geom_smooth(method="lm",se=FALSE,na.rm=TRUE,formula=y~x)`. The accuracy of the regression function can also be displayed. The function `get_regression_points()` from the moderndive package allows the deviations of the values estimated by the function from the real values to be displayed.

In the following coding, 1000 data records on motor vehicles are first divided into training and test data. This limitation to 1000 data records reduces the prediction accuracy and a deliberate reduction of the sample should not occur in reality. However, the following coding is only for the demonstration of a regression. The regression can be applied with both numerical and categorical independent variables. The following coding shows both a **Linear Regression**(1 independent variable) and a **Multiple Regression** (more than one independent variable). In the case of Multiple Linear Regression, once a numerical and a categorical variable are used as independent variables and another time two numerical variables are used.

Before the modeling begins, it is often useful to perform an exploratory data analysis. The function `ml.summary()` is suitable for viewing the variables. The price of the cars has a right-skewed distribution and several extreme outliers. To get a first idea of the dependencies of the independent variables with the dependent variable, the function `ml.summary.2()` is suitable, which displays a scatter plot and the correlation coefficient. Between age and price, there is a correlation of −0.64 and between horsepower and price a correlation of 0.66, which shows that both independent variables have a great influence on the price. The function `ml.summary.3()` is suitable for displaying the ratio of three variables. The categorical variable horsepower class has a correlation with the price between 0.34 and 0.52 for the categories low horsepower, normal horsepower, and high horsepower.

After the machine learning model mod.rg.lm has already been created, the variables can be specified using the function `recipe()` and the model with the recipe can then be connected to a workflow using the function `workflow()`. The training of the respective workflow with the model contained therein is done via the function `fit()` and predictions are made using the function `predict()`.

The coefficients of the regression formula result in the Linear Regression with 19.3 for b_0 and −1.08 for b_1 (for the variable age). This means that the price in US$ can be predicted in this model according to the formula: *Price* = *f*(*Age*) = 19,300 − 1.08 * *Age*.

In the Multiple Regression with the independent variables age and horsepower class, the coefficients are 17.9 for b_0 and −0.9 for b_1(for the variable age). In addition, there is a price discount of 2.84 for cars with low horsepower and a price surcharge of 9.2 for cars with high horsepower. This means that the price in US$ can be predicted in this model according to the formula *Price*= *f*(*Age*, *HorsepowerClass*) = 17,900 − 0.9 * *Age*− 2840 for cars with low horsepower, *Price* = *f*(*Age*, *HorsepowerClass*) = 17,900 − 0.9 * *Age* for cars with normal horsepower and *Price* = *f*(*Age*, *HorsepowerClass*) = 17,900 − 0.9 * *Age* + 9200 for cars with high horsepower.

The Multiple Regression with the independent variables age and horsepower results in the coefficients 7.24 for b_0, −0.818 for b_1(for the variable age) and 0.07 for b_2 (for the variable horsepower). This means that the price in US$ can be predicted in this model according to the formula *Price* = *f*(*Age*, *Horsepower*) = 7240 − 0.818 * *Age* + 0.07 * *Horsepower*.

The evaluation of the three models using the function `ml.eval.lr()` shows that the metrics RMSE (smaller is better) and RSQ (larger is better) improve in the Multiple Regression compared to the Linear Regression and are better in the Multiple Regression, which uses age and horsepower, than in the Multiple Regression, which uses age and horsepower class. This result is understandable, as the variable horsepower class is less accurate than the variable horsepower.

Regression: Predicting the Price of Motor Vehicles

```
# Vorbereitung--------------------------------------------------------------------
autos <- autos[1:1000,]                                # 1000 Datensätze
autos$Preis <- autos$Preis/1000                        # Preis in 1.000 US$
autos$PSKlasse<-                                       # Viel, normal und wenig PS
  case_when(autos$PS < 125 ~ "wenig PS",
    between(autos$PS,125,200) ~ "normal PS",
    between(autos$PS,200,500) ~ "viel PS")
# Trainings- und Testdaten erzeugen-----------------------------------------------
set.seed(123)
autos.split<-initial_split(autos,prop=0.80,            # Split Trainings/Testdaten 80:20
                           strata=Preis)
autos.train<-training(autos.split)
autos.test<-testing(autos.split)
rm(autos.split)                                        # Split löschen, nicht mehr benötigt
dim(autos)                                             # 100% Alle Datensätze

## [1] 1000   24

dim(autos.train)                                       # 80%  Trainingsdaten

## [1] 799  24

dim(autos.test)                                        # 20%  Testdaten

## [1] 201  24

# Visualisierung der Trainingsdaten-----------------------------------------------
ml.summary(autos.train$Preis,                          # Daten verstehen
           "Gebrauchtwagenpreis","Preis in Tausend €",box=T)
```

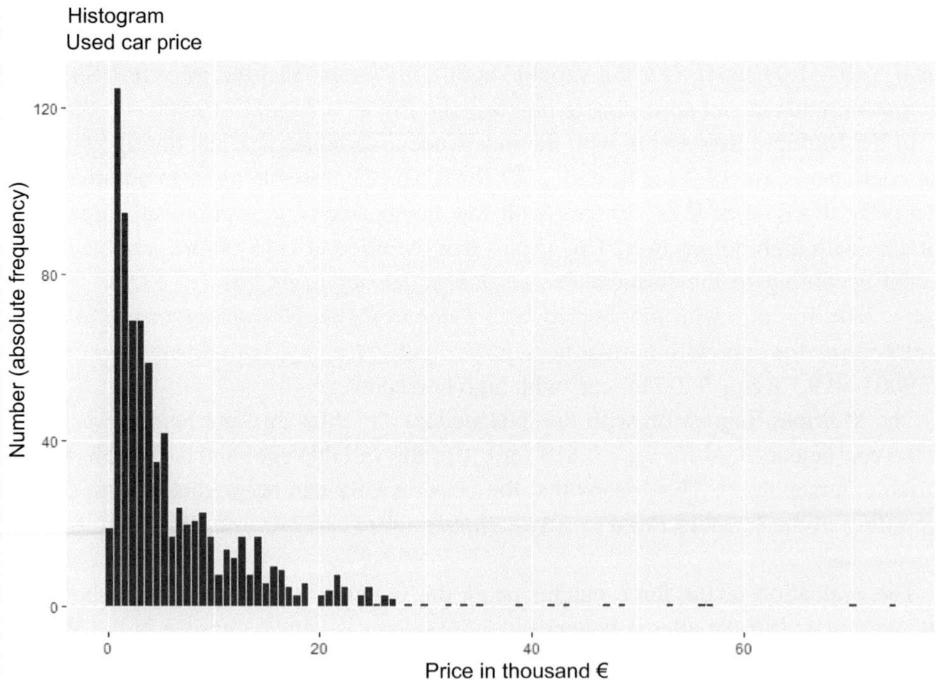

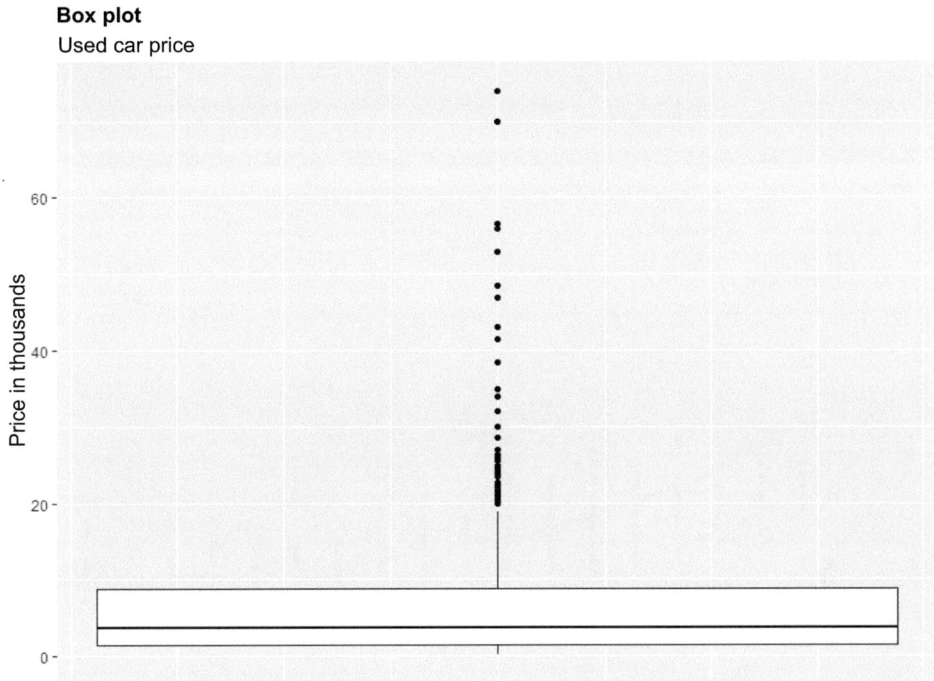


```
##                25%
## anzahl         "799"
## anzahl.na      "0"
## modus          "1.2"
## median         "3.75"
## mean           "6.53"
## min            "0.149"
## max            "73.996"
## sd             "8.12"
## q1             "1.5"
## q3             "8.8"
## iqr            "7.3"
## whisker.min    "-9.45"
## whisker.max    "19.75"
## skewness       "3.33"
## skewness.txt "Rechte Schiefe: 3.33 > 0, positive Schiefe,linkssteil,rechtsschief"
## kurtosis       "17.03"
## kurtosis.txt "Steilgipflig mit Exzess Kurtosis 17.03 > 0"
```

```
ergebnis<-ml.summary.2(autos.train[,c("Alter","Preis")],
  paste("Korrelation:",round(cor(autos.train$Alter,autos.train$Preis),2)),
  "Alter","Preis in Tausend US$",scatter=T)
```

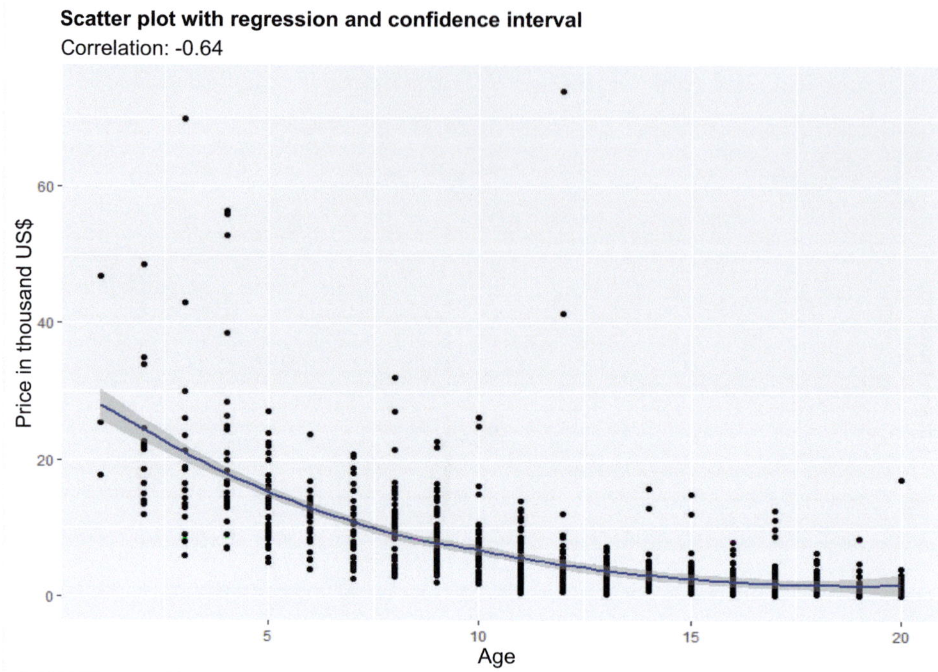


```
ergebnis<-ml.summary.2(autos.train[,c("PS","Preis")],
  paste("Korrelation:",round(cor(autos.train$PS,autos.train$Preis),2)),
  "PS","Preis in Tausend US$",scatter=T)
```

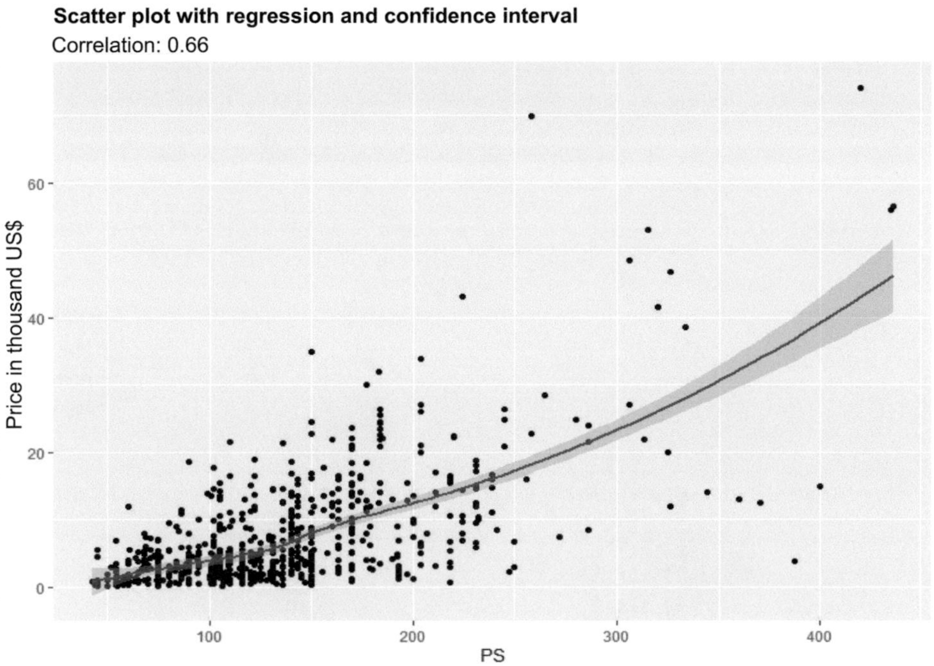


```
autos.wenig <-autos.train[autos.train$PSKlasse=="wenig PS",]
autos.normal<-autos.train[autos.train$PSKlasse=="normal PS",]
autos.viel  <-autos.train[autos.train$PSKlasse=="viel PS",]
korr.wenig  <-round(cor(autos.wenig$PS,autos.wenig$Preis),2)
korr.normal <-round(cor(autos.normal$PS,autos.normal$Preis),2)
korr.viel   <-round(cor(autos.viel$PS,autos.viel$Preis),2)
ergebnis<-ml.summary.3(autos.train[,c("PS","Preis","PSKlasse")],
  paste("Korrelation:",korr.wenig,"(wenig)",korr.normal,"(normal)",
        korr.viel,"(viel)"),"PS","Preis in Tausend US$",scatter=T)
```

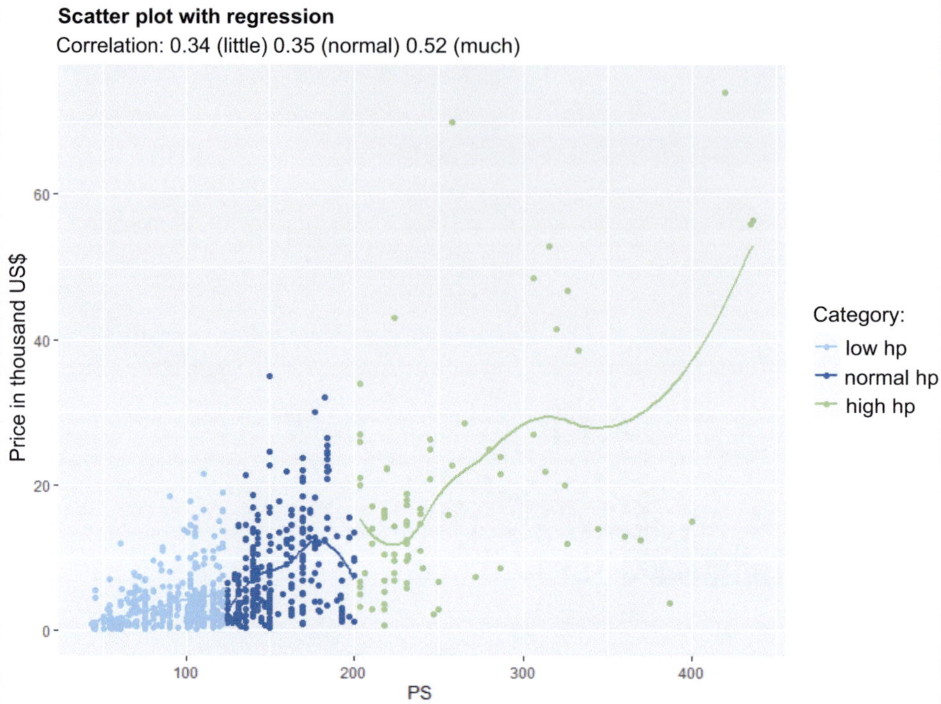


```
# Modelle bestimmen und anzeigen lassen------------------------------------------
mod.rg.lm<-linear_reg() %>% set_engine("lm")     # Modellauswahl
translate(mod.rg.lm)                             # Verwendung der Parameter

## Linear Regression Model Specification (regression)
##
## Computational engine: lm
##
## Model fit template:
## stats::lm(formula = missing_arg(), data = missing_arg(), weights = missing_arg())

## Modell mit 1 unabhängigen Variable ############################################
# Trainieren des Modells---------------------------------------------------------
rec.1<-recipe(data=autos.train,Preis~Alter)                  # Recipe
wfl.rg.lm<-workflow() %>%                                    # Workflow
  add_model(mod.rg.lm) %>%
  add_recipe(rec.1)
wfl.rg.lm

## ══ Workflow ════════════════════════════════════════════════════════════════════
## Preprocessor: Recipe
## Model: linear_reg()
##
## ── Preprocessor ────────────────────────────────────────────────────────────────
## 0 Recipe Steps
##
## ── Model ───────────────────────────────────────────────────────────────────────
## Linear Regression Model Specification (regression)
```

```
##
## Computational engine: lm

wfl.rg.lm.fit<-fit(wfl.rg.lm,data=autos.train)
# Anzeige der Koeffizienten des Modells----------------------------------------------
wfl.rg.lm.fit

## ══ Workflow [trained] ═══════════════════════════════════════════════════════════
## Preprocessor: Recipe
## Model: linear_reg()
##
## ── Preprocessor ──────────────────────────────────────────────────────────────
──────────────────────────────────────────────────────
## 0 Recipe Steps
##
## ── Model ─────────────────────────────────────────────────────────────────────
────────────────────────────────────────────────────────────
##
## Call:
## stats::lm(formula = ..y ~ ., data = data)
##
## Coefficients:
## (Intercept)        Alter
##      19.266       -1.072

tidy(wfl.rg.lm.fit)

## # A tibble: 2 × 5
##   term        estimate std.error statistic   p.value
##   <chr>          <dbl>     <dbl>     <dbl>     <dbl>
## 1 (Intercept)    19.3     0.590       32.7 3.15e-149
## 2 Alter          -1.07    0.0460     -23.3 5.51e- 92

# Anzeige der Summary für Modellparameter und p-Werte/Signifikanz----------------
extract_fit_engine(wfl.rg.lm.fit) %>% summary()

##
## Call:
## stats::lm(formula = ..y ~ ., data = data)
##
## Residuals:
##     Min      1Q  Median      3Q     Max
## -10.149  -3.375  -0.681   1.624  67.599
##
## Coefficients:
##             Estimate Std. Error t value Pr(>|t|)
## (Intercept) 19.26644    0.58972   32.67   <2e-16 ***
## Alter       -1.07243    0.04603  -23.30   <2e-16 ***
## ---
## Signif. codes:  0 '***' 0.001 '**' 0.01 '*' 0.05 '.' 0.1 ' ' 1
##
## Residual standard error: 6.263 on 797 degrees of freedom
## Multiple R-squared:  0.4052, Adjusted R-squared:  0.4044
## F-statistic: 542.9 on 1 and 797 DF,  p-value: < 2.2e-16

# Vorhersagen machen---------------------------------------------------------------
wfl.rg.lm.pred<-autos.test %>%
  select(Preis,Alter) %>%
  bind_cols(predict(wfl.rg.lm.fit,autos.test)) %>%
```

```
  bind_cols(predict(wfl.rg.lm.fit,autos.test,type="conf_int"))
head(wfl.rg.lm.pred,3)                                          # Preis/Vorhersage

## # A tibble: 3 × 5
##   Preis Alter   .pred .pred_lower .pred_upper
##   <dbl> <dbl>   <dbl>       <dbl>       <dbl>
## 1  8.99     7 11.8         11.1        12.4
## 2  1.99    18 -0.0373      -0.741       0.667
## 3  5.30     6 12.8         12.1        13.5

# Aufruf Funktion ml.eval.lr() ------------------------------------------------
ergebnis<-ml.eval.lr(wfl.rg.lm.pred[,c("Preis",".pred")],   # Evaluation
                 "Gebrauchtwagen","realer Preis",
                 "vorhergesagter Preis",scatter=T)
```

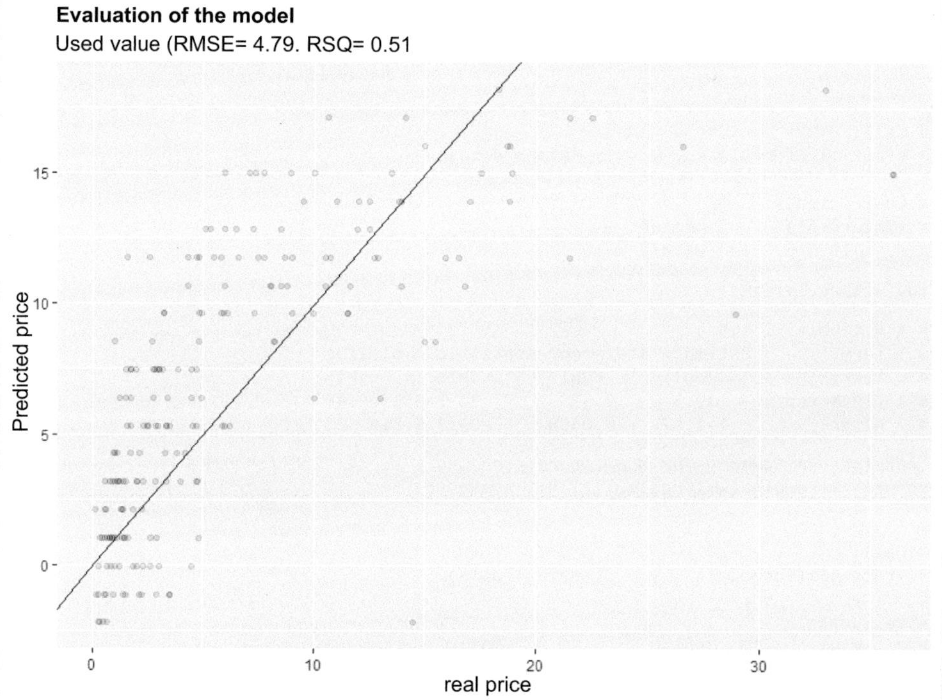


```
## Modell mit 2 unabhängigen Variablen ############################################
# Trainieren des Modells mit einer nominalen und einer kategorialen Variable------
rec.2<-recipe(data=autos.train,Preis~Alter+PSKlasse) %>%    # Recipe
  step_dummy(all_nominal_predictors())                      # Kategorial->Numerisch
wfl.rg.lm<-workflow() %>%                                   # Workflow
  add_model(mod.rg.lm) %>%
  add_recipe(rec.2)
wfl.rg.lm

## ══ Workflow ═════════════════════════════════════════════════════════════════
## Preprocessor: Recipe
## Model: linear_reg()
##
## ── Preprocessor ─────────────────────────────────────────────────────────────
```

```
## 1 Recipe Step
##
## • step_dummy()
##
## — Model ————————————————————————————————————————————
## Linear Regression Model Specification (regression)
##
## Computational engine: lm

wfl.rg.lm.fit<-fit(wfl.rg.lm,data=autos.train)
# Anzeige der Koeffizienten des Modells--------------------------------------
wfl.rg.lm.fit

## ══ Workflow [trained] ═══════════════════════════════════════
## Preprocessor: Recipe
## Model: linear_reg()
##
## ── Preprocessor ──────────────────────────────────────────
────────────────────────────────────

## 1 Recipe Step
##
## • step_dummy()
##
## ── Model ─────────────────────────────────────────────────
────────────────────────────────────────

##
## Call:
## stats::lm(formula = ..y ~ ., data = data)
##
## Coefficients:
##       (Intercept)               Alter   PSKlasse_viel.PS  PSKlasse_wenig.PS
##           17.8658             -0.8999             9.2030            -2.8374

tidy(wfl.rg.lm.fit)

## # A tibble: 4 × 5
##   term              estimate std.error statistic   p.value
##   <chr>                <dbl>     <dbl>     <dbl>     <dbl>
## 1 (Intercept)         17.9      0.533      33.5  3.68e-154
## 2 Alter               -0.900    0.0399    -22.6  1.70e- 87
## 3 PSKlasse_viel.PS     9.20     0.663      13.9  2.27e- 39
## 4 PSKlasse_wenig.PS   -2.84     0.411      -6.90 1.09e- 11

# Anzeige der Summary für Modellparameter und p-Werte/Signifikanz----------------
extract_fit_engine(wfl.rg.lm.fit) %>% summary()

##
## Call:
## stats::lm(formula = ..y ~ ., data = data)
##
## Residuals:
##     Min      1Q  Median      3Q     Max
## -16.969  -2.577  -0.430   1.870  57.727
##
## Coefficients:
##                   Estimate Std. Error t value Pr(>|t|)
## (Intercept)       17.86576    0.53322  33.505  < 2e-16 ***
## Alter             -0.89995    0.03989 -22.561  < 2e-16 ***
## PSKlasse_viel.PS   9.20302    0.66316  13.878  < 2e-16 ***
```

```
## PSKlasse_wenig.PS -2.83741    0.41146  -6.896 1.09e-11 ***
## ---
## Signif. codes:  0 '***' 0.001 '**' 0.01 '*' 0.05 '.' 0.1 ' ' 1
##
## Residual standard error: 5.226 on 795 degrees of freedom
## Multiple R-squared:  0.587,  Adjusted R-squared:  0.5854
## F-statistic: 376.6 on 3 and 795 DF,  p-value: < 2.2e-16

# Vorhersagen machen------------------------------------------------------------
wfl.rg.lm.pred<-autos.test %>%
  select(Preis,Alter,PSKlasse) %>%
  bind_cols(predict(wfl.rg.lm.fit,autos.test)) %>%
  bind_cols(predict(wfl.rg.lm.fit,autos.test,type="conf_int"))
head(wfl.rg.lm.pred,3)                                      # Preis/Vorhersage

## # A tibble: 3 × 6
##   Preis Alter PSKlasse .pred .pred_lower .pred_upper
##   <dbl> <dbl> <chr>    <dbl>       <dbl>       <dbl>
## 1  8.99     7 wenig PS  8.73        8.05       9.40
## 2  1.99    18 wenig PS -1.17       -1.80      -0.546
## 3  5.30     6 wenig PS  9.63        8.90      10.4

# Aufruf Funktion ml.eval.lr() ------------------------------------------------
ergebnis<-ml.eval.lr(wfl.rg.lm.pred[,c("Preis",".pred")],   # Evaluation
                "Gebrauchtwagen","realer Preis",
                "vorhergesagter Preis",scatter=T)
```

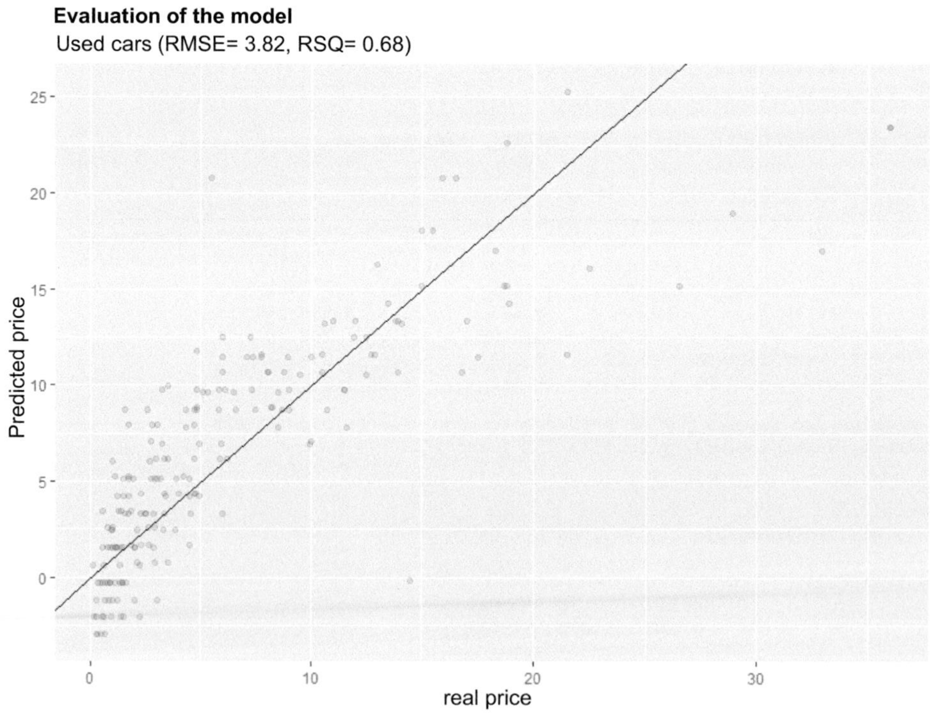


```
## Modell mit 2 unabhängigen Variablen ##########################################
# Trainieren des Modells mit zwei nominalen Variablen----------------------------
```

```
rec.3<-recipe(data=autos.train,Preis~Alter+PS)                # Recipe
wfl.rg.lm<-workflow() %>%                                      # Workflow
  add_model(mod.rg.lm) %>%
  add_recipe(rec.3)
wfl.rg.lm

## ══ Workflow ════════════════════════════════════════════════════════════════
## Preprocessor: Recipe
## Model: linear_reg()
##
## ── Preprocessor ────────────────────────────────────────────────────────────
## 0 Recipe Steps
##
## ── Model ───────────────────────────────────────────────────────────────────
## Linear Regression Model Specification (regression)
##
## Computational engine: lm

wfl.rg.lm.fit<-fit(wfl.rg.lm,data=autos.train)
# Anzeige der Koeffizienten des Modells-----------------------------------------
wfl.rg.lm.fit

## ══ Workflow [trained] ══════════════════════════════════════════════════════
## Preprocessor: Recipe
## Model: linear_reg()
##
## ── Preprocessor ────────────────────────────────────────────────────────────
────────────────────────────────────────────────────────
## 0 Recipe Steps
##
## ── Model ───────────────────────────────────────────────────────────────────
──────────────────────────────────────────────────────────────
##
## Call:
## stats::lm(formula = ..y ~ ., data = data)
##
## Coefficients:
## (Intercept)        Alter           PS
##     7.23623     -0.81752      0.07062

tidy(wfl.rg.lm.fit)

## # A tibble: 3 × 5
##   term        estimate std.error statistic  p.value
##   <chr>          <dbl>     <dbl>     <dbl>    <dbl>
## 1 (Intercept)  7.24     0.681         10.6 9.17e-25
## 2 Alter       -0.818    0.0369       -22.1 4.73e-85
## 3 PS           0.0706   0.00299       23.6 5.14e-94

# Anzeige der Summary für Modellparameter und p-Werte/Signifikanz----------------
extract_fit_engine(wfl.rg.lm.fit) %>% summary()

##
## Call:
## stats::lm(formula = ..y ~ ., data = data)
##
## Residuals:
##     Min      1Q  Median      3Q     Max
## -25.045  -2.550  -0.487   1.840  46.908
```

```
##
## Coefficients:
##              Estimate Std. Error t value Pr(>|t|)
## (Intercept)  7.236233   0.680837   10.63   <2e-16 ***
## Alter       -0.817518   0.036912  -22.15   <2e-16 ***
## PS           0.070625   0.002987   23.64   <2e-16 ***
## ---
## Signif. codes:  0 '***' 0.001 '**' 0.01 '*' 0.05 '.' 0.1 ' ' 1
##
## Residual standard error: 4.804 on 796 degrees of freedom
## Multiple R-squared:  0.6505, Adjusted R-squared:  0.6497
## F-statistic: 740.9 on 2 and 796 DF,  p-value: < 2.2e-16

# Vorhersagen machen-------------------------------------------------------------
wfl.rg.lm.pred<-autos.test %>%
  select(Preis,Alter,PS) %>%
  bind_cols(predict(wfl.rg.lm.fit,autos.test)) %>%
  bind_cols(predict(wfl.rg.lm.fit,autos.test,type="conf_int"))
head(wfl.rg.lm.pred,3)                                        # Preis/Vorhersage

## # A tibble: 3 × 6
##   Preis Alter    PS .pred .pred_lower .pred_upper
##   <dbl> <dbl> <dbl> <dbl>       <dbl>       <dbl>
## 1  8.99     7   102  8.72        8.18       9.26
## 2  1.99    18    90 -1.12       -1.67      -0.575
## 3  5.30     6    71  7.35        6.65       8.04

# Aufruf Funktion ml.eval.lr() ---------------------------------------------------
ergebnis<-ml.eval.lr(wfl.rg.lm.pred[,c("Preis",".pred")],   # Evaluation
                  "Gebrauchtwagen","realer Preis",
                  "vorhergesagter Preis",scatter=T)
```

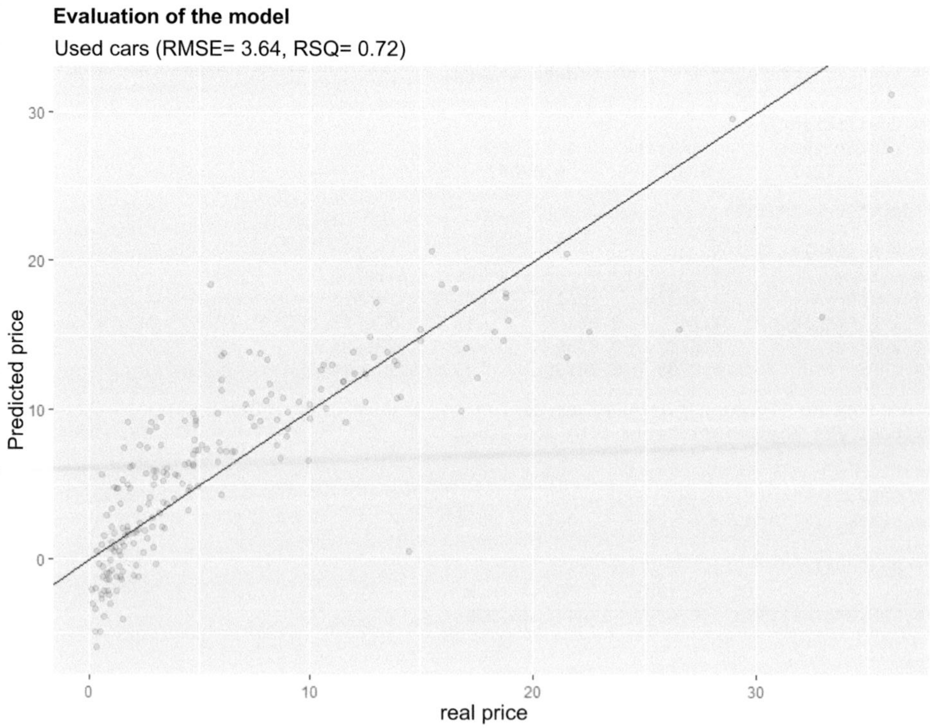

In the following example, the preparation and execution of a regression using the machine learning functions from the tidymodels package will be presented. For this purpose, the data ames from the modeldata package (Kuhn 2022) is used. The dataframe ames includes house prices of 2930 houses and 73 descriptive variables about the houses from the town of Ames in Iowa. The goal is the valuation of real estate. Based on the attributes of a house, its house price should be estimated or predicted.

First, it is checked whether there are missing data. Fortunately, all data sets are complete. The function `ml.summary()` shows a right-skewed distribution for the house price and two extreme outliers with a house price of more than 650,000 US$. To prevent these two outliers from distorting the regression formula, the two data sets are removed from the data. To eliminate the right skew of the dependent variable, this variable is logarithmized.

From the data, stratified samples are then drawn for training and test data, so that representative amounts of low, medium, and high selling prices are included in each sample. If a predictor is used for a prediction that is included with a coefficient in the regression formula, then it is also necessary that data is available for all predictors as the basis of the prediction. Therefore, it is advisable to work with a minimal number of predictors or independent variables for the regression and at the same time achieve the highest possible accuracy of the prediction. The separation into training and test data can cause rare values of categorical variables not to occur in the training data. If this is the case, then no coefficient can be determined for this value in the training process of the model. This then causes that no prediction is possible for data sets with this value. To exclude this problem, it should be examined in advance for categorical variables that no manifestations occur in the test data that are not also contained in the training data. The check in the following coding shows that this problem exists for the variable Mas_Vnr_Type, after the function `set.seed(456)` was called before the sample formation in training and test data. The function `set.seed()` determines the random numbers, which are the basis of the sample formation. After re-sampling based on `set.seed(123)`, there are no more values of categorical variables in the training data that are not also contained in the training data.

To better understand the categorical variables, the function `ml.summary.2()` from the machinelearning package can be used with a box plot. The parameter npos causes the display of the additional information N = number of records of a category and Mw = mean value for a category. It turns out that the Neighborhood has a significant influence on the price. For some neighborhoods, the average of all house prices is outside the IQR (box of the neighborhood), which suggests using this variable as one of the predictors. It is also noticeable that the neighborhoods Green Hills, Greens, and Blueste affect only very few records. If there is a very low number of records for a predictor, it is inaccurate and should be excluded. This can be taken into account in the function `recipe()` later with the function `step_other()` by giving a coefficient to manifestations of categorical variables that affect less than 1% of all records. A similar picture emerges for the variable Lot_Shape, Condition_1, Bldg_Type, House_Style, and Overall_Cond.

A better understanding of the numerical variables can also be helpful, especially to identify variables with a high correlation to the house price. First, all numerical variables are stored in the vector ames.train.num with the command `ames.train.num <- select_if(ames.train,is.numeric)`. The function `corr_cross()` from the lares package then examines the correlation among these variables, including with the house price. It is important to avoid strong correlations among the predictors. Therefore, the independent variables Year_Remod_Add, First_Flr_SF, Full_Bath, TotRms_AbvGrd, and Garage_Area are excluded as predictors. The following variables are determined as numerical predictors: Sale_Price, Gr_Liv_Area, Garage_Cars, Year_Built, Total_Bsmt_SF, Mas_Vnr_Area, Longitude, and Latitude. For all numerical variables, the correlation with the house price can initially be explicitly considered with the function `corr_var()` from the lares package. However, the function `ml.korrelation()` from the machinelearning package, which visualizes the distribution in addition to the correlation coefficients, is even more helpful. Among other things, it is noticeable that some variables are not normally distributed. Since the calculation of the coefficients for predictors works better for normally distributed variables, these variables will be logarithmized later in the function `recipe()`. The following coding shows how the distribution improves by logarithmizing, using the example of the variable Total_Bsmt_SF.

Regression: Preprocessing and Exploratory Data Analysis

```
## Laden der Daten-------------------------------------------------------------
data(ames, package = "modeldata")
ames<-mutate(ames,Sale_Price=Sale_Price/1000)   # Preis in Tausend
# Fehlende Daten prüfen--------------------------------------------------------
colSums(is.na(ames))                           # keine fehlenden Daten

##         MS_SubClass           MS_Zoning        Lot_Frontage            Lot_Area
##                   0                   0                   0                   0
##              Street               Alley           Lot_Shape        Land_Contour
##                   0                   0                   0                   0
##           Utilities          Lot_Config          Land_Slope        Neighborhood
##                   0                   0                   0                   0
##         Condition_1         Condition_2           Bldg_Type         House_Style
##                   0                   0                   0                   0
##        Overall_Cond          Year_Built      Year_Remod_Add          Roof_Style
##                   0                   0                   0                   0
##           Roof_Matl        Exterior_1st        Exterior_2nd        Mas_Vnr_Type
##                   0                   0                   0                   0
##        Mas_Vnr_Area         Exter_Cond          Foundation           Bsmt_Cond
##                   0                   0                   0                   0
##       Bsmt_Exposure      BsmtFin_Type_1        BsmtFin_SF_1      BsmtFin_Type_2
##                   0                   0                   0                   0
##        BsmtFin_SF_2         Bsmt_Unf_SF       Total_Bsmt_SF             Heating
##                   0                   0                   0                   0
##          Heating_QC         Central_Air          Electrical        First_Flr_SF
##                   0                   0                   0                   0
##       Second_Flr_SF         Gr_Liv_Area      Bsmt_Full_Bath      Bsmt_Half_Bath
##                   0                   0                   0                   0
##           Full_Bath           Half_Bath       Bedroom_AbvGr       Kitchen_AbvGr
##                   0                   0                   0                   0
##       TotRms_AbvGrd          Functional          Fireplaces         Garage_Type
##                   0                   0                   0                   0
##       Garage_Finish         Garage_Cars         Garage_Area         Garage_Cond
##                   0                   0                   0                   0
##         Paved_Drive        Wood_Deck_SF       Open_Porch_SF      Enclosed_Porch
##                   0                   0                   0                   0
## Three_season_porch        Screen_Porch           Pool_Area             Pool_QC
##                   0                   0                   0                   0
##               Fence        Misc_Feature            Misc_Val             Mo_Sold
##                   0                   0                   0                   0
##           Year_Sold           Sale_Type      Sale_Condition          Sale_Price
##                   0                   0                   0                   0
##           Longitude            Latitude
##                   0                   0

## Visualisierung der Daten----------------------------------------------------
ml.summary(ames$Sale_Price,                    # Daten verstehen:
           "Verkaufspreis des Hauses",         # 2 extreme Ausreißer > 650 +
           "Preis in 1000 US$",box=T)          # Rechte Schiefe
```

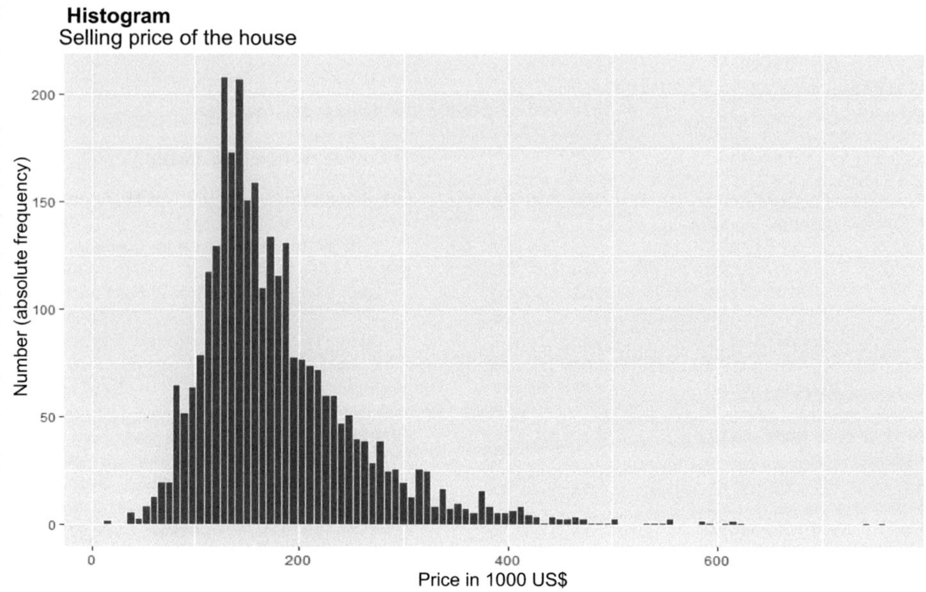

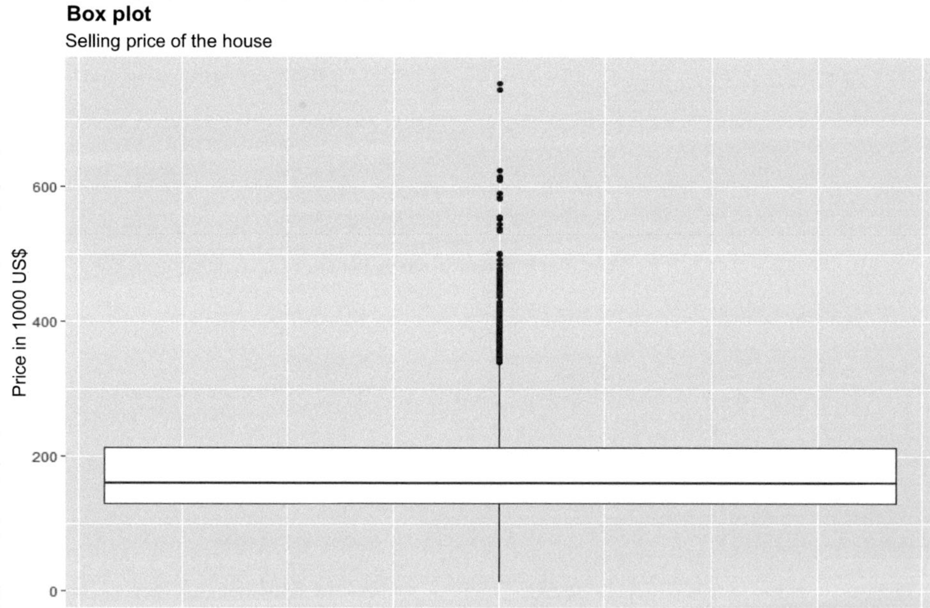


```
##                 25%
## anzahl          "2930"
## anzahl.na       "0"
## modus           "135"
## median          "160"
## mean            "180.8"
## min             "12.789"
```

```
## max          "755"
## sd           "79.89"
## q1           "129.5"
## q3           "213.5"
## iqr          "84"
## whisker.min  "3.5"
## whisker.max  "339.5"
## skewness     "1.74"
## skewness.txt "Rechte Schiefe: 1.74 > 0, positive Schiefe,linkssteil,rechtsschief"
## kurtosis     "5.1"
## kurtosis.txt "Steilgipflig mit Exzess Kurtosis 5.1 > 0"
```

```
ames<-ames[ames$Sale_Price< 650,]                # 2 extreme Ausreißer entfernen
ml.summary(log(ames$Sale_Price),                 # Daten verstehen
          "Logarithmierter Verkaufspreis des Hauses",
          "log(Preis) in 1000 US$",box=T)
```

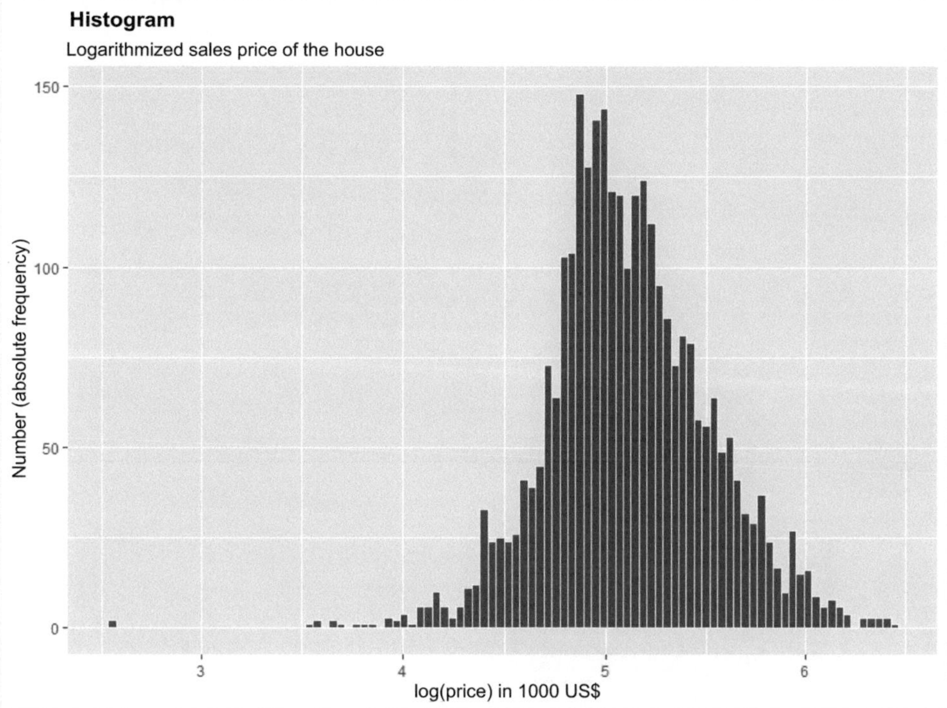

```
##              25%
## anzahl       "2928"
## anzahl.na    "0"
## modus        "4.90527477843843"
## median       "5.07517381523383"
## mean         "5.11"
## min          "2.54858542645118"
## max          "6.4377516497364"
## sd           "0.41"
## q1           "4.9"
## q3           "5.4"
## iqr          "0.5"
## whisker.min  "4.15"
## whisker.max  "6.15"
## skewness     "-0.04"
## skewness.txt "Linke Schiefe: -0.04 < 0, negative Schiefe,linksschief,rechtssteil"
## kurtosis     "1.46"
## kurtosis.txt "Steilgipflig mit Exzess Kurtosis 1.46 > 0"
```

```
ames<-mutate(ames,                               # Logarithmieren von Sale_Price
             Sale_Price=log(Sale_Price))         # für bessere Verteilung
ames<-ames[,c(sort(names(ames)))]                # Variablen alphabetisch
## Trainings- und Testdaten erzeugen--------------------------------------------
set.seed(456)
ames.split<-initial_split(ames,prop=0.80,        # Split Trainings/Testdaten 80:20
                          strata=Sale_Price)
ames.train<-training(ames.split)
```

```
ames.test<-testing(ames.split)
dim(ames.train)                                    # 80%  Trainingsdaten

## [1] 2341   74

dim(ames.test)                                     # 20%  Testdaten

## [1] 587  74

## Überprüfen ob in Testdaten Levels vorkommen, die nicht in Trainingsdaten sind--
problem<-ames.train %>%
  select(where(is.factor)) %>%
  skimr::skim() %>%
  as_tibble() %>%
  select(Variable=skim_variable,Train.Levels=factor.n_unique) %>%
  bind_cols(
    ames.test %>%
      select(where(is.factor)) %>%
      skimr::skim() %>%
      as_tibble() %>%
      select(Test.Levels= factor.n_unique)
  ) %>%
  filter(Train.Levels < Test.Levels)
if (nrow(problem) == 0) {
  print("Keine Levels in Faktorvariablen im Test-Set,
        die nicht auch im Trainings-Set vorkommen")
} else {
  print("Levels in folgenden Faktorvariablen im Test-Set,
        die nicht auch im Trainings-Set vorkommen")
  problem
}

## [1] "Levels in folgenden Faktorvariablen im Test-Set, \n        die nicht auch im
 Trainings-Set vorkommen"

## # A tibble: 1 × 3
##   Variable     Train.Levels Test.Levels
##   <chr>               <int>       <int>
## 1 Mas_Vnr_Type            4           5

## Trainings- und Testdaten erzeugen-------------------------------------------------
set.seed(123)
ames.split<-initial_split(ames,prop=0.80,          # Split Trainings/Testdaten 80:20
                          strata=Sale_Price)
ames.train<-training(ames.split)
ames.test<-testing(ames.split)
dim(ames.train)                                    # 80%  Trainingsdaten

## [1] 2341   74

dim(ames.test)                                     # 20%  Testdaten

## [1] 587  74

## Überprüfen ob in Testdaten Levels vorkommen, die nicht in Trainingsdaten sind--
problem<-ames.train %>%
  select(where(is.factor)) %>%
  skimr::skim() %>%
  as_tibble() %>%
  select(Variable=skim_variable,Train.Levels=factor.n_unique) %>%
  bind_cols(
    ames.test %>%
```

```
      select(where(is.factor)) %>%
      skimr::skim() %>%
      as_tibble() %>%
      select(Test.Levels= factor.n_unique)
  ) %>%
  filter(Train.Levels < Test.Levels)
if (nrow(problem) == 0) {
  print("Keine Levels in Faktorvariablen im Test-Set,
        die nicht auch im Trainings-Set vorkommen")
} else {
  print("Levels in folgenden Faktorvariablen im Test-Set,
        die nicht auch im Trainings-Set vorkommen")
  problem
}

## [1] "Keine Levels in Faktorvariablen im Test-Set, \n        die nicht auch im Tra
inings-Set vorkommen"

## Kategoriale Variablen besser verstehen----------------------------------------
erg<-ml.summary.2(ames.train[,c("Sale_Price","Neighborhood")],
                  "Neighborhood","Sale_Price","Neighborhood",box=T,bw=T,npos=3)
```

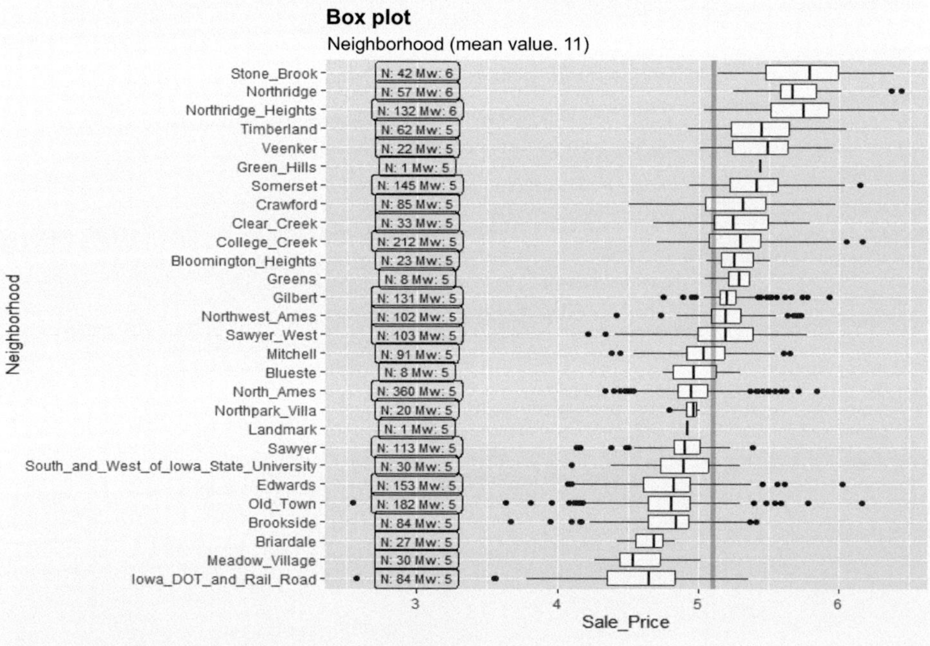


```
# Geringe Anzahl bei "Green_Hills","Greens" und "Blueste"
erg<-ml.summary.2(ames.train[,c("Sale_Price","Lot_Shape")],
                  "Lot_Shape","Sale_Price","Lot_Shape",box=T,npos=3)
```

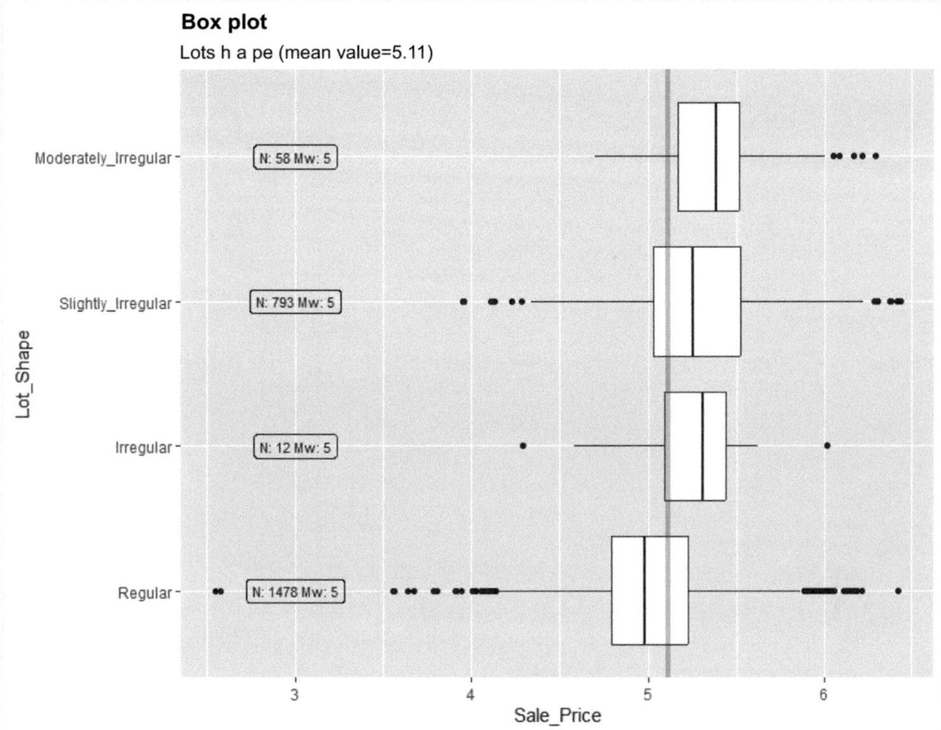


```
# Geringe Anzahl bei "Irregular"
erg<-ml.summary.2(ames.train[,c("Sale_Price","Condition_1")],
                  "Condition_1","Sale_Price","Condition_1",box=T,npos=3)
```

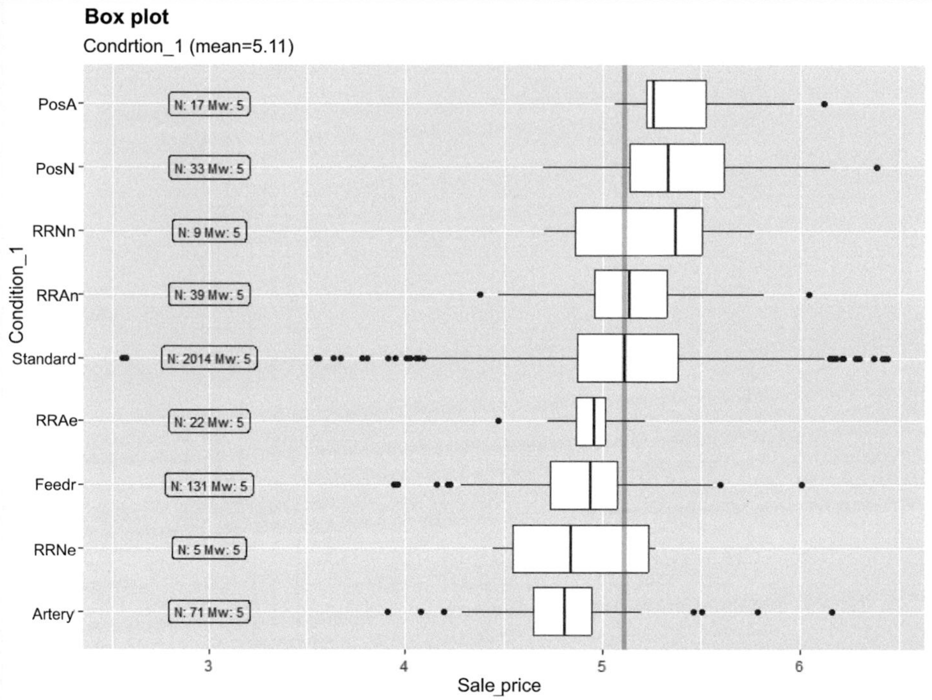


```
# Geringe Anzahl bei "RRNn" und "RRNe"
erg<-ml.summary.2(ames.train[,c("Sale_Price","Bldg_Type")],
                  "Bldg_Type","Sale_Price","Bldg_Type",box=T,npos=3)
```

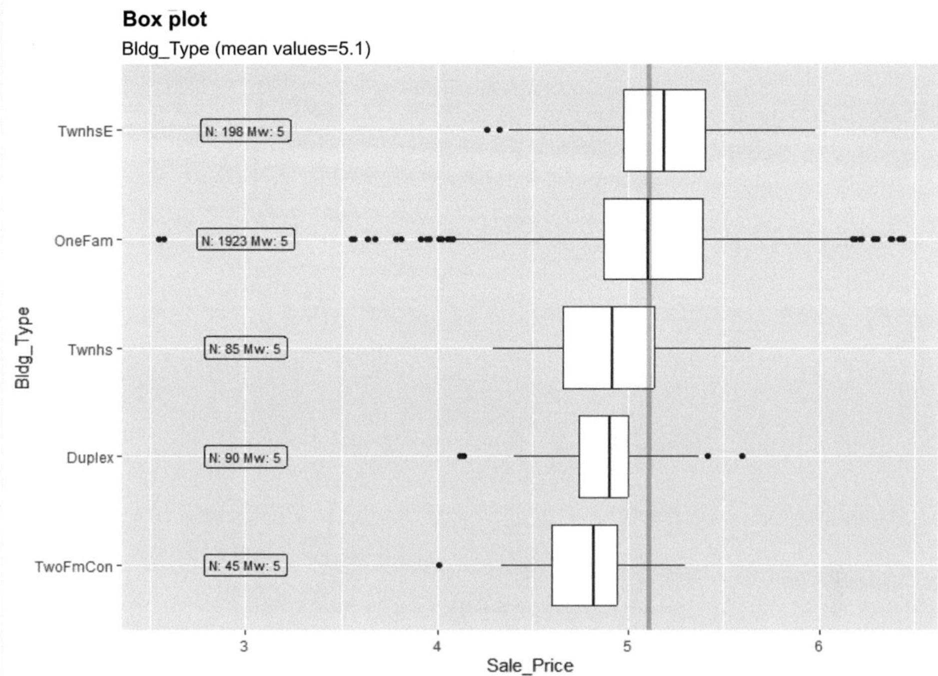


```
erg<-ml.summary.2(ames.train[,c("Sale_Price","House_Style")],
                  "House_Style","Sale_Price","House_Style",box=T,npos=3)
```

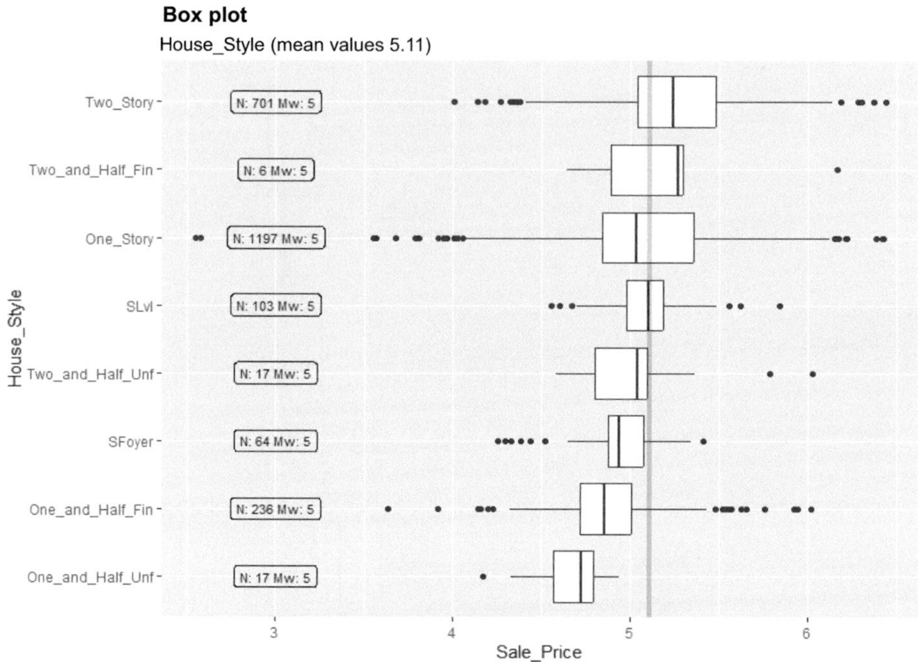


```
# Geringe Anzahl bei "Two_and_Half_Fin" und "One_and_Half_Unf"
erg<-ml.summary.2(ames.train[,c("Sale_Price","Overall_Cond")],
                  "Overall_Cond","Sale_Price","Overall_Cond",box=T,npos=3)
```

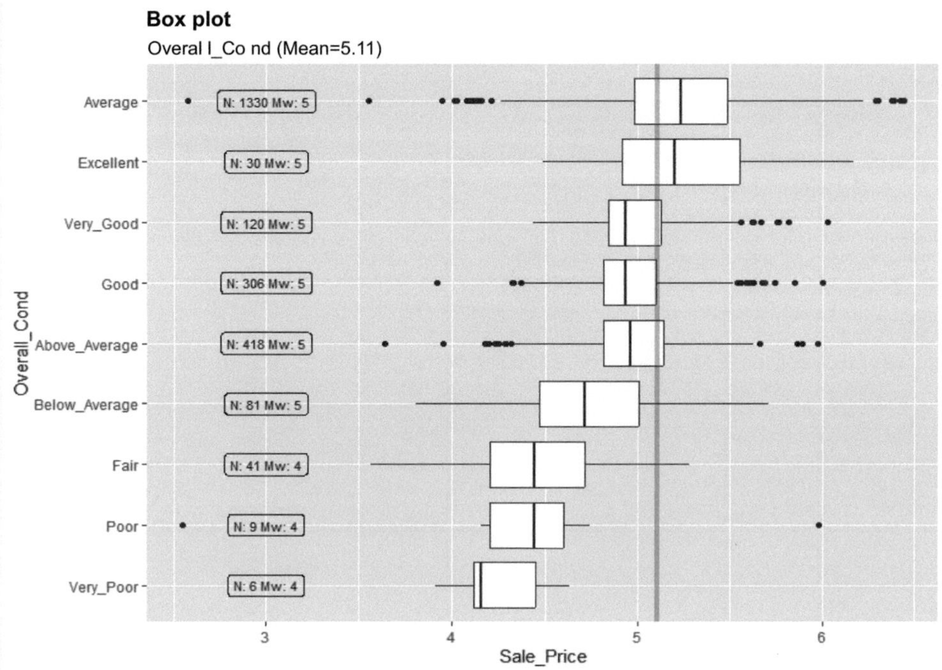


```
# Geringe Anzahl bei "Very_Poor" und "Poor"
## Numerische Variablen besser verstehen------------------------------------------
ames.train.num <- select_if(ames.train,is.numeric) # Numerische Variablen
#devtools::install_github("laresbernardo/lares")
library(lares)

lares::corr_cross(ames.train.num,
  max_pvalue = 0.05,                                # Signifikanz >95%
  top = 20                                          # TOP 20 Korrelationen
)
```

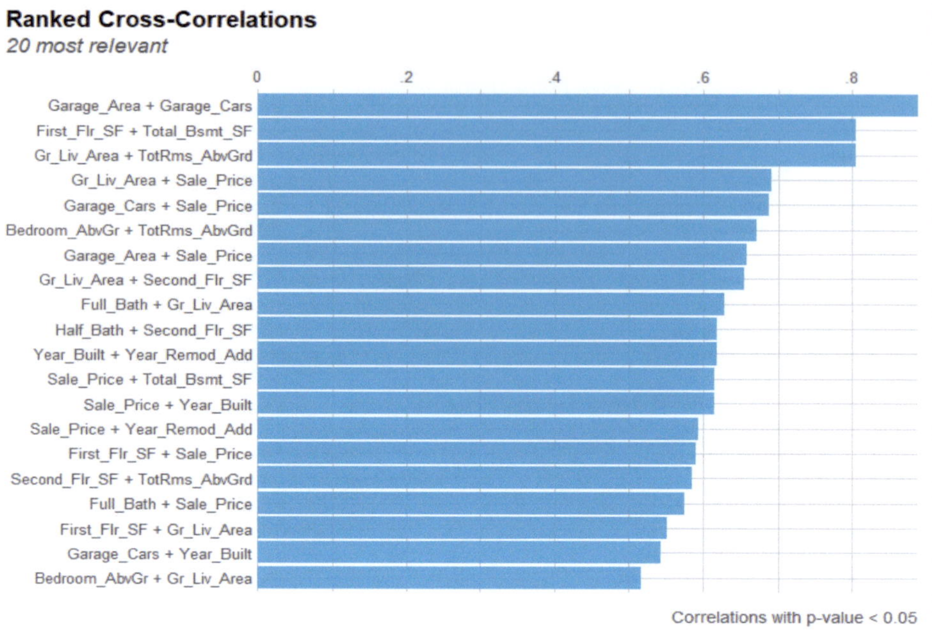


```
# Nutze Year_Remod_Add nicht, da mögliche multi-colinearity problem mit Year_Built v
on 0,62
# Nutze First_Flr_SF nicht, da mögliche multi-colinearity problem mit Total_Bsmt_SF
von 0,79)
# Nutze Full_Bath nicht, da mögliche multi-colinearity problem mit Gr_Liv_Area von 0
,630)
# Nutze TotRms_AbvGrd nicht, da mögliche multi-colinearity problem mit Gr_Liv_Area v
on 0,80)
# Nutze Garage_Area nicht, da mögliche multi-colinearity problem mit Garage_Cars von
 0,89)
var.sel<-c("Sale_Price","Gr_Liv_Area","Garage_Cars","Year_Built","Total_Bsmt_SF",
           "Mas_Vnr_Area","Longitude","Latitude")
ames.train.num.sel <- select(ames.train.num,all_of(var.sel))
lares::corr_var(ames.train.num, # name of dataset
  Sale_Price, # name of variable to focus on
  top = 10) # display top 5 correlations
```

```
ml.korrelation(ames.train.num.sel,
               "Ames Hauspreise: Vorselektierte Variablen",
               colnames(ames.train.num.sel))
```

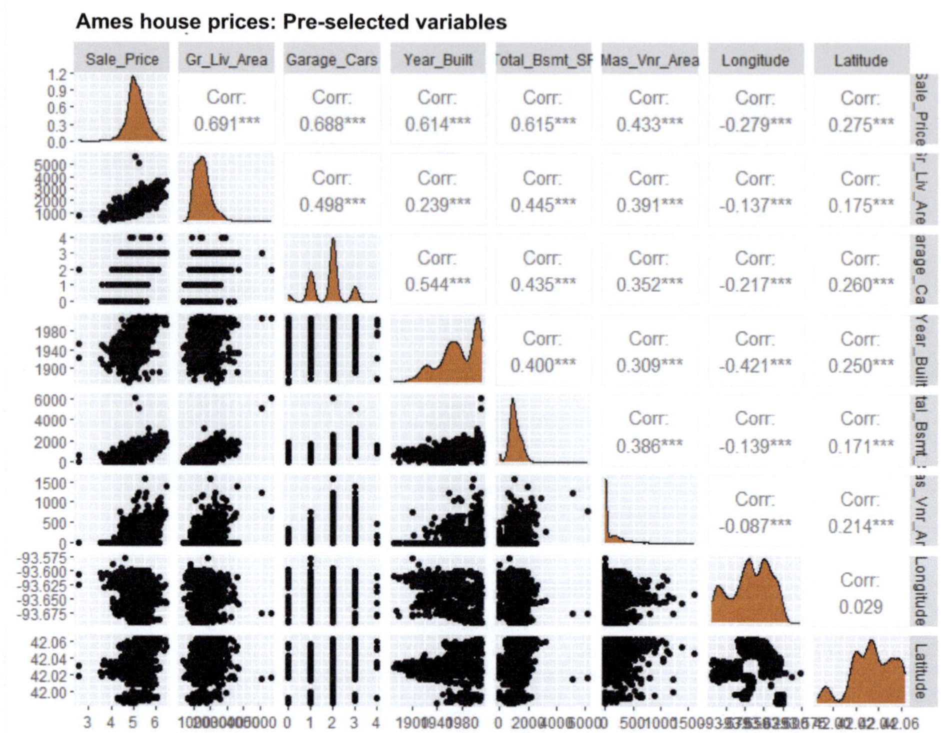


```
#- Radar-Chart -----------------------------------------------------------------
library(fmsb)
korrelation<-round(cor(ames.train.num.sel$Sale_Price,ames.train.num.sel[,-1]),2)
df<-data.frame(korrelation)
df[2,]<-1
df[3,]<-0
df[4,]<-df[1,]
df<-df[-1,]
farbe<-colour.own.nomin.2[1]                        # Farben zuweisen
farbe.alpha<-paste0(colour.own.nomin.2[1],40)    # 0.25 durchsichtig
colnames(df)<-paste(names(df),"\n",df[3,])
df[3,]<-abs(df[3,])
radarchart(df,plty=1,pcol=farbe,pfcol=farbe.alpha,lwd=1)
```

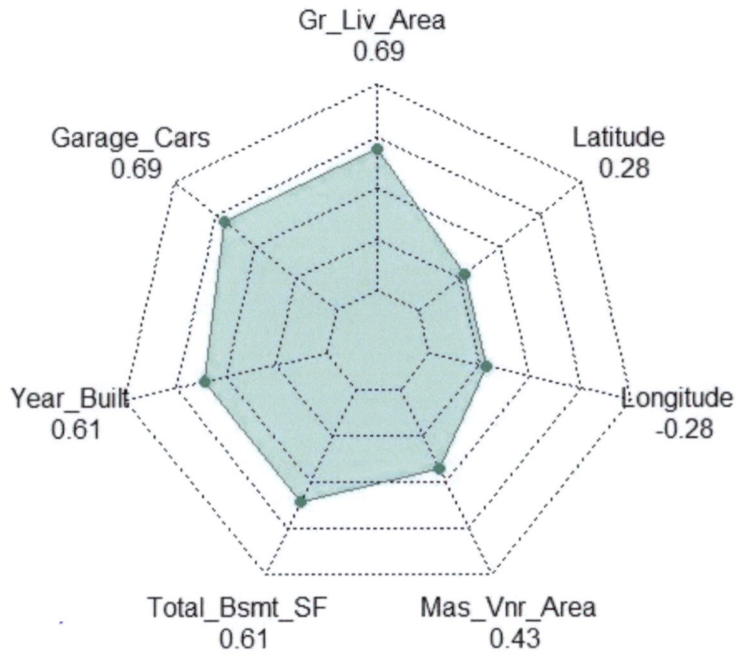


```
# Variable mit größter Korrelation nicht normalverteilt->in Recipe logarithmieren
ml.summary(ames.train.num.sel$Total_Bsmt_SF,"Total_Bsmt_SF","log(Sale_Price)")
```

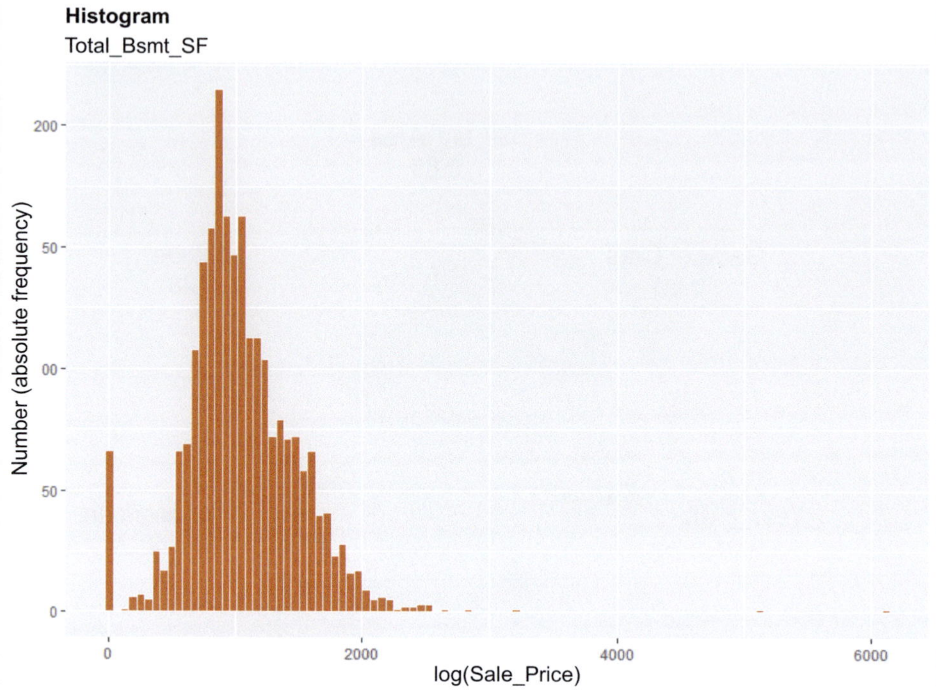


```
##                25%
## anzahl         "2341"
## anzahl.na      "0"
## modus          "0"
## median         "992"
## mean           "1049.53"
## min            "0"
## max            "6110"
## sd             "441.89"
## q1             "793"
## q3             "1298"
## iqr            "505"
## whisker.min    "35.5"
## whisker.max    "2055.5"
## skewness       "1.22"
## skewness.txt "Rechte Schiefe: 1.22 > 0, positive Schiefe,linkssteil,rechtsschief"
## kurtosis       "10.6"
## kurtosis.txt "Steilgipflig mit Exzess Kurtosis 10.6 > 0"
```

```
ml.summary(log(ames.train.num.sel$Total_Bsmt_SF),"log(Total_Bsmt_SF)","log(Sale_Pric
e)")
```

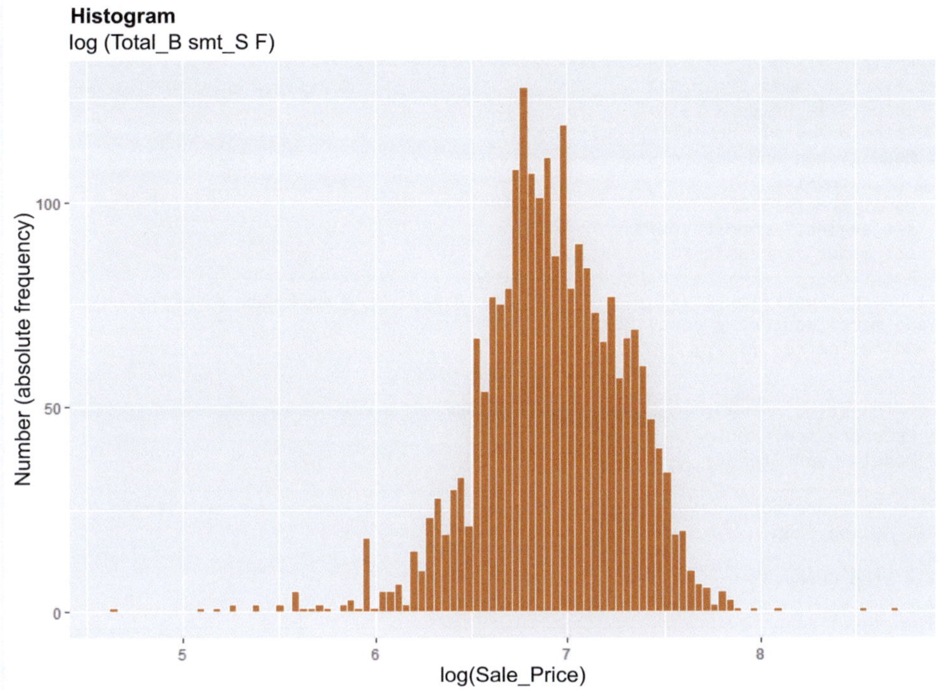


```
##               25%
## anzahl        "2341"
## anzahl.na     "0"
## modus         "-Inf"
## median        "6.89972310728487"
## mean          "-Inf"
## min           "-Inf"
## max           "8.71768205216564"
## sd            "NaN"
## q1            "6.7"
## q3            "7.2"
## iqr           "0.5"
## whisker.min   "5.95"
## whisker.max   "7.95"
## skewness      "NaN"
## skewness.txt NA
## kurtosis      "NaN"
## kurtosis.txt NA

# Keine Normalverteilung von Year_Build, Mas_Vnr_Area, Total_Bsmt_SF & Gr_Liv_Area
# -> in Recipe standardisieren
```

If all 73 independent variables are used as predictors for the dependent variable of house price, without any tuning, the result is an RMSE of 32,629 US$ with an RSQ of 0.85. As mentioned above, such an approach has the disadvantage that all 73 variables must be present for a prediction to be made.

Regression: Modeling without Tuning

```
## Recipe-----------------------------------------------------------------
rec.first <- ames.train %>%                             # Recipe
  recipe(Sale_Price ~ .) %>%
    step_dummy(all_nominal_predictors())
## Modelltypen und Engine ------------------------------------------------
mod.re.xgboost<-                                        # Modell
   boost_tree() %>%
   set_engine("xgboost") %>%
   set_mode("regression")
## Workflow---------------------------------------------------------------
wfl.re.xg <-workflow() %>%                              # Workflow
  add_model(mod.re.xgboost) %>%
  add_recipe(rec.first)
wfl.re.xg

## ══ Workflow ════════════════════════════════════════════════════════════
## Preprocessor: Recipe
## Model: boost_tree()
##
## ── Preprocessor ────────────────────────────────────────────────────────
## 1 Recipe Step
##
## • step_dummy()
##
## ── Model ───────────────────────────────────────────────────────────────
## Boosted Tree Model Specification (regression)
##
## Computational engine: xgboost

## Machine-Learning-Modell trainieren-------------------------------------
wfl.rg.xg.fit<-fit(wfl.re.xg,ames.train)                # Modell trainieren
wfl.rg.xg.fit %>% extract_recipe(estimated=TRUE)        # Anzeige Modell

## Recipe
##
## Inputs:
##
##       role #variables
##    outcome          1
##  predictor         73
##
## Training data contained 2341 data points and no missing data.
##
## Operations:
##
## Dummy variables from Alley, Bldg_Type, Bsmt_Cond, Bsmt_Exposure, BsmtFin_Type_...
 [trained]

## Evaluation ------------------------------------------------------------
vorhersage<-ames.test %>%
  select(Sale_Price) %>%
  bind_cols(predict(wfl.rg.xg.fit,ames.test)) %>%
```

```
  exp()
ml.eval.lr(vorhersage,"Modell mod.re.xgboost","Hauspreis in 1000 US$",scatter=T)
```

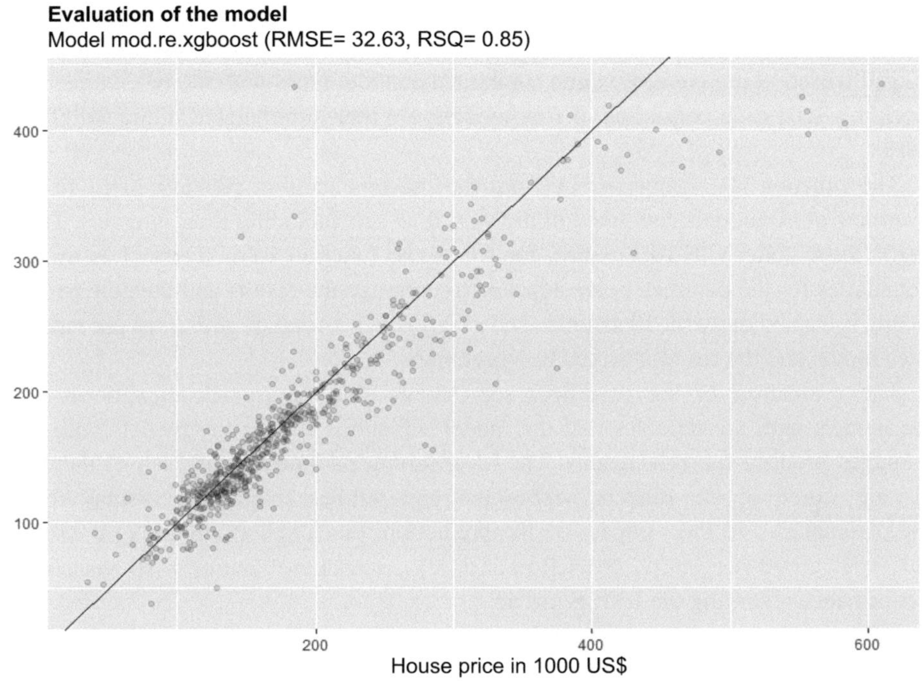


```
##    Kennzahl      Wert
## 1      rmse 32.629449
## 2       rsq  0.850048
```

After all available variables were used as predictors in the preceding example, the following coding will demonstrate how a prediction of similar or better quality can be achieved with fewer variables. For this purpose, several steps are taken to tune the machine learning.

First, two recipes rec.all with all 73 variables and rec.sel with 17 selected variables are created. Both variants are to be compared with each other and with the untuned model from before. Subsequently, four model types (many more model types could be selected here to determine the most optimal model type) and two workflow sets wflset.all and wflset.sel are created, which combine the model types with the two recipes. These workflow sets, each consisting of four models, are then summarized in the workflow set wflset.

The function `ml.tunepar()` from the machinelearning package helps to get an overview of all models contained in the workflow sets including their intervals. The tuning is to be done with 10 samples, for which the function `vfold_cv()` is used. The boundaries for the parameters are adjusted to optimize the results and then the parameter tuning is carried out with 10 samples and 10 grids, i.e. a total of 100 times per model, for each model and the result is stored in wflset.map.

The evaluation of the workflow sets shows that the model all_xgboost among the models with all variables and the model sel_cubist among the models with the 17 variables produce the best results. The ideal tuning parameters for each of these models are stored in the objects wfl.best.par and wfl.best.sel.par. To display the tuning parameters and their impact on the prediction, the function autoplot() is used, e.g. `autoplot(wflset.map,id=wflow.id[i],metric="rmse")` to visualize the performance regarding the RMSE metric.

After the best model types and their associated tuning parameters have been determined, the training and evaluation of the two best models follows.

Regression: Modeling with Tuning

```
## Recipe-----------------------------------------------------------------------------
rec.all <- ames.train %>%                                             # Alle Variablen
  recipe(Sale_Price ~ .) %>%
  step_log(all_numeric_predictors(),-all_outcomes()) %>%
  step_normalize(all_numeric_predictors()) %>%
  step_other(all_nominal_predictors(),threshold=0.01) %>%
  step_dummy(all_nominal_predictors())
rec.sel <- ames.train %>%                                             # Vorselektierte V.
  recipe(Sale_Price ~ Gr_Liv_Area + Garage_Cars + Year_Built + Total_Bsmt_SF +
         Fireplaces + Mas_Vnr_Area + Wood_Deck_SF + Half_Bath + Open_Porch_SF +
         Longitude + Latitude +  Neighborhood + Lot_Shape + Condition_1 +
         Bldg_Type + House_Style + Overall_Cond) %>%
  step_log(all_numeric_predictors(),-all_outcomes()) %>%
  step_normalize(all_numeric_predictors()) %>%
  step_other(all_nominal_predictors(),threshold=0.01) %>%
  step_dummy(all_nominal_predictors())
## Modelltypen und Engine -------------------------------------------------------------
mod.re.cubist<-
   cubist_rules(committees=tune(),neighbors=tune()) %>%
   set_engine("Cubist")
mod.re.dectreerpart<-
   decision_tree(cost_complexity=tune(),min_n=tune()) %>%
   set_engine("rpart") %>%
   set_mode("regression")
mod.re.glmnet <-
  linear_reg(penalty=tune(),mixture=tune()) %>%
  set_engine("glmnet")
mod.re.xgboost<-
   boost_tree(tree_depth=tune(),learn_rate=tune(),loss_reduction=tune(),
              min_n=tune(),sample_size=tune(),trees=tune()) %>%
   set_engine("xgboost") %>%
   set_mode("regression")
# Modell Kombinationen----------------------------------------------------------------
mod.komb<- list(
  cubist        = mod.re.cubist,
  dectreerpart  = mod.re.dectreerpart,
  glmnet        = mod.re.glmnet,
  xgboost       = mod.re.xgboost)
## Workflow-Set via Modell-Kombinationen und Preprocess Recipe definieren --------
library(rules)

wflset.all<-workflow_set(
  preproc = list(all=rec.all),
  models  = mod.komb)
wflset.all

## # A workflow set/tibble: 4 × 4
##   wflow_id          info              option     result
##   <chr>             <list>            <list>     <list>
## 1 all_cubist        <tibble [1 × 4]> <opts[0]> <list [0]>
## 2 all_dectreerpart <tibble [1 × 4]> <opts[0]> <list [0]>
## 3 all_glmnet        <tibble [1 × 4]> <opts[0]> <list [0]>
## 4 all_xgboost       <tibble [1 × 4]> <opts[0]> <list [0]>

wflset.sel<-workflow_set(
  preproc = list(sel=rec.sel),
  models  = mod.komb)
wflset.sel
```

```
## # A workflow set/tibble: 4 × 4
##   wflow_id         info             option     result
##   <chr>            <list>           <list>     <list>
## 1 sel_cubist       <tibble [1 × 4]> <opts[0]> <list [0]>
## 2 sel_dectreerpart <tibble [1 × 4]> <opts[0]> <list [0]>
## 3 sel_glmnet       <tibble [1 × 4]> <opts[0]> <list [0]>
## 4 sel_xgboost      <tibble [1 × 4]> <opts[0]> <list [0]>
wflset<-rbind(wflset.all,wflset.sel)
wflset

## # A workflow set/tibble: 8 × 4
##   wflow_id         info             option     result
##   <chr>            <list>           <list>     <list>
## 1 all_cubist       <tibble [1 × 4]> <opts[0]> <list [0]>
## 2 all_dectreerpart <tibble [1 × 4]> <opts[0]> <list [0]>
## 3 all_glmnet       <tibble [1 × 4]> <opts[0]> <list [0]>
## 4 all_xgboost      <tibble [1 × 4]> <opts[0]> <list [0]>
## 5 sel_cubist       <tibble [1 × 4]> <opts[0]> <list [0]>
## 6 sel_dectreerpart <tibble [1 × 4]> <opts[0]> <list [0]>
## 7 sel_glmnet       <tibble [1 × 4]> <opts[0]> <list [0]>
## 8 sel_xgboost      <tibble [1 × 4]> <opts[0]> <list [0]>

## Liste der Tuningparameter----------------------------------------------------
ml.tunepar(wflset.sel)                                         # Tuningparameter

##              Workflow   Paramter-Name                         Parameter Datentyp
## 1          sel_cubist      committees                        Committees  integer
## 2          sel_cubist       neighbors                 Nearest Neighbors  integer
## 3    sel_dectreerpart cost_complexity           Cost-Complexity Parameter   double
## 4    sel_dectreerpart           min_n                 Minimal Node Size  integer
## 5          sel_glmnet         penalty          Amount of Regularization   double
## 6          sel_glmnet         mixture        Proportion of Lasso Penalty   double
## 7         sel_xgboost           trees                             Trees  integer
## 8         sel_xgboost           min_n                 Minimal Node Size  integer
## 9         sel_xgboost      tree_depth                        Tree Depth  integer
## 10        sel_xgboost      learn_rate                     Learning Rate   double
## 11        sel_xgboost  loss_reduction            Minimum Loss Reduction   double
## 12        sel_xgboost     sample_size Proportion Observations Sampled   double
##    Intervall Werte
## 1      1-100
## 2        0-9
## 3     -10--1
## 4       2-40
## 5      -10-0
## 6     0.05-1
## 7     1-2000
## 8       2-40
## 9       1-15
## 10  -3--0.5
## 11  -10-1.5
## 12    0.1-1

## Bootstrap folds für Tuning---------------------------------------------------
set.seed(123)
ames.boots<-vfold_cv(ames.train,v=10,strata=Sale_Price)
## Tuning der Parameter---------------------------------------------------------
# cubist - Parametertuning
wfl.id="all_cubist"
wfl<-wflset %>% extract_workflow(id=wfl.id)
wfl.par<-extract_parameter_set_dials(wfl)
wfl.par
```

```
## Collection of 2 parameters for tuning
##
##  identifier       type    object
##  committees committees nparam[+]
##   neighbors  neighbors nparam[+]
wfl.par<-wfl.par %>%
   update(neighbors=neighbors(c(5,9)))
wflset<-wflset %>%
   option_add(id=wfl.id,param_info=wfl.par)
wfl.id="sel_cubist"
wfl<-wflset %>% extract_workflow(id=wfl.id)
wflset<-wflset %>%
   option_add(id=wfl.id,param_info=wfl.par)
# dectreerpart - Parametertuning
wfl.id="all_dectreerpart"
wfl<-wflset %>% extract_workflow(id=wfl.id)
wfl.par<-extract_parameter_set_dials(wfl)
wfl.par

## Collection of 2 parameters for tuning
##
##       identifier            type    object
##  cost_complexity cost_complexity nparam[+]
##            min_n           min_n nparam[+]

wfl.par<-wfl.par %>%
   update(cost_complexity=cost_complexity(c(-10,-4)),
          min_n=min_n(c(15,40)))
wflset<-wflset %>%
   option_add(id=wfl.id,param_info=wfl.par)
wfl.id="sel_dectreerpart"
wfl<-wflset %>% extract_workflow(id=wfl.id)
wflset<-wflset %>%
   option_add(id=wfl.id,param_info=wfl.par)
# glmnet - Parametertuning
wfl.id="all_glmnet"
wfl<-wflset %>% extract_workflow(id=wfl.id)
wfl.par<-extract_parameter_set_dials(wfl)
wfl.par

## Collection of 2 parameters for tuning
##
##  identifier    type    object
##     penalty penalty nparam[+]
##     mixture mixture nparam[+]

wfl.par<-wfl.par %>%
   update(penalty=penalty(c(-10,-2.5)))
wflset<-wflset %>%
   option_add(id=wfl.id,param_info=wfl.par)
wfl.id="sel_glmnet"
wfl<-wflset %>% extract_workflow(id=wfl.id)
wflset<-wflset %>%
   option_add(id=wfl.id,param_info=wfl.par)
#xgboost - Parametertuning
wfl.id="all_xgboost"
wfl<-wflset %>% extract_workflow(id=wfl.id)
wfl.par<-extract_parameter_set_dials(wfl)
wfl.par
```

```
## Collection of 6 parameters for tuning
##
##      identifier          type    object
##           trees         trees nparam[+]
##           min_n         min_n nparam[+]
##      tree_depth    tree_depth nparam[+]
##      learn_rate    learn_rate nparam[+]
##  loss_reduction loss_reduction nparam[+]
##     sample_size   sample_size nparam[+]

wfl.par<-wfl.par %>%
   update(trees=trees(c(1110,2000)),
          tree_depth=tree_depth(c(4,10)),
          learn_rate=learn_rate(c(-1.5,-0.5)))
wflset<-wflset %>%
   option_add(id=wfl.id,param_info=wfl.par)
wfl.id="sel_xgboost"
wfl<-wflset %>% extract_workflow(id=wfl.id)
wflset<-wflset %>%
   option_add(id=wfl.id,param_info=wfl.par)
# Grid mit 100 Durchläufen (10 Bootstrapstichproben*10 Grid) zum Parametertuning--
grid.control<-control_grid(
  save_pred      = TRUE,
  parallel_over = "everything",
  save_workflow = TRUE)
wflset.map<-wflset %>%
  workflow_map(
    seed         = 123,
    resamples    = ames.boots,
    grid         = 10,
    control      = grid.control)
wflset.map

## # A workflow set/tibble: 8 × 4
##   wflow_id         info             option     result
##   <chr>            <list>           <list>     <list>
## 1 all_cubist       <tibble [1 × 4]> <opts[4]> <tune[+]>
## 2 all_dectreerpart <tibble [1 × 4]> <opts[4]> <tune[+]>
## 3 all_glmnet       <tibble [1 × 4]> <opts[4]> <tune[+]>
## 4 all_xgboost      <tibble [1 × 4]> <opts[4]> <tune[+]>
## 5 sel_cubist       <tibble [1 × 4]> <opts[4]> <tune[+]>
## 6 sel_dectreerpart <tibble [1 × 4]> <opts[4]> <tune[+]>
## 7 sel_glmnet       <tibble [1 × 4]> <opts[4]> <tune[+]>
## 8 sel_xgboost      <tibble [1 × 4]> <opts[4]> <tune[+]>

# Workflow-Sets nach Tuning mit Resamples evaluieren----------------------------
collect_metrics(wflset.map,summarize=FALSE)

## # A tibble: 1,560 × 8
##    wflow_id   .config              preproc model id    .metric .esti…¹ .esti…²
##    <chr>      <chr>                <chr>   <chr> <chr> <chr>   <chr>     <dbl>
##  1 all_cubist Preprocessor1_Model01 recipe cubis… Fold… rmse   standa…   0.164
##  2 all_cubist Preprocessor1_Model01 recipe cubis… Fold… rsq    standa…   0.855
##  3 all_cubist Preprocessor1_Model01 recipe cubis… Fold… rmse   standa…   0.137
##  4 all_cubist Preprocessor1_Model01 recipe cubis… Fold… rsq    standa…   0.894
##  5 all_cubist Preprocessor1_Model01 recipe cubis… Fold… rmse   standa…   0.116
##  6 all_cubist Preprocessor1_Model01 recipe cubis… Fold… rsq    standa…   0.917
##  7 all_cubist Preprocessor1_Model01 recipe cubis… Fold… rmse   standa…   0.131
##  8 all_cubist Preprocessor1_Model01 recipe cubis… Fold… rsq    standa…   0.910
```

```
##  9 all_cubist Preprocessor1_Model01 recipe  cubis… Fold… rmse    standa…   0.128
## 10 all_cubist Preprocessor1_Model01 recipe  cubis… Fold… rsq     standa…   0.900
## # … with 1,550 more rows, and abbreviated variable names ¹.estimator,
## #   ².estimate
wflset.kz<-nrow(collect_metrics(wflset.map,summarize=FALSE))/2
wflset.map %>%
   rank_results() %>%
   filter(.metric == "rmse") %>%
   select(wflow_id, model, .config, rmse = mean, rank)

## # A tibble: 80 × 5
##    wflow_id    model      .config                rmse  rank
##    <chr>       <chr>      <chr>                 <dbl> <int>
##  1 all_xgboost boost_tree Preprocessor1_Model02 0.130     1
##  2 all_xgboost boost_tree Preprocessor1_Model03 0.130     2
##  3 all_xgboost boost_tree Preprocessor1_Model08 0.135     3
##  4 all_xgboost boost_tree Preprocessor1_Model10 0.136     4
##  5 all_glmnet  linear_reg Preprocessor1_Model09 0.138     5
##  6 all_glmnet  linear_reg Preprocessor1_Model01 0.138     6
##  7 all_glmnet  linear_reg Preprocessor1_Model03 0.139     7
##  8 all_glmnet  linear_reg Preprocessor1_Model04 0.139     8
##  9 all_glmnet  linear_reg Preprocessor1_Model10 0.139     9
## 10 all_glmnet  linear_reg Preprocessor1_Model05 0.139    10
## # … with 70 more rows

autoplot(wflset.map,
   rank_metric = "rmse",     # Rangordnung entsprechend Kennzahl
   metric      = "rmse",     # Visualisierte Kennzahl
   select_best = TRUE)+      # Bestes Ergebnis je Workflow anzeigen
   geom_text(aes(y = mean - 1/2, label = wflow_id), angle = 90, hjust = 1) +
   lims(y = c(-0.8,1.5)) +
   theme(legend.position = "none")
```

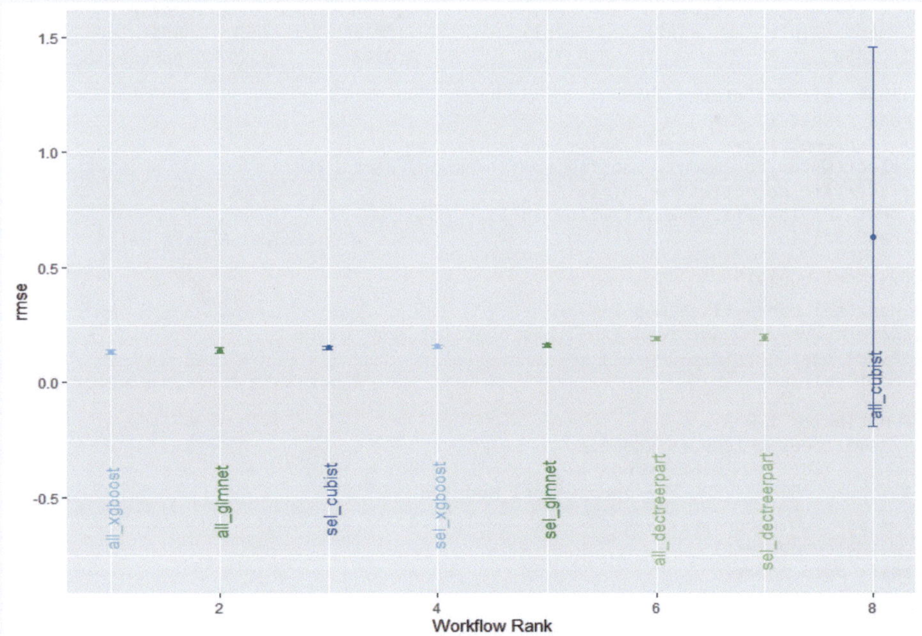

```
# Best Result
wfl.rank<-wflset.map %>%
   rank_results() %>%
   filter(.metric == "rmse") %>%
   select(wflow_id, model, .config, rmse = mean, rank) %>%
   filter(str_detect(wflow_id,"all"))
wfl.rank

## # A tibble: 40 × 5
##    wflow_id    model      .config                rmse  rank
##    <chr>       <chr>      <chr>                 <dbl> <int>
##  1 all_xgboost boost_tree Preprocessor1_Model02 0.130     1
##  2 all_xgboost boost_tree Preprocessor1_Model03 0.130     2
##  3 all_xgboost boost_tree Preprocessor1_Model08 0.135     3
##  4 all_xgboost boost_tree Preprocessor1_Model10 0.136     4
##  5 all_glmnet  linear_reg Preprocessor1_Model09 0.138     5
##  6 all_glmnet  linear_reg Preprocessor1_Model01 0.138     6
##  7 all_glmnet  linear_reg Preprocessor1_Model03 0.139     7
##  8 all_glmnet  linear_reg Preprocessor1_Model04 0.139     8
##  9 all_glmnet  linear_reg Preprocessor1_Model10 0.139     9
## 10 all_glmnet  linear_reg Preprocessor1_Model05 0.139    10
## # … with 30 more rows

wfl.best.id<-wfl.rank$wflow_id[1]
wfl.best.id                                       # Optimales Modell all_*

## [1] "all_xgboost"

wfl.best.par<-wflset.map %>%
   extract_workflow_set_result(wfl.best.id) %>%
   select_best(metric = "rmse")
wfl.best.par                                      # Optimale Tuningparameter

## # A tibble: 1 × 7
##   trees min_n tree_depth learn_rate loss_reduction sample_size .config
##   <int> <int>      <int>      <dbl>          <dbl>       <dbl> <chr>
## 1  1450     8          7     0.0389         0.0104       0.437 Preprocessor1_Mo…

wfl.rank.sel<-wflset.map %>%
   rank_results() %>%
   filter(.metric == "rmse") %>%
   select(wflow_id, model, .config, rmse = mean, rank) %>%
   filter(str_detect(wflow_id,"sel"))
wfl.best.sel.id<-wfl.rank.sel$wflow_id[1]
wfl.best.sel.id                                   # Optimales Modell sel_*

## [1] "sel_cubist"

wfl.best.sel.par<-wflset.map %>%
   extract_workflow_set_result(wfl.best.sel.id) %>%
   select_best(metric = "rmse")
wfl.best.sel.par                                  # Optimale Tuningparameter

## # A tibble: 1 × 3
##   committees neighbors .config
##        <int>     <int> <chr>
## 1         52         7 Preprocessor1_Model06

# Plot der Tuningparameter---------------------------------------------------
wflow.id<-character()
kz.best<-data.frame()
kz.plot<-list()
for(i in 1:3) {
  wflow.id[i]<-wflset.map$wflow_id[i]
```

```
  wfl.best<-wflset.map %>%
   extract_workflow_set_result(wflow.id[i]) %>%
   select_best(metric = "rmse")
  kz.best[i,1]<-paste("Modell", wflow.id[i],":",format_csv(wfl.best,eol=":"))
  px<-autoplot(wflset.map,id=wflow.id[i],metric="rmse") + labs(title=wflow.id[i])
  assign(paste("p",i,sep="."),px)
}
for(i in 1:3) {
  name <- paste("p",i,sep=".")
  inhalt<-get(name)
  grid.arrange(inhalt,nrow=1,ncol=1)
}
```

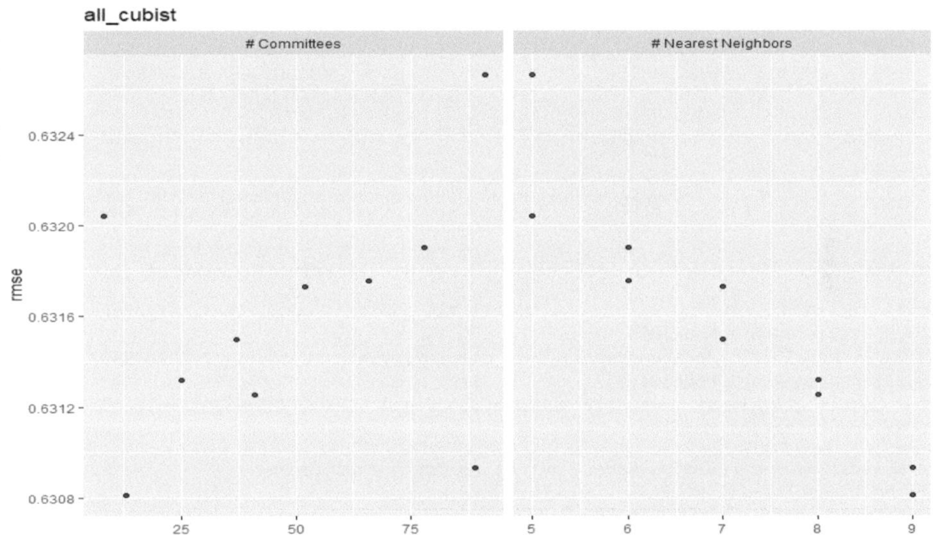

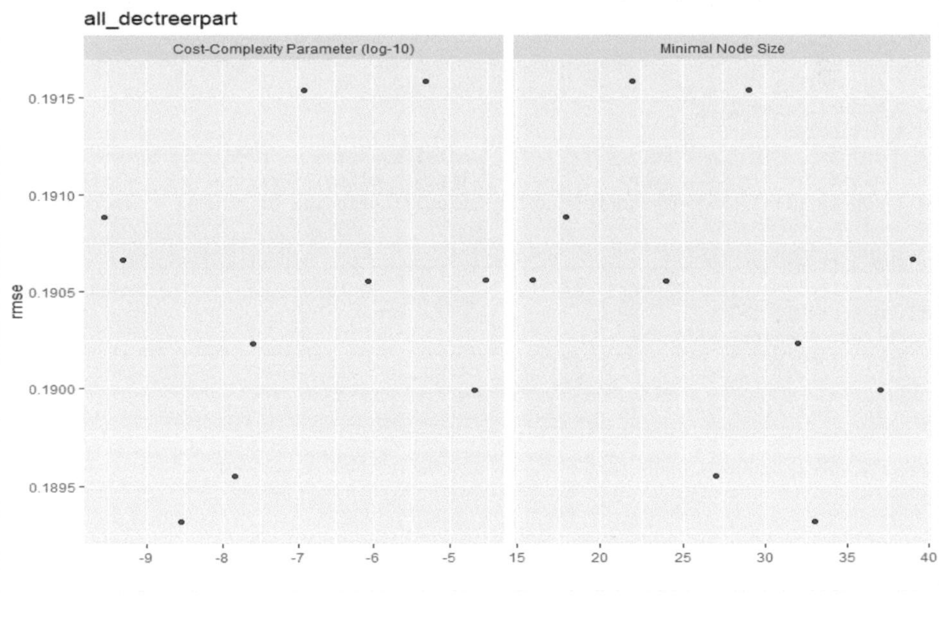

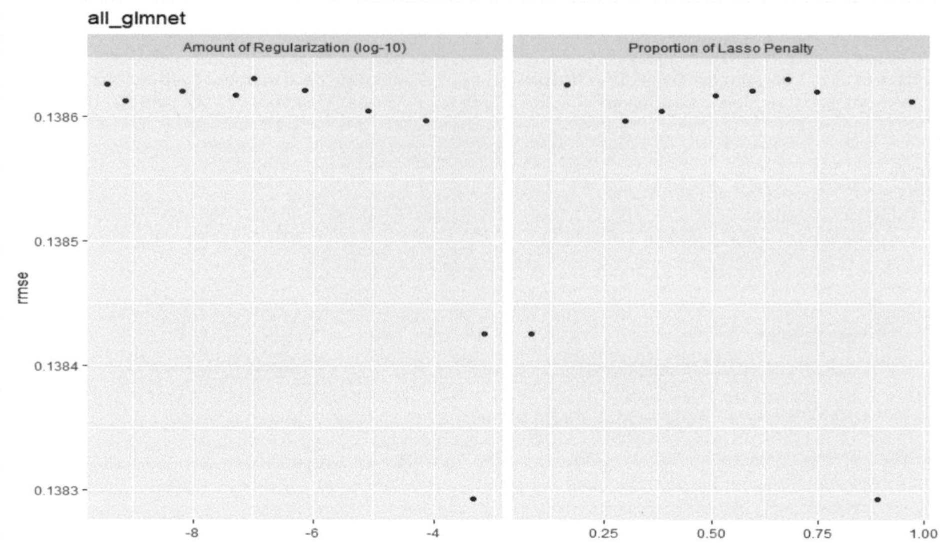


```
kz.best                                          # Beste Tuningwerte je Modell

## 1                              Modell all_cubist : committees,neighbors,.config
:13,9,Preprocessor1_Model02:
## 2     Modell all_dectreerpart : cost_complexity,min_n,.config:2.8894383577460305e
-9,33,Preprocessor1_Model01:
## 3 Modell all_glmnet : penalty,mixture,.config:4.5485865306763793e-4,0.89204747598
26902,Preprocessor1_Model09:

## Bestes Machine-Learning-Modell mit allen Daten trainieren---------------------
wfl.best.result<-wflset %>%
   extract_workflow(wfl.best.id) %>%
   finalize_workflow(wfl.best.par) %>%
   last_fit(split=ames.split)

wfl.best.result

## # Resampling results
## # Manual resampling
## # A tibble: 1 × 6
##   splits             id               .metrics .notes   .predictions .workflow
##   <list>             <chr>            <list>   <list>   <list>       <list>
## 1 <split [2341/587]> train/test split <tibble> <tibble> <tibble>     <workflow>
##

## Evaluation -------------------------------------------------------------------
best<-collect_metrics(wfl.best.result)          # Kennzahlen des besten Modells
wfl.best.result %>%
   collect_predictions() %>%
   ggplot(aes(x=Sale_Price,y=.pred)) +
     geom_abline(color="gray50",lty=2) +
     geom_point(alpha=0.5) +
```

```
    coord_obs_pred()+
    labs(x="log von realem Preis in 1000 US$",y="log von vorhergesagtem Preis in 10
00 US$",
         title=paste0("Vorhersage von ",wfl.best.id),
         subtitle=paste0(best$.metric[1],"=",
                         round(best$.estimate[1],2),",",
                         best$.metric[2],"=",
                         round(best$.estimate[2],2)))
```

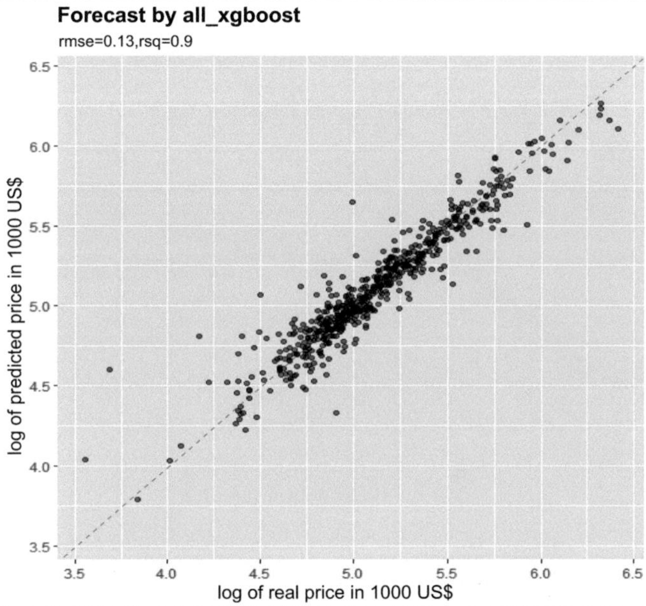


```
best.result<-wfl.best.result %>%
   collect_predictions() %>%
   select(Sale_Price,.pred) %>%
   exp()
ml.eval.lr(best.result,paste("Modell",wfl.best.id),"Hauspreis in 1000 US$",scatter=T
)
```

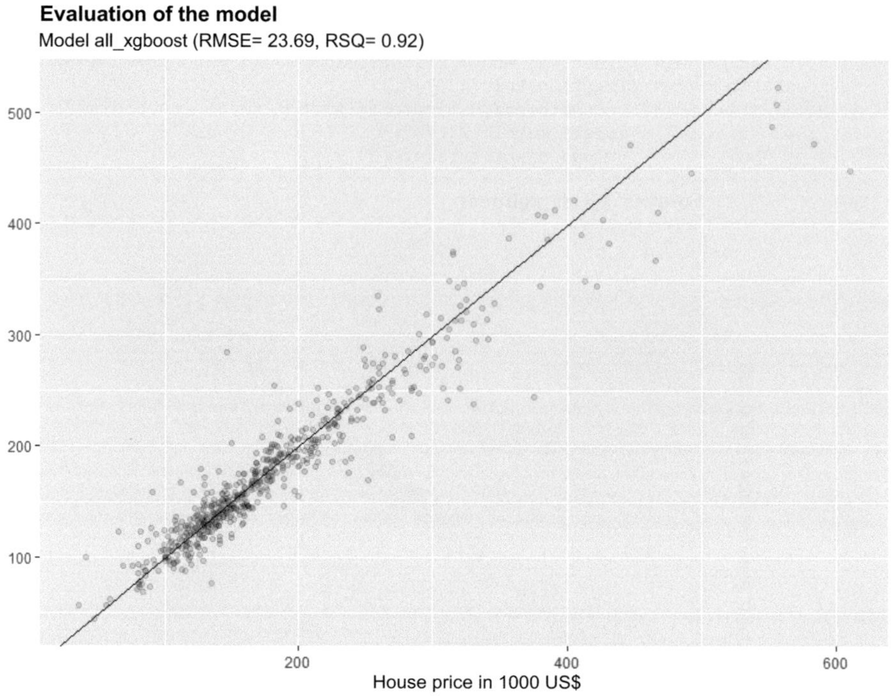


```
##   Kennzahl       Wert
## 1     rmse 23.6924902
## 2      rsq  0.9173744

## Bestes Machine-Learning-Modell mit selektiven Daten trainieren-----------------
wfl.best.result<-wflset %>%
   extract_workflow(wfl.best.sel.id) %>%
   finalize_workflow(wfl.best.sel.par) %>%
   last_fit(split=ames.split)

wfl.best.result

## # Resampling results
## # Manual resampling
## # A tibble: 1 × 6
##   splits             id               .metrics .notes   .predictions .workflow
##   <list>             <chr>            <list>   <list>   <list>       <list>
## 1 <split [2341/587]> train/test split <tibble> <tibble> <tibble>     <workflow>
##

## Evaluation -----------------------------------------------------------------------
best<-collect_metrics(wfl.best.result)             # Kennzahlen des besten Modells
wfl.best.result %>%
   collect_predictions() %>%
   ggplot(aes(x=Sale_Price,y=.pred)) +
     geom_abline(color="gray50",lty=2) +
     geom_point(alpha=0.5) +
     coord_obs_pred()+
     labs(x="log von realem Preis in 1000 US$",y="log von vorhergesagtem Preis in 10
```

```
00 US$",
         title=paste0("Vorhersage von ",wfl.best.sel.id),
         subtitle=paste0(best$.metric[1],"=",
                         round(best$.estimate[1],2),",",
                         best$.metric[2],"=",
                         round(best$.estimate[2],2)))
```

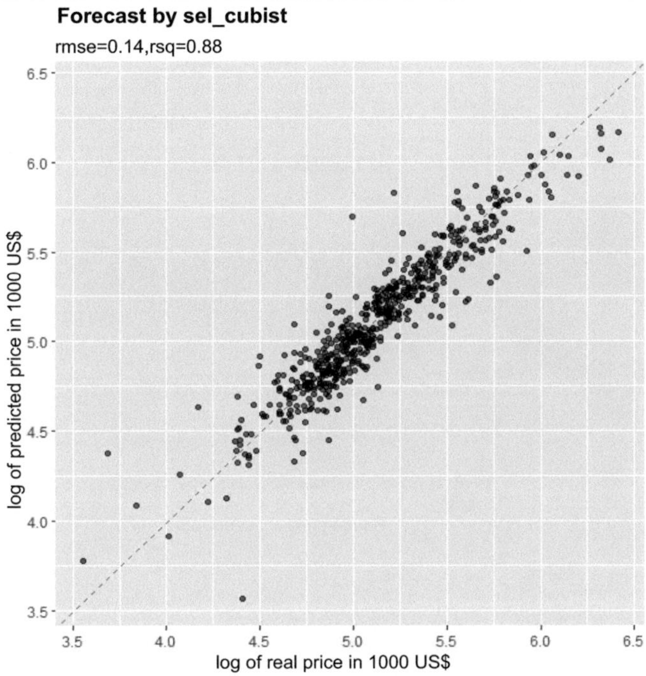


```
best.result<-wfl.best.result %>%
   collect_predictions() %>%
   select(Sale_Price,.pred) %>%
   exp()
ml.eval.lr(best.result,paste("Modell",wfl.best.sel.id),"Hauspreis in 1000 US$",scatt
er=T)
```

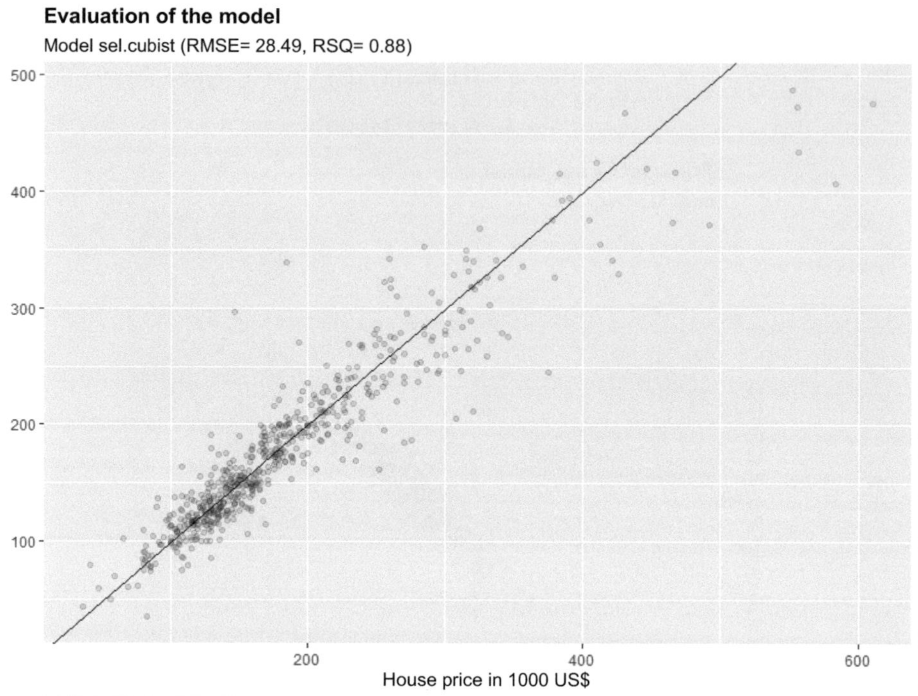


```
## Kennzahl       Wert
## 1     rmse 28.4927026
## 2      rsq  0.8788609
```

The tuning was worth it, as the key figures RMSE and RSQ both improved. The untuned model with all 73 predictors had an RMSE of 32,629 US$ and an RSQ of 0.85. The tuned model with 17 predictors was able to improve the RMSE to 28,490 US$ and the RSQ to 0.88. The tuned model with 73 predictors was slightly better with an RMSE of 23,690 and RSQ of 0.92. Therefore, the prediction error for the house price could be reduced from 32,629 to 23,690 US$. In this example, not all model types, variable combinations, and parameter optimizations were tried, but it became clear that tuning helps to improve the results.

7.1.2 Classification

The following example will present a classification of data. The data are real data on 887,379 loans with 74 variables from Kaggle (Dark_Raider 2023). The data can be found in the file loans.csv, which is read into R and stored in the dataframe loans. The goal is to be able to classify loan applications from the data, predicting whether a loan will be repaid or not. The variable loan_status indicates in the data to which class a loan is assigned. The prediction is intended to reduce losses for the credit institutions.

In the preprocessing, date variables are first converted to integer variables and character variables into factor variables to support the subsequent processing of machine learning. Variables with more than 20% NA values are removed from the dataframe as they are not suitable for analysis. Also, all variables that contain only 1 value or where each record has a different value are useless for prediction and are removed. Factor variables with many levels (different values) can greatly increase the complexity of the model. Therefore, the following code removes all variables that have such a large number of levels that on average fewer than 20 records are described by a level. Factor variables with only 2 or 3 levels are only valuable if the records of the levels differ significantly in the target variable, have a good discriminative power. The variables application_type and pymnt_plan are therefore removed due to their lack of discriminative power.

The dependent variable to be predicted is the default of a loan. The variable loan_status has several manifestations, which should be assigned to one of the two categories "good" or "bad". The variable status is created anew to mark all loans as "bad" that are "Charged Off", "Default", "Does not meet the credit policy. Status: Charged Off", "Late (16–30 days) = Considered delinquent" or "Late (31–120) = Delinquent" in loan_status. All other cases are categorized as "good". Afterwards, the variable loan_status can be deleted, as it should not be used for prediction.

In the preprocessing, it is also necessary to remove all records where the target variable is missing (status NA). The function table(loans$status) can be used to display how many of the loans are "good" or "bad". 6% of the loans were a loan default.

There are missing values (NA) in 79,488 of the records (867,328–787,750). The function `missRanger()` from the missranger package can replace missing values with values that are likely to be correct. With the help of this function, all missing data are first supplemented.

To be able to estimate which of the variables are suitable for prediction, a better understanding of the data is necessary. A display of the correlation of all numerical variables with the variable status is helpful here and can be determined with the function `cor()` after the factor variable status has been converted into the numerical variable status_num. Variables with a correlation less than 0.05 can be deleted as they will probably not be valuable for the forecast. Variables with a cross-correlation with another variable, which can be determined with the function `corr_cross()` from the lares package, are also removed. This applies to the variables collection_recovery_fee (with recoveries), issue_d (with last_pymnt_d), last_pymnt_amnt (with total_rec_prncp), last_credit_pull_d (with last_pymnt_d) and out_prncp_inv (with out_prncp). Some of the numerical variables are not normally distributed. Therefore, normalization should be done in the recipe using the function `step_normalize()`.

Classification: Preprocessing

```
## Laden der Daten---------------------------------------------------------------
loans <- read.csv("loan.csv")
loans<-loans[,c(sort(names(loans)))]                   # Variablen alphabetisch
# Datentyp ändern----------------------------------------------------------------
loans[1:7,c("issue_d","last_pymnt_d","next_pymnt_d","last_credit_pull_d",
            "earliest_cr_line")]
loans<-loans %>%                                      # Datumsvariablen -> Integer
  mutate(
    issue_d = as.numeric(my(issue_d)),
    last_pymnt_d = as.numeric(my(last_pymnt_d)),
    next_pymnt_d = as.numeric(my(next_pymnt_d)),
    last_credit_pull_d = as.numeric(my(last_credit_pull_d)),
    earliest_cr_line  = as.numeric(my(earliest_cr_line))
  )
loans[1:7,c("issue_d","last_pymnt_d","next_pymnt_d","last_credit_pull_d",
           "earliest_cr_line")]
loans <- loans %>%                                    # Charactervariable -> Faktor
  mutate_if(sapply(loans,is.character),as.factor)
# Datenqualität prüfen-----------------------------------------------------------
loans.var<-data.frame(                                # Tabelle mit Variablen
  var=names(loans),
  na=colSums(is.na(loans)),
  na.prozent=colSums(is.na(loans))/nrow(loans),
  distinct=sapply(loans, function(x) n_distinct(x)),
  nlevel=sapply(loans, function(x) nlevels(x)))
loans.var<-loans.var %>%
  mutate(distinct.prozent=distinct/nrow(loans))
# Variablen mit zu vielen NAs entfernen------------------------------------------
var.del.na<-loans.var[loans.var$na.prozent > 0.2,c("var","na","na.prozent")] %>%
  arrange(desc(na.prozent))
var.del.na
loans<-select(loans,-rownames(var.del.na))           # Var mit >20% NA löschen
loans.var<-filter(loans.var,!var %in% rownames(var.del.na))
# Variablen ohne Trennschärfe entfernen-----------------------------------------
var.del.dist<-loans.var[
                loans.var$distinct == 1 |             # Var mit 1 Wert für alle
                loans.var$distinct.prozent == 1,# Var mit keinem doppelten Wert
                c("var","distinct","distinct.prozent")] %>%
```

```
  arrange(desc(distinct))
var.del.dist
loans<-select(loans,-rownames(var.del.dist))     # Var mit 1|ungleich löschen
loans.var<-filter(loans.var,!var %in% rownames(var.del.dist))
# Variablen mit zu vielen Levels entfernen--------------------------------------
var.del.level<-loans.var[
                 loans.var$nlevel > (nrow(loans)/20),
                 c("var","nlevel")] %>%
  arrange(desc(nlevel))
var.del.level
loans<-select(loans,-rownames(var.del.level))
loans.var<-filter(loans.var,!var %in% rownames(var.del.level))
# Variablen mit geringer Trennschärfe entfernen---------------------------------
var.wenige.level<-loans.var[loans.var$nlevel %in% c(1,2,3),c("var","nlevel")] %>%
  arrange(desc(nlevel))
var.wenige.level
loans$verification_status %>% table() %>% prop.table()   # Trennschärfe gut
loans$application_type %>% table() %>% prop.table()      # Trennschärfe schlecht
loans$initial_list_status %>% table() %>% prop.table()   # Trennschärfe gut
loans$pymnt_plan %>% table() %>% prop.table()            # Trennschärfe schlecht
loans$term %>% table() %>% prop.table()                  # Trennschärfe gut
loans<-select(loans,-c("application_type","pymnt_plan")) # Schlechte Var. entf.
## Abhängige Variable-----------------------------------------------------------
table(loans$loan_status) %>% prop.table() %>% round(4)
loans$status<-
  case_when(
    loans$loan_status == "Charged Off" ~ "schlecht",
    loans$loan_status == "Current" ~ "gut",
    loans$loan_status == "Default" ~ "schlecht",
    loans$loan_status == "Does not meet the credit policy. Status:Charged Off" ~
                         "schlecht",
    loans$loan_status == "Does not meet the credit policy. Status:Fully Paid" ~
                         "gut",
    loans$loan_status == "Fully Paid" ~ "gut",
    loans$loan_status == "In Grace Period" ~ "gut",
    loans$loan_status == "Late (16-30 days)" ~ "schlecht",
    loans$loan_status == "Late (31 - 120)" ~ "schlecht")
loans<-select(loans,-loan_status)
loans$status<-as.factor(loans$status)
sum(is.na(loans$status))
loans<-filter(loans,!is.na(status))              # Datensätze ohne Zielvariable entf.
table(loans$status)                              # Risiko von 6% für Kreditausfall
table(loans$status) %>% prop.table() %>% round(2)
## Datenverständnis-------------------------------------------------------------
table(loans$purpose) %>% prop.table() %>%        # Recipe mit step_other 0.01 löst
  round(4)                                       # das Problem mit kleinen Gruppen
## NA-Werte durch Schätzwerte ersetzen------------------------------------------
varprodatensatz<-rowSums(!is.na(loans))          # Anzahl Datensätze mit Nicht-NAs
table(varprodatensatz)                           # 787750 Datensätze mit allen 46 V.
loans <- missRanger::missRanger(data=loans,
           formula = . ~ .,
           seed = 123,
           pmm.k = 4,
           num.trees = 5)
varprodatensatz<-rowSums(!is.na(loans))          # Anzahl Datensätze mit Nicht-NAs
table(varprodatensatz)                           # 867328 Datensätze mit allen 46 V.
## Relevanz von Variablen prüfen------------------------------------------------
loans$status_num<-                               # Numerische Var. für Korrelation
  case_when(loans$status == "schlecht" ~ 1,
```

```
            loans$status == "gut" ~ 0)
var.numeric <- select(loans,where(is.numeric)) %>% names()
var.name<-character()
var.korrelation<-numeric()
for(i in var.numeric){
  korrelation<-cor(x=loans[i],y=loans["status_num"])
  var.name<- append(var.name,i)
  var.korrelation<-append(var.korrelation,korrelation)
}
korrelationen<-data.frame(Korrelation=var.korrelation,Variable=var.name) %>%
  arrange(var.korrelation)
korrelationen<-korrelationen[korrelationen$Variable != "status_num",]
ergebnis<-ml.summary.2(korrelationen,"Korrelation von Variable mit Kreditausfall",
                       "Pearson Korrelation","Variable",bar=T,bw=T)
var.del.korr<-korrelationen[abs(korrelationen$Korrelation) < 0.05,]$Variable
loans<-select(loans,-all_of(var.del.korr))              # Variablen entfernen
## Multikollinearität von Variablen prüfen----------------------------------------
loans.num<-select_if(loans,is.numeric)                  # Numerische Variablen
#devtools::install_github("laresbernardo/lares")
#library(lares)
lares::corr_cross(loans.num,
  max_pvalue = 0.05,                                    # Signifikanz >95%
  top = 20                                              # TOP 20 Korrelationen
)
var.del.cross<-c("collection_recovery_fee","issue_d",
                 "last_pymnt_amnt","last_credit_pull_d",
                 "out_prncp_inv")
loans<-select(loans,-all_of(var.del.cross))             # Variablen entfernen
## Verteilung von numerischen Variablen prüfen------------------------------------
loans.num<-select_if(loans,is.numeric)               # Numerische Variablen
lares::corr_var(loans.num,
  status_num,
  top = 8)                                           # Max. Top 8 Korrelationen
loans.num<-select(loans.num,-inq_last_6mths)         # Var mit geringer Korr. entfernen
loans<-select(loans,-status_num)                     # Var status_num nicht mehr nötig
loans.num<-loans.num[,1:7]
# Variable mit größter Korrelation nicht normalverteilt->in Recipe logarithmieren
ml.summary(loans.num$recoveries,"recoveries","recoveries")
ml.summary(loans.num$last_pymnt_d,"last_pymnt_d","last_pymnt_d")
ml.summary(loans.num$out_prncp,"out_prncp","out_prncp")
ml.summary(loans.num$int_rate,"int_rate","int_rate")
# Keine Normalverteilung von recoveries, last_pymnt_d & out_prncp
# -> in Recipe standardisieren
## Kategoriale Variablen besser verstehen-----------------------------------------
erg<-ml.summary.2(loans[,c("status","home_ownership")],
                  "Home Ownership","Kreditausfall","Ownership",box=T,npos=0.5,dec=3)
# Geringe Anzahl bei "Any", Höherer Mittelwert bei "Other" und "None"
# -> in Recipe kleine Gruppen zusammenfassen
erg<-ml.summary.2(loans[,c("status","emp_length")],
                  "emp_length","Kreditausfall","emp_length",box=T,npos=0.5,dec=3)
# Geringer Mittelwert bei "10+ years"
erg<-ml.summary.2(loans[,c("status","grade")],
                  "grade","Kreditausfall","grade",box=T,npos=0.5,dec=3)
# Höherer Mittelwert bei "E", "F" und "G"
}
```

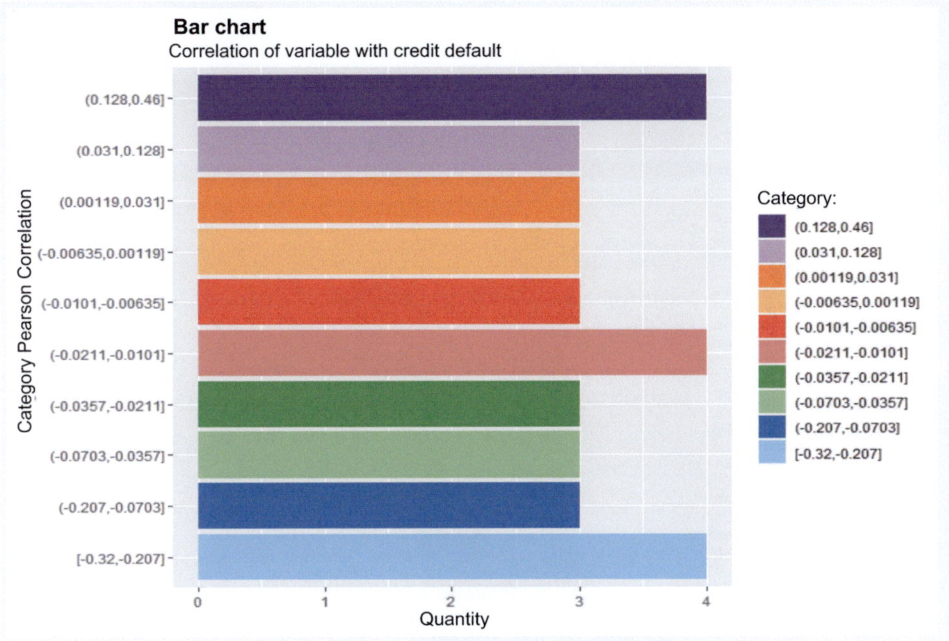
Bar chart
Correlation of variable with credit default
Category Pearson Correlation
(0.128,0.46]
(0.031,0.128]
(0.00119,0.031]
(-0.00635,0.00119]
(-0.0101,-0.00635]
(-0.0211,-0.0101]
(-0.0357,-0.0211]
(-0.0703,-0.0357]
(-0.207,-0.0703]
[-0.32,-0.207]
0
1
2
3
4
Quantity
Category:

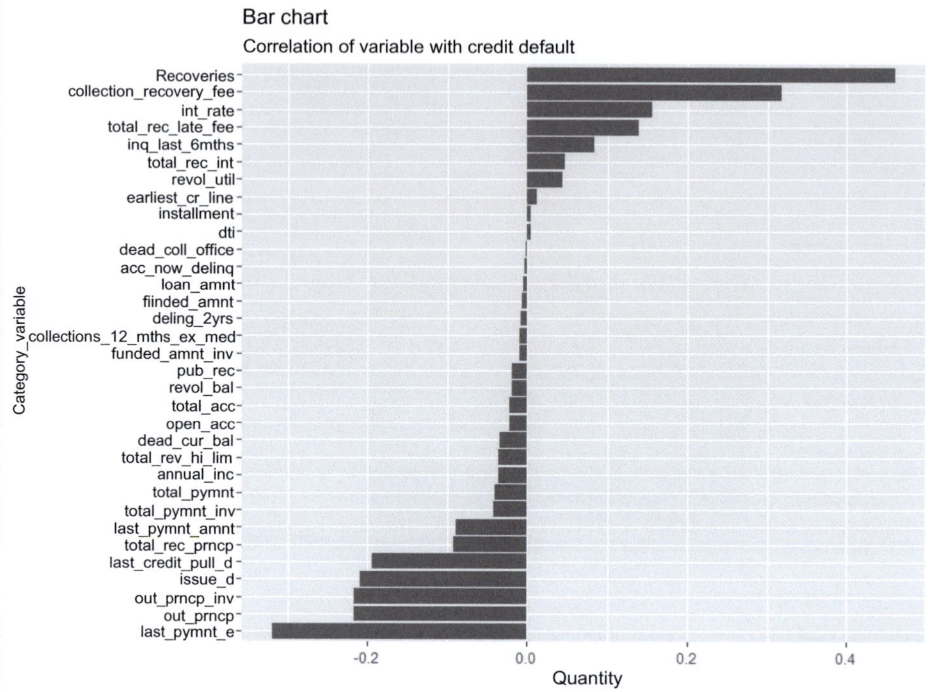
Bar chart
Correlation of variable with credit default
Recoveries
collection_recovery_fee
int_rate
total_rec_late_fee
inq_last_6mths
total_rec_int
revol_util
earliest_cr_line
installment
dti
dead_coll_office
acc_now_delinq
loan_amnt
fiinded_amnt
deling_2yrs
collections_12_mths_ex_med
funded_amnt_inv
pub_rec
revol_bal
total_acc
open_acc
dead_cur_bal
total_rev_hi_lim
annual_inc
total_pymnt
total_pymnt_inv
last_pymnt_amnt
total_rec_prncp
last_credit_pull_d
issue_d
out_prncp_inv
out_prncp
last_pymnt_e
Category_variable
-0.2
0.0
0.2
0.4
Quantity

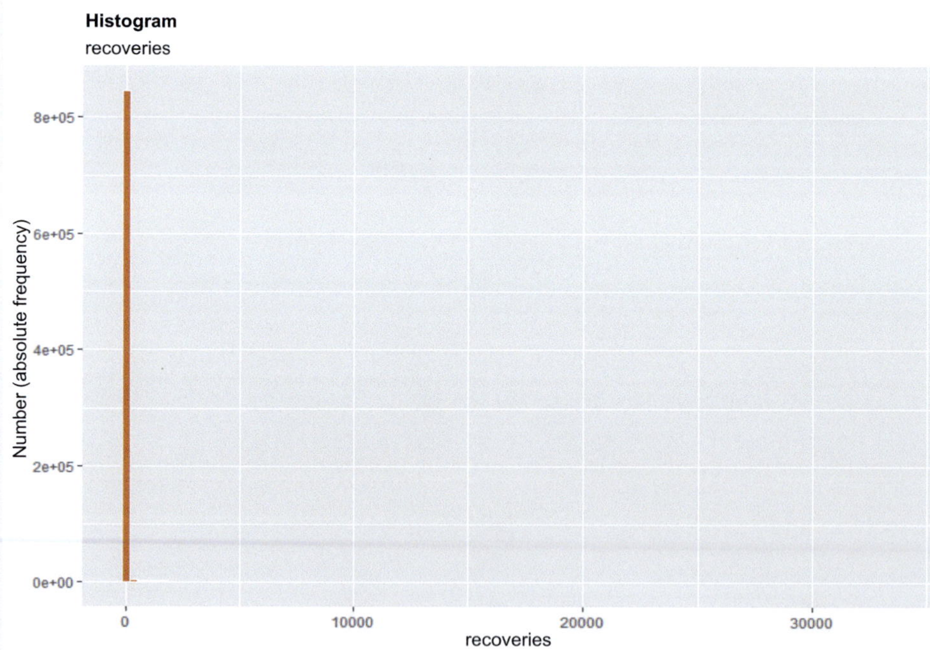
Histogram
recoveries
8e+05
6e+05
4e+05
2e+05
0e+00
Number (absolute frequency)
0
10000
20000
30000
recoveries

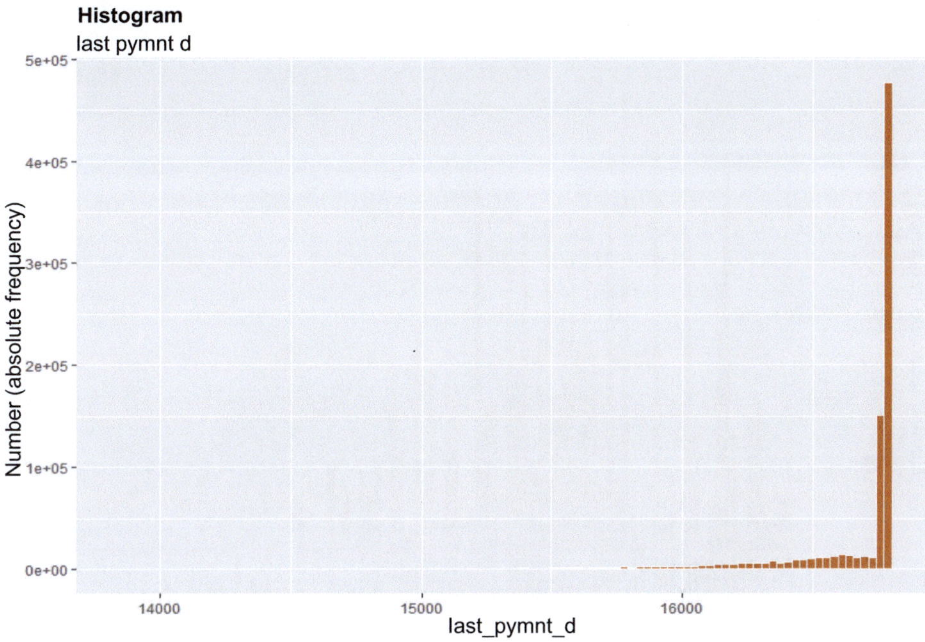
Histogram
last pymnt d
Number (absolute frequency)
5e+05
4e+05
3e+05
2e+05
1e+05
0e+00
14000
15000
16000
last_pymnt_d

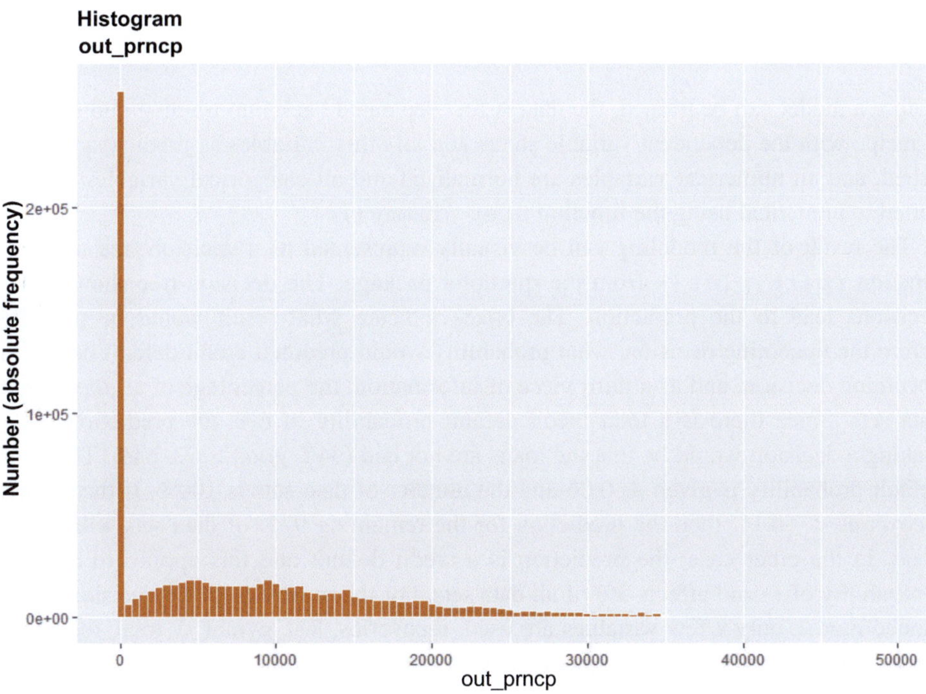
Histogram
out_prncp
Number (absolute frequency)
2e+05
1e+05
0e+00
0
10000
20000
30000
40000
50000
out_prncp

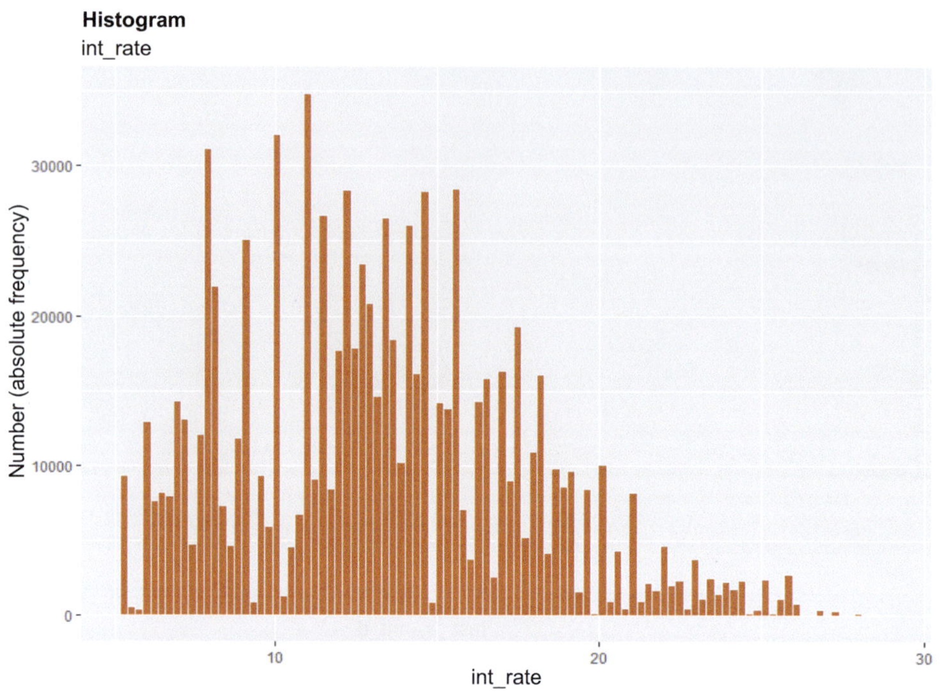

Before a model can be trained, the data is first separated back into training and test data. A recipe with the dependent variable status and all other variables as predictors is established, and all numerical variables are normalized and all categorical variables are converted to numerical using the function `step_dummy()`.

The result of the modeling can be visually represented as a decision tree using the function `rpart.plot()` from the rpart.plot package. The decision tree shows which decisions lead to the prediction. The boxes indicate what result would be predicted before the upcoming decision, what probability would predict a credit default before the upcoming decision, and as a third piece of information, the percentage of all represented data sets. Since there is a total credit default probability of 6%, the prediction before making a decision would be that the loans are not bad (94% good > 6% bad). The credit default probability is given as 0.06 and the number of data sets is 100%. If the variable recoveries < −0.11, then the prediction for the remaining 97% of data sets will also be good. In the other case, the prediction is a credit default and this applies to all cases (probability of 1) and affects 3% of all data sets. For the most significant decisions of the trained model, only a few variables are used: recoveries, last_pymnt_d, total_rec_prncp, term_X.60.months, and purpose_debt_consolidation.

Classification: Modeling

```
library(rpart.plot)

## Loading required package: rpart

##
## Attaching package: 'rpart'

## The following object is masked from 'package:dials':
##
##     prune

## Trainings- und Testdaten erzeugen-------------------------------------------------
set.seed(123)
loans.split<-initial_split(loans,prop=0.80,       # Split Trainings/Testdaten 80:20
                           strata=status)
loans.train<-training(loans.split)
loans.test<-testing(loans.split)
dim(loans.train)                                  # 80%  Trainingsdaten

## [1] 693862     19

dim(loans.test)                                   # 20%  Testdaten

## [1] 173466     19

## Recipe----------------------------------------------------------------------------
rec.all <- loans.train %>%                        # Recipe
  recipe(status ~ .) %>%
    step_normalize(all_numeric()) %>%
    step_dummy(all_nominal_predictors())
## Modelltypen und Engine -----------------------------------------------------------
mod.kl.tree<-                                     # Modell
   decision_tree() %>%
   set_engine("rpart",model=TRUE) %>%
   set_mode("classification")
## Workflow--------------------------------------------------------------------------
wfl.kl.tree <-workflow() %>%                      # Workflow
  add_model(mod.kl.tree) %>%
  add_recipe(rec.all)
wfl.kl.tree

## ══ Workflow ════════════════════════════════════════════════════════════════════
## Preprocessor: Recipe
## Model: decision_tree()
##
## ── Preprocessor ────────────────────────────────────────────────────────────────
## 2 Recipe Steps
##
## • step_normalize()
## • step_dummy()
##
## ── Model ───────────────────────────────────────────────────────────────────────
## Decision Tree Model Specification (classification)
##
## Engine-Specific Arguments:
##   model = TRUE
##
## Computational engine: rpart

## Machine-Learning-Modell trainieren-----------------------------------------------
wfl.kl.tree.fit<-fit(wfl.kl.tree,loans.train)      # Modell trainieren
wfl.kl.tree.fit %>% extract_recipe(estimated=TRUE) # Modell anzeigen
```

```
## Recipe
##
## Inputs:
##
##       role #variables
##    outcome          1
##  predictor         18
##
## Training data contained 693862 data points and no missing data.
##
## Operations:
##
## Centering and scaling for inq_last_6mths, int_rate, last_pymnt_d, out_prn... [tra
ined]
## Dummy variables from addr_state, emp_length, grade, home_ownership, initial_li...
 [trained]
```

```
rpart.plot(wfl.kl.tree.fit$fit$fit$fit)
```

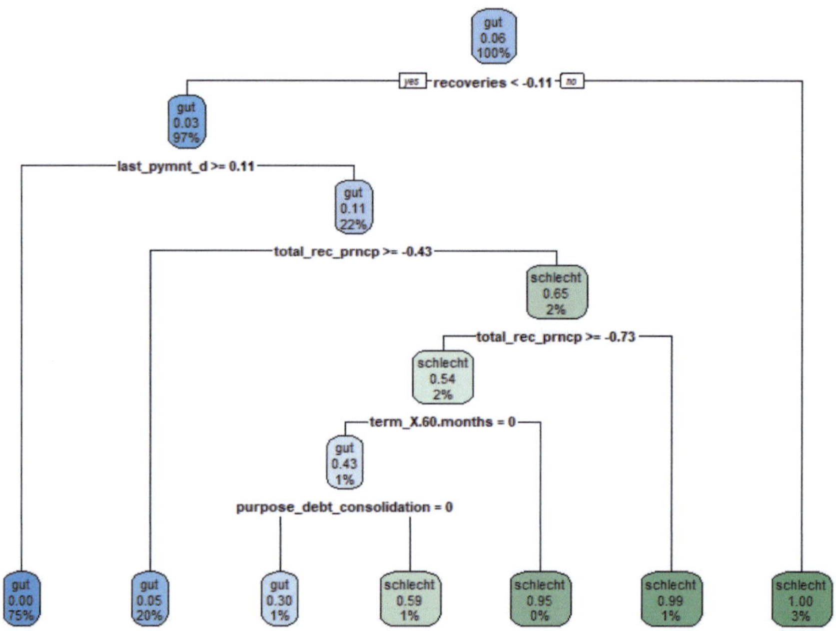


```
## Evaluation ------------------------------------------------------------------
vorhersage<-loans.test %>%
  select(status) %>%
  bind_cols(predict(wfl.kl.tree.fit,loans.test))
konfmatrix<-yardstick::conf_mat(vorhersage,          # Konfusions-Matrix
  truth=status,estimate=.pred_class)
konfmatrix
```

```
##             Truth
## Prediction    gut schlecht
##   gut       163026     2686
##   schlecht     522     7232
```

```
konfmatrix.rel<-konfmatrix$table %>% prop.table()
konfmatrix.rel

##            Truth
## Prediction          gut    schlecht
##   gut      0.939815295 0.015484302
##   schlecht 0.003009235 0.041691167

vorhersagefehler<-1
for(i in 1:nrow(konfmatrix.rel)) {
  vorhersagefehler<-vorhersagefehler - konfmatrix.rel[i,i]
}
vorhersagefehler

## [1] 0.01849354
# Aufruf Funktion accuracy() aus dem Paket yardstick---------
accuracy(vorhersage,truth=status,estimate=.pred_class)   # Genauigkeit

## # A tibble: 1 × 3
##   .metric  .estimator .estimate
##   <chr>    <chr>          <dbl>
## 1 accuracy binary         0.982
```

This model has a prediction error of 1.8% and thus an accuracy of 98.2%. A number of 0.3% of good loans are categorized as bad and 1.5% of bad loans are categorized as good. With a credit default probability of 6%, applying this model can avoid significantly more than 2/3 of bad loans. The corresponding monetary advantage can be estimated by multiplying it by the loan volume.

Now that regression and classification as applications of supervised learning have been introduced, applications of unsupervised learning will follow.

7.2 Unsupervised Learning

Two of the most significant applications of unsupervised learning are clustering and market basket analysis, which is based on association rules. These two applications will therefore be presented below.

In contrast to supervised learning, there is no target variable in unsupervised learning that can be used for training. Unsupervised learning tries to recognize completely new relationships in the data.

7.2.1 Clustering

The objective of clustering or cluster analysis is to divide objects with similar properties into homogeneous clusters. This is applied, for example, to divide customers into A, B, and C customers or similar for products etc. Category A often refers to the best customers, who order repeatedly, are loyal, and generate larger revenues. Clustering can then be useful, for example, to invite only the A customers to an event and promote the B and C customers via email.

In clustering, we distinguish between partitioning and hierarchical methods, e.g., the Ward method. The RFM and the k-means method as partitioning cluster methods are presented below.

The RFM algorithm (Recency-Frequency-MonetaryValue algorithm) performs clustering of customers based on their purchasing behavior. The following scoring values per customer are examined in more detail:

- Recency: The recency of a purchase is an important tool for identifying customers who have recently made a purchase. Customers whose purchase was not long ago are more likely to respond to a new offer than customers whose purchase was a long time ago. This is the most important factor of an RFM analysis.
- Frequency: The frequency of a purchase is less relevant than its recency. However, if customers have purchased more frequently, the probability of a positive response is higher than for customers who have purchased only rarely.
- Monetary Value: The revenue of a purchase or the monetary value refers to the total revenue of a customer. Customers with higher total revenue are more likely to respond to an offer than customers with lower total revenue.

Based on the scoring values of recency, frequency, and monetary value, customers are then divided into 5 buyer groups, with 1 being the worst and 5 being the best buyer group. This clustering is used to find out which customer groups are particularly profitable and for which customers certain campaigns or marketing offers are more or less worthwhile.

The k-means method as a partitioning cluster method is one of the most frequently used methods. The objective of this method is to form as homogeneous groups as possible. The standard deviation of the data from their assigned cluster centers serves as a measure of the achievement of the objective. The optimal number of clusters can be determined by how much the standard deviation of the data sets from the cluster centers is reduced when additional clusters are formed.

Hierarchical clustering can be achieved using the Ward method. In R, the function hclust() can be used in conjunction with the functions rect.hclust() and cutree() from the stats package. The function establishes a hierarchy and then forms a number of clusters that can be determined by the parameter k.

The data used in the following code are just over 530,000 sales transactions from a UK-based online store from 2010 and 2011, which mainly sells gift items to wholesalers (Chen et al. 2015). After the file "Online Retail.xlsx" was read into the dataframe sales, the date is first converted into a date variable and the sales amount is calculated based on the quantity and unit price. All records without sales or missing data are then deleted. From the transaction data, the information on sales and country per customer is extracted in the dataframe customers. It is noticeable that the customer ID is only unique within a country. Therefore, there are 8 duplicate customer IDs within the dataframe customers. To solve this problem, a globally unique customer ID CustomerIDGlobal is added to the dataframe, which is composed of a unique country code and the nationally unique customer ID. Most customers come from the United Kingdom, Germany, and France. In preparation for clustering, the transactions per invoice number with the details of customer, date, and invoice amount are stored in the dataframe sales. The preprocessing is now complete and the data can be analyzed.

In R, the function `findRFM()` from the didrooRFM package can be used to execute the RFM algorithm. The parameters recencyWeight, frequencyWeight, and monetaryWeight allow the weighting of the three scoring values to be determined. The smallest number of customers was assigned to the best group (Cluster 5), most of the customers are in Clusters 2 and 3. The analysis clearly shows that the average sales in euros per customer in Cluster 5 is more than 25,000, in Cluster 4 it is about 5,000, and in the other three clusters it is below 2,500. The clustering has divided the customers into 5 groups and 64% of the sales were made with 17% of the customers. This division of customers already allows for more effective marketing.

How does this compare with the k-means algorithm? To execute the k-means algorithm in R, the function `kmeans()` can be used.

In the first attempt, the variables considered relevant from the result of the RFM algorithm are initially stored in the dataframe kdata: MeanValue, NoTransaction, MonetaryScore, FrequencyScore, and RecencyScore. To determine the best number of clusters, a loop from 2 to 15 (1 is not necessary, as without clustering all data records already exist in one set) can run the algorithm and determine the standard deviation. It turns out that from about 12 clusters onwards, there is no significant improvement in the standard deviation. If the algorithm is then called with the parameter for 12 clusters, the function `aggregate()` can display the coordinates of the 12 cluster centers in the 5-dimensional space (the dataframe kdata used as the basis for the algorithm consists of 5 variables). The dataframe kdata is then supplemented by the cluster assignment for each customer using the command `kdata<-data.frame(kdata,fit12$cluster)`. The graphical analysis shows that only a few customers were assigned to clusters 7 and 10, whereas the majority were assigned to clusters 6, 3, 11, and 9. The highest average sales are found among the customers in clusters 7 and 10. The clustering has divided the customers into 12 groups and with 16% of the customers (clusters 10 + 7 + 4 + 12 + 2 + 1 + 8 + 5), it is possible to explain 55% of the sales.

In the second attempt, the variables considered relevant from the original dataframe sales are stored in the dataframe kdata: Country, Sales, SalesPerTrans, and NumberTrans. To determine the best number of clusters, a loop from 2 to 15 can again run the algorithm and determine the standard deviation. It turns out that here too, from about 12 clusters onwards, there is no significant improvement in the standard deviation. If the algorithm is then called with the parameter for 12 clusters, the function `aggregate()` can display the coordinates of the 12 cluster centers in the 4-dimensional space (this time the dataframe kdata used as the basis for the algorithm consists of 4 variables). The dataframe kdata is then again supplemented by the cluster assignment for each customer. The graphical analysis shows that only a few customers were assigned to clusters 8 and 11, whereas the majority were assigned to clusters 10, 5, and 12. The highest average sales are found among the customers in clusters 8 and 11. The clustering has divided the customers into 12 groups and this subdivision means that with 16% of the customers (clusters 11 + 8 + 7 + 4 + 9 + 2 + 1 + 6 + 3), 71% of the sales can be explained.

To output the clusters graphically, the function `clusplot()` from the cluster package is suitable. For each of the formed clusters, there is also a center, which can be specified for the cluster formed earlier with the command `fit12<-kmeans(kdata,12)` using the function `aggregate(kdata,by=list(fit12$cluster),FUN=mean)`. Since five variables were used for cluster formation in this example, the cluster is five-dimensional, even though the display with `clusplot(kdata,fit12$cluster,color=T,labels=2, lines=0)` only shows this in two dimensions. An assignment of the customers to the respective cluster in the dataframe can be achieved via the command `kdata <- data.frame(kdata, fit12$cluster)`.

Clustering

```
library(readxl)
sales<-read_excel("Online Retail.xlsx")
colnames(sales)<-c("RechnungsID","ProduktID","Beschreibung","Anzahl","Datum",
                   "Stückpreis","KundenID","Land")
sales$Datum<-as.Date(sales$Datum)
sales$Betrag<-sales$Anzahl * sales$Stückpreis
sales<-sales[sales$Betrag>0,]                         # Ohne Umsatz entf.
sales<-na.omit(sales)                                 # Unvollständige entf.
kunden<-sales %>% group_by(KundenID,Land) %>% summarize(Umsatz=sum(Betrag))
kunden[c(which(duplicated(kunden$KundenID))),]        # Kunden-ID nicht eindeutig

## # A tibble: 8 × 3
## # Groups:   KundenID [8]
##   KundenID Land         Umsatz
##      <dbl> <chr>         <dbl>
## 1    12370 Cyprus        3268.
## 2    12394 Denmark        891.
## 3    12417 Spain          436.
## 4    12422 Switzerland    417.
## 5    12429 Denmark       3312.
## 6    12431 Belgium        973.
## 7    12455 Spain          768.
## 8    12457 Switzerland   1971.

sales$Land<-as.factor(sales$Land)                     # Konvertierung in Faktor
sales$KundenIDGlobal<-as.numeric(paste0(as.numeric(sales$Land),sales$KundenID))
kunden<-sales %>% group_by(KundenIDGlobal,Land) %>% summarize(Umsatz=sum(Betrag))
kunden[c(which(duplicated(kunden$KundenIDGlobal))),]# Kunden-ID jetzt eindeutig

## # A tibble: 0 × 3
## # Groups:   KundenIDGlobal [0]
## # … with 3 variables: KundenIDGlobal <dbl>, Land <fct>, Umsatz <dbl>

ergebnis<-ml.summary(kunden$Land,"Kundenanzahl je Land","Land",bar=T)
```

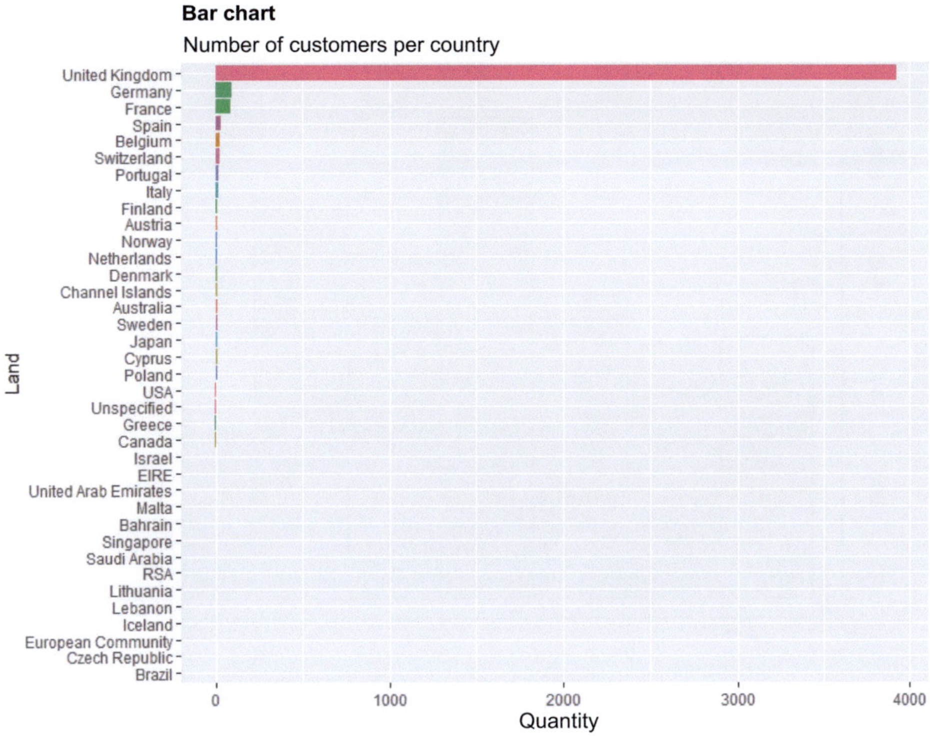


```
umsatz<-sales %>% group_by(RechnungsID,KundenIDGlobal,Datum) %>%
        summarize(Umsatz=sum(Betrag))
#- Clustering mit RFM-Algorithmus--------------------------------------------
library(didrooRFM)
RFM<-findRFM(umsatz,recencyWeight = 5,frequencyWeight = 3,monetoryWeight = 5)
```

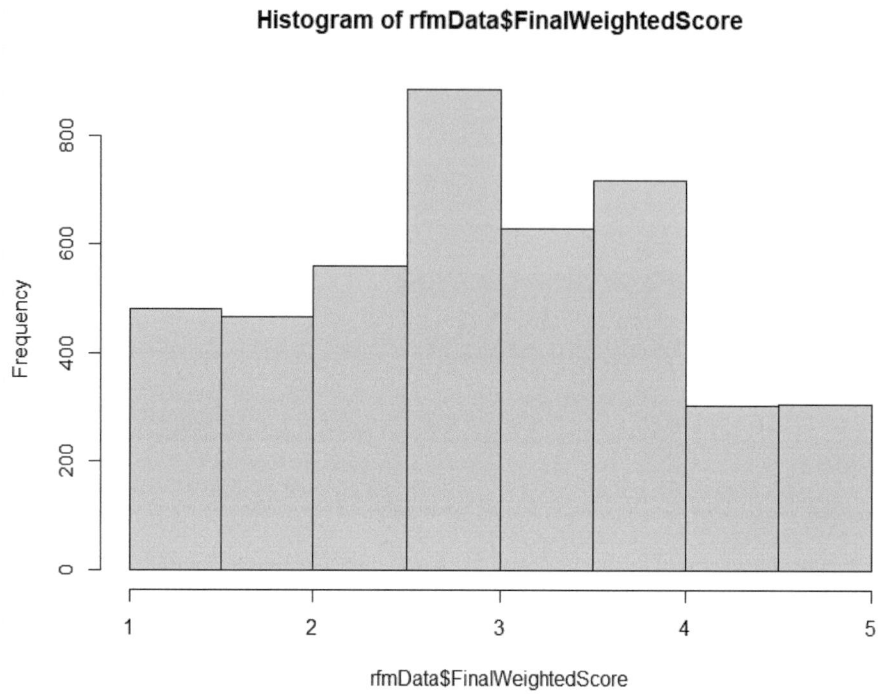


```
RFM.analyse<-merge(RFM[,c(1,8:10,16)],kunden[,c(1,3)],
                  by.x="CustomerID",by.y="KundenIDGlobal")
colnames(RFM.analyse)<-c("KundenID","Wert Monetär","Wert Häufigkeit",
                         "Wert Neuheit","Klasse","Umsatz")
RFM.analyse$KlassenID<-as.numeric(substring(RFM.analyse$Klasse,7,7))
head(RFM.analyse)                                   # Clusterzuordnung

##   KundenID Wert Monetär Wert Häufigkeit Wert Neuheit  Klasse  Umsatz KlassenID
## 1  1012367            2               1            5 Class-2  168.90         2
## 2  1012372            4               3            2 Class-3 1298.04         3
## 3  1012394            5               1            3 Class-3  891.40         3
## 4  1012403            4               1            3 Class-2  427.70         2
## 5  1012406            5               2            4 Class-3 3415.30         3
## 6  1012412            4               3            2 Class-3 1226.88         3

RFM.tab<-RFM.analyse %>% group_by(KlassenID) %>% summarize(Umsatz=mean(Umsatz))
ergebnis<-ml.summary(RFM.analyse$KlassenID,"RFM Klassifikation der Kunden",
                     "Klasse",bar=T)
```

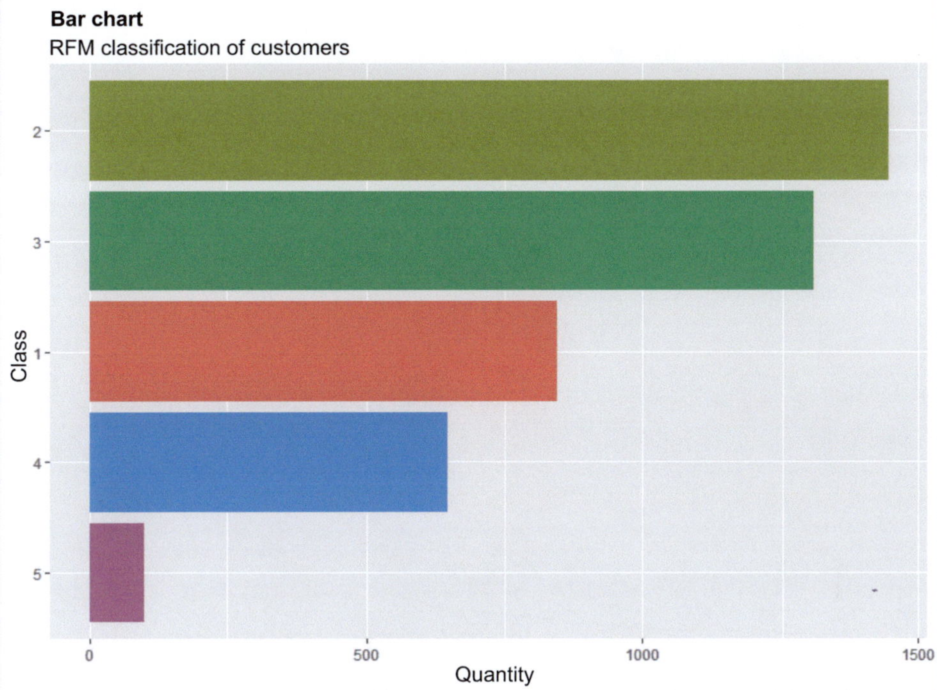


```
ergebnis<-ml.summary.2(RFM.tab,
                       "Durchschnittsumsatz je RFM Klassifikation der Kunden",
                       "Klasse","Umsatz",line=T)
```

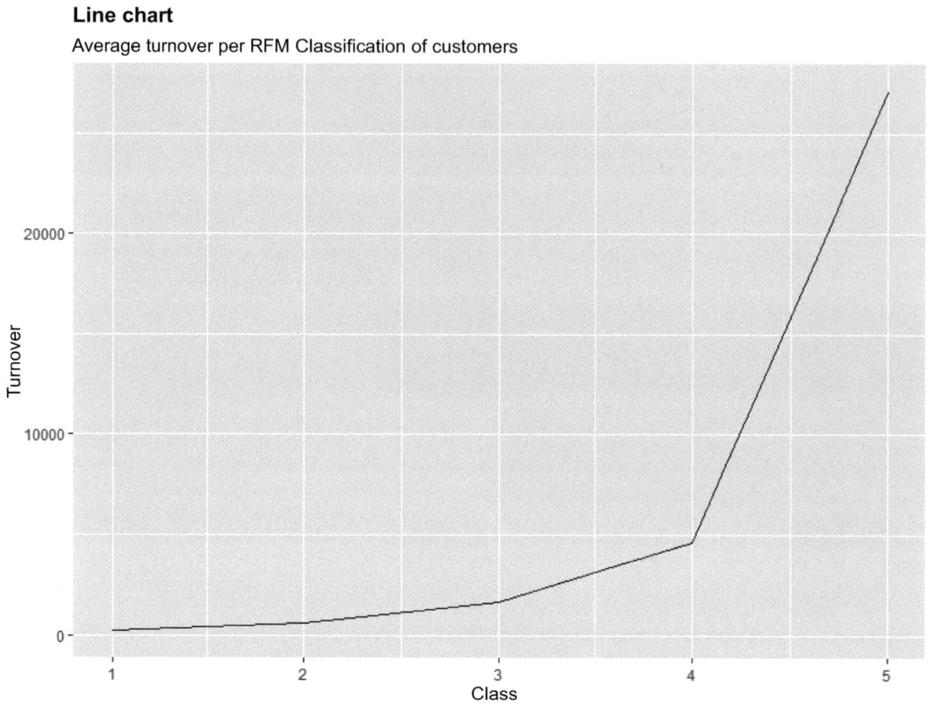


```
gesamtumsatz<-sum(RFM.analyse$Umsatz)
kunden.anteil<-RFM.analyse %>% group_by(KlassenID) %>%
       summarize(Umsatz=sum(Umsatz),
                 AnteilUmsatz=sum(Umsatz)/gesamtumsatz*100,
                 Kunden=n(),
                 AnteilKunden=n()/nrow(RFM.analyse)*100)
kunden.anteil

## # A tibble: 5 × 5
##   KlassenID   Umsatz AnteilUmsatz Kunden AnteilKunden
##       <dbl>    <dbl>        <dbl>  <int>        <dbl>
## 1         1  177901.         2.00    846        19.5
## 2         2  892485.        10.0    1446        33.3
## 3         3 2163455.        24.3    1308        30.1
## 4         4 2996600.        33.6     647        14.9
## 5         5 2680967.        30.1      99         2.28

cluster<-c(5,4)                              # Selektion der TOP Cluster
sum(kunden.anteil[cluster,]$AnteilKunden)    # Anteil der Kunden

## [1] 17.16521

sum(kunden.anteil[cluster,]$AnteilUmsatz)    # Anteil des Umsatzes

## [1] 63.71122

#- Clustering mit k-Means-Algorithmus basierend auf RFM-Ergebnis------------------
kdata<-data.frame(dplyr::select(RFM,c(2,4,8:10)))# Relevante Variablen selektieren
rownames(kdata)<-RFM$CustomerID              # Zeilenname = KundenIDGlobal
kdata.sd<-data.frame()
```

```
for (i in 2:15) {                                  # withinss=SumOfSquares je Cluster
  kdata.sd[i,1]<-i                                 # sd = Standardabweichung
  set.seed(123)                                    # Reproduzierbarkeit
  kdata.sd[i,2]<-sqrt(sum(kmeans(kdata,centers=i)$withinss)/nrow(kdata))
}
ergebnis<-ml.summary.2(kdata.sd,
                       "Standardabweichung vom nächstem Clustermittelpunkt",
                       "Anzahl Cluster","Standardabweichung",line=T)
```

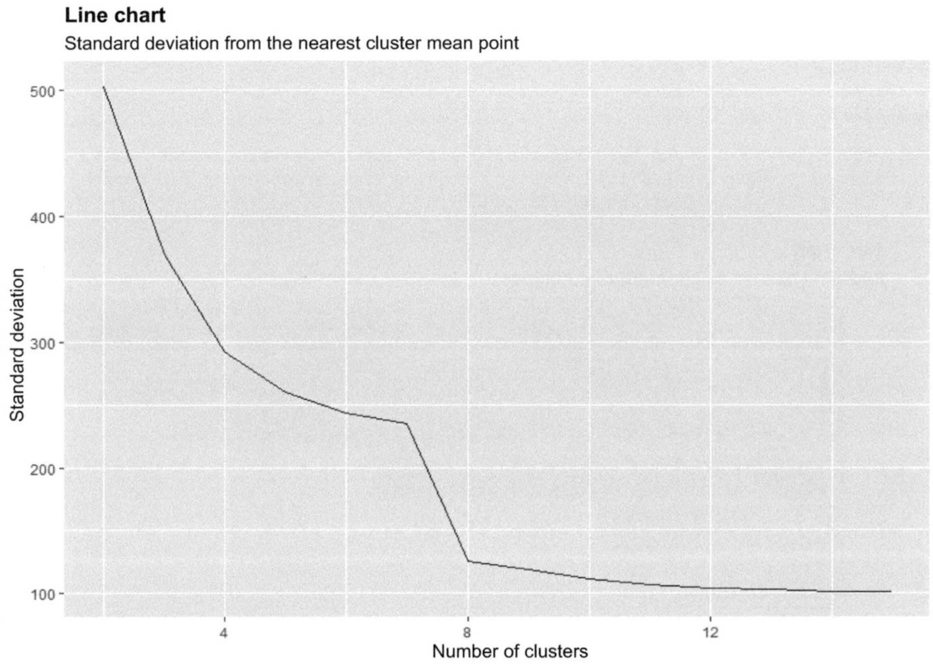


```
set.seed(123)                                      # Reproduzierbarkeit
fit12<- kmeans(kdata,12)                           # K-Means Cluster: 12 Cluster
aggregate(kdata,by=list(fit12$cluster),FUN=mean) # Cluster Zentrum im n-dim. Raum

##     Group.1  MeanValue NoTransaction MonetoryScore FrequencyScore RecencyScore
## 1         1  1116.6686      6.193878      5.000000       2.642857     2.887755
## 2         2  1623.4686      3.537037      5.000000       2.518519     2.962963
## 3         3   325.7933      4.146825      3.387897       2.727183     3.075397
## 4         4  4298.0662     13.473684      5.000000       2.842105     3.421053
## 5         5   605.8328      6.081571      5.000000       2.858006     3.211480
## 6         6   212.4803      4.247919      2.233117       2.724329     3.121184
## 7         7 14075.1333      3.000000      5.000000       3.000000     1.000000
## 8         8   827.7366      4.533708      5.000000       2.780899     3.022472
## 9         9   446.3913      4.808777      4.250784       2.879310     3.136364
## 10       10 80709.9250      1.500000      5.000000       1.500000     3.000000
## 11       11   114.7810      2.845055      1.043956       2.090110     2.539560
## 12       12  2346.3806      8.800000      5.000000       2.760000     3.120000

kdata <- data.frame(kdata, fit12$cluster)          # Cluster Zuordnung in Dataframe
head(kdata)                                        # Clusterzuordnung
```

```
##         MeanValue NoTransaction MonetoryScore FrequencyScore RecencyScore
## 1012367    168.90             1             2              1            5
## 1012372    432.68             3             4              3            2
## 1012394    891.40             1             5              1            3
## 1012403    427.70             1             4              1            3
## 1012406   1707.65             2             5              2            4
## 1012412    408.96             3             4              3            2
##         fit12.cluster
## 1012367             6
## 1012372             9
## 1012394             8
## 1012403             9
## 1012406             2
## 1012412             9

kdata.tab<-kdata %>% group_by(fit12.cluster) %>% summarize(Umsatz=mean(MeanValue))
ergebnis<-ml.summary(kdata$fit12.cluster,"k-Means Clusterzuordnung der Kunden",
                     "Cluster",bar=T)
```

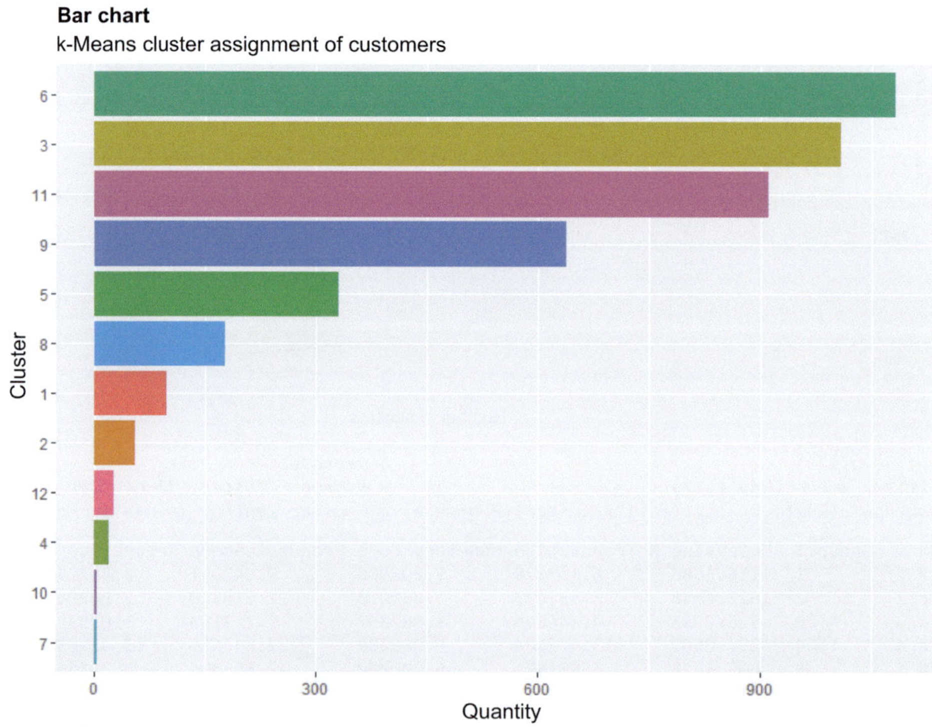


```
ergebnis<-ml.summary.2(kdata.tab,
                       "Durchschnittsumsatz je Kunde (12 Cluster)",
                       "Cluster","Umsatz",line=T)
```

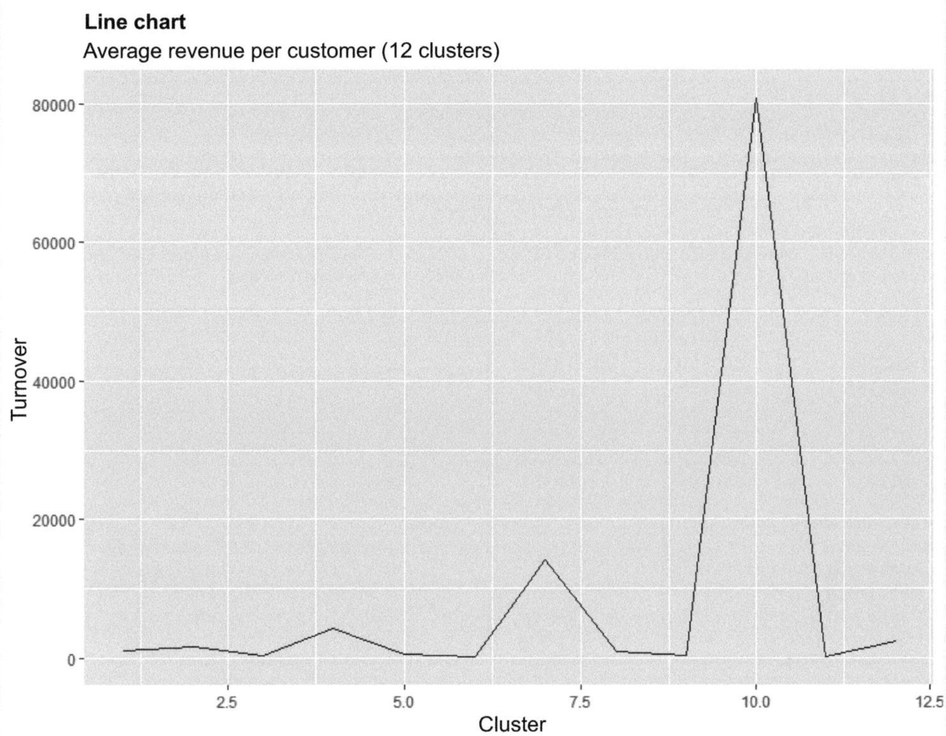


```
kdata$Umsatz<-kdata$MeanValue * kdata$NoTransaction
gesamtumsatz<-sum(kdata$Umsatz)
kunden.anteil<-kdata %>% group_by(fit12.cluster) %>%
        summarize(Umsatz=sum(Umsatz),
                  AnteilUmsatz=sum(Umsatz)/gesamtumsatz*100,
                  Kunden=n(),
                  AnteilKunden=n()/nrow(kdata)*100)
kunden.anteil

## # A tibble: 12 × 5
##    fit12.cluster   Umsatz AnteilUmsatz Kunden AnteilKunden
##            <int>    <dbl>        <dbl>  <int>        <dbl>
##  1             1  707672.        7.94      98       2.25
##  2             2  309672.        3.48      54       1.24
##  3             3 1366896.       15.3     1008      23.2
##  4             4 1081338.       12.1       19       0.437
##  5             5 1260052.       14.1      331       7.62
##  6             6  975415.       10.9     1081      24.9
##  7             7   84451.        0.948      2       0.0460
##  8             8  668130.        7.50     178       4.10
##  9             9 1370331.       15.4      638      14.7
## 10            10  245656.        2.76       2       0.0460
## 11            11  314341.        3.53     910      20.9
## 12            12  527453.        5.92      25       0.575

cluster<-c(10,7,4,12,2,1,8,5)                   # Selektion der TOP Cluster
sum(kunden.anteil[cluster,]$AnteilKunden)       # Anteil der Kunden
```

```
## [1] 16.31385

sum(kunden.anteil[cluster,]$AnteilUmsatz)         # Anteil des Umsatzes

## [1] 54.81091

#- Clustering mit k-Means-Algorithmus basierend auf Basisdaten--------------------
kdata<-sales %>% group_by(KundenIDGlobal,Land) %>%
  summarize(Umsatz=sum(Betrag),UmsatzJeTrans=sum(Betrag)/n(),
            AnzahlTrans=n()) %>% as.data.frame()
kdata$Land=as.numeric(kdata$Land)
rownames(kdata)<-kdata$KundenIDGlobal             # Zeilenname = KundenIDGlobal
kdata<-kdata[,-1]                                 # KundenIDGlobal nicht verwenden
head(kdata)

##        Land    Umsatz UmsatzJeTrans AnzahlTrans
## 112386    1    401.90      40.19000          10
## 112388    1   2780.66      27.80660         100
## 112393    1   1582.60      24.72813          64
## 112415    1 124914.53     174.95032         714
## 112422    1    386.20      18.39048          21
## 112424    1   1760.96      58.69867          30

kdata.sd<-data.frame()
for (i in 2:15) {                                 # withinss=SumOfSquares je Cluster
  kdata.sd[i,1]<-i                                # sd = Standardabweichung
  set.seed(123)                                   # Reproduzierbarkeit
  kdata.sd[i,2]<-sqrt(sum(kmeans(kdata,centers=i)$withinss)/nrow(kdata))
}

ergebnis<-ml.summary.2(kdata.sd,
                       "Standardabweichung vom nächstem Clustermittelpunkt",
                       "Anzahl Cluster","Standardabweichung",line=T)
```

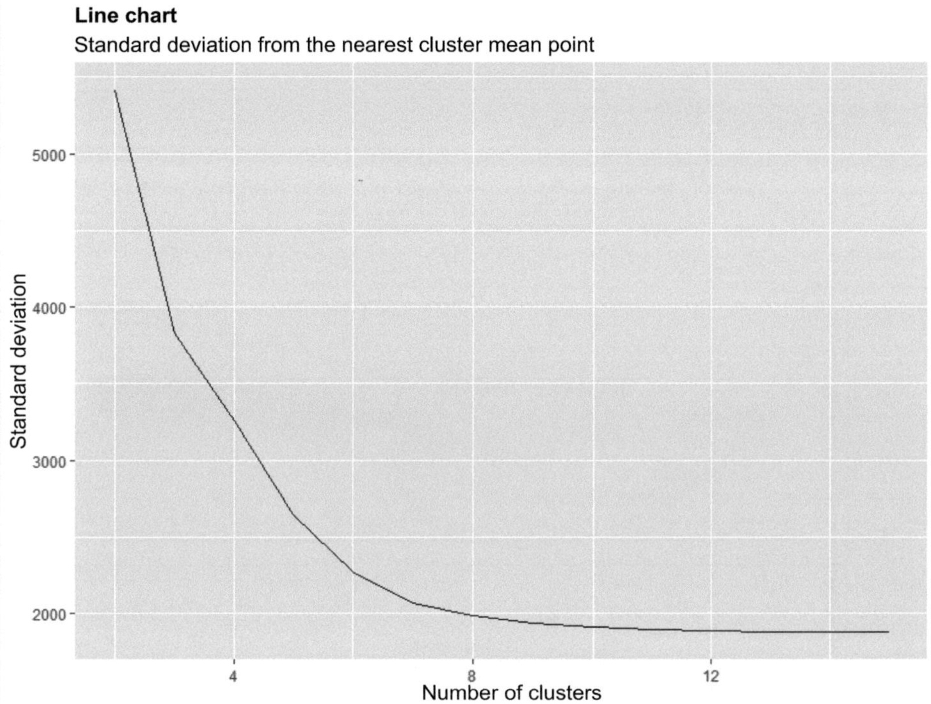


```
set.seed(123)                                          # Reproduzierbarkeit
fit12<- kmeans(kdata,12)                               # K-Means Cluster: 12 Cluster
library(cluster)                                       # Paket mit Funktion clusplot
clusplot(kdata,fit12$cluster,color=T,labels=2,lines=0) # Graphische Ausgabe
```

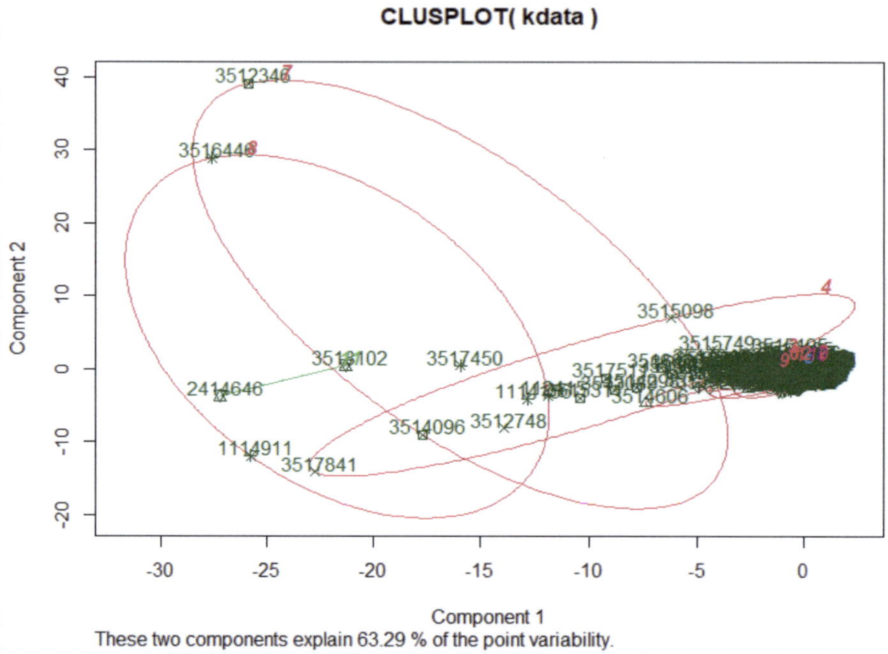


```
aggregate(kdata,by=list(fit12$cluster),FUN=mean) # Cluster Zentrum im n-dim. Raum

##    Group.1     Land      Umsatz UmsatzJeTrans AnzahlTrans
## 1        1 31.66667   7457.5763      78.05010   347.45556
## 2        2 31.80392  11653.0027      84.26700   369.23529
## 3        3 32.37647   3108.5266      54.17548   185.54118
## 4        4 34.82353  33833.3706    1211.87712   964.88235
## 5        5 33.25380    913.0552      27.76070    66.81464
## 6        6 32.01136   4879.5662      45.15705   275.46591
## 7        7 35.00000  64387.6131    6119.98902  1091.38462
## 8        8 18.60000 149828.5020   11403.78772  1625.80000
## 9        9 31.43478  18532.2413      82.89563   362.91304
## 10      10 33.68082    294.6495      24.74139    24.55594
## 11      11 29.50000 269931.6600     368.71357  1253.50000
## 12      12 33.18957   1830.1637      36.02449   116.85565

kdata <- data.frame(kdata, fit12$cluster)         # Cluster Zuordnung in Dataframe
head(kdata)                                       # Clusterzuordnung

##        Land    Umsatz UmsatzJeTrans AnzahlTrans fit12.cluster
## 112386    1    401.90      40.19000          10            10
## 112388    1   2780.66      27.80660         100             3
## 112393    1   1582.60      24.72813          64            12
## 112415    1 124914.53     174.95032         714             8
## 112422    1    386.20      18.39048          21            10
## 112424    1   1760.96      58.69867          30            12

kdata.tab<-kdata %>% group_by(fit12.cluster) %>% summarize(Umsatz=mean(Umsatz))
ergebnis<-ml.summary(kdata$fit12.cluster,"k-Means Clusterzuordnung der Kunden",
                     "Cluster",bar=T)
```

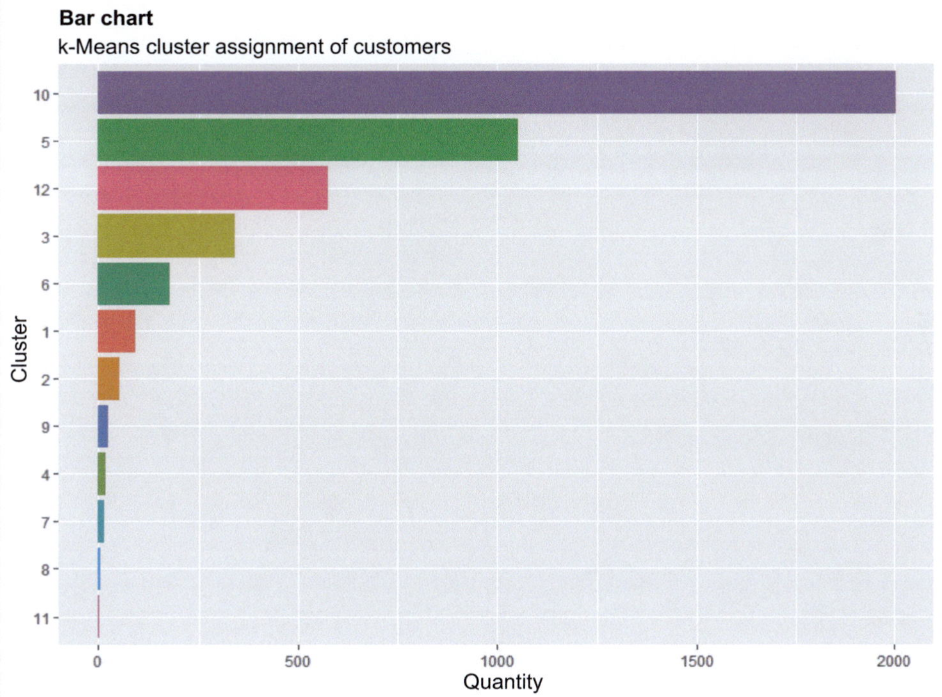


```
ergebnis<-ml.summary.2(kdata.tab,
                       "Durchschnittsumsatz je Kunde (12 Cluster)",
                       "Cluster","Umsatz",line=T)
```

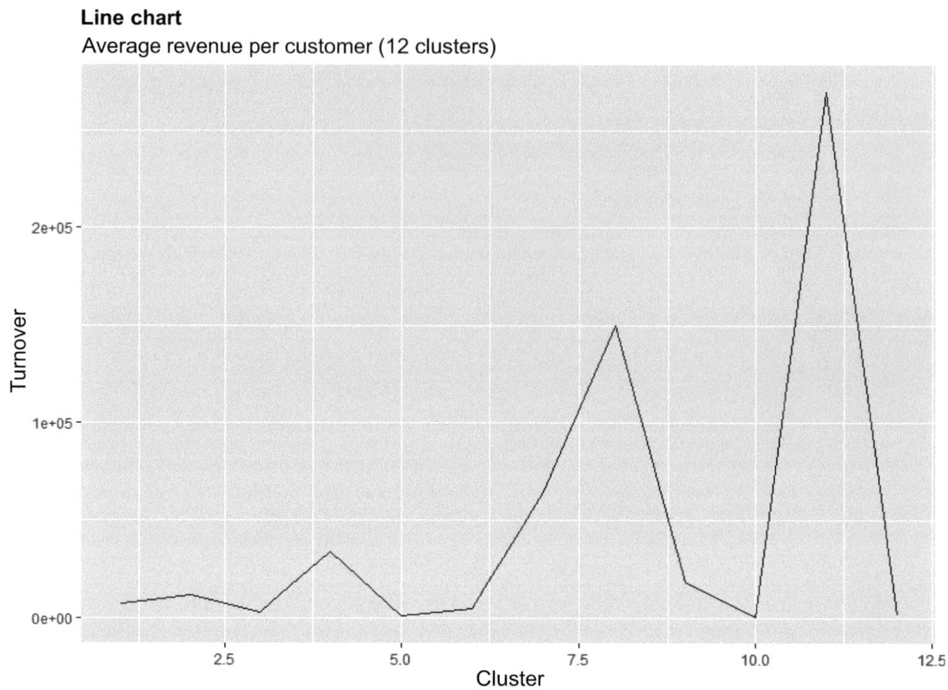


```
gesamtumsatz<-sum(kdata$Umsatz)
kunden.anteil<-kdata %>% group_by(fit12.cluster) %>%
        summarize(Umsatz=sum(Umsatz),
                  AnteilUmsatz=sum(Umsatz)/gesamtumsatz*100,
                  Kunden=n(),
                  AnteilKunden=n()/nrow(kdata)*100)
kunden.anteil

## # A tibble: 12 × 5
##    fit12.cluster   Umsatz AnteilUmsatz Kunden AnteilKunden
##            <int>    <dbl>        <dbl>  <int>        <dbl>
##  1             1  671182.         7.53     90       2.07
##  2             2  594303.         6.67     51       1.17
##  3             3 1056899.        11.9     340       7.82
##  4             4  575167.         6.45     17       0.391
##  5             5  960534.        10.8    1052      24.2
##  6             6  858804.         9.64    176       4.05
##  7             7  837039.         9.39     13       0.299
##  8             8  749143.         8.41      5       0.115
##  9             9  426242.         4.78     23       0.529
## 10            10  589888.         6.62   2002      46.1
## 11            11  539863.         6.06      2       0.0460
## 12            12 1052344.        11.8     575      13.2

cluster<-c(11,8,7,4,9,2,1,6,3)                        # Selektion der TOP Cluster
sum(kunden.anteil[cluster,]$AnteilKunden)              # Anteil der Kunden

## [1] 16.49793

sum(kunden.anteil[cluster,]$AnteilUmsatz)              # Anteil des Umsatzes
## [1] 70.79287
```

The process of clustering allowed the RFM algorithm to form clusters in which 17% of customers contribute 64% to the turnover. In the first run of the kMeans algorithm, it was 16% of customers who contributed 55% to the turnover, and in the second run, an improvement was achieved where 16% of customers contributed 71% to the turnover. The clearer the clusters are formed, the more effective their use is possible.

7.2.2 Association Rules

Association rules describe the relationship between variables, e.g., which products are often bought together, which factors contribute to lower machine downtime, or which symptoms or complications often accompany a therapy or medication.

Association rules are not primarily used for prediction, but to discover previously unknown relationships. However, they can also be used to make predictions, e.g., which product a customer is likely to buy if they have previously bought another product. The newly acquired knowledge of the relationships can then be used to offer customers products with the help of **recommendation engines**that other customers have purchased, who have previously viewed or bought similar products. This so-called cross-selling serves to increase sales. If the transaction data can be assigned to specific customers, then customer profiles can also be created based on purchases over time, which make the recommendation engines even more successful. Association rules are also used for product placement in stores and on websites and are also suitable as the basis for marketing campaigns. If product A often leads to the purchase of product B, then these should not be advertised at the same time, as it is already sufficient to advertise product A. The association rules can also be used for the classification of products that contribute particularly to sales (ABC classification).

To determine the association rules, the Apriori algorithm from 1994 is usually used. Since the determination of the association rules is one of the methods of unsupervised learning, the algorithm does not need to be trained. Therefore, the data is not divided into training and test data. The disadvantage of this method, however, is that the association rules cannot be validated to the same extent via test data. It remains for human examination and interpretation to assess their quality. An advantage of the method lies in its sheer ability to analyze large amounts of data and to recognize relationships within it that would have remained hidden to a human.

A long-established area of application for association rules is shopping basket analysis. The basis for this is the transaction data of sales, which includes the individual positions of the transaction with the purchased products for each transaction. Without prior knowledge of the relationships of the transactions, shopping basket analysis allows combinations of frequently purchased products to be recognized based on the transaction data. The data basis of the transactions is usually in a de-normalized database table, which has a column for each possible object (**Item**), a so-called **Sparse Matrix**. The rows each represent a transaction, with items contained in the transaction marked with a 1, and those not contained marked with a 0. The relationships of **Item-Sets** (set of objects/items, where a set can also consist of just one item) are stored in the **association rules**, which describe the relationship between item sets, e.g., that customers who buy jam and honey (object set: jam + honey) also buy bread (object: bread) with an x-percent probability.

Depending on the number of items, there can be a very large number of association rules to determine, which can even be a powerful task for a computer. With a number of k objects, it is possible to form 2^k object sets (item sets) which can be the basis for an association rule. However, since many of the combinations have no or little significance, the analysis can be limited to the cases that occur with a certain probability using the Apriori algorithm. Rare cases are thus ignored by the analysis. This is also the premise (the Apriori) of the algorithm, that only frequently occurring item sets are of interest. In this way, after determining the frequently occurring objects, the search can be limited to item sets that contain these objects. The search for frequent sets starts with 1-element item sets and is iteratively continued with n-element item sets until no more item sets are found that occur frequently and the algorithm stops the search and has determined all relevant item sets. In the second step of the Apriori algorithm, association rules are searched for each of the relevant item sets. This starts with the shortest possible (1-element) conclusions, which are iteratively increased.

The quality of an examined association rule is calculated by the following statistical parameters, the support, the confidence, and the lift. The **Support** is calculated as the frequency of an object combination among the transactions divided by the number of all transactions, i.e., the probability that this item set will occur in a transaction. The **Confidence** is calculated as the frequency of an object combination among the transactions divided by the frequency of the occurrence of the examined object. Confidence measures for what proportion of transactions in which X occurs, Y also occurs. The confidence of an association rule indicates its accuracy or predictive power, that the presence of item set X in a transaction allows the prediction that item set Y will also be contained in the transaction. Another parameter is the **Lift**, which is calculated as the support of the object combination divided by the support of each individual object of the object combination. Lift indicates how much more likely item set Y will be purchased if item set X is also included in the transaction. Unlike confidence, this applies reciprocally, i.e., Lift {X}=>{Y} is identical to Lift {Y}=>{X}. A Lift {X}=>{Y} of 1 means that the probability that Y will be purchased is not greater than it would be by chance, just because X was purchased. If, on the other hand, the value of Lift is greater than 1, this means that the probability is higher than it would be by chance. A Lift of 3, for example, means that the probability is 3x higher than by chance. The formulas for calculating the parameters are shown in Fig. 7.2.

The packages arules and arulesviz contain essential functions for determining association rules and transaction data. The Groceries list from the arules package contains 169 items and 9836 transactions. Own transaction data can be imported with the function

Association rule {Item-Set X} => {Item-Set Y}

- Support (X) = $\frac{Number(X)}{N}$

  Support (Y) = $\frac{Number(Y)}{N}$

  Support (X,Y) = $\frac{Number(X,Y)}{N}$

- Confidence{X}=>{Y} = $\frac{Support(X,Y)}{Support(X)}$

  Confidence{Y}=>{X} = $\frac{Support(X,Y)}{Support(Y)}$

- Lift {X}=>{Y} = Lift{Y}=>{X} = $\frac{Support\ (X,Y)}{Support(X) * Support(Y)}$

Fig. 7.2 Association rule if Item-Set X, then also Item-Set Y

`read.transactions()` or synthetically generated with the function `random.transactions()`. The function `transcations()` shows the number of transactions and items and the function `inspect()` allows the transaction data to be displayed in detail or also the association rules sorted by support, confidence or lift. The command `inspect(sort(head(rules),by="lift"))` shows, for example, the association rules with the greatest lift. If a conversion of the transaction data into a dataframe is desired, this can be achieved with the function `as(x,"data.frame")`. Details about the sparse matrix with the transactions can be visualized with the function `image()`. The function `itemFrequency()` shows items with their frequency and the function `itemFrequencyPlot()` allows the most frequent items to be visualized in a bar chart, whereby a selection of the relevant items can be made through the topN and support parameters. As can be seen in the following code, Whole Milk obviously appears in more than 25% of all transactions in the Groceries dataset.

Association rules can be determined with the help of the function `apriori()`, where the parameters support, confidence and minlen/maxlen specify the search. A support of 0.001, for example, indicates that only rules with a frequency of 0.001 * 9836 should be considered, which therefore occur at least 9 times. A confidence of 0.9 means that only rules with a confidence of 90% are determined. The minlen parameter specifies a minimum number of items that must underlie a rule and maxlen the maximum number. Since Whole Milk is included in 25% of all transactions, specifying minlen = 2 makes sense, so as not to generate a large number of rules that only refer to Whole Milk. After the association rules have been generated, the function `summary()` allows an overview to be displayed, showing how many rules were generated and how support, confidence and other variables are distributed.

If you apply the apriori algorithm with a confidence of 0.01 and a support of 0.0005 to the data, more than 175,000 rules are determined. A confidence of 0.9 and a support of 0.00125, on the other hand, leads to a manageable number of less than 50 rules. The choice of these two parameters should therefore always be well considered.

The determined rule {liquor, red/blush wine} => {bottled beer} 0.00193 0.905 0.00214 11.24 states that this rule affects 0.193% of all transactions and is correct for 90.5% of the affected transactions. The lift of 11.24 states that bottled beer is purchased with an 11 times higher than random probability if the transaction includes liquor and wine.

It can also be useful to search for rules that contain a certain item, e.g. tropical fruits. This can be used to estimate the effect of a marketing campaign for tropical fruits on the sales of other products (Item in lhs = left hand side -> left side of the association rule, rhs = right hand side -> right side of the association rule). The function `subset()` is suitable for this, to select from all rules only those that contain a certain item, e.g. `rule.sub<-subset(rules,items %in% "tropical fruit")`, to store the association rules concerning tropical fruits in the object rule. sub. If, on the other hand, you want to know which items enhance the sale of tropical fruits, the function `apriori()` with the appearance parameter can be used, e.g. `apriori(data=sales,parameter=list(support=0.00125,confidence=0.6),appearance=list(default= "lhs",rhs="tropical fruit"))`. One of the association rules states, for example, that the object set {root vegetables, other vegetables, butter, yogurt } => { tropical fruit } has a lift of 5.89. The probability of buying tropical fruits is therefore almost 6 times as high as purely random.

A visualization of the association rules is possible through the plot() function, e.g. `plot(rule.tropical,method="graph",measure="confidence",shading="lift")`.

Association Rules

```
#- Transaktionsdaten laden und verstehen--------------------------------------------
library(arules)                                    # Funktionen für Assoziationsregeln

library(arulesViz)                                 # Funktionen für Visualisierung
data(Groceries,package="arules")
sales<-Groceries; rm(Groceries)
str(sales)

## Formal class 'transactions' [package "arules"] with 3 slots
##   ..@ data       :Formal class 'ngCMatrix' [package "Matrix"] with 5 slots
##   .. .. ..@ i       : int [1:43367] 13 60 69 78 14 29 98 24 15 29 ...
##   .. .. ..@ p       : int [1:9836] 0 4 7 8 12 16 21 22 27 28 ...
##   .. .. ..@ Dim     : int [1:2] 169 9835
##   .. .. ..@ Dimnames:List of 2
##   .. .. .. ..$ : NULL
##   .. .. .. ..$ : NULL
##   .. .. ..@ factors : list()
##   ..@ itemInfo   :'data.frame':  169 obs. of  3 variables:
##   .. ..$ labels: chr [1:169] "frankfurter" "sausage" "liver loaf" "ham" ...
##   .. ..$ level2: Factor w/ 55 levels "baby food","bags",..: 44 44 44 44 44 44 44
42 42 41 ...
##   .. ..$ level1: Factor w/ 10 levels "canned food",..: 6 6 6 6 6 6 6 6 6 6 ...
##   ..@ itemsetInfo:'data.frame':  0 obs. of  0 variables

transactions(sales)                                # Anzahl Transaktionen + Items

## transactions in sparse format with
##  9835 transactions (rows) and
##  169 items (columns)

transaktionen.df<- as(sales,"data.frame")         # Transaktionen in Dataframe
head(transaktionen.df)                             # Erste Transaktionen

##                                                                    items
## 1               {citrus fruit,semi-finished bread,margarine,ready soups}
## 2                                         {tropical fruit,yogurt,coffee}
## 3                                                           {whole milk}
## 4                           {pip fruit,yogurt,cream cheese ,meat spreads}
## 5 {other vegetables,whole milk,condensed milk,long life bakery product}
## 6                         {whole milk,butter,yogurt,rice,abrasive cleaner}

inspect(head(sales))                               # Erste Transaktionen

##     items
## [1] {citrus fruit,
##      semi-finished bread,
##      margarine,
##      ready soups}
## [2] {tropical fruit,
##      yogurt,
##      coffee}
## [3] {whole milk}
## [4] {pip fruit,
##      yogurt,
##      cream cheese ,
##      meat spreads}
## [5] {other vegetables,
##      whole milk,
##      condensed milk,
##      long life bakery product}
```

```
## [6] {whole milk,
##      butter,
##      yogurt,
##      rice,
##      abrasive cleaner}
image(head(sales))                          # Sparse Matrix der ersten Transakt.
```

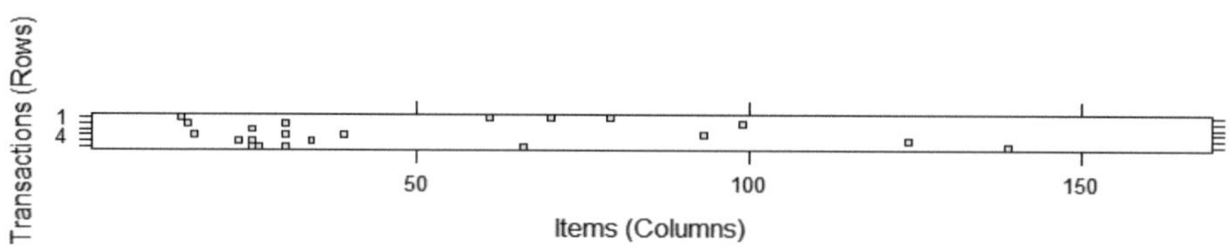


```
itemFrequency(sales)[1:5]                   # Erste 5 Items mit Häufigkeit
## frankfurter     sausage  liver loaf         ham        meat
## 0.058973055 0.093950178 0.005083884 0.026029487 0.025826131
itemFrequencyPlot(sales,topN=10,horiz=TRUE,    # Top10 Items
                  xlab="Relative Häufigkeit",
                  main="TOP 10 Items nach Häufigkeit")
```

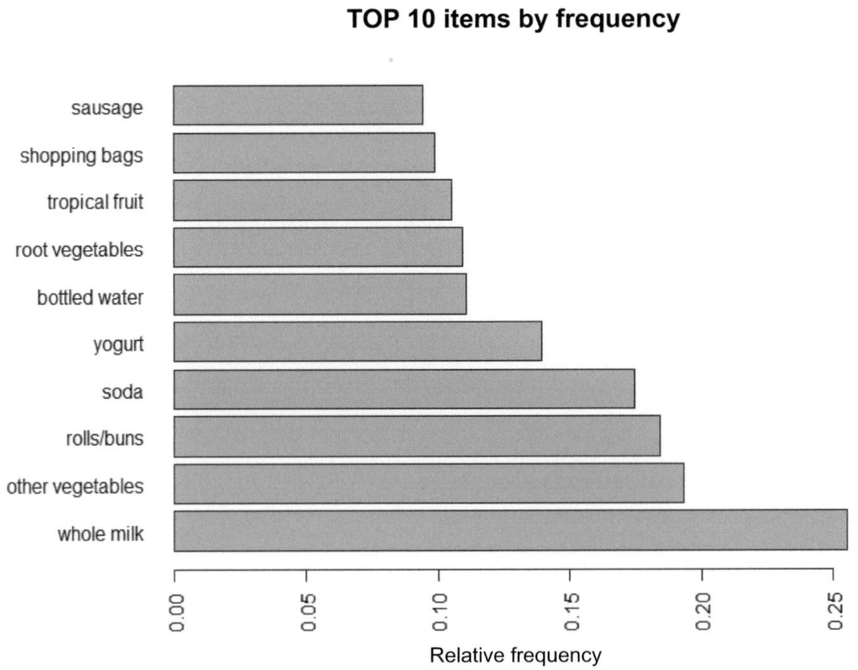


```
itemFrequencyPlot(sales,support=0.1,horiz=TRUE,# Items mit Support>0.1
                  xlab="Relative Häufigkeit",
                  main="Items mit Support > 0,1")
```

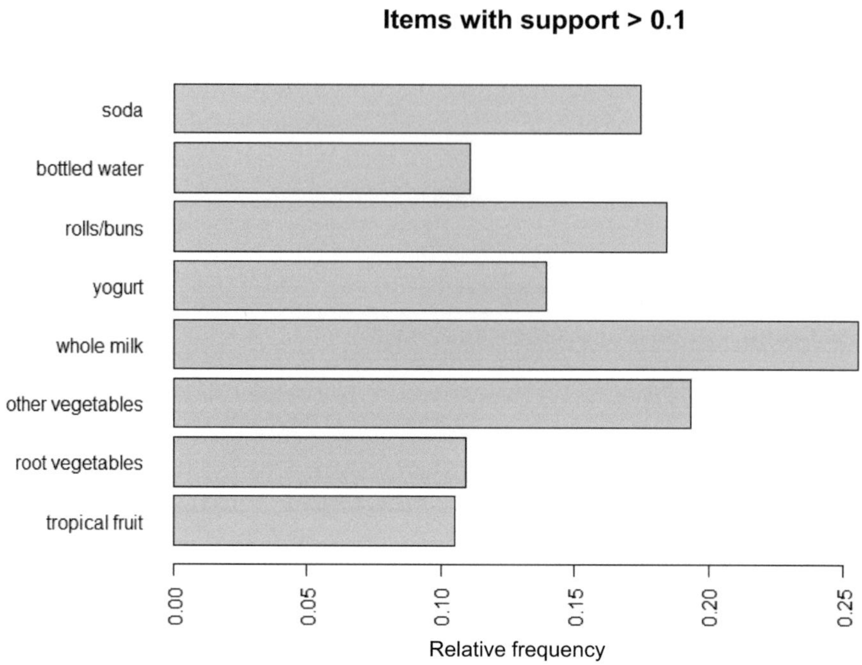


```
kreuztab<-crossTable(sales)                        # Kreuztabelle der Items
kreuztab[1:5,1:5]                                  # Anzeige erste 5 der Kreuztabelle

##             frankfurter sausage liver loaf ham meat
## frankfurter         580      99          7  25   32
## sausage              99     924         10  49   52
## liver loaf            7      10         50   3    0
## ham                  25      49          3 256    9
## meat                 32      52          0   9  254

kreuztab["sausage", "ham"]                         # Selektive Anzeige der Kreuztab.

## [1] 49

#- Assoziationsregeln ermitteln: Geeignete Parameter ausprobieren-----------------
para.support <- c(0.9, 0.1, 0.01, 0.001, 0.0005)
para.support.text <- c("0.9", "0.1", "0.01", "0.001", "0.0005")
para.konfidenz <- c(0.01, 0.1, 0.2, 0.3, 0.4, 0.5, 0.6, 0.7, 0.8, 0.9, 0.99)
rules.para<-data.frame(Konfidenz=0,Regelanzahl=0,Support="")
# Konfidenz von 0.01 und Support von 0.0005 -> mehr als 175.000 Regeln
for (i in 1:length(para.support)) {
  for (j in 1:length(para.konfidenz)) {
    rules.para[i+(j-1)*length(para.support),1] <- para.konfidenz[j]
    rules.para[i+(j-1)*length(para.support),2] <- length(
      apriori(data=sales,
              parameter=list(sup=para.support[i], conf=para.konfidenz[j],
                             target="rules"),
              control = list(verbose=FALSE)))
    rules.para[i+(j-1)*length(para.support),3] <- para.support.text[i]
```

```
  }
}
ergebnis<-ml.summary.3(rules.para,
                       "Assoziationsregeln je Kategorie (Support) und Konfidenz",
                       "Konfidenz","Anzahl Regeln",line=T)
```

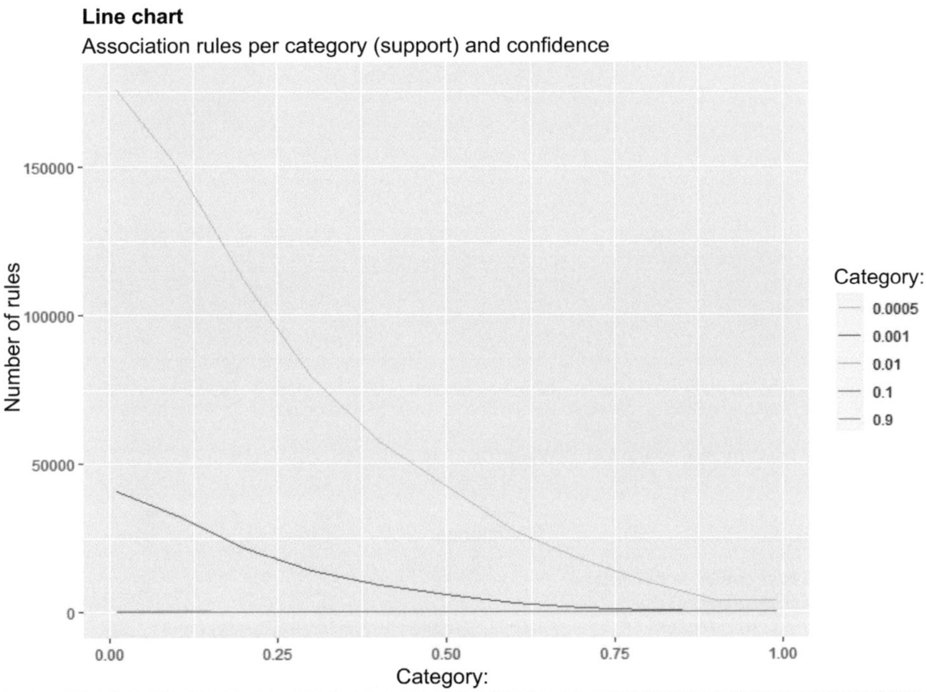


```
para.support <- c(0.002, 0.0015, 0.00125, 0.001)
para.support.text <- c("0.002", "0.0015", "0.00125", "0.001")
para.konfidenz <- c(0.90, 0.92, 0.94, 0.96, 0.98)
rules.para<-data.frame(Konfidenz=0,Regelanzahl=0,Support="")
# Konfidenz von 0.9 und Support von 0.00125 -> gute Anzahl an Regeln
for (i in 1:length(para.support)) {
  for (j in 1:length(para.konfidenz)) {
    rules.para[i+(j-1)*length(para.support),1] <- para.konfidenz[j]
    rules.para[i+(j-1)*length(para.support),2] <- length(
      apriori(data=sales,
              parameter=list(sup=para.support[i], conf=para.konfidenz[j],
                             target="rules"),
              control = list(verbose=FALSE)))
    rules.para[i+(j-1)*length(para.support),3] <- para.support.text[i]
  }
}
ergebnis<-ml.summary.3(rules.para,
                       "Assoziationsregeln je Kategorie (Support) und Konfidenz",
                       "Konfidenz","Anzahl Regeln",line=T)
```

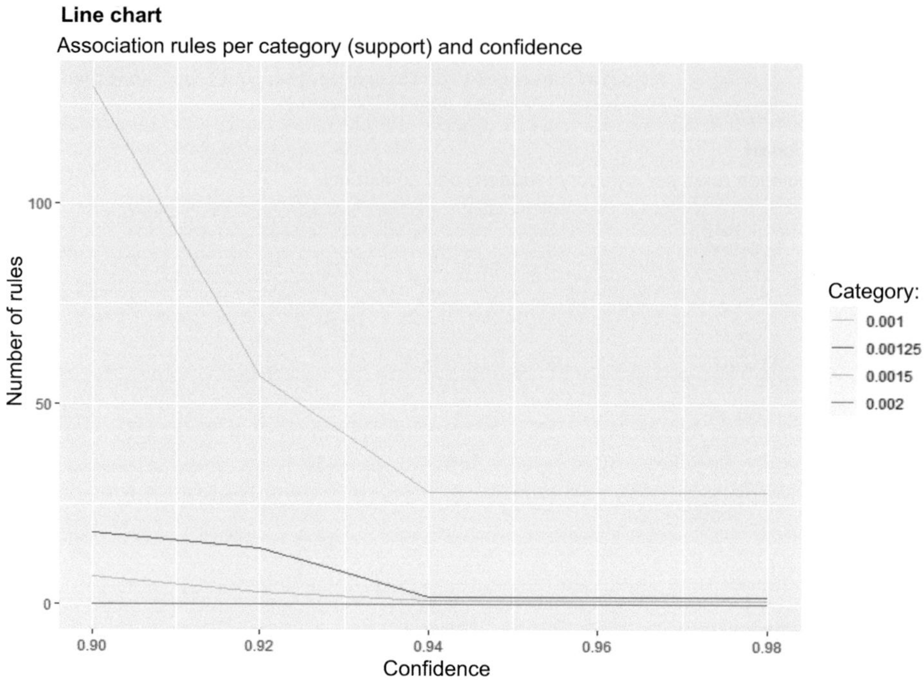


```
#- Assoziationsregeln ermitteln------------------------------------------------
regeln<-apriori(data=sales,
                parameter=list(sup=0.00125,conf=0.9,minlen=2,maxlen=4),
                control = list(verbose=FALSE))
summary(regeln)

## set of 11 rules
##
## rule length distribution (lhs + rhs):sizes
##  3  4
##  1 10
##
##    Min. 1st Qu.  Median    Mean 3rd Qu.    Max.
##   3.000   4.000   4.000   3.909   4.000   4.000
##
## summary of quality measures:
##     support           confidence        coverage             lift
##  Min.   :0.001322   Min.   :0.9000   Min.   :0.001423   Min.   : 3.522
##  1st Qu.:0.001373   1st Qu.:0.9048   1st Qu.:0.001474   1st Qu.: 3.643
##  Median :0.001525   Median :0.9286   Median :0.001627   Median : 3.669
##  Mean   :0.001599   Mean   :0.9272   Mean   :0.001729   Mean   : 4.643
##  3rd Qu.:0.001830   3rd Qu.:0.9333   3rd Qu.:0.002034   3rd Qu.: 4.737
##  Max.   :0.001932   Max.   :1.0000   Max.   :0.002135   Max.   :11.235
##      count
##  Min.   :13.00
##  1st Qu.:13.50
##  Median :15.00
##  Mean   :15.73
##  3rd Qu.:18.00
```

```
## Max.   :19.00
##
## mining info:
##   data ntransactions support confidence
##  sales          9835 0.00125        0.9
##
                                               call
##  apriori(data = sales, parameter = list(sup = 0.00125, conf = 0.9, minlen = 2, ma
xlen = 4), control = list(verbose = FALSE))
```

```
inspect(sort(head(regeln),by="support"))
```

```
##     lhs                           rhs                  support confidence    coverage
   lift count
## [1] {liquor,

##      red/blush wine}        => {bottled beer} 0.001931876  0.9047619 0.002135231 11
.235269    19
## [2] {root vegetables,

##      whipped/sour cream,

##      flour}                 => {whole milk}   0.001728521  1.0000000 0.001728521  3
.913649    17
## [3] {other vegetables,

##      cream cheese ,

##      sugar}                 => {whole milk}   0.001525165  0.9375000 0.001626843  3
.669046    15
## [4] {citrus fruit,

##      domestic eggs,

##      sugar}                 => {whole milk}   0.001423488  0.9333333 0.001525165  3
.652739    14
## [5] {yogurt,

##      domestic eggs,

##      sugar}                 => {whole milk}   0.001423488  0.9333333 0.001525165  3
.652739    14
## [6] {pip fruit,

##      whipped/sour cream,

##      cream cheese }         => {whole milk}   0.001321810  0.9285714 0.001423488  3
.634103    13
```

```
inspect(sort(head(regeln),by="confidence"))
```

```
##     lhs                           rhs                  support confidence    coverage
   lift count
## [1] {root vegetables,

##      whipped/sour cream,

##      flour}                 => {whole milk}   0.001728521  1.0000000 0.001728521  3
.913649    17
## [2] {other vegetables,
```

```
##      cream cheese ,

##      sugar}              => {whole milk}   0.001525165   0.9375000 0.001626843  3
.669046    15
## [3] {citrus fruit,

##      domestic eggs,

##      sugar}              => {whole milk}   0.001423488   0.9333333 0.001525165  3
.652739    14
## [4] {yogurt,

##      domestic eggs,

##      sugar}              => {whole milk}   0.001423488   0.9333333 0.001525165  3
.652739    14
## [5] {pip fruit,

##      whipped/sour cream,

##      cream cheese }      => {whole milk}   0.001321810   0.9285714 0.001423488  3
.634103    13
## [6] {liquor,

##      red/blush wine}     => {bottled beer} 0.001931876   0.9047619 0.002135231 11
.235269    19
inspect(sort(head(regeln),by="lift"))

##     lhs                          rhs                  support confidence    coverage
   lift count
## [1] {liquor,

##      red/blush wine}     => {bottled beer} 0.001931876   0.9047619 0.002135231 11
.235269    19
## [2] {root vegetables,

##      whipped/sour cream,

##      flour}              => {whole milk}   0.001728521   1.0000000 0.001728521  3
.913649    17
## [3] {other vegetables,

##      cream cheese ,

##      sugar}              => {whole milk}   0.001525165   0.9375000 0.001626843  3
.669046    15
## [4] {citrus fruit,

##      domestic eggs,

##      sugar}              => {whole milk}   0.001423488   0.9333333 0.001525165  3
.652739    14
## [5] {yogurt,

##      domestic eggs,

##      sugar}              => {whole milk}   0.001423488   0.9333333 0.001525165  3
```

```
.652739    14
## [6] {pip fruit,

##      whipped/sour cream,

##      cream cheese }       => {whole milk}   0.001321810  0.9285714 0.001423488  3
.634103    13

#- Assoziationsregeln für spezielles Item in lhs ermitteln-----------------------
regel.sub<-subset(regeln,items %in% "tropical fruit")
summary(regel.sub)

## set of 2 rules
##
## rule length distribution (lhs + rhs):sizes
## 4
## 2
##
##    Min. 1st Qu.  Median    Mean 3rd Qu.    Max.
##       4       4       4       4       4       4
##
## summary of quality measures:
##     support           confidence        coverage             lift
##  Min.   :0.001830   Min.   :0.9000   Min.   :0.002034   Min.   :3.522
##  1st Qu.:0.001856   1st Qu.:0.9012   1st Qu.:0.002059   1st Qu.:3.811
##  Median :0.001881   Median :0.9024   Median :0.002084   Median :4.099
##  Mean   :0.001881   Mean   :0.9024   Mean   :0.002084   Mean   :4.099
##  3rd Qu.:0.001906   3rd Qu.:0.9036   3rd Qu.:0.002110   3rd Qu.:4.388
##  Max.   :0.001932   Max.   :0.9048   Max.   :0.002135   Max.   :4.676
##      count
##  Min.   :18.00
##  1st Qu.:18.25
##  Median :18.50
##  Mean   :18.50
##  3rd Qu.:18.75
##  Max.   :19.00
##
## mining info:
##   data ntransactions support confidence
##  sales          9835 0.00125        0.9
##
                                               call
##  apriori(data = sales, parameter = list(sup = 0.00125, conf = 0.9, minlen = 2, ma
xlen = 4), control = list(verbose = FALSE))

inspect(sort(head(regel.sub),by="lift"))

##     lhs                          rhs                         support confidence    cover
age     lift count
## [1] {tropical fruit,

##      whipped/sour cream,

##      fruit/vegetable juice} => {other vegetables} 0.001931876  0.9047619 0.002135
231 4.675950    19
## [2] {tropical fruit,

##      whipped/sour cream,
```

```
##       domestic eggs}          => {whole milk}       0.001830198  0.9000000 0.002033
554 3.522284    18

#- Assoziationsregeln für spezielles Item in rhs ermitteln------------------------
regel.tropical<-apriori(data=sales,
                        parameter=list(support=0.00125,confidence=0.6),
                        appearance=list(default="lhs",rhs="tropical fruit"),
                        control = list(verbose=FALSE))
summary(regel.tropical)

## set of 11 rules
##
## rule length distribution (lhs + rhs):sizes
## 3 4 5 6
## 2 2 6 1
##
##    Min. 1st Qu.  Median    Mean 3rd Qu.    Max.
##   3.000   4.000   5.000   4.545   5.000   6.000
##
## summary of quality measures:
##     support           confidence        coverage              lift
##  Min.   :0.001322   Min.   :0.6000   Min.   :0.002034   Min.   :5.718
##  1st Qu.:0.001373   1st Qu.:0.6079   1st Qu.:0.002135   1st Qu.:5.793
##  Median :0.001525   Median :0.6190   Median :0.002339   Median :5.900
##  Mean   :0.001479   Mean   :0.6293   Mean   :0.002357   Mean   :5.997
##  3rd Qu.:0.001525   3rd Qu.:0.6356   3rd Qu.:0.002542   3rd Qu.:6.057
##  Max.   :0.001729   Max.   :0.7000   Max.   :0.002847   Max.   :6.671
##      count
##  Min.   :13.00
##  1st Qu.:13.50
##  Median :15.00
##  Mean   :14.55
##  3rd Qu.:15.00
##  Max.   :17.00
##
## mining info:
##   data ntransactions support confidence
##  sales          9835 0.00125        0.6
##

 call
##  apriori(data = sales, parameter = list(support = 0.00125, confidence = 0.6), app
earance = list(default = "lhs", rhs = "tropical fruit"), control = list(verbose = FA
LSE))

inspect(head(sort(regel.tropical,by="lift")))

##     lhs                        rhs                    support confidence    coverage
  lift count
## [1] {root vegetables,

##      other vegetables,

##      yogurt,

##      bottled water}      => {tropical fruit} 0.001423488  0.7000000 0.002033554 6.
671027    14
## [2] {pip fruit,

##      whole milk,
```

```
##       frozen meals}       => {tropical fruit} 0.001525165  0.6818182 0.002236909 6.
497754    15
## [3] {root vegetables,

##       whole milk,

##       yogurt,

##       bottled water}      => {tropical fruit} 0.001525165  0.6521739 0.002338587 6.
215243    15
## [4] {pip fruit,

##       misc. beverages}    => {tropical fruit} 0.001321810  0.6190476 0.002135231 5.
899548    13
## [5] {pip fruit,

##       other vegetables,

##       whole milk,

##       brown bread}        => {tropical fruit} 0.001321810  0.6190476 0.002135231 5.
899548    13
## [6] {root vegetables,

##       other vegetables,

##       butter,

##       yogurt}             => {tropical fruit} 0.001321810  0.6190476 0.002135231 5.
899548    13

#- Assoziationsregeln visualisieren---------------------------------------------
plot(regel.tropical,method="graph",              # Visualisierung der Regeln
     measure="confidence",shading="lift")
```

```
plot(regel.tropical, method="paracoord")
```

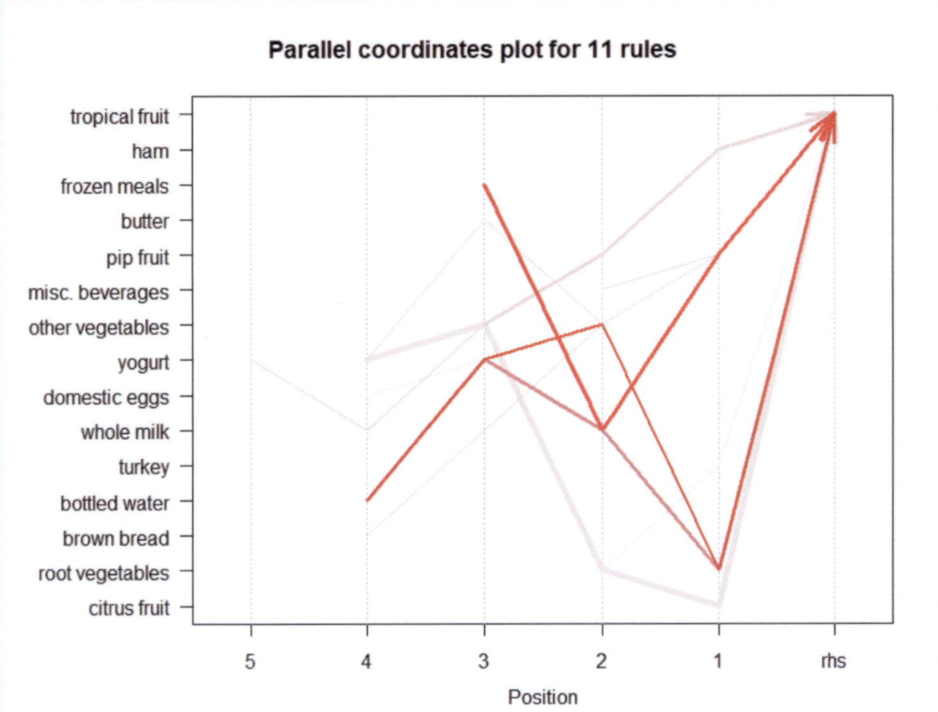


```
#- Assoziationsregeln in Dataframe speichern-------------------------------------
regel.tropical.df<- as(regel.tropical,"data.frame")
head(regel.tropical.df)

##                                                                  rules
## 1                        {turkey,root vegetables} => {tropical fruit}
## 2                     {pip fruit,misc. beverages} => {tropical fruit}
## 3             {pip fruit,whole milk,frozen meals} => {tropical fruit}
## 4                {ham,pip fruit,other vegetables} => {tropical fruit}
## 5 {pip fruit,other vegetables,whole milk,brown bread} => {tropical fruit}
## 6    {root vegetables,other vegetables,butter,yogurt} => {tropical fruit}
##       support confidence    coverage     lift count
## 1 0.001525165  0.6000000 0.002541942 5.718023    15
## 2 0.001321810  0.6190476 0.002135231 5.899548    13
## 3 0.001525165  0.6818182 0.002236909 6.497754    15
## 4 0.001626843  0.6153846 0.002643620 5.864639    16
## 5 0.001321810  0.6190476 0.002135231 5.899548    13
## 6 0.001321810  0.6190476 0.002135231 5.899548    13
```

The association rules determined in this way can then be used for cross-selling in stores as well as online shops.

References

Chen, D., Sain, S. L., & Guo, K. (2015). *UCI Machine Learning Repository: Online Retail Data Set.* Abgerufen am 06. Januar 2023 von https://archive.ics.uci.edu/ml/datasets/online+retail

Dark_Raider. (15. Januar 2023). *Credit Risk Analysis [Data set].* https://doi.org/10.34740/KAGGLE/DSV/2327131

Kuhn, M. (2022). *modeldata: Data Sets Useful for Modeling Examples*. Abgerufen am 17. November 2022 von https://modeldata.tidymodels.org/

Outlook

8

In the preceding sections, we introduced how machine learning works and how it can be very valuable for predictions, analyses, and much more.

I would now like to highlight again how limited machine learning can be. Let's take the fear of dying. Suppose I receive a database with many millions of data sets, which includes the following information from deceased people: first name, age, nationality, date of death, time of death, age at death, country of residence at death, nearest city at death, detailed location of death (e.g., street, airplane, tank, sports field, forest, field, sea, lake, chair, bed…). I use this information to train a model with the highest possible predictive accuracy. Since most people die in bed, the recommendation of the machine learning model will be to avoid beds. This is, of course, nonsense, but the data was insufficient. It takes a human to validate the results of machine learning models when it comes to newly generated knowledge. Artificial intelligence is therefore only limitedly applicable. The automation of decisions is very limited here.

On the other hand, if many millions of data sets with a comprehensive number of relevant variables are used as the basis for machine learning, for example, to determine the ideal price to optimize profit in online stores, then this works perfectly and a human would not be able to analyze the data equally quickly and well. Machine learning and artificial intelligence are perfectly applicable when the machine learning models have been sufficiently validated and have proven themselves in use. The automation of decisions is better possible in these situations.

As a conclusion, it can be derived that machine learning and artificial intelligence can be used quickly and reliably in proven use cases, but their applicability can change in new use cases or with other data. Even established machine learning models should therefore be validated again at appropriate intervals.

Regarding machine learning and artificial intelligence, there are both skeptics and enthusiasts. Both groups often express their opinions very emotionally and sometimes

B. Heesen, *Artificial Intelligence and Machine Learning with R*,
https://doi.org/10.1007/978-3-658-45392-3_8

with little understanding for the other group, the topic is polarizing. I would like to illustrate this using the example of the chatbot ChatGPT (GPT = Generative Pre-Trained Transformer), which was made available for free by OpenAI for the general public in November 2022 (Hoffmann 2023). Within a few days, the chatbot had reached 1 million users. Enthusiastic users considered the innovation as fundamental, as the trained machine learning model for natural language processing proved that it is capable of solving partially complex tasks. The problem was also the data with which the model was trained. The model was apparently extensively trained in areas such as programming and algorithms and can therefore produce impressive results. But when I asked the system what courses Ansbach University offers, the chatbot had no better answer than to refer me to the university's website and look there. The same applied to a current ranking of the best tennis players of the German Tennis Federation. The model was simply not trained with this data. The best algorithm and the most powerful computer system can therefore produce the most valuable results in one situation and worthless results in another.

Therefore, it is necessary to assess the opportunities and risks of machine learning and artificial intelligence depending on the situation and context and to select application areas that are suitable for these innovations.

In any case, there are undoubtedly many wonderful areas of application such as early cancer detection, spam filters, navigation devices, and much more. As current and future fields of application of particular importance, IDG (IDG 2021), sorted by importance, has identified the following areas: speech recognition, image recognition and analysis, text analysis, face recognition, machine translation, optical character recognition, classification, assistance systems, autonomous systems, big data analytics, robotics, and behavior pattern recognition. The biggest challenge to the use of machine learning is also seen in the IDG study as understanding the data, cleaning the data, and selecting the data. In addition, the selection of suitable machine learning models and the transfer of applications into productive operation also pose a challenge. The Fig. 8.1 shows how diverse data types can be used as input and output, that statistics, algorithms, and models also form a basis of machine learning and can only achieve the desired results in combination.

Encouraging is the fact that there are increasingly powerful computers (local hardware), virtual computing power (cloud), and suitable software tools that are increasingly easy to use for machine learning and artificial intelligence. R and Python as free tools are particularly worth mentioning here.

The integration of open-source tools into management information systems also shows that their acceptance and use are increasing. For example, Microsoft offers the integration of R and Python code in its product PowerBI (Microsoft 2023). According to the Magic Quadrant for Analytics and Business Intelligence Platforms (Gartner 2022), Microsoft Power BI is the leading business intelligence software. In Fig. 8.2, you may recognize the two lower images created with R from Sect. 7.1.1 again.

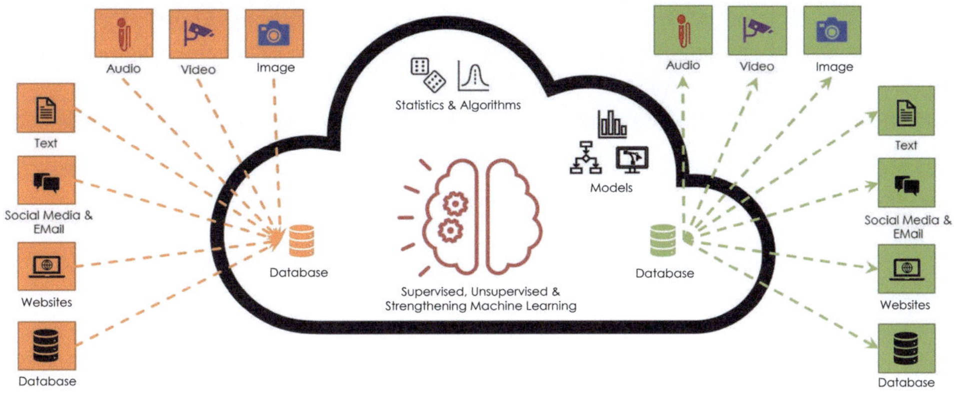

Fig. 8.1 Machine Learning

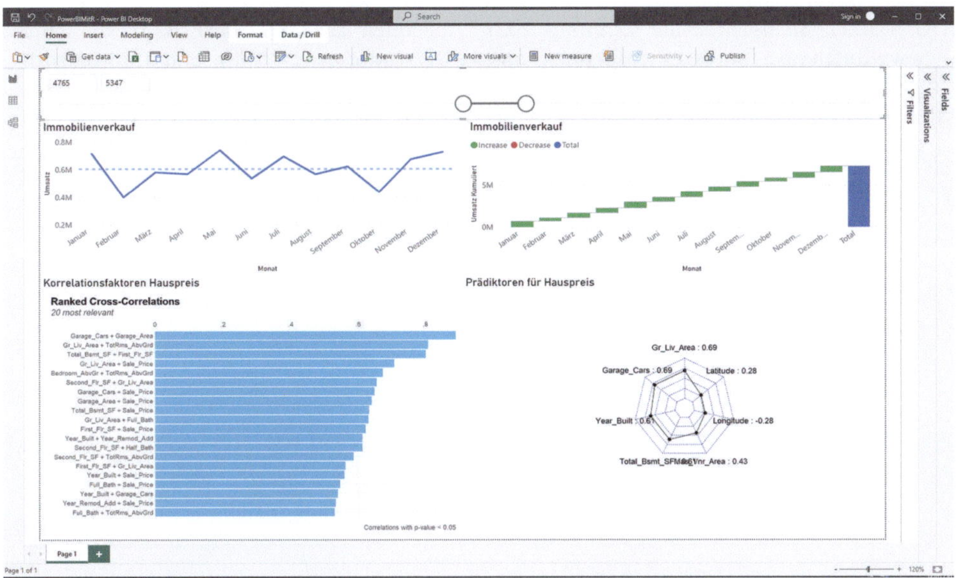

Fig. 8.2 Power BI

The integration and joint use of several innovative software solutions follows the best-of-breed approach, in which the most suitable software is used for each function. This concept reduces the dependency on a single software manufacturer.

Well, then we can start implementing and using machine learning and artificial intelligence, right? In principle, the answer to this question is probably YES. However, organizations see challenges ahead. The most significant hurdles identified by companies are

(IDG 2021): lack of skilled personnel, insufficient programming knowledge, lack of mathematical and statistical foundations, incomprehensibility of machine learning algorithms, and poor data quality.

It is therefore obviously necessary to invest both in the recruitment of data scientists and in the further training of existing employees.

There are plenty of opportunities to take advantage of these innovations.

References

Gartner. (2022). *Magic Quadrant for Analytics and Business Intelligence Platforms.* Abgerufen am 24. Januar 2023 von https://info.microsoft.com/ww-landing-2022-gartner-mq-report-on-bi-and-analytics-platforms.html?LCID=EN-US

Hoffmann, L. (20. Januar 2023). *ChatGPT im Hochschulkontext.* Abgerufen am 24. Januar 2023 von https://hochschulforumdigitalisierung.de/de/blog/chatgpt-im-hochschulkontext-kommentierte-linksammlung#Was%20ist?

IDG. (2021). *Studie Machine Learning 2021.* Abgerufen am 24. Januar 2023 von https://www.lufthansa-industry-solutions.com/de-de/studien/idg-studie-machine-learning-2021?gclid=EAIaIQobChMIwMCC9_-D9QIViPhRCh3E3wFTEAAYASAAEgLRkfD_BwE

Microsoft. (2023). *Microsoft Power BI.* Abgerufen am 24. Januar 2023 von https://powerbi.microsoft.com/

References

Agrawal, R., & Srikant, R. (1994). *Fast Algorithms for mining association rules.* Abgerufen am 04. Januar 2023 von http://rakesh.agrawal-family.com/papers/vldb94apriori.pdf

bitkom. (2017). *Künstliche Intelligenz: Wirtschaftliche Bedeutung, gesellschaftliche Herausforderungen, menschliche Verantwortung.* Abgerufen am 27. Dezember 2021 von https://www.bitkom.org/sites/default/files/file/import/171012-KI-Gipfelpapier-online.pdf

Heesen, B. (2022). *Data Science und Statistik mit R: Anwendungslösungen für die Praxis.* Heidelberg: Springer-Gabler

Price Waterhouse Coopers. (2019). *PwC Global CEO Survey 2019.* Abgerufen am 30. September 2019 von https://www.pwc.de/de/ceosurvey.html

R Core Team. (2022). *Using Rtools42 on Windows.* Abgerufen am 05. Juli 2022 von https://cran.r-project.org/bin/windows/Rtools/

B. Heesen, *Artificial Intelligence and Machine Learning with R,*
https://doi.org/10.1007/978-3-658-45392-3

MIX
Papier aus verantwortungsvollen Quellen
Paper from responsible sources
FSC® C105338

If you have any concerns about our products,
you can contact us on
ProductSafety@springernature.com

In case Publisher is established outside the EU,
the EU authorized representative is:
Springer Nature Customer Service Center GmbH
Europaplatz 3, 69115 Heidelberg, Germany

Printed by Libri Plureos GmbH
in Hamburg, Germany